HEAVY METALS
IN THE AQUATIC ENVIRONMENT

INTERNATIONAL ASSOCIATION ON WATER POLLUTION RESEARCH

President
DR. G. J. STANDER

Vice-Presidents
PROFESSOR B. B. BERGER and MR. E. KUNTZE

Secretary-Treasurer
MR. P. E. ODENDAAL

Executive Editor
DR. S. H. JENKINS

Conference Officer
PROFESSOR P. A. KRENKEL

SUPPLEMENT TO
PROGRESS IN WATER TECHNOLOGY

Heavy Metals in the Aquatic Environment

Proceedings of the International Conference held in Nashville, Tennessee, December 1973

Edited by P. A. KRENKEL

Sponsored by

The International Association on Water Pollution Research,
The Sport Fishing Institute, and
The Department of Environmental and Water Resources Engineering at
Vanderbilt University, Nashville, Tennessee

PERGAMON PRESS

OXFORD · NEW YORK · TORONTO
SYDNEY · BRAUNSCHWEIG

Pergamon Press Ltd., Headington Hill Hall, Oxford
Pergamon Press Inc., Maxwell House, Fairview Park, Elmsford,
New York 10523
Pergamon of Canada Ltd., 207 Queen's Quay West, Toronto 1
Pergamon Press (Aust.) Pty. Ltd., 19a Boundary Street,
Rushcutters Bay, N.S.W. 2011, Australia
Pergamon Press GmbH, Burgplatz 1, Braunschweig 3300, West Germany

Copyright © 1975 Pergamon Press Ltd.

All Rights Reserved. No part of this publication may be reproduced, stored in a retrieval system, or transmitted, in any form or by any means, electronic, mechanical, photocopying, recording or otherwise, without the prior permission of Pergamon Press Ltd.

First edition 1975
Library of Congress Catalog Card No. 75-649

Printed and bound in Great Britain by
A. Wheaton & Co., Exeter

ISBN 0 08 018068 X

CONTENTS

Preface ix

SESSION I TOXICOLOGY AND PHYSIOLOGY

A Review of Dose-Response Relationships Resulting from Human Exposure to Methylmercury Compounds. T. W. CLARKSON, J. CRISPIN SMITH, D. O. MARSH, and M. D. TURNER, University of Rochester School of Medicine and Dentistry, Rochester, New York. 1

Discussion by LEONARD J. GOLDWATER, Department of Community Health Sciences, Duke University, Durham, North Carolina. 13

Toxicity of Cadmium—Mechanism and Diagnosis. KAZUO NOMIYAMA, Professor, Department of Environmental Health, Jichi Medical School, Minamakawachi-Machi, Tochigi-Ken 329-04, Japan. 15

The Effects of Heavy Metals on Fish and Aquatic Organisms. MAX KATZ, Parametrix Inc., Seattle, Washington. 25

Discussion by DONALD I. MOUNT, US Environmental Protection Agency, National Water Quality Laboratory, Duluth, Minnesota. 31

Mercury Concentrations in Humans and Consumption of Fish Containing Methylmercury. HAROLD E. B. HUMPHREY, Environmental Epidemiologist, Michigan Department of Public Health, Lansing, Michigan. 33

SESSION II ANALYTICAL TECHNIQUES

Methylmercury Analysis (A Review and Some Data). KIMIAKI SUMINO, Kobe University School of Medicine. 35

Discussion by GUNNEL WESTÖÖ, Food Laboratory, Swedish National Food Administration, Stockholm, Sweden. 47

A Review of the Status of Total Mercury Analysis. W. DICKINSON BURROWS, Associated Water and Air Resources Engineers, Inc., Nashville, Tennessee. 51

Discussion by JAMES H. FINGER and TOM B. BENNETT, Environmental Protection Agency, Region IV, Surveillance and Analysis Division, Chemicals Services Branch, Athens, Georgia 30601. 63

Analytical Techniques for Heavy Metals other than Mercury. HERBERT A. LAITINEN, Professor of Chemistry, University of Illinois at Urbana-Champaign, Urbana, Illinois 61801. 73

Discussion by R. K. SKOGERBOE, Department of Chemistry, Colorado State University, Fort Collins, Colorado 80521. 81

SESSION III TRANSPORT MECHANISMS (1)

Field Observations on the Transport of Heavy Metals in Sediments. A. J. DE GROOT, Head, Department of Chemistry, Institute for Soil Fertility, Haren (Groningen), Netherlands, and E. ALLERSMA, Head, Department of Hydrodynamics and Morphology, Delft Hydraulics Laboratory, Delft, Netherlands, and Department of Fisheries and Wildlife, Michigan State University, East Lansing, Michigan 48824. 85

Discussion by I. R. JONASSON, Geological Survey of Canada, Ottawa, Canada, and M. H. TIMPERLEY, Chemistry Division, DSIR, Petone, New Zealand. 97

Some Remarks by A. J. DE GROOT. 103

Metabolic Cycles for Toxic Elements in the Environment. A Study of Kinetics and Mechanism. J. M. WOOD, Director, Freshwater Biological Research Institute, University of Minnesota, College of Biological Sciences, St. Paul, Minnesota. 105

Discussion by J. J. BISOGNI JR. and A. W. LAWRENCE, Department of Environmental Engineering, Cornell University, Ithaca, NY. 113

Sorption Phenomenon in the Organics of Bottom Sediments. ROBERT S. REIMERS, Battelle, Columbus Laboratories, Columbus, Ohio, PETER A. KRENKEL, Director of Environmental Planning Division, Tennessee Valley Authority, Chattanooga, Tennessee 37401, and MARGARET EAGLE and GREGORY TRAGITT, Vanderbilt University, Nashville, Tennessee. 117

Discussion by E. A. JENNE, US Geological Survey, Menlo Park, California. 131

Contents

SESSION IV TRANSPORT MECHANISMS (2)

Role of Hydrous Metal Oxides in the Transport of Heavy Metals in the Environment. G. FRED LEE, Institute for Environmental Sciences, University of Texas, Dallas, Texas.   137
Discussion by J. D. HEM, US Geological Survey, Menlo Park, California 94025.   149
The Accumulation and Excretion of Heavy Metals in Organisms. JORMA K. MIETTINEN, Department of Radiochemistry, University of Helsinki, Helsinki, Finland.   155
Discussion by ROBERT A. GOYER, University of North Carolina, Chapel Hill, North Carolina.   163
Transport and Biological Effects of Molybdenum in the Environment. WILLARD R. CHAPPELL, Department of Physics and Astrophysics, and The Molybdenum Project, University of Colorado, Boulder, Colorado 80302.   167
Discussion by CORALE L. BRIERLEY, Chemical Microbiologist, New Mexico Bureau of Mines and Mineral Resources, New Mexico Institute of Mining and Technology, Socorro, N.M. 87801.   189
Biological and Nonbiological Transformations of Mercury in Aquatic Systems. W. P. IVERSON and C. HUEY, Corrosion Section, National Bureau of Standards, Washington DC 20234, and F. E. BRINCKMAN, K. J. JEWETT, and W. BLAIR, Inorganic Chemistry Section, National Bureau of Standards, Washington DC 20234.   193

SESSION V DISTRIBUTION OF HEAVY METALS IN THE ENVIRONMENT

The Distribution of Mercury in Fish and its Form of Occurrence. RIKUO DOI, Tokyo Metropolitan Research Institute for Environmental Protection, Health Division, and JUN UI, University of Tokyo, Tokyo, Japan.   197
Discussion by FRANK M. D'ITRI, Associate Professor of Water Chemistry, Institute of Water Research, and Department of Fisheries and Wildlife, Michigan State University, Each Lansing, Michigan 48824, and PATRICIA A. D'ITRI, Associate Professor of Thought and Language, Michigan State University, East Lansing, Michigan 48824.   223
Comment by JUN UI.   229
Environmental Lead Distribution in Relation to Automobile and Mine and Smelter Sources. GARY L. ROLFE, Assistant Professor of Forest Ecology and Environmental Studies, University of Illinois, and J. CHARLES JENNETT, Associate Professor of Civil Engineering, University of Missouri–Rolla.   231
Discussion by S. R. KOIRTYOHANN, Project Director, Environmental Trace Substances Center, The University of Missouri, Columbia, Missouri, B. G. WIXSON, Professor of Environmental Health, Civil Engineering Department, Environmental Research Center, The University of Missouri–Rolla, Rolla, Missouri, and H. W. EDWARDS, Associate Professor, Department of Mechanical Engineering, Colorado State University, Fort Collins, Colorado.   243
Investigations of Heavy Metals and Other Persistent Chemicals, Westernport Bay, Australia. M. A. SHAPIRO, Professor, Environmental Health Engineering, University of Pittsburgh, and Director, Westernport Bay Environmental Study, Melbourne, and D. W. CONNELL, Marine Studies Coordinator, Westernport Bay Environmental Study, Melbourne.   247
Mercury Distribution in the Chesapeake Bay. F. E. BRINCKMAN, K. L. JEWETT, and W. R. BLAIR, Inorganic Chemistry Section, National Bureau of Standards, Washington DC 20234, and W. P. IVERSON and C. HUEY, Corrosion Section, National Bureau of Standards, Washington DC 20234.   251

SESSION VI SOURCE REDUCTION METHODOLOGY

Canadian Experience with the Reduction of Mercury at Chlor-alkali Plants. F. J. FLEWELLING, Canadian Industries Ltd., Montreal.   253
Discussion by FRED T. OLOTKA, Superintendent Environmental Control, Hooker Chemical Corporation, Niagara Falls, New York.   257
Physical–Chemical Methods of Heavy Metals Removal. JAMES W. PATTERSON, Associate Professor and Chairman, Department of Environmental Engineering, Illinois Institute of Technology, Chicago 60616, and ROGER A. MINEAR, Associate Professor of Environmental Engineering, Department of Civil Engineering, University of Tennessee, Knoxville 37916.   261
Discussion by JOHN D. WEEKS, US Environmental Protection Agency, National Environmental Research Center, Cincinnati, Ohio 45268.   273

Contents

The Effects and Removal of Heavy Metals in Biological Treatment. CARL E. ADAMS JR., W. WESLEY ECKENFELDER JR., and BRIAN L. GOODMAN, Associated Water and Air Resources Engineers Inc.   277
Discussion by E. F. BARTH, Advanced Waste Treatment Research Laboratory, National Environmental Research Center, Cincinnati, Ohio 45268.   293

SESSION VII   CORRECTIVE MEASURES FOR EXISTING PROBLEMS

The Feasibility of Restoring Mercury-contaminated Waters. ARNE JERNELÖV and BO ÅSÉLL, Swedish Water and Air Pollution Research Laboratory, Stockholm, Sweden.   299
Discussion by FREDERICK G. ZIEGLER, Associated Water and Air Resources Engineers, Nashville, Tennessee, and PETER A. KRENKEL, Director of Environmental Planning Division, Tennessee Valley Authority, Chattanooga, Tennessee 37401.   311
Experience with Heavy Metals in the Tennessee Valley Authority System. W. R. NICHOLAS, Chief, Water Quality Branch, TVA, and BRUCE A. BRYE, Assistant Chief, Water Quality Branch, TVA.   315
The Use of Synthetic Scavengers for the Binding of Heavy Metals. GEORGE FEICK, EDWARD E. JOHANSON, and DONALD S. YEAPLE, JBF Scientific Corporation, 2 Ray Avenue, Burlington, Massachusetts 01803.   329
Discussion by D. G. LANGLEY, I.E.C. International Environmental Consultants Ltd., Toronto, Canada.   333

SESSION VIII   LEGISLATION, STANDARDS, SURVEILLANCE, AND MONITORING

Current Regulations and Enforcement Experience by the Environmental Protection Agency. CARL J. SCHAFER, Chief, and DR. MURRAY P. STRIER, Chemical Engineer; Permit Assistance Branch, Permit Assistant and Evaluation Division, Office of Water Enforcement, EPA, Washington DC 20460.   335
Risk–Benefit Analysis and the Economics of Heavy Metals Control. GEORGE PROVENZANO, Research Economist, Institute for Environmental Studies, University of Illinois, Urbana, Illinois 61801.   339

Author Index   347

Subject Index   349

PREFACE

These papers resulted from an International Conference held in Nashville, Tennessee, USA, from December 4 to December 7, 1973. The meeting was co-sponsored by the International Association on Water Pollution Research, the Sport Fishing Institute, the American Fishing Tackle Manufacturers Association, and the Department of Environmental and Water Resources Engineering at Vanderbilt University.

More than 4 years have elapsed since the announcement that some Canadian fish contained mercury in concentrations up to 10 parts per million and the subsequent discovery of excessive concentrations of mercury in some American fish. Many investigations have been undertaken since that time; however, the "state of the art" of the effects of all metals on the environment is still in its infancy.

A plethora of statements concerning the hazards of heavy metals have been made with little data to substantiate or refute them. Limits have been placed on mercury in fish that are somewhat arbitrary and highly questionable. In spite of the lack of data, the entire swordfish industry has been eliminated in the United States and many people have been subjected to economic loss.

Recent proposed limitations on mercury discharges into the aquatic environment have reflected the entry of unknowledgeable people into the environmental field inasmuch as the ambient water limits are above natural background concentrations in many instances and are not within practical analytical capabilities. In addition, efforts are currently underway to prohibit the use of all organic mercurials in the United States, an objective that, if attained, will preclude many beneficial uses of these materials without a demonstrable benefit.

No one can disagree with a philosophy of protecting the environment; however, overreaction is no better than the "do nothing" attitude that prevailed prior to the environmental movement. The instigation of unreasonable treatment requirements with no commensurate benefits is totally irrational. Treatment for "treatment's sake" and the application of arbitrary standards will only result in economic disaster. For example, cooling towers are being constructed where they are not needed, sophisticated wastewater treatment is being required where it is not necessary, and automobiles are being subjected to air pollution controls in all geographic areas when serious problems only exist in certain metropolitan ones. Furthermore, the numbers of personnel of regulatory agencies at the local, state, and Federal level are being unreasonably increased in order to enforce and promulgate these standards. The end result of these activities will be significantly increased consumer costs, increased taxes, and a lower economic growth.

It is therefore time to place the environmental movement in its proper perspective, i.e., to determine the true effect of various materials contributed to the environment, to determine the economic impact of removing these materials from waste effluents, and to compare these costs with the benefits attained by their removal. In addition, the social costs should be evaluated. For example, priorities should be assigned to various societal problems and monies directed towards the areas dictated by these priorities in their assigned order.

In an environmental class that I have taught for several years to liberal arts students, I listed the various problems facing society and asked which were the most serious. These problems included: housing, mass transportation, air pollution, water pollution, noise, solid waste disposal, vector and rodent control, and radiation. On the first day of class, air and water pollution were always the number one and two problems. However, at the end of the semester, after gaining at least a cursory knowledge of the various problems, housing and mass transportation always occupied the most serious problem categories.

This meeting was directed towards the attainment of a rational approach towards the control of heavy metals in the aquatic environment. Recognized scientists were gathered from all over the world in order to discuss their investigations and conclusions regarding their areas of expertise in heavy metals. The results, as presented herein, are significant and timely. It is hoped that this "state of the art" document will aid all countries in pursuing a rational approach to the control of heavy metals in the environment.

Several acknowledgements should be made with regards to the conference planning, execution, and the publication of this book. Mrs. Margaret Eagle played a major role in all categories, and without her sincere efforts this publication would not have been possible. Miss Donna Nowels contributed in many aspects, her secretarial assistance being particularly outstanding. Finally, the financial assistance of the American Fishing Tackle Manufacturers Association through the Sport Fishing Institute is gratefully acknowledged.

PETER ASHTON KRENKEL

SESSION I

TOXICOLOGY AND PHYSIOLOGY

A REVIEW OF DOSE–RESPONSE RELATIONSHIPS RESULTING FROM HUMAN EXPOSURE TO METHYLMERCURY COMPOUNDS†

T. W. CLARKSON, J. CRISPIN SMITH, D. O. MARSH, and M. D. TURNER

University of Rochester School of Medicine and Dentistry, Rochester, New York

INTRODUCTION

Published studies of dose–response relationships in several populations exposed to methylmercury fall into three general categories, distinguished according to intensity and duration of exposure. One group consists of populations in Minamata and Niigata, Japan, where people were exposed to high concentrations of methylmercury in fish and many cases of poisoning occurred.[1,2] The second group comprises persons who have a high daily intake of fish in areas of the world where methylmercury concentrations in the fish are high but no cases of poisoning have been reported.[3] The third population is in Iraq, where people received methylmercury in contaminated bread, resulting in the exposure of many thousands of persons, a large number of cases of poisoning, and many fatalities.[4] These three populations differ with respect to the intensity and duration of exposure to methylmercury. A comparison of the effects of methylmercury in these three populations should lead to a better understanding of the effects of methylmercury in man.

The effects of methylmercury on mammals may be neurotoxic, teratologic (or embryotoxic), and genetic (for details see Berglund[5]). The neurotoxic effects are the most widely studied and the most obvious. These arise essentially from damage to the central nervous system, especially the visual cortex and the cerebellum. Many other areas of the brain may be affected, depending upon the intensity of the exposure. The usual effects, which have been well described, include ataxia, constriction of the visual field (in some cases leading to blindness), difficulties in hearing, and difficulties in speech. Another neurological effect, possibly connected with the effects on the peripheral nervous system, is the loss of sensation in the hands and feet and in the perioral region. Teratogenic and embryotoxic effects have been described in animals and have been claimed to have occurred in the population in Minamata Bay. Genetic effects have been described only in animals or lower forms of life. These include a dominant lethal mutation in male rats due to exposure to methylmercury (Khera[6]) and effects on onion root.[7] Possible effects in human populations are indicated by Skerfving *et al.*,[8] who claim a higher frequency of chromosome breakage in individuals having high exposure to methylmercury. Except for the work on chromosome breakage, all the reports on dose–response relationships have been concerned exclusively with the neurotoxic effects.

Before discussing dose–response relationships, it would be well to point out the importance of such studies and the attendant difficulties. Methylmercury and the other short chain alkyl mercurials produce effects unique among the mercury compounds. However, some—if not all—of these effects can be caused by agents other than mercury or by certain disease states. Thus, in examining the population for neurological changes, it must be borne in mind that there could be, at least in theory, many causes of neurological changes other than methylmercury itself. The dose–response relationships we will discuss imply a cause-effect relationship. In fact, the only proof we have that methylmercury causes certain effects is that (1) these effects coincide in time with exposure to methylmercury, and (2) that the frequency of these effects in a given population increases with increasing exposure to methylmercury. One of the key problems in these studies, thus, is to distinguish between the background frequency

---

†This work was sponsored by grants from the NSF (RANN) (G-300978), NIGMS (GM 15190) (GM 01791), NIAMDD (AM-05177) NINDS (NS-5084), and by the Tuna Research Foundation.

of a sign or symptom and the increase in that frequency due to increased exposure to methylmercury.

JAPANESE POPULATIONS

Data on dose–effect relationships in the Japanese populations were analyzed by a Swedish Expert Committee.[5] After admission to the hospital, the blood levels of total mercury of each patient declined, with a half-time of approximately 70 days (Fig. 1). The blood level at the time of onset of symptoms was estimated by back-extrapolation of the lines in the figure. It was concluded that the lowest blood level for the appearance of signs and symptoms of methylmercury poisoning was 200 parts per billion (ppb) (0.2 ppm).

Studies on the pharmacokinetics of methylmercury in man enable one to relate blood levels to daily intake of methylmercury. It was concluded from studies of volunteers taking a tracer dose of radioactive methylmercury, along with experimental work on animals, that mercury was completely absorbed from food and was distributed rapidly throughout the body, that the body behaved kinetically as a single compartment, and that the elimination of methylmercury from the body could be described by a single half-time with an average value of 70 days.[9, 10] This information allowed one to develop a simple compartment model depicting the total body burden of mercury as a function of time when a constant daily dose of methylmercury was ingested (Fig. 2). The body burden becomes essentially steady after one year. Thereafter, excretion balances intake. In this steady state the total level in the body is proportional to the daily intake. Furthermore, isotopic studies indicate that the amount in the blood is an approximately constant fraction of the amount in the body,[10] so we can assume that blood levels also become constant after one year and are proportional to the daily intake over this period.

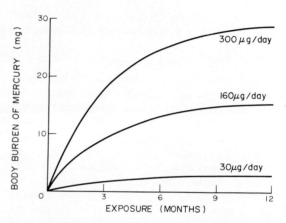

FIG. 2. Theoretical relationships between the body burden in man and the time of exposure to methylmercury compounds in food. Each curve corresponds to a constant daily intake of mercury. From Clarkson.[11]

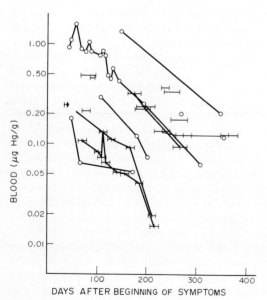

FIG. 1. The relationships between the concentration of total mercury in whole blood and the time after onset of symptoms. From Berglund et al.[5]

This assumption was confirmed by studies on fishermen having a relatively high daily intake of methylmercury (Fig. 3). Blood levels were linearly related to the daily intake in this population, which is believed to be in steady state—i.e., the fishermen had been on more or less the same diet of fish for at least a year. Two relationships are shown in Fig. 3. The upper line is believed to be the correct one; the lower is believed to be in error due to difficulties in the measurement of dietary intake. Obviously, the conclusion on linearity of either of these relationships depends a great deal on the datum point for the individual with the highest mercury intake. Nevertheless, this was the best information available, and it was concluded that a blood level of 200 ppb (the lowest level associated with symptoms) would be attained by a minimum daily intake of approximately 300 µg of mercury present as methylmercury in the diet.

Strictly speaking, the results of these studies do not amount to a dose–response relationship. The analysis was directed toward determining that individual who was most sensitive to methylmer-

THE IRAQ POPULATION

The large epidemic of methylmercury poisoning in Iraq provided another opportunity to study dose–response relationships.[4] Approximately 3000 males and slightly more females were admitted to hospitals throughout the country (Table 1). The total number of cases admitted was 6530. The death rate was six to seven per hundred hospitalized patients. The rate was higher in pregnant patients, but there were too few such patients for statistical accuracy. It is not known to what extent these data on hospital admissions reflect the full extent of the epidemic. Iraq has a population of 10 million people divided equally between urban and rural areas. The poisoning resulted from the consumption of homemade bread prepared in farming families from wheat treated with a methylmercury fungicide. The wheat had been distributed to virtually all the agricultural areas in the south, center, and north of the country.

Of special importance for future studies in this population is the fact that large numbers of women of child-bearing age were exposed to methylmercury. As will be discussed later, this population will allow the study of the prenatal effects of methylmercury. However, the dose–response relationships studied to date have been based on exposure of adults.

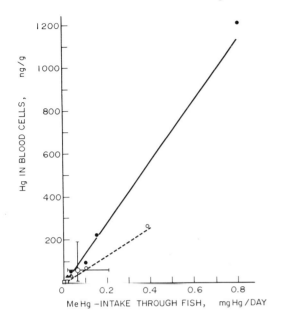
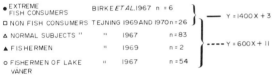

FIG. 3. The relationship between the concentrations of mercury in the red blood cell and the intake of methylmercury from fish. From Berglund et al.[5]

cury in the exposed Japanese population. The evidence that his symptoms were due to methylmercury is (1) that they appeared simultaneously with, or shortly after, exposure to methylmercury, and (2) that he belonged to a highly exposed population whose members exhibited signs and symptoms which were at once similar to his and typical of those observed in previous cases of methylmercury poisoning. The difficulty with this approach is that we do not know to what extent this particular individual has a sensitivity common to other persons. We do not know, in other words, the probability of finding other individuals as sensitive as this one in the Japanese population as a whole, or in any other population. We have no rational basis for extending this type of observation to the whole population. In setting standards for a whole population, it is customary to apply a safety factor (usually 10) to the data. In this case it would mean that the maximum safe intake would be 10 times lower than the level that would produce 200 ppb of mercury in blood, or no more than 30 μg of mercury in the form of methylmercury.

TABLE 1
Total distribution of cases of poisoning admitted to hospitals in Iraq according to age[a]

Patients	Number	Deaths	
		No.	Percent of cases
Males	3144	197	6.3
Nonpregnant females	3353	248	7.4
Pregnant females	33	14	42.0
Total	6530	459	7.0

[a] Data from Bakir et al.[4]

The time course of the body burden of methylmercury in this population is diagrammed in Fig. 4. The first ingestion of the contaminated bread took place in late 1971 and continued for about 2 months, the rate of intake of methylmercury being very high in many cases. The ingestion period is indicated by the hatched areas in the figure. At the end of consumption, the body burden of methylmercury would be expected to decline, with a half-time of about 70 days, so that at the end of one year mercury levels should be close to normal. Thus this

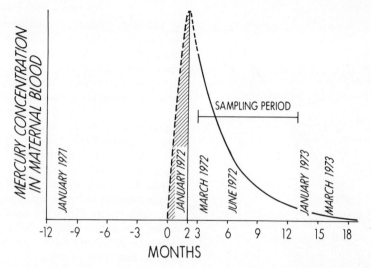

FIG. 4. A diagrammatic representation of methylmercury in maternal blood (or the body burden of methylmercury) throughout the Iraq epidemic. According to a previous report,[4] the average period of consumption of the contaminated bread was 2 months. The blood concentration of methylmercury before (broken line) and during (solid line) the sampling period was calculated assuming a 70 day half-time clearance from blood from Amin-Zaki et al.[16]

population, with exposure time of less than a year, differs from that of Japan, where contaminated fish were consumed for many years by some of the population. It also differs from the fishing populations to be discussed next, in which exposure may have been for a lifetime. Dose–response relationships studied so far in the Iraq population were carried out in the time period shortly after maximum exposure. The blood samples for these studies and for clinical examinations were obtained in March, April, and May 1972, some time after the point of maximum body burden. It was therefore necessary to find some means of recapitulating exposure and of determining the dose received by the individual.

It was possible to relate the concentrations of mercury in blood samples collected in this post-exposure period to the ingested dose of mercury (Fig. 5). This relationship was established for a limited number of patients who were able to give information on the number of loaves they had ingested per day and the period over which they had consumed the contaminated bread. The total dose of methylmercury was then calculated from the weight and flour content of the loaves and analysis of samples of the contaminated wheat and the flour prepared from that wheat. The methylmercury content of the first blood samples was linearly related in adults (Fig. 5, lower graph) and in the 10–15 year age group (upper graph) as indicated by

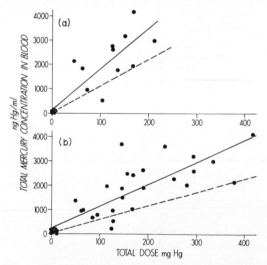

FIG. 5. The relationship between the concentration of total mercury in blood and the amount of mercury ingested from contaminated bread. The solid line was drawn from linear regression analysis. The broken line is the predicted concentration in the blood estimated from data published by Miettinen[10] on 15 volunteers given a single oral dose of labeled methylmercury. (a) Patients aged 10–15 years; (b) patients more than 18 years of age. From Bakir et al.[4]

the least square linear regression line. The steeper slope in the younger age group would be expected since body weight (and therefore volume of distribution of methylmercury) was smaller in this group. The average body weight in the adult group was 51 kg and in the younger group 31 kg.

The observed relationship between blood levels and estimated dose compares favorably with that established for data on volunteers given trace doses of methylmercury.[10] Specifically, the parameters used from Miettinen's study were that (1) methylmercury was completely absorbed from food, (2) approximately 1% of the body burden of methylmercury was found in one liter of blood, (3) this fraction changed slightly with time, decreasing with a half-time of approximately 170 days, and (4) mercury was eliminated from the body with a half-time of approximately 70 days. From this information, we were able to plot the predicted blood level (broken line, Fig. 5) under the circumstances of this exposure. Although the predicted and observed lines are not identical, they appear to be remarkably close and do not differ to a statistically significant extent. In fact, subsequent studies on hair samples indicate that the discrepancy between the two lines may be due to errors in patients' recapitulation of the period of ingestion of the contaminated bread.[4]

The success of the Miettinen line on Fig. 5 in simulating blood levels under the exposure conditions of Iraq indicates an interesting property of methylmercury. Miettinen's studies were carried out with tracer quantities of radioactive methylmercury given as a single dose. The conditions in Iraq included wide ranges of doses of methylmercury and repeated daily doses over a period of 2 months or more. The fact that observations made on a single dose of tracer quantities of methylmercury can give parameters that describe a situation under chronic dosing indicates that each dose of methylmercury behaves independently of the other. That is to say, during chronic or repetitive exposures as in Iraq, we can regard each daily dose as behaving identically to that described by Miettinen. This property of independence of successive doses greatly assists us in the prediction of the pharmacokinetic behavior of methylmercury in man under conditions of long exposure. It is noteworthy that the same property has been demonstrated in experimental animals and that it is a property shared by the other compounds of mercury, including metallic mercury vapor (for a detailed discussion, see Clarkson[11]). Clearly this generalization will hold only to the extent that the damage inflicted by extremely high doses of methylmercury does not influence the kinetics of distribution and elimination of mercury from the body.

The linear regression line and the Miettinen line in these figures were then used to convert blood levels to daily intakes of patients who could not supply sufficient information on the ingested dose. In these cases, the problem was that the patients had either washed the wheat or had mixed the contaminated wheat in an unknown proportion with noncontaminated wheat. Thus by using the determination of methylmercury in the first blood sample, we were able to estimate the total ingested dose according to the relationships shown in Fig. 5.

Two dose–response relationships are shown in Fig. 6. The frequencies of signs and symptoms were plotted against the body burden of methylmercury (logarithmic scale) at the time of cessation of exposure (lower graph) and at the time of onset of symptoms (upper graph). The body burden was calculated from the total ingested dose, as described previously.[4] The patients were divided into groups according to their estimated body burden; each group consisted of approximately 20 patients or subjects. The abscissa has two scales, the upper

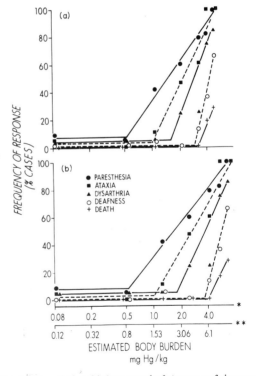

FIG. 6. The relationship between the frequency of signs and symptoms and the estimated body burden of methylmercury. (a) At the time of onset of symptoms, (b) at the time of cessation of ingestion of methylmercury in bread. Both scales on the abscissa are for body burdens of methylmercury calculated as described by Bakir et al.[4] from the linear regression lines of Fig. 5(b). From Bakir et al.[4]

based on the linear regression line, the lower based on the Miettinen line for estimating dose from blood levels. In each group the frequency of the specified signs and symptoms is reported. For example, the solid line connecting the solid circles refers to the frequency of paresthesia in this population. The two groups having the lowest body burdens gave a frequency of about 1 in 20, or 5%. The frequency increased in proportion to the log of the body burden above this point. Thus, the effects of body burden of mercury become evident at the point of intersection of these two lines; this would be in the range of 0.5–0.8 mg Hg/kg, depending on which scale one uses. These figures would correspond to 40–60 mg Hg for a 70 kg man.

The other signs and symptoms of poisoning became evident at higher doses of methylmercury, but the general picture is the same as with paresthesia. There was a background frequency of the sign or symptom, which was not related to the mercury level, and then an increase in frequency related to the mercury level at high doses of mercury. The deaths associated with methylmercury became apparent at doses of the order of 4–6 mg/kg. The upper graph, based on the body burden at the time of the appearance of symptoms, gave virtually the same data. The results of this study have been taken to indicate that an effect of mercury can be detected at approximately 50 mg for a 70 kg man in the sense that there is an increase over the background frequency of complaints of paresthesia in this population. In other words, there is a 5% probability of mercury causing paresthesia at this body burden. By greatly extending the number of patients studied and by more critical examination of patients claiming to have paresthesia, it might be possible to increase our sensitivity so that we could discuss probabilities of 1%. However, the limit on this type of dose–response analysis seems to be the background frequency of the subjective symptoms. Since there are many causes of paresthesia, a background frequency in any population is to be expected. The effects of mercury cannot be detected until this background frequency is significantly exceeded.

FISH-EATING POPULATIONS

Several populations having high fish consumption have been studied. Birke et al.[12] reported results of examinations of 12 subjects having unusually high intakes of methylmercury from fish. We are studying fishermen in American Samoa who are exposed to high fish intake for periods of up to 2 years or more. The concentration of mercury and its chemical state in blood samples from these individuals were determined by two independent methods. Selective atomic absorption was used to measure total mercury and inorganic mercury, and the concentration of organic mercury was calculated as the difference between total and inorganic mercury. Gas chromatography was used to determine methylmercury. Figure 7 compares the values for organic mercury with those for methylmercury. The line has a gradient near unity and passes through the origin. The correlation coefficient between the two sets of numbers is 0.97.

The distribution of methylmercury in blood in 163 persons is depicted in the histogram in Fig. 8. Most of this population lies in the 20–100 ppb range, significantly above the normal levels (usually taken as less than 10 ppb). Furthermore, several individuals can be found in the 100–150 range and one or two persons in excess of this.

These subjects were given a thorough neurological examination, and some of the results are given in Fig. 9, plotted in comparison with the Iraqi data. The data from Iraq were taken from Fig. 6 but plotted in the form of a histogram. The line depicting the frequency of paresthesia in Fig. 6 would pass through the center of the tops of each column in the histogram. The Iraqi and Samoan

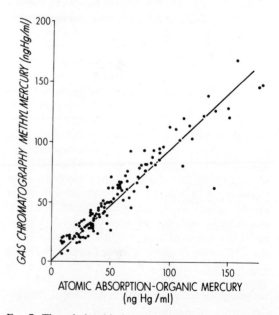

FIG. 7. The relationship between organic mercury determined by atomic absorption and methylmercury determined by gas chromatograpy in blood samples from Korean contract fishermen based in American Samoa. The equation of the line is: $y = 0.95x + 0.79$. $r = 0.97$. Data from Smith et al.[13]

increased frequency of signs and symptoms in comparison with the Iraqi population.

Skerfving[3] reported a dose-response curve calculated from data from Japan and from studies of Swedish fisherman having high intakes of methylmercury from fish (Fig. 10). The frequency of signs and symptoms were related to concentrations of mercury in hair. The approximate blood level can be obtained by dividing the hair concentrations by 300. The apparent threshold effect in Fig. 10 corresponds to 60–90 ppm in hair and to a blood level of 200–300 ppb. Using Miettinen's metabolic data referred to above, this blood level would indicate a body burden of 20–30 mg.

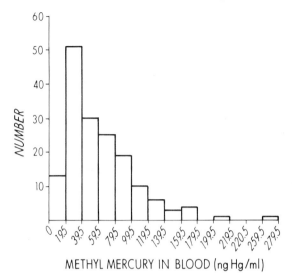

FIG. 8. A histogram indicating the blood concentrations of methylmercury in 163 Korean contract fishermen based in American Samoa. Data from Smith et al.[13] and Marsh et al.[14]

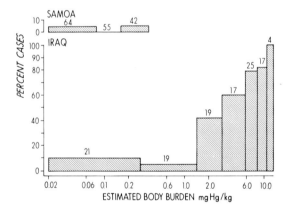

FIG. 9. A histogram indicating the frequency of symptoms of paresthesia in Korean contract fishermen based in American Samoa and in the Iraqi population plotted as a function of the estimated body burden. Data from Bakir et al.[4] and Marsh et al.[14]

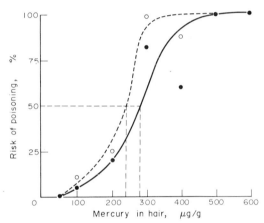

FIG. 10. The dose-response relationships according to hair levels of total mercury estimated from the Japanese epidemic and from the Swedish families having a high intake of fish. From Skerfving.[3]

dose-response relationships are in agreement. The range indicated by the Iraqi dose-response as background frequency is also the range in which the Samoan results fall and where there appears to be no correlation with mercury exposure. Background frequency in the Samoan group runs about 5% and is similar to the Iraqi findings. The importance of these results lies in the fact that the Samoan population represents a much longer exposure to methylmercury, and yet shows no

In summary, the results of studies on populations having brief exposure (Iraq), an exposure of more than one year (Japan), and extended exposure over many years (fish-eating populations) yield more or less similar dose-response relationships. The adverse effects of methylmercury become detectable at body burdens of 0.3–0.9 mg Hg/kg, with corresponding blood concentrations of 200–600 ppb. The length of exposure does not appear to be the critical factor, since the dose-response curves in the populations studied agree despite the greatly different periods of exposure. The important factor determining the damage thus appears to be the maximum level of methylmercury obtained, presumably in the brain.

Observations of the Minamata epidemic and more recent reports on experimental animals suggest that prenatal life is the most sensitive to attack by methylmercury.[6] If this is the case, then the dose–response curves in adults have limited value in determining safe levels of methylmercury for human beings. Our ignorance about the dose–relationship in prenatal and early postnatal life in humans is one reason why a Swedish expert committee included a safety factor of ten in extrapolating the dose–response curve in adults to the general population.[5] This safety factor is of critical importance from both the public health and the economic points of view. Economically, this safety factor reduces the maximum allowable intake of methylmercury in food to about 30 μg/day, and has led to restriction in the sale of fish having high concentrations of methylmercury. On the other hand, an arbitrary safety factor is hardly the most rational way to protect people's health. Of more importance would be an effort to relieve our ignorance in the area that this safety factor is supposed to cover. Future studies on dose–response relationships should be directed toward the question of effects on early stages in the human life cycle.

The population of women exposed to methylmercury in the Iraq epidemic offers a promising source for this type of information. The infants born before the epidemic (see Fig. 4) would be exposed only postnatally, through ingestion of methylmercury in the maternal milk. Those born after the beginning of the epidemic would be exposed both prenatally and postnatally from the milk. Preliminary findings on these two groups of infants have been reported by Amin-Zaki et al.[15,16]

The data in Fig. 11 show declining blood and milk levels of mercury in an infant–mother pair in which the infant was exposed only postnatally. The first measurements were made in July 1972, about 6 months after the mother had consumed contaminated bread. At the time of sampling the infant was one year old and was therefore about 6 months old at the time the mother ingested the mercury. Exposure of this infant was entirely from milk, since the infant was fully breast-fed without any additional food.† The maternal blood level was higher in the early samples than that in the infant, but in August the infant blood levels exceeded the maternal levels and remained higher from that time on. One reason for the slower decline in the infant blood level is the continued ingestion of methylmercury from the mother's milk. Concentrations of

---

†It is customary in rural Iraq to feed infants in this way. After the age of one year they begin to receive some solid food, and breast feeding is stopped at about the end of the second year.

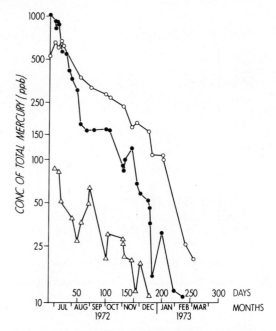

FIG. 11. The concentration of total mercury in infant blood (open circles) and mother (solid circle) blood and in milk (triangles) when the infant's exposure to methylmercury was only via the mother's milk. The date of birth was June 1971. From Amin-Zaki et al.[15]

methylmercury in milk are considerably below those in blood, as reported previously.[4] Nevertheless, these levels are sufficient to produce substantial levels of methylmercury in infant blood. For example, the maximum infant blood level in Fig. 11 is 500 ppm, a value within the so-called threshold range of concentration associated with symptoms in adults.

Unfortunately, very few infant–mother pairs could be sampled for the prolonged period of time indicated in Fig. 11. However, it was possible to get single samples, and in some cases a few samples, from several infant–mother pairs, and to determine whether the ratios of infant-to-mother blood levels of mercury changed as in this figure. The ratio was considerably less than unity for March through June, indicating that the infant blood level was less than that of the mother (Fig. 12). In July and August the ratio was about unity, and thereafter the infant blood level was higher than the maternal level. This build-up of mercury in infant blood is what one would expect when mercury is transferred from the mother to the infant through milk. One would expect in this two-compartment model a slower build-up in the infant as mercury moves from the mother to the infant, reaching a maximum in the infant well after

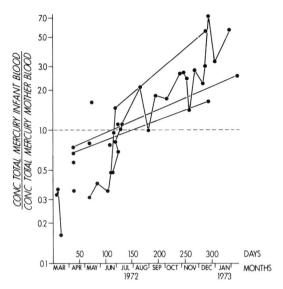

FIG. 12. The ratio of the concentration of total mercury in infant blood to the concentrations in mother blood in infants having postnatal exposure. The Iraqi infant–mother pairs were sampled over the period March 1972 to January 1973. Data points for the same infant–mother pair are connected by straight lines. From Amin-Zaki et al.[15]

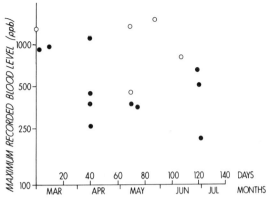

FIG. 13. The maximum recorded concentrations of mercury in infant blood plotted according to the date at which the samples were collected. The solid circles are from infants having exposure to methylmercury only via mother's milk. The open circles refer to infants who may also have ingested some contaminated bread. From Amin-Zaki et al.[15]

the mother's maximum is attained. The infant blood would attain a lower maximum level and would decline at a rate slower than that of the mother, since the infant continues to receive methylmercury as long as there are significant levels in the mother. Thus the infant with postnatal exposure from milk would carry significant body burdens of methylmercury for a longer period of time than the mother. But if maximum levels of methylmercury determine effects, rather than period of exposure, the infant would appear to face less hazard than the mother. However, we do not know whether the infant's tissues may be less sensitive, or more sensitive, to methylmercury than are the adult tissues.

Infants aged 18 months or less who were examined in the summer of 1972 (approximately 6 months after the outbreak of the epidemic) had maximum levels shown in Fig. 13. At the time of examination no adverse effects were apparent in these infants; they seemed to be perfectly happy, healthy, normal babies. This finding was surprising in view of the high blood levels in some of them. All the infants but one had maximum blood levels about 250 ppb and at least three had blood levels in excess of 1000 ppb. According to the dose–response curve from the Iraq data, such a blood level in adults would give a 50% probability of neurological deficits.

Amin-Zaki et al.[16] reported on studies of 15 infant–mother pairs exposed prenatally to methylmercury. In Fig. 14 the ratio of mercury in infant blood to that in maternal blood is plotted against the age of the infant in months at the time of sampling. The samples taken from the same infant–mother pair are connected by a line. The small numbers adjacent to the points indicate the month of gestation in which consumption of the contaminated bread ceased as reported by the mother. The infants were examined from March to November of 1972; all were less than one year of age and most were less than 6 months.

From Fig. 14 it is obvious that, unlike the ratios in the postnatal exposures, most of the blood level ratios are substantially above unity at the time of birth. The mercury levels in these infants remained generally higher than the maternal levels for several months after birth.[16] It has previously been reported that levels of mercury in cord blood were about 20% higher than those in maternal blood (Tejning, cited by Berglund et al.[5]), and animal experiments have indicated values in the fetus twice as high as those in the mother under conditions of low dosage of methylmercury.[17]

No detailed kinetic analysis has been attempted in these infant–mother pairs. However, one reason for the high mercury levels in the infant at birth may be the differences in hematocrit between maternal and infant blood. Many of the mothers in Iraq were anemic and had low hematocrits; values of 30% were not uncommon. Infants, on the other hand, are

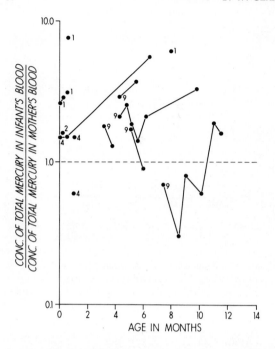

FIG. 14. The ratio of the concentration of total mercury in infant blood to the concentration in mother blood in infant–mother pairs according to the age of the infant. Points from the same mother–infant pair are connected by straight lines. The number adjacent to the points indicates the estimated month of gestation when the mother stopped eating the contaminated bread. From Amin-Zaki.[16]

known to have high hematocrits at birth—about 60%. Since most of the methylmercury in blood is attached to red blood cells, this difference in hematocrit could account for a twofold difference between mercury levels in infant and maternal blood. An infant and mother could thus have the same plasma levels of methylmercury, but the whole blood levels could differ by a factor of 2. If tissue levels are determined by the concentration of mercury in plasma, the infant would not necessarily face a greater hazard than the mother even with blood level ratios as high as 2. On the other hand, these high ratios persist in this population and may even increase with time, undoubtedly due to the consumption of methylmercury in the mother's milk, whereas the high hematocrits in infants persist for only a few days. The role of hematocrit differences between mother and infant is therefore not established. Furthermore, differences in hematocrit could not explain elevations of more than a factor of 2 in the infant blood level ratio, and as can be seen in Fig. 14, in many cases the ratio was greater than 2.

In the 15 infant–mother pairs studied by Amin Zaki et al.[16] symptoms of methylmercury poisoning were observed as follows. Five mothers had paresthesia, four had changes of vision (usually constriction of the visual fields), two had motor weakness, and other signs and symptoms occurred occasionally in other mothers. Ataxia, auditory changes, and dysarthria were not frequent. Five of the 15 infants had signs of poisoning; four of these were severely affected, with generalized spasticity, hyperactive reflexes, blindness, and deafness, and the fifth had milder manifestations, such as poor head control and poor grasp reflex. The relationship of signs and symptoms of poisoning to the observed blood levels is shown in Fig. 15. Although the number of cases are clearly insufficient to establish any statistical dose–response relationship, the results are interesting because they show that the signs and symptoms in the infant do not appear at remarkably different levels from those of the mothers. All mothers who had blood levels in excess of 400 ppb exhibited symptoms of poisoning; seven mothers who had levels of 5–400 ppb had no such signs and symptoms. As pointed out earlier, these are by no means the maximum blood levels, since the samples were collected several months after exposure. The lowest concentration of mercury in blood associated with signs and symptoms of poisoning in the mothers was 300 ppb, recorded on August 7, 1972. The next highest level, 416 ppb, was recorded on June 1. Allowing for differences in time of collection, these two levels would be more or less equivalent.

An exhaustive evaluation of the methylmercury poisoning in the infants is not possible at this time. Some infants were examined only 3 or 4 days after birth; others were examined at the age of 6 months. The infants examined in March and April 1972 who had blood levels over 3000 ppb were severely affected. One infant having a blood level of 1053 when examined on August 6 was also severely affected; assuming a clearance half-time of 70 days, this blood level would have been about 3000 ppb in March and April. The lowest blood level associated with signs and symptoms of poisoning was 565 ppb in an infant examined in August 1972. The symptoms were mild. It is noteworthy that seven infants had blood levels in the range of 122–636 ppb and were free of signs of poisoning.

Some mothers had signs of methylmercury poisoning and their infants did not. For example, a mother having a blood level of 416 ppb on the first of June had signs of poisoning, but her infant, aged $1\frac{1}{2}$ months, had a blood level of 636 ppb without signs of

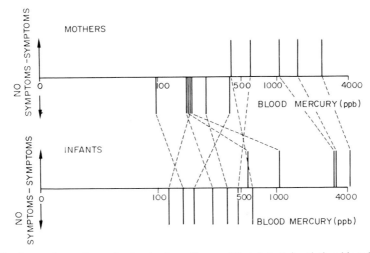

FIG. 15. The signs and symptoms of poisoning in mothers and their infants in relationship to blood levels of total mercury. Blood concentrations are plotted horizontally on a logarithmic scale. A vertical line indicates a single patient. A vertical line placed above the horizontal line indicates a patient having signs and symptoms of poisoning. A vertical line starting below the horizontal line indicates a patient without signs and symptoms of poisoning. Each infant–mother pair is connected by a broken line. Blood samples were collected from the mothers and infants at the same time. The samples from the 15 infant–mother pairs described in this figure were collected over the period of March to November 1972. Data from Amin-Zaki.[16]

poisoning. In another pair, examined on July 15, 1972, the mother had signs and symptoms of poisoning and a blood level of 602 ppb, whereas her 6-month-old infant had a blood level of 476 but no signs of poisoning.

It is clear that these preliminary data from Iraq are not adequate to permit conclusions to be drawn about the dose–response relationships for prenatal exposure. Besides the small number of cases involved, blood levels are not an accurate means of indicating exposure when the samples have been taken many months after the exposure to methylmercury. To establish dose–response relationships for prenatal exposure, one would have to (1) greatly increase the number of infants on the study who have been prenatally exposed, (2) collect hair samples from each mother to permit a recapitulation of mercury exposure throughout the life of the fetus and the first year of postnatal life, and (3) follow up clinically a large population of these infants for many years. Collection and analysis of hair samples would give the most reliable data on exposure and would help answer such important questions as: How much methylmercury can the mother ingest before damage to the unborn child results? At what stage in gestation is the fetus most sensitive to methylmercury? Are there sex differences in the sensitivity to prenatal exposure to methylmercury?

Does early postnatal exposure to methylmercury produce similar damage to that occurring prenatally?

REFERENCES

1. KUTSUNA, M. (ed.), *Minamata Disease*, Study Group of Minamata Disease, Kumamoto Univ., Japan, 1968.
2. TSUBAKI, T., Clinical and epidemiological aspects of organic mercury intoxication. In *Mercury in Man's Environment*, p. 131, Royal Society of Canada, Ottawa, 1971.
3. SKERFVING, S., Mercury in fish—some toxicological considerations, *Food Cosmet. Toxicol.* **10**, 545–556 (1972).
4. BAKIR, F., DAMLUJI, S. F., AMIN-ZAKI, L., MURTADHA, M., KHALIDI, A., AL-RAWI, N. Y., TAKRITI, S., DAHIR, H. I., CLARKSON, T. W., SMITH, J. C., and DOHERTY, R. A., Methyl mercury poisoning in Iraq, *Science* **181**, 230–241 (1973).
5. BERGLUND, F., MERLIN, M., BIRKE, G., CEDERLOF, R., VON EULER, U., FRIBERG, L., HOLMSTEDT, B., JONSSON, E., LUNING, K. G., RAMEL, C., SKERFVING, S., SWENSSON, A., and TEJNING, S., Methyl mercury in fish, *Nord. Hyg. Tidskr.*, Suppl. 4 (1971).
6. KHERA, K. S., Reproductive capability of male rats and mice treated with methylmercury. *Toxicology and Applied Pharmacology* **24**, 167–177 (1973).
7. RAMEL, C., Genetic effects of organic mercury compounds, 1. Cytological investigations on allium roots, *Hereditas* **61**, 208–230 (1971).

8. SKERFVING, S., HANSSON, A., and LINDSTEN, J., Chromosome breakage in human subjects exposed to methylmercury by fish consumption, *Arch. Environ. Hlth.* **21**, 133–139 (1970).
9. ABERG, B., EKMAN, L., FALK, R., GREITZ, U., PERSSON, G., and SNIHS, J. O., Metabolism of methyl mercury (^{203}Hg) compounds in man; excretion and distribution, *Arch. Environ. Hlth.* **19**, 478–484 (1969).
10. MIETTINEN, J. K., Absorption and elimination of dietary mercury (Hg^{++}) and methyl mercury in man. In *Mercury, Mercurials and Mercaptans* (M. W. Miller and T. W. Clarkson, eds.), pp. 233–243, Charles C. Thomas, Springfield, Ill., 1973.
11. CLARKSON, T. W., Recent advances in the toxicology of mercury with emphasis on the alkyl mercurials. *CRC Critical Reviews in Toxicology* (L. Goldberg, ed.), Vol. 1, No. 2, pp. 203–233, 1972.
12. BIRKE, G., JOHNELS, A. G., PLANTIN, L. O., SJOSTRAND, B., SKERFVING, S., and WESTERMARK, T., Studies on humans exposed to methyl mercury through fish consumption, *Arch. Environ. Hlth.* **15**, 77–91 (1972).
13. SMITH, J. C., CLARKSON, T. W., TURNER, M. D., and MARSH, D. O., Total, inorganic, and methylmercury levels in hair and blood, I. American Samoa. In preparation.
14. MARSH, D. O., TURNER, M. D., SMITH, J. C., CHOI, J. W., and CLARKSON, T. W., Methylmercury in human populations eating large quantities of marine fish, I. American Samoa; cannery workers. In preparation.
15. AMIN-ZAKI, L., EL-HASSANI, S., MAJEED, M. A., CLARKSON, T. W., DOHERTY, R. A. and GREENWOOD, M., Studies of infants postnatally exposed to methyl mercury, *J. Pediat.* **85**, 81–84 (1974).
16. AMIN-ZAKI, L., EL-HASSANI, S., MAJEED, A., CLARKSON, T. W., DOHERTY, R. A., and GREENWOOD, M., Intrauterine methyl mercury poisoning in Iraq, *Am. J. Pediat.*, in press (1974).
17. CHILDS, E. A., Kinetics of transplacental movement of mercury fed in a tuna fish matrix to mice, *Arch. Environ. Hlth.* **17**, 50–52 (1973).

A Review of Dose–Response Relationships Resulting from Human Exposure to Methylmercury Compounds
(T. W. Clarkson et al.)

DISCUSSION by Leonard J. Goldwater
Department of Community Health Sciences, Duke University, Durham, North Carolina

First I should like to pay homage to Dr. Clarkson for his many contributions to our present knowledge of the toxicology of mercury compounds and particularly to his 1961 monograph "The General Pharmacology of the Heavy Metals" written in collaboration with Rothstein and Passow, published in *Pharmacological Reviews*. This work has served as a sort of "bible" for me over the years and can be read with profit by anyone concerned with heavy metal toxicology. Second, I should like to commend Dr. Clarkson and his associates at the University of Rochester and the University of Baghdad for the vigor, skill and devotion with which they have investigated the recent outbreak of methylmercury poisoning in Iraq. Their efforts and achievements can only be appreciated by one who has attempted to conduct epidemiological and clinical studies in the field. The problems are quite different from those facing the "rat and mouse fraternity." Dr. Clarkson aptly pointed out the difficulties the research team faced when he mentioned "guessing, estimations and extrapolations," particularly the latter, with curves based on a single point.

The Iraq experience illustrates, once more, the contrast of mass poisoning which results from gross contamination of food (previously seen in Japan and elsewhere) and the absence of effect from the ingestion of presumably "contaminated" fish as observed in the Lake St. Clair region, the Aleutian Islands, and the Cape Fear basin in North Carolina. Dr. Clarkson's preliminary findings among the Samoan tuna fishermen and those in Peru further underline this contrast.

I was interested in the mention of paresthesias as a frequently reported symptom among the Iraqis. I found a similar situation in the persons studied in the North Carolina investigation. Dr. Clarkson questions the significance of this as do we, particularly since, in our studies, there was no correlation between the occurrence of this symptom and the amount of local fish and game consumed or the levels of mercury in urine or blood. Clinical toxicologists know that conclusions based on reported subjective symptoms must be interpreted with the greatest of caution. Lack of correlation between mercury concentrations in blood and urine, on the one hand, and clinical manifestations of toxicity, on the other, has been demonstrated repeatedly in human studies. This is corroborated in Dr. Clarkson's findings.

Of great interest in Dr. Clarkson's studies is the difference in the reaction of fetuses and neonates to the mercury intake of the mothers. It appears that mercury entering through placental transfer is more dangerous than that coming through maternal milk. This suggests, among other things, that there may be some form of binding in the milk which to some extent detoxifies the mercury. (Passing mention may be made of the fact that maternal milk was once used as a vehicle for treating congenital syphilis with mercury.) Another possible explanation is that those fetuses which survive to the time of delivery have acquired a tolerance based on repeated uptake of small amounts of mercury. The Iraq studies so far do not shed any light on the question of chromosomal injury of a type that could result in true genetic damage. In this connection it is worth mentioning that recent follow-up studies of the Minamata victims have not revealed any evidence of chromosomal abnormalities in excess of those found in control populations. Fetal abnormalities appear to be a result of injury to the embryo after rather than before fertilization of the ovum by the sperm. The Minamata findings confirm what has long been suspected: that humans do not necessarily behave like onion tips or fruit flies.

The term "dose–response relationship" receives major attention in Dr. Clarkson's paper as well as in many articles in the literature. Unfortunately the term is rarely defined by those who use it. All too often it seems to refer to the relationship between known or (more frequently) assumed intake and amounts of something found in tissues or body fluids. As a clinician I find this definition unacceptable if for no other reason than the well-known phenomenon of biochemical individuality, sometimes referred to as individual susceptibility or host factors, in response to potentially harmful agents. To me, response means a demonstrable or measurable aberration in body structure or function. If the only "abnormality" is an elevation of the level of mercury (or other material) above some arbitrary "normal" value the "response" has little or no clinical significance as far as any individual exposed person is concerned.

Dr. Clarkson made a point to the effect that each daily or repetitive dose behaves independently of previous doses. I presume that by this he meant to imply that this is true in a chemical sense only. After the intake of any foreign substance the body is no longer the same as it was before the intake, there being a number of protective or adaptive mechanisms which alter the body's response to subsequent stresses. In extreme cases an initial dose can be highly damaging or even lethal, making subsequent doses more dangerous or impossible. (Note: In a post-meeting discussion Dr. Clarkson expressed agreement with my interpretation.)

Another assumption appearing in Dr. Clarkson's calculations (and elsewhere) had to do with the 70 day clearance half-time for methylmercury. For the most part there seems to be an idea that in applying this figure the body behaves as a single, homogeneous compartment. On purely anatomical and biochemical grounds, such an assumption is patently questionable since all tissues are not chemically alike. The differential uptake of methylmercury in different organs and even in different parts of the central nervous system has been demonstrated by Berlin and by others. Differences in the rate of elimination from different organs have also been demonstrated experimentally and studies on human tissues performed at Kumamoto University have confirmed this. This means that calculations based on a "single compartment" theory are open to serious question, at least as far as clinical significance is concerned. Most important is the questionable validity of using blood levels as an index of tissue levels.

So much important information has been presented by Dr. Clarkson that one highly significant point, mentioned in passing, nearly escaped attention. I refer to his statement that he found evidence of demethylation of methylmercury in the liver. This certainly deserves further exploration and emphasis, particularly in light of the recent demonstration of biological demethylation in aquatic ecosystems.

In closing I should like to "make it perfectly clear" that the few questions I have raised should not be interpreted as detracting in any way from the value and significance of the information being collected by Dr. Clarkson and his associates. Their observations constitute an invaluable addition to the understanding of methylmercury toxicity.

TOXICITY OF CADMIUM—MECHANISM AND DIAGNOSIS

Kazuo Nomiyama

Professor, Department of Environmental Health, Jichi Medical School, Minamakawachi-Machi, Tochigi-Ken 329-04, Japan

INTRODUCTION

The so-called Minamata and itai-itai diseases are serious conditions caused apparently by heavy metals in the aquatic environment. Minamata disease was already mentioned by Dr. Clarkson under the title of "toxicity of mercury"; and I will discuss here the toxicity of cadmium in relation particularly to itai-itai disease.

Itai-itai disease was endemic between 1940 and 1965 in the Zinzu river area in Japan. Most patients were middle-aged or elderly females, who suffered from severe pains all over the body and died in pain. Itai-itai disease was named by Dr. Hagino after patients' cries "itai-itai". Itai-itai means "ouch-ouch" in English (Fig. 1).[1] The first symptom of the disease was usually lumbago, followed by the later development of pseudofractures and a waddling gait. In May of 1968, Japan Ministry of Health and Welfare took an official position on the identification and cause of itai-itai disease.[2] "The etiology of itai-itai disease is related to dietary cadmium and malnutrition during and after the Second World War." Cadmium has been known, as will be discussed in detail later, to inhibit renal function. Increased loss of calcium due to depression of proximal reabsorption was thought to lead finally to osteomalacia. Slight movements set up fractures, which brought the patients intolerable pains. The etiological agent of itai-itai disease, cadmium, was thought to be derived from the Mitsui Mine, a source of zinc, lead, and cadmium approximately 50 km upstream on the Zinzu river.

On the other hand, cadmium poisoning among cadmium workers has been characterized by a triad of signs—emphysema of the lungs, dysfunction of the kidneys, and low molecular weight proteinuria. Emphysema apparently resulted from cadmium deposits in the lungs following inhalation of cadmium dust or fumes. Cadmium nephropathy has been characterized by polyuria, hypercalciuria, and aminoaciduria. Finally, characteristic proteins in urine were of low molecular weight (20,000–30,000) with electrophoretic mobility of α- or β-globulins.[3] The mechanism of cadmium poisoning, however, was little understood when cadmium pollution was recognized as a serious hazard in Japan in 1965. Zinc mines and refineries and cadmium-plating factories caused pollution in many parts of Japan. Inhabitants of cadmium-polluted areas became fearful of itai-itai disease. At that time researches on health effects of cadmium started in Japan. Most of the data presented here were obtained from our laboratory.

MECHANISM OF CADMIUM ACCUMULATION IN THE KIDNEYS

Cadmium in the aquatic environment is generally taken up into the human body via the gastrointestinal tract as drinking water and food; the absorption rate from the gastrointestinal tract has been demonstrated to be 3–6%, while uptake of cadmium through the lung is said to be as high as 10–40%.[4] One-third of the cadmium taken up by the body is accumulated in the kidneys. The mechanism of cadmium accumulation in the kidneys, however, still remains unsolved. Most cadmium remains in the plasma for the first few hours after administration. As to the transport of cadmium, it has been believed that some cadmium-binding protein, such as metallothionein of molecular weight 7000, might play an important role on the transport from liver to kidneys.[5] However, the mechanism of transport still remains unsolved because a large amount of metallothionein or theionein has not been detected in the urine of cadmium workers. Furthermore, the independent mechanisms of proteinuria and cadmiuria do not fully support the metallothionein transport theory, but they might play, to a certain extent, a role in the transport of cadmium in the body. It is assumed that reabsorption of cadmium occurs in the renal tubules in normal rabbits because the cadmium/inulin clearance ratio was as small as 0.0002–0.0004.[6] However, the mechanism remains unknown.

MECHANISM OF URINARY EXCRETION OF CADMIUM

Urinary cadmium levels have been used as an index of cadmium exposure. The mechanism of urinary excretion of cadmium is, however, still not fully elucidated. Little cadmium is excreted in

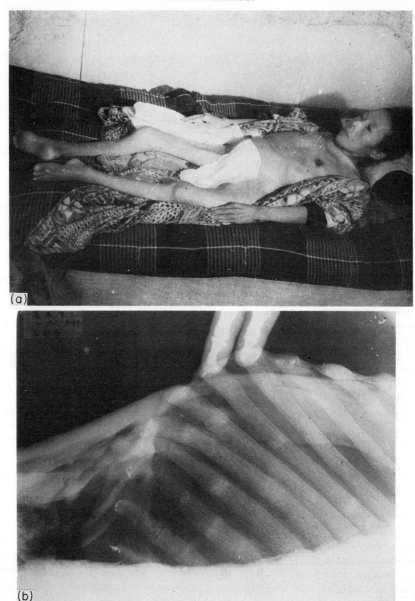

Fig. 1. Itai-itai disease patient and her roentgenogram. (*a*) A severe case of itai-itai disease, 55 years old, female; (*b*) roentgenogram of multiple milkman's pseudofracture in ribs (osteomalacia).[1]

normal persons or animals. The biological half-time of cadmium in humans is estimated as 13 years on the basis of the total body burden of cadmium in autopsied Japanese.[7] After successive exposures to cadmium, urinary cadmium excretion increases. Subsequently it decreases very slowly after cessation of cadmium exposure. This has been attributed to cadmium nephropathy, interfering with normal reabsorption of filtered cadmium. These suggestions need experimental confirmation.

Seventeen rabbits were given 300 ppm cadmium chloride as pelleted food for up to 54 weeks.[8] Urinary cadmium excretion was directly correlated with renal cortex cadmium level. This finding agrees well with the fact that urinary cadmium excretion parallels the cadmium level of the renal cortex as it increases with age in humans, as shown in Fig. 2.[7] To clarify the mechanism of urinary excretion of cadmium in rabbits, the following experiment was performed.[9] Cadmium chloride was given to

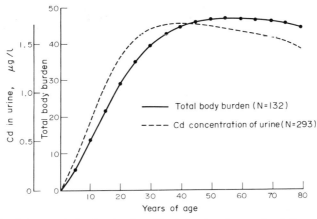

FIG. 2. Urinary cadmium level and total body burden of normal male Japanese.[17]

20 rabbits subcutaneously on the back at a dose level of 1.5 mg/kg/day up to 5 weeks. A very high correlation was observed between cadmium level of the renal cortex and urinary excretion of cadmium or cadmium clearance ($r = .85$) as shown in Fig. 3. This finding suggested that the determination of urinary cadmium excretion would give some indication of the cadmium level of the renal cortex. The latter, in turn, reflects the total body burden of cadmium because cadmium in the kidneys constitutes one-third of the total body cadmium.[4] Further, a strong inverse correlation was observed between maximal tubular capacity for excreting p-aminohippurate (PAH) and cadmium clearance.[9] The depressed normal function may play an important role in the increase of urinary cadmium excretion. However, as shown in Table 1, cadmium clearance in uranyl intoxicated rabbits with reversibly damaged kidneys was as low as in normal rabbits.[9] This observation is at variance with the conclusion that the accumulation of cadmium in the renal cortex leads to renal dysfunction followed by an increase of urinary cadmium excretion. The mechanism of urinary excretion clearly remains unknown.

TABLE 1
Relationship between renal function and clearance ratio of cadmium^{115m} to inulin[9]

Treatment	Number of rabbits	C_{in} (ml/min)	Tm_{PAH} (µg/kg/min)	$C_{Cd^{115m}}/C_{in}$
Normal	3	13.3	3.76	0.002
Cadmium	4	6.9	1.69	0.141
Uranium	2	10.6	1.19	0.002

Cadmium chloride: s.c., 1.5 mg/kg/day for 21 days.
Uranyl acetate: i.v., 0.2 mg/kg, single.

After the cessation of exposure to cadmium in humans, the urinary cadmium level decreases very slowly while the blood cadmium level decreases rapidly.[10] This fact agrees with our finding that urinary cadmium excretion parallels cadmium of the renal cortex, as the latter remains high. On the other hand, the blood cadmium level seems to parallel ongoing exposure to cadmium, and it decreases rapidly after the cessation of cadmium exposure. However, further detailed experiments are required to study the significance of these findings.

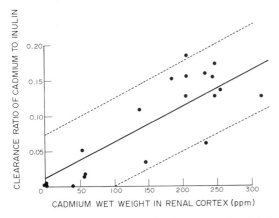

FIG. 3. Relationship between the clearance ratio of cadmium^{115m} to inulin and the renal cadmium level (ppm wet weight) in 20 rabbits given successive subcutaneous injections of cadmium. Solid and dotted lines represent the correlation line and the confidence limits at 10% level, respectively.[19]

MECHANISM OF CADMIUM NEPHROPATHY

Cadmium nephropathy is characterized by polyuria, aciduria, hypercalciuria, and aminoaciduria. Renal function as measured by glomerular filtration rate and urine concentrating ability is depressed as well.[3] These findings suggest that cadmium affects not only proximal tubules but also other parts of the kidneys such as glomeruli, Henle's loop, and distal tubules. Histological examination revealed atrophy and denaturation of renal tubular cells, especially of proximal tubular cells. As to aminoaciduria, it is characteristic of intoxication with cadmium as well as other heavy metals.[4] Further, aminoaciduria is one of the earliest signs of cadmium intoxication,[18] and the determination of amino acids in urine is helpful for the detection of cadmium nephropathy. This will be discussed in detail in the section on Early Symptoms of Cadmium Intoxication (p. 20).

As a first step in the study of mechanism of cadmium nephropathy, the relationship between the degree of toxic nephropathy and cadmium level in renal cortex was tested.[12] Nineteen male rabbits were given subcutaneous injections of cadmium chloride at dose levels of 15, 6, and 1.5 mg/kg. Body weight, renal functions, and urinary excretion of enzymes were periodically examined. Signs of intoxication were found to be closely correlated with the accumulated dose of cadmium. The cumulative median lethal dose was estimated to be 40–44 mg/kg. Further, the relationship between renal cadmium level and renal functions was critically examined.[9] To 20 rabbits were administered successive subcutaneous injections of cadmium chloride on the back at a dose level of 1.5 mg/kg for up to 5 weeks. There was seen a strong negative correlation between renal cadmium level and glomerular filtration ($r = -0.61$) and between renal cadmium level and maximal tubular capacity for secreting p-aminohippurate ($r = -0.58$).

Some researchers reported that a high prevalance of proteinuria was often seen when renal cadmium levels exceeded 200 μg per wet g weight in human autopsy or biopsy specimens.[4] It seems true that the higher the exposure to cadmium, the higher the renal cadmium level. This is also supported by our autopsy data as shown in Table 2.[13] However, the critical level of renal cadmium has still not been ascertained. Histological study did not show any changes in the kidneys with a renal cadmium level of 264 μg/g.[13]

The renal cadmium level of itai-itai disease patients were curiously lower than those of normal people.[14] Some researchers suggested that itai-itai disease patients suffered from cadmium nephropathy, which led to loss of cadmium by malabsorption of cadmium at renal tubules.[4] However, there are difficulties in such a view, because I already referred to the fact that cadmium excretion is quite independent of renal function but varies rather with renal cadmium level. The question of the etiology of itai-itai disease will be discussed later.

IDENTIFICATION OF LOW MOLECULAR WEIGHT PROTEINS IN URINE OF CADMIUM WORKERS AND ITAI-ITAI DISEASE PATIENTS

Since Friberg and Olhagen[15] first observed the curious urinary proteins of cadmium workers, proteins have been shown to have molecular weights of 20–30 thousand and electrophoretic mobilities of

TABLE 2
Tissue cadmium level and renal dysfunction in residents in cadmium-polluted area[13]

Sex	Age	Cadmium exposure		Cadmium level (μg/g)			Renal injury and proteinuria	Pathological findings in the kidneys
				Renal cortex	Renal medulla	Liver		
Female	77	211 μg/day	33 years	31.7	19.1	1.6	−	Arteriosclerotic atrophic kidneys
Female	80	Slight	34 years	41.3	35.9	11.2	+	Arteriosclerotic atrophic kidneys
Female	84	Slight	7 months	134.1	48.7	29.4	?	Arteriosclerotic atrophic kidneys
Male	57	350 μg/day	34 years	263.9	128.8	31.7	−	Slight arteriosclerosis
Male	45	211 μg/day	34 years	28.5	23.3	34.8	+ +	Chronic glomerulonephritis
41 cases	>60	No known exposure		133	104	12	(Ishizaki)	
25 cases	>60	No known exposure		50	20	10	(Kitamura)	
160 cases	>60	No known exposure		100	20	5	(Tsuchiya)	(Accidental death)

α_2-, β-, and post-γ-globulin. While renal tubular dysfunction has been diagnosed in Japan by the detection of β-globulin in urine with electrophoresis and of urinary low molecular weight proteins with column gel filtration, the urinary proteins of cadmium workers and itai-itai disease patients have so far not been identified.

A urine sample of itai-itai disease patient F.O. was analyzed for molecular size distribution of proteins by column gel filtration (Fig. 4). The major low molecular weight proteins were found to be of 12, 21, 45, and 54 thousand molecular weights. Another urine of cadmium worker M.S. contained proteins with molecular weights of 16, 24, 45, and 51 thousand.

Secondly, counterimmunoelectrophoresis indicated that the protein fraction of molecular weight approximately 54,000 was composed of immunoglobulin G, albumin, transferrin, α_1-acidoglycoprotein, and α_1-β-glycoprotein. The protein fraction of approximately 45,000 molecular weight contained Zn-α_2-glycoprotein, α_2-HS-glycoprotein, and Bence–Jones protein (types λ and κ). The lighter fraction (molecular weight 21,000) showed maximum light absorption at 330 nm, and fluorescence excitation and emission spectra with maxima at 344 and 470 nm, respectively: these are properties of retinol. Single radial immunodiffusion also suggested that this protein fraction might be retinol-binding protein (RBP). The protein fraction of molecular weight 12,000 probably consisted mainly of β_2-microglobulin because the amino acid composition of the present protein agreed quite well with the fraction described by Berggård and Bearn.[17] This protein fraction rapidly disappeared from urine samples, probably due to denaturation upon freezing and thawing or upon storage at 4°C. Identification and purification of the protein should therefore be performed shortly after the collection of urine samples of cadmium workers or itai-itai disease patients.

Finally, we studied protein bands by disc electrophoresis. Urinary proteins formed five bands. Electrophoresis of each protein sample fractionated by Sephadex G-100 suggested that band 1 was composed of prealbumin, band 2, albumin, band 3, apo-RBP (RBP free of retinol), band 4, holo-RBP (RBP plus retinol), and β_2-microglobulin, and band 5, transferrin, respectively.

These results suggest that immunoassay of urinary RBP and β_2-microglobulin could prove a simple method for diagnosis of renal tubular lesions, as will be further discussed in detail in the following section.

SIMPLE DETERMINATION OF LOW MOLECULAR WEIGHT PROTEINS

Low molecular weight proteins have been detected by gel filtration. Column gel filtration is not suitable for screening large numbers of samples in areas polluted with cadmium as it is an elaborate and time-consuming procedure. On the other hand, several samples can be analyzed simultaneously by thin-layer gel filtration, although time is required to concentrate urine samples for gel filtration. Disc electrophoresis does not give us the molecular weight of proteins in urine although it is simple and easy to perform. Retinol-binding protein and β_2-microglobulin, however, can be semiquantitatively determined by bands 3 and 4 in disc electrophoresis, and, furthermore, RBP is easily qualitatively determined by its fluorescence in ultraviolet light. Sodium dodecyl sulfate disc electrophoresis permits determination of molecular weights of urinary proteins.

As already mentioned above, low molecular weight proteins in urine are mainly composed of RBP, β_2-microglobulin, and lysozyme.[16, 18] We produced rabbit antisera to human RBP, β_2-

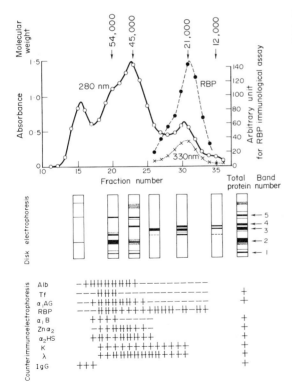

FIG. 4. Characteristics of fractioned urinary proteins of itai-itai disease (patient F.O.) with Sephadex G-100.[16]

microglobulin, and lysozyme. Anti-albumin sera were obtained from Hoechst Japan. Preliminary tests with single radial immunodiffusion permitted us to discriminate among normal, glomerular, and tubular disorders. A large amount of albumin (more than 1000 mg/l) was detected in nephritis and nephrosis, 30–150 mg/l of β_2-microglobulin, and 4–20 mg/l lysozyme in cadmium workers, cadmium nephropathy, Wilson disease, and itai-itai disease.[19] We are now determining low molecular weight proteins in urine sample from areas polluted with cadmium.

MECHANISM OF LOW MOLECULAR WEIGHT PROTEINURIA

Two kinds of pores are known to occur in renal glomerular membranes in a ratio of 10,000 pores of 2.0–2.8 nm diameter to 1 pore of 8 nm diameter.[20] Substances of molecular weight less than 15,000 pass freely through glomerular membranes, while substances of molecular weight greater than 50,000 are essentially unable to pass through glomerular membranes.[21] However, the etiology of low molecular weight proteinuria also should be studied in relation to the reabsorption of filtered proteins in renal tubules.

Six rabbits were divided into three groups;[22] to the first group was administered cadmium chloride subcutaneously on the back at a dose level of 1.5 mg/kg/day for 21 days; the second group was given a single intravenous administration of uranyl acetate at a dose level of 0.2 mg/kg to cause proximal dysfunctions; the third group served as control. The rabbits were given a continuous infusion of labeled dextrans and proteins of molecular weight 17,000 and 69,000 and their clearances were determined. Further, the uptake of iodine-labeled proteins into liver and kidneys was determined. The results are shown in Table 3. In normal rabbits, dextran clearance of proteins of molecular weight 69,000 was smaller than that of proteins of 17,000 molecular weight by a factor of 5. Most of the large proteins were found to be reabsorbed by the tubules, whereas only about one-third of the smaller proteins were reabsorbed. In cadmium-treated rabbits, no significant difference in protein reabsorption was found between larger and smaller proteins. However, the fact, that the uptake by the kidneys of smaller proteins depressed more than that of larger protein, suggests the possibility of low molecular weight proteinuria in cadmium nephropathy. In uranyl-treated rabbits with tubular nephropathy, clear-cut results were obtained showing that the reabsorption of low molecular weight

TABLE 3
Mechanism of low molecular weight proteinuria[22]

Renal functions	Molecular weight	Normal	Cadmium	Uranium
C_{in}		8.74	4.53	3.26
$C_{dextran}/C_{in}$	17,000	0.90	0.76	0.70
	69,000	0.16	0.16	0.15
Reabsorption of				
myoglobin	17,000	0.36	0.25	0.00
Albumin	69,000	0.97	0.93	0.89
Renal cortex/ plasma				
myoglobin	17,000	5.90	4.33	1.72
Albumin	69,000	0.13	0.13	0.12

Cadmium chloride; s.c., 1.5 mg/kg/day for 21 days.
Uranyl acetate; i.v., 0.2 mg/kg, single.
Number of animals; 2 rabbits for each group.

protein at the tubules was completely inhibited. Therefore, low molecular weight proteinuria in cadmium nephropathy might be caused by the depressed reabsorption of low molecular weight proteins at the renal tubules. Dr. Tsuchiya[10] observed that proteinuria in cadmium workers disappeared shortly after the cessation of cadmium exposure while urinary cadmium excretion continued at a high concentration. This suggests that proteinuria and urinary cadmium excretion may be of different etiologies, in good agreement with our results.

On the other hand, Chiappino et al.[23,24] suggested that low molecular weight proteinuria in cadmium intoxication might be caused by the depression of protein catabolism in the tubular cells rather than the depression of protein reabsorption by the tubules, because in cadmium-intoxicated rats, there was observed a depression of leucine aminopeptidase in epithelial cells, and after administration of peroxidase, unmetabolized enzyme could be detected in cells of proximal tubules. It is our conclusion, however, that the mechanism of low molecular weight proteinuria in cadmium intoxication is still unsolved.

EARLY SYMPTOMS OF CADMIUM INTOXICATION

Cadmium chloride was subcutaneously administered to seven rabbits daily at a dose level of 1.5 mg/kg/day.[12] On the 2nd day of the experiment the urine volume decreased. Temporal proteinuria was also observed. On the 21st day, proteinuria and aminoaciduria were observed again, accompanied by glycosuria and a decrease of maximal capacity for secreting p-aminohippurate. No change in urine osmolality and pH could be observed. Urinary enzyme activities rose significantly far earlier than the usual indices of renal function. Alkaline phosphatase activity rose on the 6th day of the experiment, and glutamic–pyruvic transaminase and acid phosphatase on the 9th day and 14th day, respectively. These findings suggest that urinary activity might be a good index of renal tubular injury and that the cadmium nephropathy might be diagnosed earlier by urinary enzyme assay.

Cadmium is taken into the body through the gastrointestinal tract in the case of environmental cadmium pollution. Seventeen rabbits were given pelleted food containing 300 ppm cadmium chloride for up to 54 weeks;[8] 16 rabbits served as control. As shown in Table 4, the earliest signs of intoxication were aminoaciduria and enzymuria, both of which were detected after 14–16 weeks of cadmium feeding. Anemia was observed after 27 weeks, and later proteinuria and glycosuria appeared. Loss of body weight and appetite were also seen after 42 weeks of cadmium feeding. Osteomalacia was not observed at all. These data suggested that early cadmium intoxication can be detected by determining urinary excretion of amino acids and enzymes. Proteinuria and glycosuria indicate a later stage of cadmium intoxication.

The determination of urinary enzymes of cadmium workers revealed its usefulness because of sensitivity and the fact that urinary enzyme excretion parallels urinary cadmium excretion.

Next we examined whether the elevation of urinary enzyme activity in toxic nephropathy is really due to enzyme migration from renal cells. Rabbits were given a single intravenous injection of uranyl acetate at a dose level of 0.2 mg/kg, and enzyme activity in serum and urine were assayed 2 days after the administration.[25] Significant increases occurred in the activities of alkaline phosphatase, glutamic-oxalacetic transaminase, and glutamic-pyruvic transaminase at an earlier stage than pathological changes indicated by urinalysis or renal function tests. The increased activity of urinary enzymes was found to be almost proportional to the dose 24 h after uranyl administration. Urinary enzyme levels soon returned to normal levels, whereas renal function tests remained depressed. These findings suggest that the increased activity of enzymes in the urine results from renal cellular injury. Clearance studies indicate that most urinary enzymes after uranyl administration were released from renal tubular cells. These findings suggest that the assay of urinary enzymes may be useful in the detection of early injury to renal tubules. Specifically, proximal tubular injury is indicated by increased alkaline phosphatase activity; distal tubular injury by an elevation of lactic dehydrogenase.[26]

DIAGNOSTIC TECHNIQUES OF CADMIUM POISONING

It is very neccessary to know the degree of cadmium exposure and total body burden of cadmium. Environmental cadmium should be checked first of all. The blood cadmium level probably indicates the present degree of cadmium exposure, and total body burden of cadmium is roughly proportional to urinary cadmium excretion.[10] Recently developed techniques such as anodic stripping analysis or flameless atomic absorption spectrophotometry have made it possible to determine 50–100 samples per technician per day. As to urinary cadmium level, intra-day variation was found to be greater than inter-day variation.[27] Therefore, a whole day urine sample should be used for the determination of urinary cadmium excretion in order to avoid such an effect. However, when it is difficult to obtain a whole day urine sample, a spot sample in the morning is recommended for the screening test because morning urine contains more cadmium than afternoon or evening urine.[27]

For diagnosing toxic nephropathy, urinary excretion of total amino acids and low molecular weight

TABLE 4
Effects of dietary cadmium (300 ppm) on rabbits[8]

Signs		Weeks of cadmium feeding
Body weight	Decreased	After 42 weeks
Food consumption	Decreased	After 41 weeks
Urine volume	Unchanged	Throughout experiment
Urinalysis		
Protein	Increased	After 38 weeks
Glucose	Increased	After 42 weeks
Amino acid	Increased	After 16 weeks
Enzymuria		
ACP	Increased	After 49 weeks
ALP	Increased	16–28 weeks
GOT	Increased	16–28 weeks
GPT	Increased	16–38 weeks
Renal function		
C_{in}	Unchanged	Until 24 weeks
Tm_{PAH}	Unchanged	Until 24 weeks
Hemoglobin	Decreased	After 27 weeks
Osteomalacia	Cannot be observed	Throughout experiment

proteins should be determined as already mentioned. Total amino acids in urine can be determined easily and accurately by our modified procedures; the urine sample is mixed with Tsuchiya's reagent for deproteinization and the supernatant is used for amino acid determination by the colorimetric method with trinitrobenzenesulfonic acid reagent; no color is developed with peptides, urea, and ammonium in urine.[28] Urinary low molecular weight proteins can be detected with thin layer gel filtration or sodium dodecyl sulfate disc electrophoresis, and determined quantitatively by single radial immunodiffusion with the use of albumin, RBP, β_2-microglobulin, and lysozyme antisera.[19]

In order to detect early cadmium nephropathy, the determination of urinary enzyme activities, especially of alkaline phosphatase, is very helpful. The urine sample should be dialyzed for more than 2 h against water just after collection; enzyme activities in urine will then remain stable for 20 days at 4°C.[28]

QUESTIONS ON CADMIUM AS AN ETIOLOGICAL AGENT OF ITAI-ITAI DISEASE

The Japan Ministry of Health and Welfare concluded officially in 1968 that the etiological agent of itai-itai disease is cadmium intoxication superimposed upon malnutrition.[1] Osteomalacia in itai-itai disease was thought to be caused by cadmium-induced tubular nephropathy. However, several questions arose after the announcement of the Japan Ministry of Health and Welfare.[29]

Firstly, clinical pathology and autopsy data did not reveal any serious renal dysfunctions in itai-itai disease before the vitamin D treatment. In other words, osteomalacia seemed not to be caused by cadmium nephropathy. This fact is uncompatible with the views of Japan Ministry of Health and Welfare. Secondly, more than 40,000 inhabitants in areas polluted with cadmium have been tested for cadmium poisoning and itai-itai disease, but no tubular-nephropathic osteomalacia could be found so far except in Zinzu river area. Thirdly, although calcium balance study revealed the loss of calcium in cadmium-treated animals, no typical osteomalacia could be produced so far.

Although many questions remain, we cannot assert that cadmium did not play any role in the etiology of itai-itai disease. No new cases of itai-itai disease have been reported since 1965, and the patients are becoming older year by year. The form of the disease is changing with treatment. It is urgently necessary to study further the etiology of itai-itai disease now; otherwise the etiology may remain unsolved.

CONCLUSION

Because of the shortage of my time, the discussion has been focused on toxic nephropathy and the etiology of itai-itai disease. We have some data on the hypersensitivity to cadmium emphysema[30,31] and several critical comments on the proposal by the 16th report of the Joint FAO-WHO Expert Committee on Food Additives in 1972.

The mechanism and early diagnosis of cadmium intoxication has been elucidated little by little as already mentioned. However, the studies on the health effects of cadmium have been focused too much on the toxic nephropathy because the toxic nephropathy has been believed to relate to itai-itai disease. It is necessary now to perform epidemiological surveys relating long term exposure to low levels of cadmium to hypertension, coronary disease, and malignant tumors of prostate, stomach, and lungs. Further, in order to establish firmly a tolerable weekly intake of cadmium, more health surveys on inhabitants in areas polluted with cadmium are required, as well as are more animal experiments involving longterm feeding with low levels of cadmium.

REFERENCES

1. MURATA, I., HIRONO, T., SAEKI, Y., and NAKAGAWA, S., *Bull. Soc. Int. Chir.* **2**, 34 (1970).
2. Japan Ministry of Health and Welfare (1968) *Etiology of Itai-Itai Disease.*
3. NOMIYAMA, K., *Igaku no Ayumi* **83**, 121 (1972).
4. FRIBERG, L., PISCATOR, M., and NORDBERG, G., *Cadmium in the Environment*, CRC Press, Cleveland (1971).
5. PISCATOR, M., *Nord. Hyg. Tid.* **45**, 76 (1964).
6. NOMIYAMA, K., *Proceedings of the 2nd Symposium on Cadmium Poisoning*, p. 78, 1971.
7. TSUCHIYA, K., SUGITA, M., and SEKI, Y., *Proceedings of the 17th International Congress on Occupational Health*, 1972.
8. NOMIYAMA, K., SUGATA, Y., YAMAMOTO, A., and NOMIYAMA, H., *Toxicol. Appl. Pharmacol.* (in press).
9. NOMIYAMA, K., SUGATA, Y., NOMIYAMA, H., and YAMAMOTO, A., *Proceedings of the 46th Annual Meeting of the Japan Association of Industrial Health*, p. 336, 1973.
10. TSUCHIYA, K., personal communication, 1973.
11. CLARKSON, T. W. and KENCH, J. E., *Biochem. J.* **62**, 361 (1956).
12. NOMIYAMA, K., SATO, C., and YAMAMOTO, A., *Toxicol. Appl. Pharmacol.* **24**, 625 (1973).
13. NOMIYAMA, K., *Jap. J. Hyg.* **28**, 45 (1973).

14. ISHIZAKI, A., FUKUSHIMA, M., and SAKAMOTO, M., *Jap. J. Hyg.* **26**, 268 (1971).
15. FRIBERG, L., *Acta med. scand.* **138**, Suppl. 240, 1 (1950).
16. NOMIYAMA, K., SUGATA, Y., MURATA, I., and NAKAGAWA, S., *Environ. Res.* **6**, 373 (1973).
17. BERGGÅRD, I. and BEARN, A. G., *J. Biol. Chem.* **243**, 4095 (1968).
18. OHSAWA, M. and KIMURA, M., *Experientia* **29**, 556 (1973).
19. NOMIYAMA, K., KIMURA, M., OHSAWA, Y., and SUGATA, Y., *Jap. J. Hyg.* **28**, 42 (1973).
20. ARTURSON, G., GROTH, T., and GROTTE, G., *Clin. Sci.* **40**, 137 (1971).
21. HARDWICKE, J. and SOOTHILL, J. F., *Renal Disease*, 2nd ed. (D. A. K. Black, ed.), p. 261, Blackwell Sci. Publ., Oxford, 1968.
22. NOMIYAMA, K., YAMAMOTO, A., and NOMIYAMA, H., *Proceedings of the 46th Annual Meeting of the Japan Association of Industrial Health*, p. 288, 1973.
23. CHIAPPINO, G., REPETTO, L., and PERNIS, B., *Med. Lab.* **59**, 584 (1968).
24. CHIAPPINO, G. and PERNIS, B., *Med. Lab.* **62**, 424 (1970).
25. NOMIYAMA, K., YAMAMOTO, A., and SATO, C., *Toxicol. Appl. Pharmacol.* **27**, 484 (1974).
26. POLLAK, V. E. and MATTENHEIMER, H., *Arch. Intern. Med.* **109**, 473 (1962).
27. SUDO, Y. and NOMIYAMA, K., *Jap. J. Ind. Hlth.* **14**, 117 (1972).
28. NOMIYAMA, K., *Jap. Med. J.* **2587**, 132 (1973).
29. TAKEUCHI, J., *Jap. J. Clin. Med.* **31**, 2048 (1973).
30. NOMIYAMA, K., NOMIYAMA, H., and MATSUI, H., *Bull. Wld. Hlth. Org.* **45**, 253 (1971).
31. NOMIYAMA, K., NOMIYAMA, H. and KOBAYASHI, H., *Allerg. Immunol.* **18**, 81 (1972).

THE EFFECTS OF HEAVY METALS ON FISH AND AQUATIC ORGANISMS

MAX KATZ

Parametrix Inc., Seattle, Washington

I was asked by Peter Krenkel to prepare a talk regarding the toxicity of metals to aquatic organisms, in particular, fish. This is certainly a reasonable assignment for a paper to be delivered at a symposium regarding heavy metals in the environment. And it is not an especially arduous task, for there has been a lot of work done on the toxicity of metals to aquatic organisms and there is little difficulty in finding toxicological papers and compilations of toxicity data.

Yet the toxicity of heavy metals is not the topic that I would really like to talk about. The real problem today is not to determine if heavy metals like copper, zinc, cadmium, etc., are toxic to fish and other aquatic organisms: we know that they are. The problem today that we are being called upon to solve, or at least try to make informed judgements on, is what concentrations of heavy metals, as well as other substances, are permissible in water without having any harmful effect on the uses and users of this water. This is information that we must have because many of the newer environmental enthusiasts inside and outside of the regulatory agencies insist that there should be zero levels of almost any given substance that you choose to name in all of the water supplies of the world. Yet, on the other hand, those of us who have picked up some experience and some bitterly gained knowledge in our struggles to survive as scientific workers are aware that several of the substances that are toxic at low concentrations, at even lower concentrations are necessary for the maintenance of populations of desirable organisms. An example is copper sulfate, which is one of the best molluscicides known to man and is effectively used to control the vectors of schistosomiasis. Yet, some years ago, I was intrigued by a paper of which I no longer have the reference, which describes how a snail population was limited by a lack of copper in its environment. There were obviously insufficient amounts of this metal to supply the copper snails' respiratory pigment, hemocyanin.

Thus, the removal of all copper from all water supplies is one example of a demand made in all sincerity by well-intentioned enthusiasts, and is typical of some of the suggested regulations which have come perilously close to being enacted into statutes by enthusiastic legislators. The enforcement of such regulations would have been technologically wellnigh impossible but, in addition, does not make ecological good sense.

Another important and timely argument for an understanding of the action of heavy metals and the levels of heavy metals that aquatic organisms can tolerate, or even require, was stimulated by a recent estimate by an economist in regard to the energy requirements of American agriculture. The relationship caught my eye in this paper was that it took 5 cal of fossil energy to bring a calorie of food energy to the table of the American consumer. I am not an agricultural specialist; I am not sure if the United States agricultural practices are efficient or if they are unusually wasteful, but in the light of the energy crisis in the United States and other countries this is an alarming figure.

We can argue that, inefficient or not, we must have food for our people, and until we learn how to produce and distribute more efficiently we will have to spend 5 cal in order to get the one food calorie that we need to keep us alive. This inefficiency is one I am forced to go along with. Yet, as I look at the technical literature in the environmental areas, I am alarmed at the vast expenditures that are advocated to clean up some of our environmental problems, and I am even more alarmed when I study the requirements proposed by regulatory agencies upon municipalities and industries to correct imaginary environmental problems.

I am willing to agree that to solve certain environmental problems we must be willing to spend large sums of money and allocate vast amounts of energy to gain a certain tangible or intangible environmental benefit. Conversely, the limits to our funds, time, and energy make it imperative that the problems that we are intending to solve are actually problems and not a requirement to meet a bureaucratic numerical standard that often is based on incomplete, insufficient, and incorrect knowledge.

I will admit that there was a time, just a year or two ago, when we could spend several millions of dollars, many, many man years, tie up a lot of material and energy, and correct a water quality problem only to find out in short order that we had solved a nonexistent problem. I am sure that many of us have been involved in some of these projects and we were able to shrug off these expenditures as a learning experience.

I do not believe that we can afford to be so cavalier in our attitude today, at least if we wish to maintain our self-respect and the respect of the many people who have entrusted us to maintain our environment in a logical and sensible manner.

The remarks I have made indicate that, as biologists, it is important for us to do the necessary research so that data will be available to insure that recommendations and the treatment facilities based upon these recommendations achieve an improvement in the environment which may justify in part the expenditure of time, funds, and energy. And, in certain cases, this research can be done in the library, for much of the information we could be using is already available but is not being used.

There are some of you, of course, who may agree or disagree with my philosophy, but are attending this meeting to get some factual material for your files. Therefore, to satisfy those who have put up their hard-earned money, I will proceed. There are several useful tabulations listing the results of studies regarding the acute toxicity of heavy metals to fish and other organisms. These start with the old review, Doudoroff and Katz,[1] and the more recent compilations of McKee and Wolf.[2] A more recent review which covers the literature regarding the toxicity of substances on aquatic life is that of the US Environmental Protection Agency.[3]

A most recent compilation is by Becker and Thatcher,[4] who checked the literature through 1972. Most of their references deal with the toxicity of copper and zinc salts to aquatic organisms.

For those whose needs go beyond the need to know what concentrations of heavy metals kill or do not kill aquatic organisms, the review of Jones[5] in his textbook will form a useful introduction to the other effects of heavy metals in the aquatic environment. Jones, in his two chapters, points out that heavy metal pollution and its biological effects have been a concern for at least 100 years, and cites the fifth report of the River Pollution Commission in 1874.[6]

These reports concerned themselves with the polluting effects of the lead mines in Cardiganshire, Wales. The extended studies by Carpenter[7–10] of the Cardiganshire rivers has set the foundation of our current knowledge regarding the effects of heavy metals in the freshwater environment.

The rest of my paper will be devoted to a discussion of just a little of the vast amount of information that is being made available on a daily basis, which can be used to help us decide what levels of heavy metals can be tolerated in our waters with good assurance that valuable aquatic populations will not be harmed. There are different ways that different workers have chosen to collect information that can be applied to this practical problem. Of special interest is the excellent work of Sprague and associates,[11,12] in regard to the effects of mining activities on the Miramichi river in Nova Scotia and their effect on the Atlantic salmon that utilize this stream as a rearing and spawning area. I might remark that Sprague's studies are of particular interest and value because they show how he develops his study from the necessary basic toxicity bioassays to the interesting and very practical field observations that give an understanding of the effects of the wastes upon the fish in the river and which give a basis for regulations that can be used to serve useful purposes.

Sprague[12] conducted constant flow bioassays with copper and zinc sulfides in a soft water at pH about 7.3 and water temperatures of 5°, 15°, and 17°. Using a logarithmic scale, Sprague demonstrated that the concentration of metal and the survival time could be described by a straight line until, at a certain low concentration, the survival became indefinitely long, or expressed in simpler terms, the fish did not die. This sharp break was designated the incipient lethal level (ILL), which was 48 μg/l of copper iron and 500 μg/l of zinc. Sprague also observed that survival time in any given concentration of zinc was four times as long at 5°C as at 15°C, and the ILL was at least 1.5 times higher. For the person responsible for waste treatment at a plant, the latter information could allow him to make significant savings in money and energy in waste treatment during the colder periods of the year when energy resources are apt to be limited.

Sprague's paper alone would provide a certain amount of pertinent data for an agency that had to regulate this industry. His ILLs could be used by a regulatory agency to set standards for streams receiving heavy metal wastes. Expenditures of time, money, and, above all, energy, could be justified on the basis of Sprague's data to reduce the heavy metal contents in the effluents to just below the ILL.

Below these levels, expenditures for treatments could be viewed with skepticism by metal-producing industries. The decreased toxicity and the reduced ILL at 5°C, as compared to the toxicity at 15°C, indicate that a lesser degree of treatment can be justified during the colder winter months. This is worthy of consideration because energy demands for many of man's necessary activities increase at the lower seasonal temperatures and there is as little justification for waste of energy as there is for the waste of fish.

Sprague and his associates extended his studies to the field and made some interesting observations in regard to the effects of heavy metals on significant aspects of fish behavior that were most pertinent to the Miramichi problem. A series of tagging experiments demonstrated that concentrations of dissolved copper and zinc somewhat below the ILL served as a barrier to the adult salmon which were attempting to migrate upstream to spawn. At a salmon-counting weir it was observed that 10–22% of mature salmon went back downstream without spawning during the 4 year period in which pollution from the mine was acute. In "normal" years, only 1–3% demonstrated this aberrant behavior. A tagging program indicated that one-third of these fish reascended later, but that most of the fish that had descended were not observed in any other river system and are believed not to have reproduced. Laboratory studies conducted to confirm these results showed that small salmon avoided less than one-tenth the ILLs. Further, the avoidance threshold was 0.09 ILL of zinc, 0.05 ILL of copper, and 0.02 ILL of a mixture of the salts that were equally toxic. In nature, however, the avoidance thresholds were higher, and abnormal downstream movements past the counting weir only occurred when heavy metal content exceeded 0.35–0.43 ILL. As a consequence, young salmon populations were severely depressed in the Miramichi as a result of insufficient egg deposition by spawning adults and poor survival rates.

Weir and Hine[13] determined the effect of low concentrations of mercury, arsenic, lead, and selenium on conditioned responses of goldfish. These fish were subject in training experiments to light and electric shock stimuli. They were then transferred to exposure tanks for 24 h where they were subjected to heavy metals, retested for their response, exposed to the heavy metals for a second 24 h, retested and returned to the holding tank.

The lowest concentrations of heavy metals which gave a significant impairment of the learning response and the fraction that these concentrations were of the 50% and 1% lethal concentrations (LC_{50} and LC_1) are given in Table 1.

TABLE 1
Impairment of conditioned avoidance responses by metal ions[a]

Compound	LC_{50} (mg/l)	LC_1 (mg/l)	Lowest conc. of signif. imp. (mg/l)	LC_{50} (%)	LC_1 (%)
As	32.0	1.5	0.1	1/320	1/15
Pb (hard water)	110.0	60.0	0.07	1/570	1/857
Pb (soft water)	6.6	1.5			
Hg	0.82	0.36	0.03	1/273	1/120
Se	12.0	1.0	0.25	1/48	1/4

[a] Lowest concentrations of ions which significantly impaired conditioned avoidance responses and their fractions of the 50% and 1% lethal concentrations.

These results, although not directly parallel to those of Sprague, support his observation of disturbed behavior of fish due to metallic wastes. This disturbed behavior is certainly of importance in affecting the well-being of a fish population.

There are those who correlate heavy metal pollution with fish diseases. Rippy and Hare[14] investigated an epidemic involving Atlantic salmon and suckers in the Miramichi river. The bacterium *Aeromonas liquefacians* was isolated from the affected fish. High water temperatures and increased content of copper and zinc in the summer of 1967 and 1968 were also involved. Although the heavy metal may not have been directly involved, it is known that disease epidemics of fish in nature are unusual, and it is entirely proper to investigate the heavy metals as one of the contributing factors in regard to the epidemic.

Some of you may feel that correlating disease problems with sublethal heavy metal concentrations is belaboring the issue. I will admit that this does fit into the "prophet of doom" approach, yet Christensen et al.[15] indicate that changes in the blood picture of brown bullhead result from exposure to low level concentrations of copper. It is well known that blood, because of the phagocytic properties of some of the leucocytes and the antibodies of the plasma, is the first line of defense against disease. The documented changes observed by Christensen et al. add credence to the observations of Rippy and Hare.[14]

For an understanding of the histological effects of heavy metals to fish the best single source of information is the review by Skidmore.[16] Of particular interest is Skidmore's discussion on the mode of toxic action of zinc. Skidmore reviewed the observations of earlier workers and mentioned that both Carpenter[9] and Jones[17] believed that gill damage and copious secretions of mucus restricted the respiration and were responsible for deaths in freshwater fish.

Lloyd[18] reported that in 20 mg/l of zinc a cytological breakdown of the gill epithelium of trout occurred in about $2\frac{1}{2}$ h. At about 4 mg/l zinc, the gill lamellae became swollen before death. Cytological changes were detected in fish exposed to 3 mg/l for about 48 h. In a report by the Department of Scientific and Industrial Research, probably by Lloyd,[19] some additional details are given regarding the pathological effects of heavy metals. In toxic but unspecified concentrations of zinc, copper, and lead salts, a specific action on the gill epithelium was observed. At about one-half of the expected survival time, about half of the gill epithelium had been sloughed off the lamellae, and at death about three-quarters of the epithelium had sloughed off.

Morphological and growth changes in guppies were observed by Crandall and Goodnight.[20, 21] These workers exposed newborn guppies to tap-water containing 1.15 mg/l zinc. The 79 guppies in the zinc solution grew less rapidly than the 54 controls in zinc-free water. The challenged group had more mortalities and exhibited less sexual dimorphism. After 90 days the median weight of the test fish was 23 mg and of the control fish 52 mg. Mortalities were 41% for the test fish and 9% for the control.

Only one male of the 79 test fish developed a male reproductive organ as contrasted to about 30–40% of the male fish in the control.

In a second series of experiments, these authors repeated the experiments and made histopathological examination of newborn guppies that had been exposed for long periods of time after birth to 1.15 and 2.3 mg/l zinc. Histological examinations of fish that had been exposed to these concentrations of zinc for over 55 days showed that the blood vessels in the liver were poorly developed, and that the mesenteries were lacking in fat. The tubules and glomeruli of the kidneys were distended, the lymphoid tissues of these organs were reduced, and the gonads were underdeveloped. After 95 days the liver contained large vacuoles, and granulocytes had accumulated in the heart muscle. The kidney tubules were even more expanded, the spleen was underdeveloped, and only one-fourth of the fish were mature. The controls at 95 days were sexually mature and showed no gross or microscopic abnormalities. The authors remarked that in none of the fish examined was there any gill damage.

Sangalang and O'Halloran[22] made some observations that throw light on the reproductive failure of the guppies of Crandall and Goodnight. Brook trout, which had been exposed to 25 ppb of cadmium, showed marked damage of the testicular tissues. *In vitro* tests of testicular tissue exposed to 25 ppb showed no markedly reduced levels of 11-ketosterone, 11B-hydroxytestosterone, and testosterone.

These observations are significant because they indicate the possible mechanism of depletion and possible extinction of a fishery in a stream that receives an insufficiently treated heavy metal waste. Although there may be no fish kill episodes, the population would disappear and decline to an unproductive level by the deterioration of the fish brought about by the destruction of several organ systems which would not allow the individual to cope efficiently with his environment. The retarded growth, if it did not lead to excess mortalities, would perhaps result in a population of fish which would be unattractive to fishermen and, of course, the failure to achieve sexual maturity would mean an abnormally low recruitment to the population. It would be of interest if some of you in the audience could perhaps present examples of retarded growth in fish populations which you have studied, which had been subjected to heavy metals.

An interesting paper describing the pathological effects of zinc salts and the recovery of fish that had been replaced in clean water is described by Mathiessen and Bradfield.[23] At a concentration of 0.5–1.0 mg Zn/dm^3, sticklebacks in distilled water were killed within 1–3 days. The pathological effects were the detachment and sloughing of epithelial cells and the coalescing of adjacent secondary lamellar epithelia. Cytoplasmic abnormalities included extensive vacuolation, followed by the swelling of nuclei and mitochondria leading to cellular disintegration. Many acutely poisoned fish recovered in zinc-free fresh water. The recovery of the epithelium was accompanied by a temporary appearance of chloride cells in the secondary lamellae. These authors also observed that concentrations of 2.0–6.0 mg Zn/dm^3 were not lethal over periods up to 700 h. Extensive cytoplasmic abnormalities appeared, however, including the formation of membrane-bounded vesicles and dense accumulations of metabolites. The most pronounced effect was the appearance of active chloride cells in the secondary lamellae.

It should be noted that the title of my paper is the toxicity of metals to the biota, and it is reasonable that I finally get around to speaking on the subject. We all know that there is a good deal of material on the toxicity of metals to fish. But with the exception of several papers on the toxicity of heavy metals to *Daphnia*, most of the recent papers written in English regarding the toxicity of heavy metals to aquatic animals other than fish have been prepared by members of Dr. Mount's staff in Duluth. Warnick and Bell[24] determined the 96 h TLm of the metals Cu^{++}, Zn^{++}, Cd^{++}, Pb^{++}, Fe^{++}, Ni^{++}, Co^{++}, Cr^{++}, and Hg^{++} to a stone fly *Acroneuria*, a may fly *Ephemerella*, and a caddis fly *Hydropsyche*. Those that did not die in 96 h in the highest concentrations used, 64 mg/l, were retested and the survival times at various metal concentrations were recorded. The results indicated that the aquatic insects were more tolerant to heavy metals than were fish. From the practical point of view, this means the insects might survive an industrial accident of brief duration that would wipe out the fish population of an area. When and if the area would be repopulated by fish, there would be, perhaps, a thriving population of aquatic insects to serve as a food resource.

Ephemerella was the most sensitive of the three insects and copper was the most toxic substance followed by Fe^{++}, cadmium, chromium, and mercury.

Arthur and Leonard[25] determined the toxicity of heavy metals to three other invertebrates, the amphipod *Gammarus* and the snails *Campeloma* and *Physa* under continuous flow conditions. The responses used to measure the effects of the toxicants were survival, growth, reproduction, and feeding. The 96 h TLm for *Campeloma*, *Physa* and *Gammarus* was 1.7, 0.039, and 0.020 mg/l, respectively. In extended bioassays it was determined that the total concentration of copper that had no effect after 6 weeks' exposure was between 8.0 and 14.8 ppb (μg/l).

Gammarids reproduced in copper concentrations of 8 μg/l or less during both 6-week trials. After 9 weeks' additional exposure the newly hatched amphipods reached adult size in copper concentration of 4.6 μg/l or less. In their discussion, Arthur and Leonard state that the safe concentration of copper is about the same as that already existing in many freshwater supplies. This means that there is no safety margin for copper wastes in water supplies, and indicates the burden that our copper-dependent civilization must bear in regard to its water supplies and the desirable organisms that grow in them.

Beisinger and Christensen[26] made an interesting study of the effects of heavy metals to *Daphnia magna*. This study included a determination of the acute toxicity and the effects of continued exposures to metals on reproduction and growth and on metabolism. The metallic ions tested were sodium, calcium, magnesium, potassium, strontium, barium, iron, manganese, arsenic, tin, chromium, aluminum, zinc, nickel, lead, copper, platinum, cobalt, mercury, and cadmium.

To me, the most interesting summary is in their Table 7 which lists the safe concentrations of heavy metals as determined by testing aquatic animals through a complete generation under laboratory conditions. These animals were *Daphnia*, brook trout, fat head minnow, and *Gammarus*. The reproductive impairment for fish and crustacea was about the same for chromium, copper, and zinc. Nickel and cadmium were more toxic to *Daphnia magna* in soft water than to fat head minnow in hard water.

If these experiments indicate what happens in the field as a result of metal waste discharges, then fish and the crustacean fish food organisms will be equally affected by heavy metal wastes.

REFERENCES

1. DOUDOROFF, P. and KATZ, M., Critical review of literature on toxicity of industrial wastes and their components to fish: II, Metals as salts, *Sewage Industr. Wastes* **25**, 802–839 (1953).
2. MCKEE, J. E. and WOLFE, H. W., *Water Quality Criteria*, 2nd edn., Water Resources Control Bd., State of Calif. Publ. No. 3-A, 548 pp., 1963.
3. Environmental Protection Agency, *Water Quality Criteria Data Book*, Vol. 3, *Effects of Chemicals on Aquatic Life*, selected data from the literature through 1968, Environmental Protection Agency, Project No. 18050 AWV, Battelle Columbus Laboratories, 100 pp. + 4 appendices, 1971.
4. BECKER, C. D. and THATCHER, T. O., *Toxicity of Power Plant Chemicals to Aquatic Life*, US Atomic Energy Commission, Washington DC, 1249 UC-11, 1973.
5. JONES, J. R. E., *Fish and River Pollution*, Butterworths, London, VIII, 203 pp., 1964.
6. River Pollution Comm., River Pollution Commission appointed in 1868, Fifth Report (1874), Vol. I, *Report + maps*, Vol. 2, *Evidence* (in ref. 5), 1874.
7. CARPENTER, K. E., On the biological factors involved in the destruction of river fisheries by pollution due to lead mining, *Ann. Appl. Biol.* **12**, 1–13 (1925).
8. CARPENTER, K. E., The lead mine as an active agent in river pollution, *Ann. Appl. Biol.* **13**, 399–401 (1926).
9. CARPENTER, K. E., The lethal action of soluble metallic salts on fishes. *Br. J. Exp. Biol.* **4**, 378–390 (1927).

10. CARPENTER, K. E., Further researches in the action of metallic salts on fishes, *J. Exp. Zool.* **56**, 407–422 (1930).
11. SPRAGUE, J. B., ELSON, P. F., and SANDERS, R. L., Sublethal copper and zinc pollution in a salmon river, a field and laboratory study. *Advances in Water Pollution Research*, Vol. I, pp. 61–82, Pergamon Press, 1963.
12. SPRAGUE, J. B., Lethal concentration of copper and zinc for young Atlantic salmon, *J. Fish. Res. Bd. Can.* **21** (1), 17–26 (1964).
13. WEIR, P., and HINE, C. H., Effects of various metals on behavior of conditioned goldfish, *Arch. Environ. Hlth.* **20** (1), 45–51; *Chem. Abst.* **72**, 64079b (1970).
14. RIPPY, J. H. and HARE, G. M., Relation of river pollution to bacterial infection in salmon (*Salmo salar*) and sucker (*Catostomus commersonii*), *Trans. Am. Fish. Soc.* **98** (4) (1969).
15. CHRISTENSEN, G. M., MCKIM, J. M., BRUNGS, W. A., and HUNT, E. P., Changes in the blood of the brown bullhead (*Ictalurus nebulosus* Le Sueur) following short and long term exposure to copper, *Toxicol. Appl. Pharmacol.* **23**, 417–427 (1972).
16. SKIDMORE, J. F., Toxicity of zinc compounds to aquatic animals with special reference to fish, *Q. Rev. Biol.* **39** (3), 227–248 (1964).
17. JONES, J. R. E., The relative toxicity of salts of lead, zinc, and copper to the stickleback (*Gasterosteus aculeatus* L.) and the effect of calcium on the toxicity of lead and zinc salts, *J. Exp. Biol.* **15**, 399–497 (1938).
18. LLOYD, R., The toxicity of zinc sulphate to rainbow trout, *Ann. Appl. Biol.* **48**, 84–94 (1960).
19. LLOYD, R., Department of Scientific and Industrial Research, 1958, Great Britain Dept. of Scientific and Industrial Research, *Water Pollution Research, 1958, 1960*, p. 83, 1960.
20. CRANDALL, C. A. and GOODNIGHT, C. J., Effects of sublethal concentration of several toxicants on growth of the common guppy, *Lebistes reticulatus. Limnol. Oceanography* **7**, 233–234 (1962).
21. CRANDALL, C. A. and GOODNIGHT, C. J., The effects of sublethal concentrations of several toxicants to the common guppy, *Lebistes reticulatus, Trans. Am. Micros. Soc.* **82**, 59–73 (1963).
22. SANGALANG, L. B. and O'HALLORAN, M. J., Cadmium induced testicular injury and alterations of androgen synthesis in brook trout, *Nature* **240**, 470–471 (1972).
23. MATTHIESSEN, P. and BRADFIELD, A. E., The effects of dissolved zinc on the gills of the stickleback, *Gasterosteus aculeatus* (L.)., *J. Fish. Biol.* **5**, 607–613 (1973).
24. WARNICK, S. L., and BELL, H. L., The acute toxicity of some heavy metals to different species of aquatic insects, *J. Wat. Pollut. Control Fed.* **41** (2), Part I, 280–284 (1969).
25. ARTHUR, J. W. and LEONARD, E. N., Effects of copper on *Gammarus pseudolimnaeus*, *Physa Integra* and *Campeloma decisum* in soft water, *J. Fish. Res. Bd. Can.* **27**, 1277–1283 (1970).
26. BEISINGER, K. E. and CHRISTENSEN, C. M., Effects of various metals on survival, growth, reproduction, and metabolism of *Daphnia magna, J. Fish. Res. Bd. Can.* **29**, 1691–1700 (1972).

The Effects of Heavy Metals on Fish and Aquatic Organisms
(M. Katz)

DISCUSSION by DONALD I. MOUNT
US Environmental Protection Agency, National Water Quality Laboratory, Duluth, Minnesota

The stress of this paper, the need for measuring sublethal effects of heavy metals in addition to the lethal effects, reflects the current intensive effort to provide habitats for aquatic life where it can not only survive, but thrive. This change in concern from survival to well-being reflects the advancement that has occurred in the field of aquatic toxicology over the past 5-10 years, a change that has greatly increased the work needed in aquatic toxicology. In the case of heavy metals that are essential to the well-being of the aquatic organism, zero is not the most desirable concentration in the water, and the biologist must therefore elucidate the full range of effects of metals on aquatic life from effects that are highly desirable to those that are detrimental.

Using the observation that many metals are desirable at some low concentration, the paper discusses waste treatment from the point of view that uniform treatment should be required of all dischargers. Such uniform treatment requirements do appear to be the intent of the new Federal Water Quality Legislation, but scientists should remain detached from involvement in this controversial area lest they lose their credibility as scientists and aquatic toxicologists. By law, it is for the public to decide the degree of treatment that it wishes to finance. The job of the scientist remains one of assisting the public, the private sector, and the regulatory agency in understanding the consequences of varying degrees of treatment by careful research that elucidates the full gamut of effects of pollutants on aquatic organisms. A part of this responsibility is to make scientific information available to those who need it, in a usable form. To toss our data to busy administrators and expect them to apply it unassisted is to expect too much. In fact, the problem of applying research data to the real world problems is a challenging and difficult one, perhaps even more so than the challenge of obtaining the data at the research bench.

Some discussion ensues in the paper regarding the importance of changes in behavior as endpoints from the exposure of organisms to pollutants. The point is made that behavioral changes must be biologically significant before they can be used as significant endpoints of effect in pollution control. Certainly the avoidance of reaches of streams, such as Sprague found in Canada, is a change in behavior that is biologically significant. Avoidance such as this is also likely to play a very important role in mixing zones, and yet the effects of conditions permissible in mixing zones, such as varying concentrations above the safe levels for varying periods of time, are effects that are virtually unstudied and little effort is being directed toward determining them. Research effort is sorely needed to unravel the meaning to a system of allowing a percentage of its area to be devoted to water of less than liveable quality, i.e. mixing zones.

An issue indirectly raised in the paper but of direct importance to the field is the balance of effort needed between understanding the cause or mode of action of some insult on an organism and work describing what happens, without concern for how or why it happens. Undoubtedly, the best mix includes some of both. The remarkable success of cancer chemotherapy illustrates the value of observing what happens, without knowing why. In this case, it would seem that the observations of what happens led eventually to some understanding of why it happens. On the other hand, knowledge of the "what" without the "why" is of limited usefulness. The problematical question is how much time can be devoted to time-consuming "why" research, while discharges go on and problems of pollution control exist. After years of study, we know little of the mode of action of a much-studied chemical, DDT. Can we expect quicker solutions for metals?

Let us turn now to five key areas of research need that are of prime importance to the field of aquatic toxicology. First, we can no longer treat metals each as a single toxicant. The chemistry of all metals in water is complex at best and one really never tests the toxicity of a metal as a single form, but rather as a mixture of a half a dozen or more. The effects of a few water characteristics, such as pH and hardness, have long been realized, but various molecular and ionized forms, as well as equilibrium times and organic chelates, may be, and probably are, even more significant. We cannot stop toxicity testing until all chemical behavior is understood, as some would have us do, but a competent chemist should always assist in test design and conduct and bring to bear what information is known about the chemical behavior in water of the toxicants under study. Chemists in their field, too, have a responsibility to pursue their research in a biologically meaningful way.

A second area of need involves extrapolation, and this is where knowing "why" is especially useful. We will never have sufficient resources to study many species or systems or combinations of toxicants. We must therefore select species of organisms and experimental conditions to maximize the utility of the information for extrapolation of data from one animal to another, as well as from one set of conditions to another. Those involved in regulatory activity know the critical need for this ability. If you can accept that we must extrapolate, then it is clear that we need not use the most sensitive animals in our tests because it is as logical to extrapolate from a resistant animal to a sensitive one as the converse and it is easier to test certain animals. Recognition and use of this approach can save important resources as laboratory research is conducted.

A third area of concern regards models. It seems that nearly everyone is working on models to the extent that they appear to be the end rather than a means to an end. Many appear to have no real interest or intent to determine if a theoretical model filled with shaky assumptions has any real-world validity. In the late 1950s and early 1960s physiological studies were the rage among workers in our field. Physiological tests are a tool, but so far they have played a relatively minor role in regulatory activity. I hope that models are not going to follow a like path. Clearly, they too are a tool, but we are years away from valid complex models useful to pollution control agencies and with most or all of the important assumptions validated. Validation efforts to date might better be termed "chance correlations" than proof of the validity of the model. I wonder if workers will be willing to do the drudgery required to build useful models because they have not been willing to do it for the application of physiological effects.

The fourth area of concern relates to the need for field methods that can measure effects on entire water bodies. It has been said that the perch population in Lake Michigan would have to change by 50% before man could detect it. If such is typical, we can never expect to evaluate anything but effects of grossly polluting discharges unless our quantitative methods are improved. With the advent of treated discharges, the value of classical "oxygen sag studies" and the "rat-tailed maggot" studies has nearly disappeared, and studies of combined effects of many discharges on whole water bodies are necessary. Unfortunately, such data are not available at present. Site studies have their place, but not for evaluating effects of single discharges on whole water bodies. As treatment is instituted on more and more discharges, and the effects of existing discharges become more subtle, only highly refined whole-water-body studies will answer the questions that need to be answered as to whether or not sufficient clean-up has been achieved to protect the aquatic environment.

Finally, I would suggest a fifth point—one which it may be unnecessary to make. You may have noticed throughout this paper that I have not used the word "ecosystem" in my comments. The meaning of this word has been all but destroyed by abusive, indiscriminate use. Many, who 4 years ago, did not know the word's meaning now are "ecological engineers" or "ecological chemists." Departments and agencies have put ecological in their names but have changed nothing in the content of their programs or activities. More to the point, there is a common feeling that all research must be "ecological research." It would seem, from observing titles of papers and grant proposals, that ecological research is a name applied to almost any research that involves studies on more than a single species. Are studies involving a large number of unknown organism interactions really more useful? Studies where cause and effect are rarely known? Why is it that laboratory microcosm experiments are thought to be more nearly "natural" than a test system involving a single species? Both are artificial and the former is far less controlled and understood than the latter. Have you ever asked yourself critically whether data can be extrapolated from one ecosystem to another? If the ability to extrapolate is a necessary use of our data, as I have suggested earlier, maybe we are not ready for ecosystem research today. The ecology bandwagon of the last 4 years is covered with scars made by futile efforts to climb aboard. None of us should try to get on that bandwagon until he is sure that the time is right and that his efforts will be fruitful.

MERCURY CONCENTRATIONS IN HUMANS AND CONSUMPTION OF FISH CONTAINING METHYLMERCURY

HAROLD E. B. HUMPHREY

Environmental Epidemiologist, Michigan Department of Public Health, Lansing, Michigan

Certain species of fish from Lake St. Clair contain a questionably toxic concentration of mercury, over 0.5 ppm. Accordingly, the relationship between whole blood total mercury concentrations and consumption of fish has been studied in this and another area of Michigan. Randomly sampled adult residents of Algonac who eat less than 6 lb of fish annually from the St. Clair River and Lake St. Clair (noneaters) were contrasted with residents eating 26 lb or more of such fish annually (eaters). These were compared with similarly defined noneaters and eaters in South Haven where Lake Michigan fish contain less than 0.5 ppm of mercury.

Blood mercury levels for 65 Algonac noneaters averaged 5.7 ppb and ranged from 1.1 to 20.6 ppb. Comparative values for 42 South Haven noneaters were 5.2 ppb and 1.6–11.5 ppb, respectively. Mercury levels for persons who ate fish were higher in both communities. Blood mercury levels for 42 Algonac fish eaters averaged 36.4 ppb and ranged from 3.0 to 95.6 ppb. Comparative levels for 54 eaters from South Haven were lower, averaging 11.8 ppb and ranging from 3.7 to 44.6 ppb.

The data show that a direct relationship between the quantity of fish consumed and the concentration of mercury observed in human blood exists. Preliminary tests for methylmercury in the same human blood samples indicate that relationships similar to those observed for total mercury occur. Eating of sport fish from the St. Clair waters apparently leads to some absorption of methylmercury, albeit at lower than known toxic levels.

SESSION II
ANALYTICAL TECHNIQUES

METHYLMERCURY ANALYSIS
(A REVIEW AND SOME DATA)

Kimiaki Sumino
Kobe University School of Medicine

MINAMATA DISEASE AND METHYLMERCURY

In Japan, following the outbreak of Minamata disease which occurred in 1953 on the coast of Minamata Bay, Kumamoto Prefecture, it was reported in 1965 that several individuals had developed symptoms similar to those of Minamata disease, near the mouth of the Agano river, Niigata Prefecture. After clinical, analytical, and epidemiologic investigations it was confirmed that this was a second outbreak of Minamata disease caused by waste discharge from a factory, including a small amount of methylmercury (MM). Both incidents were of exactly the same type. Soon after notice of the first outbreak (1955), much research was performed by Kumamoto University staff members. In 1956 it was found that the disease appeared to be due to the victims having eaten large quantities of fish from the bay. In 1959 the Kumamoto University research group found fish and sea mud contaminated with much mercury in Minamata Bay. They also found that all of the victims were suffering from symptoms similar to those of MM poisoning mentioned by Hunter and Russel.[1] In fact, MM crystals were extracted from waste sludge in an acetaldehyde factory in 1963 in the Minamata Bay area.

However, unfortunately this experience and knowledge had not been utilized in order to prevent a second incident. The reasons were many, mainly negligence of the Japanese Government. Also at that time analytical methods for measuring MM were undiscovered, though this was a small reason.

Micro-quantitative determination of MM using gas chromatography (GC) was presented at first by the author and co-workers in the 3rd Kinki Industrial Medicine Society in November 1965.[2] Thereafter, this technique was improved, and research reports on it were read at medical meetings and published.[3-5] At just about the same time, coincidentally, almost the same method, using GC, was discovered and published by Professor Gunnel Westöö in Sweden following mercury problems there.[6-8]

At the time that the second outbreak was discovered in Japan, this gas chromatographic method and a thin layer chromatographic (TLC) technique developed just previously, played an important role in the survey of the cause of the disease by tracing out the contamination process of MM.[9] A definite diagnosis was established only after the identification of MM by GC in patients' hair and in fish (Table 1). After this application, the method using GC became widely popularized in Japan for measuring MM in fish and in waste water. Many laboratories and institutes in Japan have been able to analyze MM in recent times using this method.

This paper will report the analytical methods coming into wide use for measuring MM throughout the world, will discuss, somewhat, the problems in this field, and will also show application data.

ANALYTICAL METHOD OF METHYLMERCURY

Concerning the analysis of organic mercury compounds or MM, several methods have been reported in the past, employing distribution between solvents,[10,11] paper chromatography,[12] TLC,[6,13] and others. But nowadays it is usual that GC is used in order to determine methyl, ethyl, methoxyethyl, phenyl mercury, and so on, especially to clarify the chemical form.[5] Though TLC can be used for pretreatment of the sample in part, it may safely be said that only GC is used in the end for micro-determination. Therefore the problems in each process in GC analysis are reviewed. Reviews like this have already been published,[14,15] so that only recently discussed points, mainly in Japan, are described in this report.

TABLE 1
Materials related to Minimata disease in Niigata prefecture—mercury content in patients' hair and contaminated fish

Material	Sex	Age	MMC (ppm as Hg)	Total Hg (ppm)
K.I	♂	31	56.0	425.0
F.O	♂	38	64.6	160.0
K.O	♂	28	28.7	59.0
M.K	♂	42	33.2	78.0
C.K	♀	25	23.0	66.0
T.E	♀	58	120.0	180.0
H.C	♂	48	149.0	305.0
RAIGYO[a] (*Channa argus*)			99.0	119.0

Sampling: June–October 1965.
[a]Dried sample.

The fundamentals of the analytical method for determining and measuring the presence of MM have been established: after the addition of halogenhydrogen acid to a homogenated sample, the MM must be extracted with some organic solvent, thereafter purified with a sulfhydryl compound and reextracted with benzene. Then the final extract is analyzed quantitatively by GC with an electron capture detector.

The remaining problem is whether a true value is obtained or not. Much effort and attention are required for exact determination. The author describes some particular problems in the analysis of MM.

One is a problem in pretreatment and the other is in the GC process itself. In the former, the main point is whether the final extract includes all the MM without any loss and degradation.

Extraction

(1) There are few problems in homogenizing a sample of fishes, organs, and other living materials. It is necessary to tear the sample to pieces and mix it thoroughly with water. Though a stainless homogenizer (Ultra-Turrax in my laboratory) is used, loss of MM in this stage is not by chemical adsorption of MM but by adhesion of material on the surface of the homogenizer. It can easily be removed to be added to the homogenized sample by washing the homogenizer with water.

Hair is adequately prepared if it is cut uniformly with scissors. Sea mud or sludge is usually analyzed as it is. Though the homogenized sample is usually acidified without washing, it is better to wash it with benzene or acetone before homogenizing if it contains fat-rich or colored material.

(2) Some acid solution is added in order to separate CH_3Hg— from the bonded form. Many workers use hydrochloric acid; very few use hydrobromic acid.[16] In any case, the bonds are theoretically cleaved and MM is changed to the halide in conditions below a pH of one. Usually under acid conditions 2 N to 3 N HCl, MM bonded to the organic material is changed to MM chloride (MMC). In order to promote the change to the chloride, excess sodium chloride is added.[8] This process is not effective in changing bisdimethyl mercuric sulfide to MMC. Before acidification with hydrochloric acid, the papain digestion method in living materials has been adopted in Standard Method of Analysis for Hygienic Chemists in Japan.[17] In fact, this method has not been used widely in Japan, for loss of MM may occur if the digestion time is long.

(3) Benzene is usually used for the extraction solvent. Free MM can theoretically be transferred to the benzene layer by one extraction. According to Westöö, after the homogenized sample is extracted with an equal amount of benzene once, about 80% of clear benzene can be recovered after centrifugation. But in our method the added benzene is lost during centrifugation, and sometimes not all the MM can be transferred to the benzene in one extraction, depending on amount and type of material in suspension in the water, so that three extractions are carried out to get complete recovery of MM and to avoid more complicated calculations. In this process, great care must be taken to recover the added benzene.

In Westöö's method 5% $HgCl_2$–HCl solution is added to the homogenized sample or to the first benzene phase in order to avoid the recombination of MM and sulfur compounds in the extracting solvent in the distillation process, especially when analyzing liver and egg yolk. The present author also found that the recovery of MM is decreased because of the recombination and the volatilization when the benzene extract is left for a long time or especially if it is distilled. The problem is eliminated if clean up is carried out without distillation and within a few hours. Moreover, the Hg^{++} in benzene causes a ghost peak similar to MM, which probably comes from the MM which adheres to the column packing or the column wall in the presence of Hg^{++}. Therefore, no Hg^{++} is added in our method. Concerning the determination of MM in aqueous solutions containing sulfides or other extraction-disturbing sulfur compounds, Nishi et al.[18] have reported that by adding excess amounts of $HgCl_2$, CH_3Hg, which was bonded to the sulfur compounds, was replaced by inorganic mercury, resulting in the liberation of CH_3HgCl. After the removal of HgS or other inorganic sulfur compounds, the excess of $HgCl_2$ was also removed as $HgNH_2Cl$ by adding aqueous NH_3. On acidification, CH_3HgCl thus liberated was readily extractable in organic solvents. But in reality they have said it must be considered that the excess of $HgCl_2$ might be converted to MM in the sample solution or in the column.

In the presence of sulfide compounds or thiocyanate, Japanese Industrial Standards has suggested that $CuCl_2$ powder instead of Hg^{++} should be added to the homogenized sample to avoid interference in extraction of MM from these compounds. The author has found that the MM in a 1% Na_2S solution could be extracted with benzene in strongly acidic conditions.

(4) L-cysteine or cysteine acetate is used for the clean-up process. MM is readily combined with cysteine, and it becomes water soluble at a neutral pH. But if hydrochloric acid remains in the benzene as an emulsion, of course no MM can transfer to the cysteine solution, so that the washing with sodium chloride solution is a necessity till the sample becomes neutral. If there are any substances which oxidize cysteine in the benzene solution, e.g. H_2S in mud or I_2 compounds, cysteine is oxidized and MM is unable to connect with cysteine, so that the recovery rate becomes extremely low. But common biological materials, even egg yolk and liver, are recovered sufficiently by the clean up with cysteine. However, Prof. Westöö has presented the fact that MM from the benzene layer can also be extracted using NH_4OH—Na_2SO_4 solution. This ammonium method is useful in extraction from mud.

In any case, these extractions are repeated at least twice, and this repeated extraction is a very important process, which may increase the recovery of MM. In spite of these careful procedures, if the recovery of MM is still low in accordance with the properties of the sample, recovery tests of small additional amounts of MM are needed to determine what percentage is actually extracted from the sample.

(5) It is quite a bit less trouble to extract from cysteine solution into benzene. Good results are obtained under 2N to 3N hydrochloric acid conditions with only one extraction. This final benzene extract is ready for GC analysis or for flameless atomic absorption after changing the form of the mercury from MM to Hg^{++}.[19]

(6) Throughout the pretreatment process, the greatest attention must be paid to see that MM is extracted with benzene under acid conditions, and that the MM in the benzene extract is transferred to the cysteine solution. For example, when hydrochloric acid is added to bottom mud, H_2S will be generally produced, causing the evaporation of the MM and obstructing the combination of MM and cysteine in the benzene. The former, the evaporation of MM, cannot be controlled, even though the latter, obstruction of the combination, can be controlled using metal such as Hg^{++} or Cu^{++} and using an ammonium solution instead of the cysteine solution. Therefore, the loss of MM will be impossible to avoid in such materials. However, in the case of usual biological or experimental materials, if the extraction process is repeated and the separation of the benzene layer from the water is carefully done, the final benzene extract will contain the MM at the recovery rate of 92–95% of the original amount.

The flow chart, Fig. 1, shows Westöö's method and our method for extraction of biological samples.

Estimation

The one remaining problem is in the GC technique, namely, GC conditions for estimation of MM. However, this problem is not clearly mentioned by other workers. The description below is mainly based on our data.

(1) The conditions of the GC should be adjusted so that the MM peak has sharpness, high sensitivity, and good reproducibility.

Column packings: DEGS and 1.4-BDS are mostly used in Japan; Carbowax-2000 or PEG-20M in Sweden; and sometimes OV-17 in the United States, Canada, and some other countries.[16] However, poor results are obtained with a nonpolar stationary phase such as OV-17. Recently, column packings such as DEGS-Hg or 1.4-BDS-Hg with the support previously coated with 5–10% NaCl, have been put on sale in Japan. This packing gives good sharpness for alkyl mercury.

Column. An all-glass system is desirable, and the direct on-column injection technique should be employed. As to the column length, it is determined by the stationary liquid phase concentration, e.g. 20%—0.5 m, 10%—1 m, or 5%—1.5 m.

Instrument conditions. Column temperature is usually kept within 140–180°C, and detector temperature and injection temperature are held somewhat higher than column temperature.

Detector. Electron capture detector (ECD) with ^3H or ^{63}Ni is usually employed. However, Bache *et*

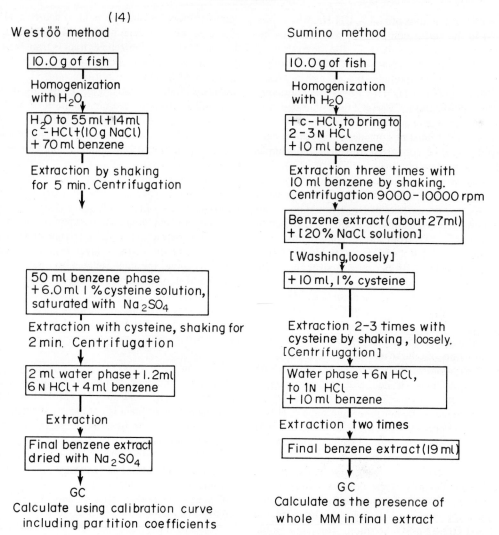

Fig. 1. Flow chart of extraction method for methylmercury in fish. Square brackets indicate sometimes omittable process.

al.[20] have reported that almost the same results were obtained by using emission spectrometry in a helium plasm. In the Japanese official method, the ECD with discharge electrode is also adopted. A chromatogram of MM and other alkyl mercury compounds under these conditions is given in Fig. 2.

(2) Pure MMC is usually used for the standard solution of MM. MMC synthesized from methyl magnesium bromide (Grignard's reagent) sometimes includes MM bromide. Of course some bias occurs in using impure MMC. In our laboratory, MMC synthesized from Hg^{++} and tetra methyl lead is used. Quantitative determination can usually be made by comparing the peak height of the sample with that of a standard solution of about the same MM concentration after confirming the linearity and dynamic range of MM. In some laboratories, real amounts have been estimated after simple supplementary calculation involving recovery rate; however, this is not a sufficiently accurate method.

Though methods using internal standards should be considered in cases of low recovery of MM, no report of this sort has yet been published. In The Standard Method of Analysis for Hygienic Chemists in Japan, p-nitrobenzylchloride as an internal standard is used in order to avoid the error resulting from injection technique.

After these procedures, in order to ascertain whether the peak of MM is real MM or not, the

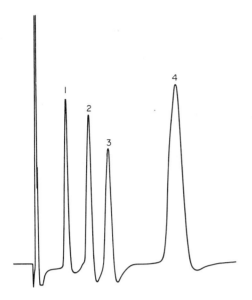

FIG. 2. Chromatogram of organic mercury compounds. 1, Methylmercury chloride, 0.1 ng; 2, ethylmercury chloride, 0.1 ng; 3, butylmercury chloride, 0.1 ng; 4, methoxyethylmercury chloride, 1 ng. GC conditions: 10% DEGS on Chromosorb W 60–80 mesh, 1.2 m × 3 mm ϕ glass column; column temperature, 160°C; N_2, 50 ml/min; ECD with ^{3}H at 200°C.

elimination method by glutathione is usually used. Nishi et al.[21] have reported that peak substraction occurred by inserting a short column of metal powders between the chromatographic column and detector. Mass fragmentgraphy (spectroscopy) is most clearcut for this purpose as mentioned later.

Precision and accuracy

In order to obtain the true value, first of all, high sensitivity and good separation in GC is necessary. In our laboratory, the detection limit for MM of 10^{-12} g was achieved, the coefficient of variation was within 5% after eight analyses of the sample in one tuna can, and good recovery was obtained (Table 2).

Since there were some problems in the accuracy of MM analysis among different laboratories, during the last year the cross-checking of MM in tuna processed at the same time was carried out among three countries; the United States, Canada, and Japan. However, these results have been unpublished.

Before this analysis (1971), there were some data among Japanese laboratories, shown in Table 3. Definite differences were found, and after that the official method was established. However, these results have shown that the percentage of MM in the total mercury (TM) was about 70%, and this fact may be different from that in the fish in Sweden.

Tentative analytical method for determining methylmercury in bottom mud (from Environmental Agency of Japan)

(1) Method of taking up mud (omitted).
(2) Analytical method.
  (a) Reagents:
    $CuSO_4$-sol. (1 w/v%).
    NaCl-sol. (20 w/v%).
    L-Cysteine-sol. (0.1 w/v%)⎫ freshly pre-
    Glutathione-sol. (1 w/v%) ⎭ pared.
    HCl.
    Methylmercury standard sol. in benzene.
    Benzene.
    Anhydrous Na_2SO_4.
  (b) Procedure up to sample solution. Ten grams of sample material are measured and transferred into separatory funnel; 20 ml of HCl (1:1) and 5 ml of $CuSO_4$-sol. are added, and the mixture is shaken vigorously for 5 min; 40 ml of benzene is added and shaken for 5 min. After phase separation, the same volume of benzene is added into the water layer and shaken. The benzene layers are added together and washed slowly with NaCl-sol. until the solution becomes neutral. After washing, 8 ml of cysteine sol. are added into the benzene extract and shaken vigorously for 10 min. The same extraction is again repeated. After adding all of the cysteine sol., 2 ml of HCl and 5 ml of benzene are added and shaken for 5 min and separated. After the benzene layer is dehydrated with anhydrous Na_2SO_4, this benzene extract is ready for GC analysis.
  (c) Measuring.
    GC conditions:
    Column: glass (3–4 mm ϕ × 1–2 m).
    Packing: 5–25% DEGS or 1.4-BDS.
    Column temp.: 150–160°C.
    Detector: ECD.
    Carrier gas: N_2; 30–80 ml/min.

A few microliters of the sample solution are injected with a micro-syringe into the GC set up under the conditions mentioned above. Peak height or area is measured and then the amount of MM is calculated by comparison with the calibration curve of MMC made previously. After measuring, in order to confirm whether the peak obtained by this method is produced by MM or not, 1 ml of

TABLE 2
Accuracy and recovery for mercury in fish cans

	Skipjack		Yellow fin		Albacore	
	TM[a]	MM[b]	TM	MM	TM	MM
Found (μg/g as Hg)	0.282	0.21	0.569	0.41	0.519	0.34
	0.286	0.21	0.562	0.42	0.519	0.35
	0.286	0.20	0.559	0.41	0.515	0.34
	0.297	0.18	0.559	0.42	0.515	0.36
	0.282	0.20	0.578	0.42	0.523	0.35
	0.282	0.20	0.569	0.43	0.519	0.38
	0.297	0.20	0.559	0.42	0.519	0.35
	0.286	0.20	0.559	0.42	0.526	0.34
Mean	0.287	0.20	0.564	0.42	0.519	0.35
σ	0.0059	0.0093	0.0066	0.0060	0.0035	0.013
c.v.[c]%	2.1	4.7	1.2	1.4	0.67	3.7
MM/TM	70		75		67	
Added Hg (μg/g as Hg)	0.30	0.20	0.50			0.30
Found (μg/g as Hg)	0.589	0.40	1.08			0.62
	0.592	0.39	1.05			0.64
	0.596	0.38	1.08			0.64
	0.585	0.39	1.05			0.63
Mean	0.591	0.39	1.07			0.63
Recovery (%)	101	95	101			94

[a] Total mercury estimated by flameless atomic absorption.
[b] Methylmercury estimated by GC with our method.
[c] Coefficient of variation was obtained $(\sigma \times 100)/\text{mean}$.

TABLE 3
Interlaboratory comparisons of mercury analysis of two fish cans in Japan

	Albacore			Swordfish		
	TM (μg/g)	MM (μg/g)	MM/TM (%)	TM (μg/g)	MM (μg/g)	MM/TM (%)
Range	0.32–0.65	0.24–0.37	52–76	0.86–1.3	0.60–0.81	55–72
Mean	0.47	0.30	65	1.1	0.71	65
s.d.	0.10	0.048	9.9	0.16	0.075	6.4
c.v. (%)	21	16	15	15	11	9.8

No. of laboratories: 9 in TM and 8 in MM.

TM, total mercury was estimated by wet digestion and reductive vaporization in five laboratories and by combustion digestion and vaporization in four laboratories.

MM, methylmercury was estimated using GC by benzene extraction three times and by cysteine extraction two times in all laboratories.

These data have not been previously published. The author obtained them from Prof. Kitamura, who was a member of the Committee of Mercury Analysis in Japan.

glutathione solution is added to 1 ml of sample solution and shaken for 2 min. The upper benzene layer is injected into the GC. If the peak of MM disappears, the peak given above is confirmed as MM. The measured value of MMC is converted to that of mercury on a dry weight basis.

APPLICATION DATA

Because of the environmental pollution conditions caused by mercury around the world, the necessity for reliable analytical methods for determining and measuring MM is increasing.

In Japan, the procedure for determination of MM using GC is specified in the official method. And there have been presented many analytical reports on environmental pollution surveys, such as the amounts of MM in industrial wastes, soils, and many fishes. However, the analytical procedure has rarely been applied to biological and chemical studies. This paper describes below some new data obtained in these fields.

The amounts of total mercury and methylmercury in bodies of Japanese[22]

TM and MM in the organs were determined as follows: 30 Japanese (20–70 years old—15 males and 15 females) were subjected to this investigation. TM was determined by a flameless atomic absorption method developed in our laboratory,[23] and the determination of MM was performed by the method as shown in Fig. 1. The results are shown in Table 4. In the following, a numeral is μg/g based on wet weight; $\gg$ means significant at $p = 0.01$; $>$ is $p = 0.05$.

TM: There was a decreasing tendency in the order of kidney $(1.1) \gg$ liver $(0.47) \gg$ adrenal $(0.14) >$ cerebellum, cerebrum (0.10), and the amount in each of all other organs was less 0.1. The value in blood was 0.059 ± 0.026. This value was clearly at a lower level than other organs, but was a little higher than other data reported in the past.[14,24] The upper limit of the 95% confidence interval for blood concentration was 0.10. No remarkable difference was found between hair (4.1 ± 2.6) and pubic hair (3.8 ± 2.5), so that external adhesion might be ignored.

On the difference in the sexes, each average in the lung, cerebrum, and spleen tends to be higher in the male, but the average in all other organs or parts tends to be higher in the female, especially in the heart and trachea ($p = 0.01$). However, the fact that a high value was found in the male hair may indicate a difference in the metabolic pathway for mercury between the sexes.

The difference among the age groups was not clear, but the level in the intestinal tract was a little higher in the older groups in both sexes. The whole body burden of mercury was more than 3–4 mg (Table 5). About two-thirds of this amount was present in the muscle (1.4 mg) and liver (0.7 mg), and in the brain (0.13 mg).

MM: The level of MM in the organs and parts was in the order of liver $(0.044) \gg$ kidney (0.023), cerebellum (0.019), cerebrum (0.016), about 0.01 in blood and pancreas, and less than 0.01 in the other organs studied. A characteristic difference compared with the case of TM was that MM concentration in the liver was higher than that in the kidney, and the small intestine higher than the large intestine ($p = 0.01$). The level in the blood was 0.011 ± 0.0073. The amount in the hair was 2.6 ± 2.1, and only three female subjects among 14 showed less than 1.0.

About the difference between the sexes, the levels in all viscera tend to be higher in the male subjects in contrast to the case of TM. The difference in the kidney was statistically significant ($p = 0.05$). Its difference in the old age group was not remarkable, but in the middle age group significant differences were found in the kidney, small intestine, muscle, and hair, so we may say that the content of MM is the highest in the middle age group.

More than 400 μg of MM was present in the whole body. Of these 400 μg of MM, about 50% was present in the muscles, about 20% in the liver, more than 10% in the blood, and 5% in the brain.

MM/TM: From the fact that the level of TM tends to be higher in the female and that of MM tends to be higher in the male, it will naturally be concluded that the MM/TM is higher in the male.

The organs which have high ratios of MM/TM are arranged as follows: (1) in the male: cerebellum (21.4%), blood (20.7%), small intestine, heart (19.7%); (2) in the female: blood (15.4%), pancreas (15.1%), small intestine (14.2%).

Some differences are found between the sexes, however; in both sexes the MM/TM ratio is high in the blood and small intestine and low in the kidney (2.8%) and colon (8.2%). This fact may indicate that the MM is decomposed in both excretion organs. In any case, the MM/TM ratio for nonexposed man is less than 20%. Unfortunately, there are no other data with which to compare these results.

Biological half-life of methylmercury in rat organs

The labeled mercury compounds have been used in almost all animal experiments of MM. In animal

TABLE 4
Mercury content of human organs/parts in Japanese

Organ or part	Sex	N	Total mercury (μgHg/g wet tissue)				Sex	N	Methylmercury (μgHg/g wet tissue)				MM/TM (%)		
			Average	Range	Mean ± s.d.	Median			Average	Range	Mean ± s.d.	Median	♂	♀	♂+♀
Cerebrum	♂	10	0.11	0.039–0.17	0.10 ± 0.042	0.097	♂	9	0.022	0.0015–0.069	0.016 ± 0.014	0.012	19.4	11.1	15.0
	♀	10	0.099				♀	11	0.010						
Cerebellum	♂	10	0.11	0.048–0.23	0.10 ± 0.045	0.093	♂	9	0.028	0.0015–0.096	0.019 ± 0.020	0.014	21.4	12.7	16.3
	♀	11	0.089				♀	11	0.012						
Trachea	♂	12	0.036	0.015–0.11	0.047 ± 0.029	0.036									
	♀	4	0.079												
Lung	♂	15	0.081	0.015–0.30	0.080 ± 0.054	0.070	♂	5	0.0083	0.0023–0.015	0.0065 ± 0.0034	0.0060	12.0	10.8	11.3
	♀	13	0.078				♀	6	0.0050						
Heart	♂	15	0.054	0.023–0.13	0.069 ± 0.028	0.069	♂	7	0.011	0.0030–0.027	0.0092 ± 0.0066	0.0070	19.7	9.3	14.9
	♀	14	0.085				♀	6	0.0067						
Liver	♂	15	0.42	0.16–1.3	0.47 ± 0.26	0.42	♂	15	0.048	0.012–0.080	0.044 ± 0.019	0.042	13.7	8.7	11.2
	♀	15	0.52				♀	15	0.041						
Pancreas	♂	15	0.077	0.023–0.29	0.083 ± 0.048	0.077	♂	14	0.010	0.0013–0.033	0.010 ± 0.0078	0.0083	15.8	15.1	15.4
	♀	15	0.090				♀	15	0.010						
Spleen	♂	13	0.073	0.021–0.14	0.068 ± 0.028	0.062									
	♀	15	0.064												
Kidney	♂	15	0.97	0.18–2.6	1.11 ± 0.67	0.98	♂	15	0.029	0.010–0.080	0.023 ± 0.015	0.019	3.6	1.9	2.8
	♀	15	1.24				♀	14	0.018						
Suprarenal gland	♂	12	0.12	0.03–0.33	0.14 ± 0.073	0.15									
	♀	12	0.16												
Small intestine	♂	12	0.057	0.024–0.19	0.069 ± 0.037	0.064	♂	12	0.016	0.0030–0.069	0.014 ± 0.017	0.0082	19.7	14.2	16.7
	♀	13	0.080				♀	12	0.012						
Colon	♂	14	0.078	0.032–0.16	0.083 ± 0.037	0.075	♂	9	0.0086	0.0018–0.026	0.0065 ± 0.0061	0.0044	10.6	6.2	8.2
	♀	13	0.090				♀	11	0.0047						
Testicles	♂	14	0.067												
Ovary	♀	14	0.069												
Muscle	♂	13	0.056	0.018–0.15	0.060 ± 0.027	0.057	♂	6	0.0090	0.0041–0.019	0.0078 ± 0.0043	0.0064	13.2	13.8	13.5
	♀	14	0.064				♀	6	0.0065						
Skin	♂	15	0.051	0.017–0.15	0.059 ± 0.034	0.048									
	♀	12	0.066												
Blood	♂	9	0.054	0.016–0.11	0.059 ± 0.026	0.058	♂	6	0.012	0.0036–0.026	0.011 ± 0.0073	0.0092	20.7	15.4	17.8
	♀	10	0.064				♀	6	0.010						
Hair	♂	14	5.4	1.4–15	4.1 ± 2.6	3.4	♂	14	3.4	0.63–10.4	2.6 ± 2.1	2	59.9	56.1	57.9
	♀	15	3.0				♀	14	1.8						

TABLE 5
The contents of mercury in Japanese human bodies

Organ or part	Weight (g)	TM[a] (mg)	MM[b] (mg)
Muscle	24,000	1.44	0.19
Bone	8500	—	—
Fat	6600	—	—
Blood	4500	0.27	0.050
Skin	4200	0.25	—
Connective tissue	1800	—	—
Liver	1500	0.71	0.066
Brain	1300	0.13	0.023
Intestine	1000	0.076	0.010
Lung	900	0.072	0.0058
Heart	300	0.021	0.0028
Kidney	250	0.28	0.0058
Spleen	150	0.010	—
Pancreas	100	0.0083	0.0010
Total	55,000	3.3	0.35

[a]TM, total mercury.
[b]MM, methylmercury.

experiments using labeled MM, however, it cannot be clarified whether the MM may be decomposed in animal bodies. In this study, the GC method was used to demonstrate the behavior of MM in rats.

The female rats had 500 μg/head of MMC orally administered. After the administration, the amounts of MM and TM in the brain, liver, kidney, spleen, muscle, intestine without contents, contents of intestine, blood, and fur, were determined at a given interval. Each organ was divided in two equal parts for measurement of MM and TM, and two rats were used at one point.

Part of the results are shown in Fig. 3 in terms of averages. From these results, the biological half-life of MM in each organ was calculated as follows: blood about 8 days; brain about 9 days; liver and kidney about 10 days. Therefore no marked difference was found in any organ examined. The decrease curve of TM showed also almost the same biological half-life as MM in each organ, respectively. Only the curve of TM in the kidney was a little slow, so that MM/TM in the kidney tends to become smaller as the days go by.

Chemical reaction of Hg^{++} with methyl cobalamine

Besides the synthesis of MM using Grignard's reagent in the laboratory, the following transformations of inorganic mercury to MM have been found:

(1) Methylation of catalytic inorganic mercury in an acetaldehyde production plant.

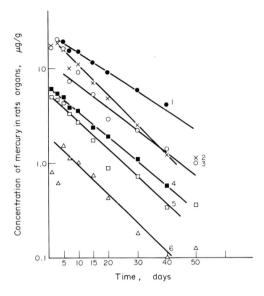

FIG. 3. Decrease of mercury in organs of rats, which were administered 500 μg/head of MMC, orally. The values were obtained as average of two rats and as mercury. 1, TM in kidney; 2, MM in kidney; 3, MM in blood; 4, TM in liver; 5, MM in liver; 6, MM in brain.

(2) Simple chemical reaction with Hg^{++} and CH_3 donors.[26]
(3) Methylation of inorganic mercury by microorganisms such as fungi and bacteria.[27,28]
(4) Methylation of inorganic mercury by exposure to ultraviolet light.[29,30]
(5) Methylation resulting from chemical reaction of inorganic mercury with methyl cobalamine (Met-Co).

Some reports on these mechanisms have been published,[31,32] but in this report the results of item (5) obtained by using GC and GC–MS (mass spectrometry) are presented.

The Met-Co was added to $HgCl_2$ solutions (some of which contained cysteine; some, MM; some neither, as shown in Table 6) and MM in the mixed solution was determined at various subsequent times.

As is apparent from Table 6, the MM produced was found to be more than 20% of the orginal added Hg^{++}. This reaction reached the steady state within 10 min. but synthesized MM decreased after a few hours. The additional MM soon decreased and was equilibrated at a certain ratio. When cysteine was added, no MM was found, and when fish were present in the water in tanks in which concentrations of 0.1 ppm of Hg^{++} and 10 ppm Met-Co were maintained, the MM amounts in fish were not

TABLE 6
Amounts of methylmercury synthesized by chemical reaction of Hg^{++} and methyl cobalamine

Added (μM)				Reaction time (min)	Found (μM) MM	
Hg^{++}	Met-Co	Cysteine	MM			
1	1	—	—	10	0.20	
5	5	—	—	10	0.62	1*,
10	10	—	—	10	1.8	
1	1	—	—	60	0.06	
5	5	—	—	60	1.4	2*,
10	10	—	—	60	2.5	
5	5	—	4	60	1.4	
5	5	—	40	60	2	3*, 4*,
5	5	1	—	60	(−)	
5	5	100	—	60	(−)	

Incubations were performed in brown flask and 10 ml of water solution were extracted. MM was determined by ECD and mass fragmentgraphy before and after purification of cysteine

Values in this table were obtained by ECD. (−) indicates levels below 0.01 μM. Mass fragmentgraphies of 1*, 2*, 3*, 4* are shown in Fig. 4.

increased compared with the control fish. These detailed results will be reported separately, but they may not always show that the MM in the natural environment is synthesized as in the reaction of the laboratory.

By the way, the synthesized MM in each benzene extract was confirmed by GC–MS and the same values were obtained before or after purifying (Fig. 4). This mass fragmentgraphic method had almost the same sensitivity as the GC with ECD, and it may be utilized in the very near future.

COMMENT

This report has pointed out that the GC technique for MM analysis had a very important role in the past, especially in Japan, and that though the determination of MM has usually been performed only by GC, nowadays there are still some problems in the analytical process. However, this method is useful enough in the survey of mercury pollution or in biological and chemical experiments.

Some application data were measured, such as mercury content in human organs, biological half-lives of MM in the rat organs, and the synthesized MM from inorganic mercury in the presence of Met-Co. Mass fragmentgraphy using GC–MS may become popular for MM analysis in the future.

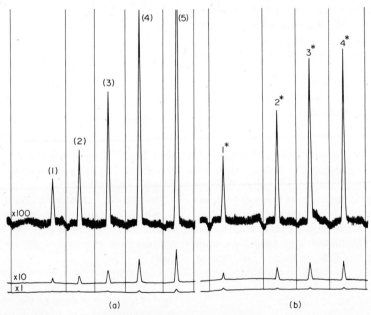

FIG. 4. Determination of methylmercury by mass fragmentgraphy (m/e 252) using GC–MS (Shimadzu LKB-9000). (a) Calibration of standard MMC: (1) 0.5 ng of MMC, (2) 1 ng, (3) 2 ng, (4) 3 ng, (5) 4 ng, respectively. (b) Estimation of MM in the solution of reaction by mass fragmentgraphy 1*, 2*, 3*, 4*, are shown in Table 6, 5 μl injection of benzene extract each. GC conditions: 20% DEG-Hg on Chromosorb W 60–80 mesh; 0.3 m × 3 mm ϕ glass column; Column temp. 150°C; carrier gas He 40 ml/min. MS conditions: temp. IS 330°C; temp. separ. 260°C; elec. energy 20 eV; trap curr. 60 μA; Acc. HV 3.5 kV; slits 0.1/0.3.

Acknowledgement—The author wishes to express his deep gratitude to Professor Kitamura, Department of Public Health, Kobe University School of Medicine, for helpful advice and suggestions.

REFERENCES

1. HUNTER, D. and RUSSELL, D. S., *J. Neurol. Neurosurg. Psychiat.* **17**, 235 (1954).
2. KITAMURA, S. and SUMINO, K., *Jap. J. Ind. Hlth.* **8**, 299 (1966).
3. KITAMURA, S., SUMINO, K., HAYAKAWA, K., SHIBATA, T. and TSUKAMOTO, T., *Med. Biol.* **72**, 274 (1966).
4. KITAMURA, S., SUMINO, K. and HAYAKAWA, K., *Jap. J. P. Hlth.* **15**, 379 (1968).
5. SUMINO, K., *Kobe J. Med. Sci.* **14**, 115 (1968).
6. WESTÖÖ, G., *Acta chem. scand.* **20**, 2131 (1966).
7. WESTÖÖ, G., *Acta chem. scand.* **21**, 1790 (1967).
8. WESTÖÖ, G., *Acta chem. scand.* **22**, 2277 (1968).
9. Special Research Group of the Ministry of Welfare, Report of special research on the mercury poisoning in Niigata, Japan, April 1967.
10. POLLEY, D. and MILLER, V. L., *Anal. Chem.* **24**, 1622 (1952).
11. GAGE, J. C., *Analyst* **86**, 457 (1961).
12. BARTLETT, J. N. and CURTIS G. W., *Anal. Chem.* **34**, 80 (1962).
13. TATTON, J. O'G. and WAGSTAFFE, P. J., *J. Chromatogr.* **44**, 284 (1969).
14. Report from an expert group, Methylmercury in fish, *Nord. Hyg. Tidskr. Suppl.* 4, Stockholm, 1971.
15. FRIBERG, L. and VOSTAL, J. (eds.), *Mercury in the Environment*, CRC Press, Cleveland, Ohio, 1972.
16. SPANGLER, W. J., SPIGARELLI, J., ROSE, J. M. and MILLER, J. L., *Science* **180**, 192 (1973).
17. Pharmaceutical Society of Japan, *Standard Method of Analysis for Hygienic Chemists*, Kinbara Shuppan Co., Tokyo, 1973.
18. NISHI, S., HORIMOTO, Y. and UMEZAWA, Y., *Jap. Analyst.* **19**, 1646 (1970).
19. RIVERS, J. B., PEARSON, J. E. and SHULTZ, C. D., *Bull. Environment. Contam. Toxicol.* **8**, 257 (1972).
20. BACHE, C. and LISK, J., *Anal. Chem.* **43**, 950 (1971).
21. NISHI, S. and HORIMOTO, Y., *Jap. Analyst.* **20**, 16 (1971).
22. KITAMURA, S., SUMINO, K., HAYAKAWA, K. and SHIBATA, T., *Jap. J. Hyg.* **28**, 108 (1973).
23. TAGUCHI, Y., *Jap. J. Hyg.* **25**, 553 (1971).
24. JOSELOW, M. M., GOLDWATER, L. J. and WEINBERG, S. B., *Arch. Environ. Hlth.* **15**, 64 (1967).
25. KITAMURA, S., SUMINO, K., HAYAKAWA, K. and SEBE, E., *Jap. J. Pharmacol.* **63**, 228 (1967).
26. SEBE, E., KITAMURA, S., HAYAKAWA, K. and SUMINO, K., *Formosan Sci.* **25**, 64 (1971)
27. JENSEN, S. and JERNELÖV, A., *Nature* **223**, 753 (1969).
28. KITAMURA, S., SUMINO, K. and TAIRA, M., *Jap. J. Hyg.* **24**, 76 (1969).
29. KITAMURA, S. and SUMINO, K., *Jap. J. Hyg.* **27**, 123 (1972).
30. AKAGI, H., FUJITA, M. and SAKAGAMI, Y., *J. Hyg. Chem. Jap.* **18**, 309 (1972).
31. WOOD, J. M., KENNEDY, F. S. and ROSEN, C. G., *Nature* **220**, 173 (1968).
32. IMURA, N., SUKEGAWA, E., PAN, S. K., NAGAO, K., KIM, J. Y., KWAN, T. and UKITA, T., *Science* **172**, 1248 (1971).

Methylmercury Analysis (K. Sumino)

DISCUSSION by GUNNEL WESTÖÖ
Food Laboratory, Swedish National Food Administration, Stockholm, Sweden

Since 1967 there has been some concern about the different results obtained in Japan and Sweden as to the percentage of total mercury (TM) in fish present as methylmercury (MM). In Sweden almost all the mercury was identified as MM, whereas in Japan the figures found have increased from an average of 25% MM in 1967[1] to about 65% today.[2] I hope that the present discussion will eliminate the remaining difference.

In 1966–8 Dr. Sumino and I published methods, based on similar principles, for the determination of MM in certain foods. Our most recently published procedures are seen in Fig. 1.[3-9] The first of these methods is the one mainly used in my laboratory. Dr. Sumino has now presented a modified method in which he has increased the amounts of hydrochloric acid added before the first extraction so that he reaches the acidity originally proposed by Gage in 1961. He has also increased the amount of hydrochloric acid added before the last extraction. Furthermore, instead of a very dilute glutathione solution (0.01%) he now uses a 1% cysteine solution (Fig. 2).

After these essential improvements in the Japanese method, there are still two important differences between our procedures:

(1) Dr. Sumino attempts to transfer all the MM in the sample into the final extract by repeated extractions, whereas I rely on partition coefficients, perform only one extraction in each step, and use aliquots of the extracts.

(2) In the last extraction, using 2×10 ml benzene, Dr. Sumino has 1 N HCl solution in his chart, instead of the more efficient 2 N acid solution in my procedure. However, he writes in his paper that good results are obtained using 2 N–3 N HCl solution and only one extraction. This indicates that the analysis may sometimes be performed in this way in Japan.

The principle of complete extraction should be abandoned only if there are very good reasons for doing so. In the case of routine MM determination, the reasons are the following.

Efforts to transfer the MM completely from the biological material into the final extract by repeated extractions make the analyses time consuming. There is also a strong tendency to lose MM because of the difficulty in getting complete separation between the solvent phases, especially when stubborn emulsions are formed. Special difficulties are encountered when hydrochloric acid is used to liberate the MM from the SH-groups in the proteins, peptides, and amino acids in the first and the last extraction steps. The reason is the comparatively small partition coefficient of MM chloride between the benzene and the hydrochloric acid phases.

Because of these difficulties, I recommend simple procedures (Fig. 1) based on partial extraction of MM (80% is extracted). The extraction yield can be raised to more than 95% by increasing the volumes of the benzene phases compared to the water phases as shown in Fig. 2, but this does not enhance the reliability of the method significantly. A high concentration of hydrochloric acid is used for complete liberation of MM from the proteins and amino acids. The foods are diluted so much that the partition coefficients are not influenced markedly by the nature of the samples analyzed. Only one extraction is performed at each step, and aliquots of the extracts are used. Recoveries of added MM are good when the aliquots and the partition coefficients are taken into consideration in the calculations by using a calibration curve or a conversion factor equal to the slope of the calibration curve. The average recovery, using my methods shown in Fig. 1, was found to be $98 \mp 3\%$, when 0.2–0.6 mgHg/kg had been added to 10 samples of fish flesh.[5]

When hydrobromic acid is used instead of hydrochloric acid in the first extraction, thiosulfate instead of cysteine in the second extraction, and iodide instead of hydrochloric acid in the last extraction,[10,11] a complete transfer of MM is more easily achieved than in the procedures shown in Fig. 1. The reason is that the MM salts formed have more favorable partition coefficients. Especially MM iodide has a high partition coefficient. How-

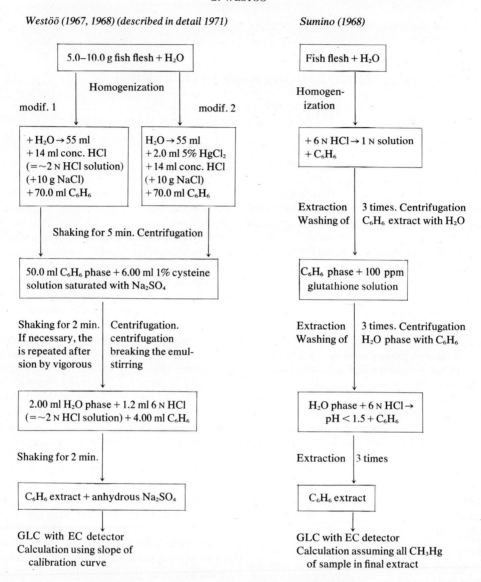

Fig. 1. Analysis of methylmercury in fish flesh.

ever, the use of iodide leads to instability, and the extract cannot be stored. Sometimes, for instance when liver is analyzed, decomposition may occur as early as during the extraction. Accordingly, procedures based on iodide give rise to problems in routine analysis, in spite of the excellent extraction figures.

The principle of complete extraction is sound. However, it is not achieved by just repeating the extractions in each stage once or twice. The solvent

---

†Extraction three times with 10.00 ml 1% cysteine solution.

volumes and the pH must be properly chosen. In my opinion, they are not well chosen in the Japanese method. A rough estimation of the extraction efficiency (based on partition coefficients) made me suspicious, and my suspicions were fully confirmed when I ran MM standards through the procedure.

The Japanese procedure in Fig. 2 uses such a low concentration of hydrochloric acid (1 N) in the last extraction step, and such small volumes of benzene, compared to the volumes of acid solution, in the first and last extraction stages, that only about 65% of the MM present in the samples can be extracted.† If 2–3 N HCl solution and only one

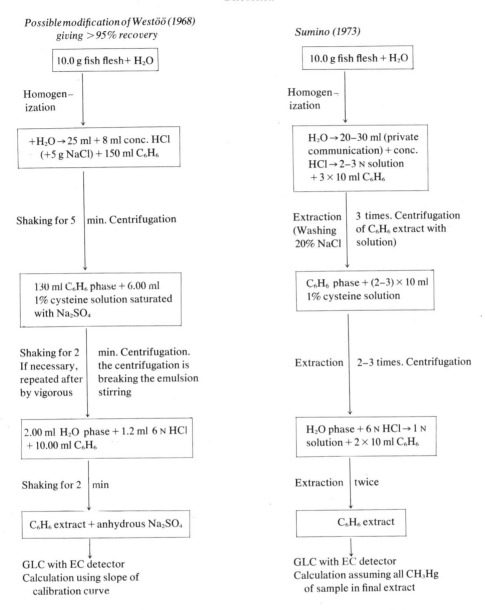

FIG. 2. Modified methods for the analysis of methylmercury in fish flesh.

extraction is used in the last step,[2] about 65% of the MM can also be extracted. The figure obtained depends on the benzene volume used. The incomplete extraction without using a calibration curve or a conversion factor seems to be the reason why in Japan an average of only 65% of the TM in fish is recovered as MM.[2] In my laboratory we have analyzed several samples of Japanese fish and found almost all the mercury in the fish to be in the form of MM, exactly as in Swedish fish.[1] A low percentage of MM can of course exist in special fish in special areas. An example is the blue marlin from Hawaii, described by Rivers et al.[12]

Dr. Sumino reports a recovery of 92–95% of MM added to fish samples.[2] This recovery is theoretically impossible, using his method of 1973, described in Fig. 2.

A few minor points mentioned by Dr. Sumino should also be discussed. The addition of mercuric chloride, under the partition conditions described in my Fig. 1, produces a MM ghost peak corresponding to 1–2 μgHg/kg sample. In most samples this contribution is insignificant. Furthermore, it can be incorporated into the calibration curve or compensated for by subtraction of a blank of 1 (for 10 g sample) or 2 (for 5 g sample) μgHg/kg. The ghost peak will be larger only if the phase separations after the extractions are not performed properly. One must always transfer clear, homogeneous phases. Several hundred fish samples have been analyzed by the first two methods in my Fig. 1 without significant differences in the results. In about 60 fish samples, analyzed during the last 2 years, 94 ± 11% and 93 ± 15% of the TM was recovered as MM according to the procedures with and without addition of mercuric chloride. Thus the addition of mercuric chloride—which is seldom made in fish analysis—does not contribute to the different MM percentages found in fish in Sweden and Japan.

I do not believe that an extraction from the final extract with glutathione[2] can reveal the contribution from impurities having the same retention time as MM chloride unless the impurity has been introduced by the analyst after the second extraction step. If an impurity in the original sample passes the cysteine step, it will no doubt be extracted also by the glutathione solution. A metal powder column would probably remove many compounds besides MM and likewise would not confirm the presence of MM.

A few details in Dr. Sumino's paper[2] should also be mentioned. No benzene is lost (p. 36) during centrifugation if stoppers are used. Sodium chloride is not added in order to transform the MM compounds to MM chloride (p. 37),[2] but simply to facilitate the separation of the phases. Bis(MM) sulfide is easily transformed to MM chloride by excess hydrochloric acid. However, the extraction of hydrogen sulfide into the benzene phase must be prevented by precipitation of the sulfide ions with mercuric ions[4] or other appropriate metal ions[13] in order to prevent the reformation of bis(MM) sulfide in the benzene phase, where there is no excess of hydrochloric acid.

Hydrogen sulfide cannot oxidize cystcine (p. 37).[2]

There is no loss of bis(MM) sulfide when mud is analyzed (p. 37),[2] if benzene is added immediately after weighing the sample.

In the tentative analytical method for mud (p. 39)[2] (Environmental Agency of Japan), the hydrochloric acid and benzene volumes added at the end of the extraction procedure are definitely too small. About half the MM will be lost in this extraction. A calibration curve is used here. A much better recovery would have been obtained if a few millilitres of 1% cysteine solution had been used, instead of a large volume of 0.1% solution.

REFERENCES

1. BERGLUND, F., BERLIN, M., BIRKE, G., CEDERLÖF, R., VON EULER, U., FRIBERG, L., HOLMSTEDT, B., JONSSON, E., LÜNING, K. G., RAMEL, C., SKERFVING, S., SWENSSON, Å., and TEJNING, S., Methylmercury in fish. Report from an expert group, Stockholm 1971, p. 53.
2. SUMINO, K., International Conference on Heavy Metals in the Aquatic Environment, Nashville, Tennessee, December 1973.
3. WESTÖÖ, G., Acta chem. scand. 20, 2131 (1966).
4. WESTÖÖ, G., Acta chem. scand. 21, 1790 (1967).
5. WESTÖÖ, G., Acta chem. scand. 22, 2277 (1968).
6. KITAMURA, S., TSUKAMOTO, T., HAYAKAWA, K., SUMINO, K., and SHIBATA, T., Med. Biol. 72(5), 274 (1966).
7. SUMINO, K., Kobe J. Med. Sci. 14, 115 (1968).
8. SUMINO, K., Kobe J. Med. Sci. 14, 131 (1968).
9. WESTÖÖ, G. and RYDÄLV, M., Vår föda 23, 177 (1971).
10. BOUVENG, H., Publication No C7A Swedish Water and Air Pollution Research Laboratory, Stockholm, 1970.
11. UTHE, J. F., SOLOMON, J., and GRIFT, B., J. Ass. Offic. Anal. Chem. 55, 583 (1972).
12. RIVERS, J. B., PEARSON, J. E., and SHULTZ, C. D., Bull. Environ. Contam. Toxicol. 8, 257 (1972).
13. FUJIKI, M., Jap. Analyst 19, 1507 (1970).

A REVIEW OF THE STATUS OF TOTAL MERCURY ANALYSIS

W. DICKINSON BURROWS

Associated Water and Air Resources Engineers, Inc., Nashville, Tennessee

INTRODUCTION

In an earlier paper we reviewed the status of total mercury analysis through the beginning of the present decade.[1] Since then there has been substantial progress in defining methodologies for water and biological samples in particular, and agreement among different laboratories has improved. Almost all analyses are performed using atomic absorption spectroscopy or neutron activation, although research continues on more exotic techniques.

FLAMELESS ATOMIC ABSORPTION

Flameless or cold vapor atomic absorption spectrometry, the method of choice for most laboratories, seems likely to become even more dominant with the appearance on the market of several relatively low-priced instruments. Most commonly used is the method of Hatch and Ott[2] as modified by Uthe et al.[3] This involves oxidative digestion ("wet ashing"), followed by reduction, aeration, and measurement of mercury vapor absorption at 253.7 nm. The detection limit at this wavelength is on the order of 1–5 ng. For the very low levels in water samples, it may be necessary to preconcentrate the mercury before analysis. This has been achieved by dithizone extraction,[4,5] by electrodeposition,[6] and by amalgamation on silver wire,[7,8] in each case permitting detection limits of 0.01–0.001 ppb. The Environmental Protection Agency has developed a procedure which involves oxidation of the sample with nitric acid, sulfuric acid, potassium permanganate, and potassium persulfate, followed by reduction with hydroxylamine and stannous sulfate, and aeration.[9] Preconcentration is not specified, and the procedure is claimed to give superior results in the range of 0.2 ppb and above. A similar but simpler procedure is reported by Cranston and Buckley.[10]

With sediment and biological samples the analyst encounters more difficulty in bringing about complete oxidation. Derivatives of benzene and other aromatic hydrocarbons which absorb strongly in the 253 nm region may be carried into the detector chamber with mercury vapor. Kopp et al. have tabulated interference from a variety of organic solvents.[11] They also note (as have others) that where mercury is organically bound, conversion to elemental mercury requires more rigorous conditions in the digestion step. Hoover et al. found boiling nitric acid to be satisfactory for oxidizing organomercurials in alfalfa meal,[12] but others advocate stronger reagents, such as persulfuric acid.[9,11] Malaiyandi and Barrette have devised a procedure for determining mercury in cereal grains contaminated with organomercurials which employs vanadium pentoxide and nitric acid in the digestion step;[13] the same reagent has been used by Deitz et al. to estimate total mercury in milk, meat, soil, and alfalfa.[14] For fish samples, Munns and Holland have devised a procedure using nitric and perchloric acids which has given good reproducibility in a collaborative study.[15] Rains and Menis have used this reagent for determining mercury in standard reference materials.[16] Nord et al. have recovered mercury from human hair after digestion with sulfuric and nitric acids and potassium permanganate.[17] For pike and cod, Omang has utilized a mixture of hydrobromic and nitric acids.[18] Thus, there is as yet no agreement on any one reagent for wet oxidation of all environmental samples.

Various items of equipment have been devised to facilitate the different steps in the Hatch and Ott procedure. Holak et al. have determined mercury in fish, perhaps the most troublesome of biological samples, by digestion with nitric acid at 150°C in a sealed Teflon reaction vessel.[19] By subjecting sediment samples to ultrasonic dispersion prior to digestion, Bonner and King have consistently achieved an enhancement of about 10% in mercury recovery.[20]

A remarkably simple digestion procedure has been devised by Bishop and Taylor[21] using a "hot block," an aluminum billet with holes bored to accept standard test tubes. With the sample contained in a mixture of sulfuric and nitric acids, the block is heated until the solution begins to reflux up the exposed sides of the tubes (about 240°C). Digestion is complete when the solution is clear. Bishop and Taylor report no volatilization losses, and analyses of fish samples containing 0.5–1.0 mg/kg of mercury are in good agreement with those obtained by neutron activation.

For transferring mercury into the detector cell, three different techniques are commonly employed. The air from the reduction flask containing mercury vapor may simply be passed through the cell and vented[3, 16] or it may be recycled back through the reaction flask by means of a peristaltic pump, as demonstrated in Fig. 1.[2, 9, 11] Mercury vapor may also be removed from the reduction flask with a syringe and injected into a closed detector cell, a method recommended by Thorpe for food products and biological fluids,[22] and by Deitz et al.[14] Loss of mercury during transfer may be limited by carrying out the reduction reaction in the syringe itself, a procedure developed by Stainton[23] and exploited by Thompson and McComas for analysis of mercury in natural waters.[5]

Because wet oxidation can be a slow and tedious procedure and because results are frequently erratic due to incomplete digestion, volatilization losses, and mercury contamination of commercial reagents, a number of investigators are exploring total combustion of samples. Well before the development of the method of Hatch and Ott, Gutenmann and Lisk had already determined the mercury content of apples in the 0.3–0.6 ppm range by combustion in an oxygen atmosphere over hydrochloric acid in a modified Schoniger flask, prior to analysis by the dithizone technique.[24] In 1966 Pappas and Rosenberg described adaptation of this procedure to determination of mercury in wheat,[25] fish, and eggs.[26] More recently, Okuno et al. have combined Schoniger combustion and silver wire amalgamation in the analysis of a variety of field samples.[27] In 1968 Lidums and Ulfvarson constructed a complete, closed combustion train for determination of mercury by atomic absorption in biological samples such as blood, urine, liver, and eggs.[28] Agreement with values obtained by neutron activation was quite good at the 1 ppb level. Because combustion is not very rapid even at high temperatures, the concentration of mercury vapor in the combustion stream is usually too low for direct measurement; the mercury must be immobilized at some point in the train for later release. Lidums and Ulfvarson trapped the mercury using a gold filter, while Anderson et al. collect the mercury after combustion on a gold-impregnated fritted disk or gold-coated asbestos.[29] Mercury is released from the amalgamator by heating and carried in an air stream into the absorption cell as a single pulse. Perhaps the most promising development of the combustion–amalgamation technique is that of Willford and co-workers, who use a high frequency induction furnace to oxidize 0.05 to 0.1 g samples, and a gold wire nest to trap the mercury vapor.[30] This system, also employed by Schulert et al.,[31] is adaptable to all biological materials, water, and sediments. A simple combustion train constructed by Reimers at Vanderbilt for large (1–10 g) samples uses a permanganate trap, from which mercury is released by the addition of hydroxylamine and tin(II) salts.[32] A system designed by Hermann et al. for direct measurement of mercury in clinical samples employs Ascarite and other chemical filters to remove interfering gases from the mercury vapor, which is analyzed without prior trapping.[33] A variation on this technique is used by Thomas et al. for measuring total mercury in very small fish samples (50–150 mg).[34] Kunkel has devised a versatile apparatus utilizing Wickbold oxyhydrogen combustion with or without a permanganate trap, depending on the nature of the sample.[35] In yet one more variation on the combustion furnace, Joensuu has used solid sodium nitrate to oxidize sulfur and

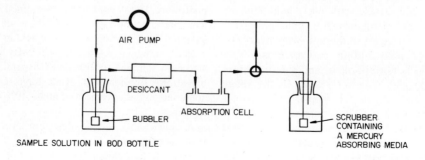

FIG. 1. Apparatus for flameless mercury determination. After Kopp et al.[11]

organic matter and a gold foil amalgamator to trap the mercury vapor carried through the train in a nitrogen stream.[36] Other total combustion methods have been reported for fish,[37] eggs,[38] sediments,[39] and minerals.[40]

The technique of trapping elemental mercury by amalgamation on gold,[28-31] silver,[7,8] or platinum[41] can advantageously be used in both wet and combustion methods. After interfering gases have been purged from the system, the mercury is vaporized by heating and passed into the detector, relatively uncontaminated. Cadmium sulfide pads may be used in the same manner.[29] Head and Nicholson have combined wet digestion–reduction–aeration with gold wire amalgamation for the atomic absorption analysis of geological samples.[42] A different approach is taken by Windham, who measures the sample absorption in a closed, recycling system, then introduces a palladium chloride trap into the line to remove the mercury.[43] The background absorption is measured and subtracted from the total.

Magos has devised a wholly reductive technique for the separate determination of methylmercury and total mercury in biological samples by atomic absorption spectroscopy.[44] The chemical basis for this procedure is that cysteine solutions of alkylmercury and inorganic mercury derivatives are converted to elemental mercury by a mixture of cadmium chloride and stannous chloride, while only inorganic mercury compounds are reduced by stannous chloride alone. A modification by Magos and Clarkson permits the determination of mercury at the ppb level in blood.[45] Tables 1 and 2 summarize current methods using atomic absorption spectroscopy for the determination of mercury in water and other environmental samples.

OTHER METHODS BASED ON ATOMIC SPECTROMETRY

The resonance line at 184.9 nm has found limited utility in the total analysis of mercury in environmental samples by atomic absorption, though the theoretical detection limit for the resonance line is nearly two orders of magnitude lower than for the forbidden resonance line at 253.7 nm. Because oxygen absorbs strongly in the 185 nm region, the system must be evacuated or purged with a noble gas, such as argon, thus complicating the design of whatever mercury vapor transfer system is used. Robinson et al. have measured the mercury content of air by atomic absorption spectrometry at 184.9 nm by drawing the air through a bed of hot carbon before introduction to the detector cell, thereby converting oxygen to carbon monoxide.[46] The detection limit is reported as about 0.5 $\mu g/m^3$.

Flameless emission spectrometry is another procedure yet to be substantially exploited for environmental samples. In a technique devised by April and Hume, elemental mercury generated in solution by reduction is swept by a helium stream into a chamber enclosing a radio-frequency plasma in helium.[47] The emission from the plasma is monitored at 253.7 nm, giving a detection limit of about 2 ng. Lichte and Skogerboe have described a superficially similar system based on emission from a microwave plasma generated in argon.[48] A precision of 10–12% is claimed in analysis of 10 ml water samples containing 0.01 ppb mercury. Leaf

TABLE 1
Determination of mercury in water by atomic absorption

Sample size (ml)	Mercury content ($\mu g/l$)	Standard deviation[a] ($\mu g/l$)	Recovery (%)	Method of concentration	Method of digestion	Reference
500	0.5	0.0087	>90	Dithizone	None	4
500	0.05	0.0042	>90	Dithizone	None	4
200	0.023	0.0049	>94	Dithizone	None	5
200	0.166	0.0112	>94	Dithizone	None	5
50	0.1–1.0	10%	—	Electrodeposition	None	6
100	0.1–1.5	0.04–0.2	>80	Silver amalgamation	None	8
100	0.35	0.16	—	None	Wet	9
100	1.35	0.14	89	None	Wet	9
250	0.03–3	—	—	None	Wet	10

[a] Reported precision may not be comparable among different investigators. Some use distilled water containing mercury rather than natural water, and some report precision for atomic absorption only, rather than for the whole procedure.

TABLE 2
Determination of mercury in sediment, soil, and biological samples by atomic absorption. Some typical results

Sample	Sample size (g)	Mercury content (µg/g)	Standard deviation (µg/g or %)	Recovery (%)	Method of digestion	Reference
Fish	0.5	0.5	0.051	ca. 100	Wet	3
Illite	ca. 1	0.158	—	—	Wet	10
Flour	5	0.5	0.006	100	Wet	13
Soil	≤2	0.0351	0.0046	—	Wet	14
Fish[a]	5	0.5	0.072	76	Wet	15
Fish[a]	5	0.5	0.13	92	Wet	15
Orchard leaves	2–3	0.160	0.006	—	Wet	16
Hair	0.03	0.0117	0.002	ca. 100	Wet	17
Fish	0.1–1	2.12	0.16	—	Wet	18
Fish	1	0.5	—	90–100	Wet	19
Fish	0.3–2	0.8	0.05	>100	Wet	22
Eggs	0.3–2	0.5	—	100	Wet	22
Wheat	1	0.1	—	80–90	Combustion	25
Eggs	1	0.1	—	ca. 80	Combustion	26
Duck breast	1	0.1	7–17%[b]	80–100	Combustion	27
Blood	0.2–0.4	0.003–0.02	ca. 10%	—	Combustion	28
Liver	0.025	0.5	ca. 10%	—	Combustion	28
Gelatin	0.025–0.05	1.3	0.27	—	Combustion	29
Fish	0.05–0.1	0.2	4.7%	ca. 100	Combustion	30
Coal	0.05–0.1	0.06	3.1%	ca. 100	Combustion	31
Blood	0.1	0.033	0.0025	—	Combustion	33
Fish	0.05–0.15	0.5	ca. 0.05	—	Combustion	34
Blood	0.1	0.001–0.027	—	≥100	Combustion	34
Urine	0.1	0.05–0.2	—	ca. 100	Combustion	34
Cod meal	3	0.200	—	—	Combustion	35
Sediment	0.5–2	0.160	7.8%[b]	—	Wet	42
Blood, red cells	1	0.016	0.0022	95	None	45
Blood, plasma	1	0.0022	0.0009	—	None	45

[a] Collaborative study.
[b] Coefficient of variation.

and blood samples have been analyzed with an estimated accuracy of 10%.

Kirkbright and West have recently reviewed the applications of atomic fluorescence spectrophotometry in chemical analysis.[49] The most important advantage of atomic fluorescence where mercury is concerned is that the cold vapor detection limit at 253.7 nm is 0.05 ng, nearly two orders of magnitude smaller than the limit for flameless atomic absorption. Furthermore, the broad band absorption caused by traces of organic materials, for which a correction must be made in absorption spectrometry, is not a problem in fluorescence analysis. The fluorescence signal is highly subject to quenching by many common gases, however, particularly oxygen. Robinson and Araktingi have found that quenching by nitrogen and carbon monoxide is so efficient that atomic fluorescence analysis of mercury in air is not feasible, even if the oxygen is removed by passage through hot carbon.[50] For water samples, Thompson and Reynolds have constructed a simple atomic fluorescence system with a detection limit of 2 ng.[51] Elemental mercury is generated in the usual reduction unit and swept through the detector in an argon stream. Because an open cell is employed, there are no cell windows to fog, and hence there is no need for drying columns. They note that although organic materials in general do not fluoresce, and therefore do not generate spurious signals, benzene absorbs strongly in the 253.7 nm region and will almost totally attenuate the mercury fluorescence if present at the 2% level. Muscat and Vickers have carried out water analyses by atomic fluorescence using a recirculating reduction–aeration system similar to that of Hatch and Ott, but because they used air as the carrier, the detection limit was higher than that of Thompson and

Reynolds.[52] In a later paper, Muscat et al. have described the analysis of flour, water, rock, and sediment samples using either the digestion–reduction–aeration system (for liquids) or a combustion furnace (for solids).[53] In both cases, elementary mercury is collected in a silver amalgamator, and argon is the carrier gas to the resonance cell. Detection limits and precision appear to be about the same as for atomic absorption. Shimomura and Hiroto, by taking great care in the design and geometry of the nozzle from which the mercury vapor enters the detection system, have achieved a detection limit of 0.5 ng for a 5 ml water sample (0.1 ppb), with a standard deviation of 3% for 15 ng.[54] A reduction–aeration train was employed, with nitrogen as the carrier.

The application of the Zeeman effect on atomic absorption is another promising approach to the elimination of interference from contaminating vapors. This effect is the splitting of the absorption line by a magnetic field. By pyrolyzing 10 mg tuna samples directly in the absorption cell, Hadeishi and McLaughlin have measured the mercury content to 0.04 ppm from the hyperfine Zeeman lines.[55] Recent improvements in instrumentation have increased the sensitivity a hundred-fold, permitting the detection of mercury vapor at a concentration of 0.2 $\mu g/m^3$.[56]

Two flameless procedures for analyzing a film of water picked up on a metal loop have been described. Dagnall et al. evaporate the water from a platinum loop, then rapidly heat the loop in an argon stream to release the mercury for detection by emission or absorption at 184.9 nm.[57] Stephens atomizes the entire sample in a fast current pulse through a copper loop directly in the absorption cell.[58] In each case the sensitivity in terms of concentration is at least an order of magnitude less than given by flameless atomic absorption because of the extremely small sample size (ca. 0.1 μl).

Research on flame techniques continues, in particular by investigators at Imperial College.[59,60] The disadvantages of the method compared with flameless techniques are that the detector is exposed to everything present in the sample, and the amount of mercury at the detector is limited by the rate at which the sample can be nebulized. Detection limits for water are typically several orders of magnitude higher.

NEUTRON ACTIVATION ANALYSIS

Although neutron activation analysis is used by only a few laboratories for routine estimation of mercury in environmental samples, the method is commonly the standard to which all other methods are compared. When natural mercury, a mixture of stable isotopes, is exposed to a high flux of thermal (slow) neutrons, it is converted to a mixture of radioactive isotopes, principal among which are ^{197}Hg and ^{203}Hg, with decay half-lives of 65 h and 47 days, respectively. In selecting a procedure for analysis by neutron activation, there are three principal sets of options to consider: treatment of the sample—which may be destructive or nondestructive, the choice of isotope to be measured, and the choice of detector. For example, the Sjostrand technique,[61] which has been applied to a number of environmental samples, involves destruction of the sample after neutron irradiation, followed by isolation and counting of ^{197}Hg using a NaI (Tl) detector. Digestion is usually carried out with refluxing hydrochloric acid in a Bethge apparatus, and isolation of mercury may be achieved by electrodeposition on a gold foil. By adding a precise weight of nonradioactive mercury carrier to the sample after irradiation and before digestion, and by measuring the recovery of this carrier on the gold foil, the fraction of mercury recovered from the sample can be determined. Thus isolation losses, which introduce a degree of uncertainty in atomic absorption analysis, are not a factor. Pillay et al. have used this general procedure, for which Sjostrand claims a sensitivity limit of 0.5 ppb, to estimate mercury in Lake Erie fish, sediments, and plankton,[62,63] and Ruf and Rohde have determined mercury contents of a variety of biological materials.[64] Mercury may also be separated by ion exchange,[65,66] and quantitative precipitation as the sulfide is commonly used for estimation of carrier recovery.[65–68] Weiss and Bertine have determined carrier yield without recourse to gravimetry by re-irradiating the counted sample (plus carrier) and comparing the induced γ-activity with that of carrier standards similarly irradiated.[69]

Chemical separation of mercury is required, depending on the nature of the sample, by the presence of radioactive elements such as ^{24}Na ($t_{1/2}$ = 15 h), ^{82}Br ($t_{1/2}$ = 35 h), ^{32}P ($t_{1/2}$ = 15 days), and ^{75}Se ($t_{1/2}$ = 4 months), which are also produced by neutron activation, and which would interfere with counting if allowed to remain in the sample. Sodium and bromine interference can be eliminated simply by allowing the radioactivity to decay for 1–2 weeks before processing the sample. By measuring the longer-lived ^{203}Hg rather than ^{197}Hg, interference from phosphorus can be eliminated in the same way, although this may take more than a month. (A higher neutron flux or a longer irradiation period is required to achieve the same activity using ^{203}Hg.) Heitzman and Simpson have combined the digestion–electrodeposition technique of Sjostrand with measurement of ^{203}Hg to determine the mercury contents of fish, flour, and reference samples.[70] Margosis and Tanner have determined mercury in pharmaceutical products at higher detection levels by nondestructive analysis using the NaI (Tl) detector to measure both ^{197}Hg and ^{203}Hg.[71] Becknell et al.,[72] and more recently Law,[73] have described procedures whereby total mercury is converted to $HgCl_4^{2-}$ by treatment with chlorine and hydrochloric acid and recovered from solution by passage through anionic or chelating resin-loaded papers. The paper, more or less free from interfering metals, can be analyzed for mercury by nondestructive methods.

The interference from other radioisotopes can also be overcome to some extent by using a more discriminating γ-ray detector. The high resolution spectrum provided by the Ge (Li) detector can be used to measure both ^{197}Hg and ^{203}Hg, although the sensitivity of this detector is low compared with NaI (Tl). Rook et al. have analyzed assorted environmental samples using total combustion to separate the mercury after irradiation, followed by counting on a Ge (Li) detector.[74] Where the mercury level of the sample was high enough (>1 ppm), nondestructive analysis by irradiation and direct counting of ^{197}Hg gave satisfactory precision. A nondestructive technique for blood and other biological tissues containing mercury at the 1–100 ppb level has been devised by Filby et al.[75] Irradiation is followed by 3–6 weeks' decay, and the sample is counted intact on a Ge (Li) detector. By measuring peak areas and employing suitable standards, the interference due to ^{75}Se can be eliminated from ^{203}Hg. A recent development is the low energy photon detector, which is a special Ge (Li) or Si (Li) detector maintained at liquid nitrogen temperature, and which measures decay X-rays.[76, 77] Weaver has described a nondestructive procedure for coal which uses the low energy photon detector to estimate mercury at the 100 ppb level with a precision of 10%.[77]

Sample preparation prior to irradiation depends on the method of analysis and on the temperature to which the sample will be exposed in the reactor. Both quartz[61, 64, 66, 70, 74, 77] and polyethylene[62, 63, 65, 67, 71, 75] containers are used, but there appears to be no general agreement on the efficacy of freeze-drying samples before encapsulation.[62, 70] Jervis and Tiefenbach have encountered a much greater loss of mercury on freeze-drying and irradiating fish, compared with other biological material.[78] The applications of neutron activation analysis of mercury to environmental studies have been reviewed by Westermark and Sjostrand.[79] Table 3 summarizes recent methods used in the analysis of some typical environmental samples.

COLORIMETRIC METHODS

Despite inferior sensitivity, colorimetric and photometric methods for estimating mercury continue to warrant investigation because of the low cost of equipment needed and because they provide the prospect of a simple field method. Best known is the dithizone method, wherein mercury(II) is converted to the yellow–orange organic soluble complex and extracted with chloroform. Digestion or combustion of environmental samples requires the same care as with other methods. Leong and Ong have combined the dithizone method with pyrolysis and gold amalgamation to determine mercury at the ppm level in soil and rock samples containing sulfide minerals.[80] For biological samples, Woidich and Pfannhauser apply the sulfuric acid digestate to a chromatography column pretreated with dithizone.[81] The mercury dithizonate is eluted with chloroform for photometric determination at 483 nm. The same authors have devised a thin layer chromatography procedure for canned fish and mussels using dithizone combined with an oxidative digestion process.[82] A similar procedure is reported for food and cosmetics except that total combustion with permanganate trapping is employed rather than digestion.[83] Chelate formation by 1,2-diketonebisthiobenzhydrazones is the basis of an analytical method devised by Heizman and Ballschmiter.[84] The chelate formed by mercury(II) with 2,3-pentanedionebisthiobenzhydrazone absorbs strongly at 420 nm, and can be used for estimation of mercury at the ppm level. The interference from a number of metals which also form chelates can be reduced by competitive complexing with cyanide and other ions. The complex formed by mercury(II) with 2-mercaptobenzoic acid has been used by Cresser for the estimation of mercury down to 0.4 ppm with a precision of 2.3%.[85] Between pH 2 and pH 4 the complex absorbs strongly at 264 nm. Co^{2+}, Cu^{2+}, Fe^{2+}, and Zn^{2+} interfere substantially at a hundred-fold molar excess, but masking procedures are available. Many other colorimetric reagents for mercury have been described, including 8-mercaptoquinoline,[86] methylthymol blue,[87] Ruhemann's purple,[88] and 1,5-diphenylcarbazide.[89]

TABLE 3
Activation analysis of environmental samples for mercury

Sample	Sample weight (g)	Mercury content (μg/g)	Standard deviation (μg/g or %)[a]	Method	Isotope	Decay time	Detector	Reference
Human remains	0.07	1.6	ca. 10%	Destructive	^{197}Hg	10 h	NaI (Tl)	61
Fish	1–3	1.77	<0.12	Destructive	^{197}Hg	1 h	NaI (Tl)	62
Coal	1–3	0.54	—	Destructive	^{197}Hg	1 h	NaI (Tl)	63
Plankton/algae	1–3	31.2	—	Destructive	^{197}Hg	1 h	NaI (Tl)	63
Swan feathers	—	0.161	—	Destructive	^{197}Hg	—	NaI (Tl)	64
Human hair	—	0.56	—	Destructive	^{197}Hg	—	NaI (Tl)	64
Soil	0.1	0.14	≤10%	Destructive	^{197}Hg–^{203}Hg	24 h	NaI (Tl)	65
Carrot root	0.08	0.04	≤10%	Destructive	^{197}Hg–^{203}Hg	24 h	NaI (Tl)	65
Orchard leaves	0.2	0.148	0.010	Destructive	^{197}Hg	1 week	NaI (Tl)	66
Fish	0.2	1.29	0.13	Destructive	^{197}Hg	1 week	NaI (Tl)	66
Glacial ice	200–400	0.000230	0.000018	Destructive	^{197}Hg	4 days	NaI (Tl)	67
Fish	0.2	2.24	—	Destructive	^{203}Hg	3–5 days	NaI (Tl)	70
Flour	0.2	0.012	0.003	Destructive	^{197}Hg	3–5 days	NaI (Tl)	70
Beef liver	0.5	0.0145	0.0017	Destructive	^{197}Hg	3 days	Ge (Li)	74
Whole blood	ca. 5	0.007	10%	Nondestructive	^{203}Hg	3–6 weeks	Ge (Li)	75
Lake water	ca. 30	0.004	22.7%	Nondestructive	^{203}Hg	3–6 weeks	Ge (Li)	75
Furnace oil	ca. 5	0.015	11.4%	Nondestructive	^{203}Hg	3–6 weeks	Ge (Li)	75
Coal	0.5	0.10	10%	Nondestructive	^{197}Hg	3 days	LEPD[b]	77

[a] based on replicate analyses.
[b] LEPD = low energy photon detector.

KINETIC METHODS

The application of kinetic methods to the determination of mercury has been vigorously pursued by Ke and Thibert.[90–92] The iodide-catalyzed reaction of cerium(IV) ion with arsenite [eqn. (1)], is inhibited by mercury(II) at a very low concentration:

$$2Ce^{4+} + AsO_3^{3-} \longrightarrow 2Ce^{3+} + AsO_3^- \quad (1)$$

Reduction of the yellow cerium(IV) ion to the colorless cerium(III) ion is measured spectrophotometrically at 275 nm 20 min after standard solutions of the reactants have been combined with a sample containing mercury, under carefully controlled conditions of temperature and pH. Samples containing 1 ng of mercury have been determined with a precision of 0.05 ng.[90] The uncatalyzed reaction between cerium(IV) and arsenite, which is also inhibited by mercuric ion, has been used by Ke and Thibert in mercury analysis, but with lesser precision.[91] By combining an acid digestion step with the kinetic method, environmental samples containing both organic and inorganic mercury have been analyzed.[92] Urine, serum, and natural water samples containing 0.05–2 μg/ml of total mercury have been analyzed with a precision of 5–10%. A number of ions cause substantial interference at the ppm level; these include Cu(II), Co^{2+}, H$_2$PO$_4^-$, Ni^{2+}, and Fe^{3+}, the last suggesting that whole blood samples would present special problems. Silver and iodide ion interfere at much lower concentrations, the latter in particular with the uncatalyzed reaction. Weisz and Ludwig have utilized the iodide-catalyzed cerium(IV)-arsenite reaction to determine mercury in water spectroscopically by means of a continuous flow-through cell.[93] The mercury(II) catalyzed substitution of α,α'-dipyridyl for cyanide in potassium ferrocyanide [eqn. (2)] has been exploited by Datta and Das to estimate mercury at the 0.2–20 μg/ml level:[94]

$$Fe(CN)_6^{3-} + 3\,dipy \longrightarrow Fe(dipy)_3^{3+} + 6CN^- \quad (2)$$

The rate of complex formation is monitored spectrophotometrically at 520 nm, and the precision is reported as 0.5%. Interference is noted from palladium(II) and Ag$^+$, which behave similarly to mercury(II), and ions which complex with mercury(II), such as bromide and iodide.

MISCELLANEOUS METHODS

Photometric[95] and amperometric[96] titrations of mercury(II) with sequestering agents such as EDTA have been investigated by van der Linden and co-workers. The first is reported to be capable of analyzing standard samples containing 4 ppm mercury with a precision of 2–3%; the second is an order of magnitude less sensitive. Iron and copper interfere, but can be tolerated in the second case if suitable masking agents are used.

The fluorimetric determination of mercury has been described by Holzbecher and Ryan.[97] Mercury(II) undergoes an oxidative reaction with thiamine to produce thiochrome, which fluoresces intensely at 440 nm, and can be determined thereby in water at the 0.05 µg/ml level with a standard deviation of 4.1%.

The sensitivity of the gravimetric method, frequently used in analyzing ores, is inadequate for estimating mercury in environmental samples. However, as pointed out by Tandon et al.,[98] the procedures may be useful for separating mercury for analysis by other methods, such as activation analysis. Among the gravimetric reagents under recent investigation are 2-amino-4-methylthiazole,[98] 8-quinolinethiol,[99] and phenyl substituted 1,2-dithiole-3-thiones.[100]

Gas chromatography is commonly employed for analysis of organic mercury compounds; Jones and Nickless have devised a scheme for estimating inorganic mercury in water in the 0.05–50 ppm range by gas chromatography.[101] The Peters reaction between mercury(II) salts and sulfinic acids is used to convert mercuric chloride to phenylmercuric chloride [eqn. (3)], which is separated and purified by the Weströö procedure[102] for chromatography in toluene solution:

$$C_6H_5SO_2H + HgCl_2 \longrightarrow C_6H_5HgCl + SO_2 + HCl \tag{3}$$

Fishbein has exhaustively reviewed the application of all types of chromatography to the analysis of inorganic mercury.[103]

A unique and highly sensitive procedure contrived by McNerney et al. is based on resistance changes produced in a thin gold film by adsorption of elemental mercury vapor.[104] A portable mercury detector has been developed capable of detecting 5×10^{-11} g. It is, however, necessary to present a relatively clean mercury vapor sample to the detector, as with atomic absorption.

Analysis by X-ray fluorescence continues to find utility for samples containing mercury at the ppm level and above.[105, 106] Vassos et al. have concentrated mercury from water solution by electrodeposition on graphite, thereby greatly extending the sensitivity of the method.[105]

REFERENCES

1. REIMERS, R. S., BURROWS, W. D., and KRENKEL, P. A., Total mercury analysis: review and critique, J. Wat. Pollut. Control Fed. **45**, 814–828 (1973).
2. HATCH, W. R. and OTT, W. L., Determination of sub-microgram quantities of mercury by atomic absorption spectrophotometry, Anal. Chem. **40**, 2085–2087 (1968).
3. UTHE, J. F., ARMSTRONG, F. A. J., and STAINTON, M. P., Mercury determination in fish samples by wet digestion and flameless atomic absorption spectrophotometry, J. Fish. Res. Bd. Can. **27**, 805–811 (1970).
4. CHAU, Y.-K. and SAITOH, H., Determination of submicrogram quantities of mercury in lake waters, Environ. Sci. Technol. **4**, 839–841 (1970).
5. THOMPSON, J. A. J. and MCCOMAS, F. T., Application of the Stainton syringe method to the analysis of mercury in natural waters, Environ. Lett. **5**, 189–197 (1973).
6. DOHERTY, P. E. and DORSETT, R. S., Determination of trace concentrations of mercury in environmental water samples, Anal. Chem. **43**, 1887–1888 (1971).
7. HINKLE, M. E. and LEARNED, R. E., Determination of mercury in natural waters by collection on silver screens, US Geol. Survey Prof. Paper 650-D, pp. D251–D254, 1969.
8. FISHMAN, M. J., Determination of mercury in water, Anal. Chem. **42**, 1462–1463 (1970).
9. WINTER, J. A. and CLEMENTS, H. A., Analyses for mercury in water: a preliminary study of methods, EPA-R4-72-003, National Environmental Research Center, Office of Research and Monitoring, US Environmental Protection Agency, Cincinnati, Ohio, September 1972, 58 pp.
10. CRANSTON, R. E. and BUCKLEY, D. E., Mercury pathways in a river and estuary, Environ. Sci. Technol. **6**, 274–278 (1972).
11. KOPP, J. F., LONGBOTTOM, M. C., and LOBRING, L. B., "Cold vapor" method for determining mercury, J. Am. Wat. Wks. Ass. **64**, 20–25 (1972).
12. HOOVER, W. L., MELTON, J. R., and HOWARD, P. A., Determination of trace amounts of mercury in foods by flameless atomic absorption, J. Ass. Offic. Anal. Chem. **54**, 860–865 (1971).
13. MALAIYANDI, M. and BARETTE, J. P., Wet oxidation method for the determination of submicrogram quantities of mercury in cereal grains, J. Ass. Offic. Anal. Chem. **55**, 951–959 (1972).
14. DEITZ, F. D., SELL, J. L., and BRISTOL, D., Rapid, sensitive method for determination of mercury in a variety of biological samples, J. Ass. Offic. Anal. Chem. **56**, 378–382 (1973).
15. MUNNS, R. K. and HOLLAND, D. C., Determination of mercury in fish by flameless atomic absorption: a collaborative study, J. Ass. Offic. Anal. Chem. **54**, 202–205, 466–467 (1971).
16. RAINS, T. C. and MENIS, O., Determination of submicrogram amounts of mercury in standard reference materials by flameless atomic absorption spectrometry, J. Ass. Offic. Anal. Chem. **55**, 1339–1344 (1972).
17. NORD, P. J., KADABA, M. P., and SORENSON, J. R. J.,

Mercury in human hair, *Arch. Environ Hlth.* **27**, 40–44 (1973).
18. OMANG, S. H., Trace determination of mercury in biological materials by flameless atomic absorption spectrometry, *Anal. chim. Acta* **63**, 247–253 (1973).
19. HOLAK, W., KRINITZ, B., and WILLIAMS, J. C., Simple, rapid digestion technique for the determination of mercury in fish by flameless atomic absorption, *J. Ass. Offic. Anal. Chem.* **55**, 741–742 (1972).
20. BONNER, W. P. and KING, W. C., Tennessee Technological University, personal communication, 1973.
21. BISHOP, J. N. and TAYLOR, L. A., Ministry of Environment, Etobicoke, Ontario; personal communication from J. N. Bishop, December 1973.
22. THORPE, V. A., Determination of mercury in food products and biological fluids by aeration and flameless atomic absorption spectrophotometry, *J. Ass. Offic. Anal. Chem.* **54**, 206–210 (1971).
23. STAINTON, M. P., Syringe procedure for transfer of nanogram quantities of mercury vapor for flameless atomic absorption spectrophotometry, *Anal. Chem.* **43**, 625–627 (1971).
24. GUTENMANN, W. H. and LISK, D. J., Rapid determination of mercury in apples by modified Schoniger combustion, *J. Agric. Fd. Chem.* **8**, 306–308 (1960).
25. PAPPAS, E. G. and ROSENBERG, L. A., Determination of submicrogram quantities of mercury by cold vapor atomic absorption photometry, *J. Ass. Offic. Anal. Chem.* **49**, 782–792 (1966).
26. PAPPAS, E. G. and ROSENBERG, L. A., Determination of submicrogram quantities of mercury in fish and eggs by cold vapor atomic absorption photometry, *J. Ass. Offic. Anal. Chem.* **49**, 792–793 (1966).
27. OKUNO, I., WILSON, R. A., and WHITE, R. E., Determination of mercury in biological samples by flameless atomic absorption after combustion and mercury–silver amalgamation, *J. Ass. Offic. Anal. Chem.* **55**, 96–100 (1972).
28. LIDUMS, V. and ULFVARSON, U., Mercury analysis in biological material by direct combustion in oxygen and photometric determination of mercury vapor, *Acta chem. scand.* **22**, 2150–2156 (1968).
29. ANDERSON, D. H., EVANS, J. H., MURPHY, J. J., and WHITE, W. W., Determination of mercury by a combustion technique using gold as a collector, *Anal. Chem.* **43**, 1511–1512 (1971).
30. WILLFORD, W. A., HESSELBERG, R. J., and BERGMAN, H. L., Versatile combustion–amalgamation technique for the photometric determination of mercury in fish and environmental samples, *J. Ass. Offic. Anal. Chem.* **56**, 1008–1014 (1973).
31. SCHULERT, A. R., ZEIND, A., and WITT, D., Mercury assay in human blood and tissue, paper presented at Spring Meeting of Southeast Section of American Association of Clinical Chemists, Nashville, Tenn., April 20–22, 1972.
32. REIMERS, R. S., unpublished research, Vanderbilt Univ., 1972.
33. HERMANN, W. J., Jr., BUTLER, J. W., and SMITH, R. G., in *Laboratory Diagnosis of Diseases Caused by Toxic Agents* (F. W. Sunderman and F. W. Sunderman, Jr., eds.), pp. 379–386, Warren Green, St. Louis, Mo., 1970.
34. THOMAS, R. J., HAGSTROM, R. A., and KUCHAR, E. J., Rapid pyrolytic method to determine total mercury in fish, *Anal. Chem.* **44**, 512–515 (1972).
35. KUNKEL, E., Trace analysis of mercury in Wickbold combustion and flameless atomic absorption, *Z. Anal. Chem.* **258**, 337–341 (1972).
36. JOENSUU, O. I., Mercury-vapor detector, *Appl. Spectr.* **25**, 526–528 (1971).
37. UKISHIMA, Y., NAGAI, T., NAGANO, T., UNNO, T., TERADA, S., and HATTORI, S., Rapid determination of mercury by an oxygen flask combustion method and atomic–absorption spectrophotometry, *Eisei Kagaku* **18**, 270–273 (1972); *Chem. Abstr.* **78**, 28009u (1973).
38. EICHNER, K., Determination of mercury in eggs by atomic absorption spectrometry (AAS), *Mitteilungsbl. G. D. Ch—Fachgruppe Lebensmittelchem. Gerichtl. Chem.* **26**, 240–241 (1972); *Chem. Abstr.* **78**, 82977w (1973).
39. VERNET, J. P. and THOMAS, R. L., Levels of mercury in the sediments of some Swiss lakes including Lake Geneva and the Rhone river, *Eclogae Geol. Helv.* **65**, 293–306 (1972); *Chem. Abstr.* **78**, 19011m (1973).
40. LIDUMS, V., Determination of mercury in small quantities by direct combustion combined with cold vapor atomic absorption, *Chem. Sch.* **2**, 159–163 (1972); *Chem. Abstr.* **78**, 37523j (1973).
41. THILLIEZ, G., Rapid and precise determination of traces of mercury in air and biological materials by atomic absorption, *Chim. Anal.* **50**, 226–232 (1968).
42. HEAD, P. C. and NICHOLSON, R. A., A cold vapor technique for the determination of mercury in geological materials involving its reduction with tin(II) chloride and collection on gold wire, *Analyst* **98**, 53–56 (1973).
43. WINDHAM, R. L., Simple device for compensation of broad-band absorption interference in flameless atomic absorption determination of mercury, *Anal. Chem.* **44**, 1334–1336 (1972).
44. MAGOS, L., Selective atomic-absorption determination of inorganic mercury and methylmercury in undigested biological samples, *Analyst* **96**, 847–853 (1971).
45. MAGOS, L. and CLARKSON, T. W., Atomic absorption determination of total, inorganic, and organic mercury in blood, *J. Ass. Offic. Anal. Chem.* **55**, 966–971 (1972).
46. ROBINSON, J. W., SLEVIN, P. J., HINDMAN, G. D., and WOLCOTT, D. K., Non-flame atomic absorption in the vacuum ultraviolet region: the direct determination of mercury in air at the 184.9 nm resonance line, *Anal. chim. Acta* **51**, 431–438 (1972).
47. APRIL, R. W. and HUME, D. N., Environmental

mercury: rapid determination in water at nanogram levels, *Science* **170**, 849–850 (1970).
48. LICHTE, F. E. and SKOGERBOE, R. K., Emission spectrometric determination of trace amounts of mercury, *Anal. Chem.* **44**, 1321–1323 (1972).
49. KIRKBRIGHT, G. F. and WEST, T. S., Atomic fluorescence for chemical analysis, *Chem. Br.* **8**, 429–435, 438 (1972).
50. ROBINSON, J. W. and ARAKTINGI, Y. E., Study of the application of atomic fluorescence spectrometry to the direct determination of mercury and cadmium in the atmosphere, *Anal. chim. Acta* **63**, 29–38 (1973).
51. THOMPSON, K. C. and REYNOLDS, G. D., The atomic-fluorescence determination of mercury by the cold vapor technique, *Analyst* **96**, 771–775 (1971).
52. MUSCAT, V. I. and VICKERS, T. J., Determination of nanogram quantities of mercury by the reduction–aeration method and atomic fluorescence spectrophotometry, *Anal. chim. Acta* **57**, 23–30 (1971).
53. MUSCAT, V. I., VICKERS, T. J., and ANDREN, A., Simple and versatile atomic fluorescence system for determination of nanogram quantities of mercury, *Anal. Chem.* **44**, 218–221 (1972).
54. SHIMOMURA, S. and HIROTO, R., Determination of trace amounts of mercury by non-disperse atomic fluorescence method, *Anal. Lett.* **6**, 613–618 (1973).
55. HADEISHI, T. and MCLAUGHLIN, R. D., Hyperfine Zeeman effect atom absorption spectrometer for mercury, *Science* **174**, 404–407 (1971).
56. HADEISHI, T., CHURCH, D. A., MCLAUGHLIN, R. D., ZAK, B. D., and NAKAMURA, M., Total mercury monitor for ambient air: the IZAA spectrometer, Lawrence Berkeley Laboratory—1593 preprint, 16 pp., February 1973.
57. DAGNALL, R. M., MANFIELD, J. M., SILVESTER, M. D., and WEST, T. S., Atomic absorption and emission spectrometry of mercury at 184.9 nm, *Spectr. Lett.* **6**, 183–189 (1973).
58. STEPHENS, R., The application of a fast pulse atom reservoir to the determination of mercury, *Anal. Lett.* **5**, 851–861 (1972).
59. KIRKBRIGHT, G. F., WEST, T. S., and WILSON, P. J., The direct determination of mercury by atomic-absorption spectrophotometry at 184.9 nm by using a nitrogen-separated nitrous oxide–acetylene flame, *Analyst* **98**, 49–52 (1973).
60. KIRKBRIGHT, G. F., WEST, T. S., and WILSON, P. J., The direct determination of mercury by atomic fluorescence spectrometry in a nitrogen-separated air–acetylene flame with excitation at 184.9 nm, *Anal. chim. Acta* **66**, 130–133 (1973).
61. SJOSTRAND, B., Simultaneous determination of mercury and arsenic in biological and organic materials by activation analysis, *Anal. Chem.* **36**, 814–818 (1964).
62. PILLAY, K. K. S., THOMAS, C. C., Jr., SONDEL, J. A., and HYCHE, C. M., Determination of mercury in biological and environmental samples by neutron activation analysis, *Anal. Chem.* **43**, 1419–1425 (1971).
63. PILLAY, K. K. S., THOMAS, C. C., Jr., SONDEL, J. A., and HYCHE, C. M., Mercury pollution of Lake Erie ecosphere, *Environ. Res.* **5**, 172–181 (1972).
64. RUF, H. and ROHDE, H., Determination of mercury contents of diverse samples of fish and other biological materials by neutron activation, *Z. Anal. Chem.* **263**, 116–120 (1973).
65. CHATTOPADHYAY, A., BENNETT, L. G. I., and JERVIS, R. E., Activation analysis of environmental pollutants, *Can. J. Chem. Eng.* **50**, 189–193 (1972).
66. TANNER, J. T., FRIEDMAN, M. H., LINCOLN, D. N., FORD, L. A., and JAFFE, M., Mercury content of common foods determined by neutron activation analysis, *Science* **177**, 1102–1103 (1972).
67. WEISS, H. V., KOIDE, M., and GOLDBERG, E. D., Mercury in the Greenland ice sheet: evidence of recent input by man, *Science* **174**, 692–694 (1971).
68. ARUSCAVAGE, P. J., Neutron activation analysis procedure for the determination of mercury in soil and rock samples, *US Geol. Surv. Prof. Pap.* No. 880-C, 209–214, 1972.
69. WEISS, H. V. and BERTINE, K. K., Simultaneous determination of manganese, copper, arsenic, antimony and mercury in glacial ice by radioactivation, *Anal. chim. Acta* **65**, 253–259 (1973).
70. HEITZMAN, M. W. and SIMPSON, R. E., Neutron activation analysis of mercury in fish, flour, and standard reference orchard leaves by electrodeposition radiochemistry, *J. Ass. Offic. Anal. Chem.* **55**, 960–965 (1972).
71. MARGOSIS, M. and TANNER, J. T., Determination of mercury in pharmaceutical products by neutron activation analysis, *J. Pharm. Sci.* **61**, 936–938 (1972).
72. BECKNELL, D. E., MARSH, R. H., and ALLIE, W., Jr., Use of anion exchange resin-loaded paper in the determination of trace mercury in water by neutron activation analysis, *Anal. Chem.* **43**, 1230–1233 (1971).
73. LAW, S. L., Resin-loaded papers for methyl mercury and inorganic mercury determination, *Am. Lab.* 91–93, 96–97, July 1973.
74. ROOK, H. L., GILLS, T. E., and LAFLEUR, P. D., Method for determination of mercury in biological materials by neutron activation analysis, *Anal. Chem.* **44**, 1114–1117 (1972).
75. FILBY, R. H., DAVIS, A. I., SHAH, K. R., and HALLER, W. A., Determination of mercury in biological and environmental materials by instrumental neutron activation analysis, *Mikrochim. Acta* **1970**, 1130–1136.
76. KAY, M. A., MCKOWN, D. M., GRAY, D. H., EICHOR, M. E., and VOGT, J. R., Neutron activation analysis in environmental chemistry, *Am. Lab.* 39–48, July 1973.
77. WEAVER, J. N., Determination of mercury and

selenium in coal by neutron activation analysis, *Anal. Chem.* **45**, 1950–1952 (1973).
78. JERVIS, R. E. and TIEFENBACH, B., Trace mercury determinations in a variety of foods, *Symposium on Nuclear Methods in Environmental Research, Univ. of Missouri, August 1971*, pp. 188–196.
79. WESTERMARK, T. and SJOSTRAND, B., Activation analysis of mercury in environmental studies, *Advan. Activ. Anal.* 1972, 57–88.
80. LEONG, P. C. and ONG, H. P., Determination of mercury by using a gold trap in samples containing considerable sulfide minerals, *Anal. Chem.* **43**, 940–941 (1971).
81. WOIDICH, H. and PFANNHAUSER, W., Rapid separation of mercury from biological materials by extraction chromatography with dithizone, *Z. Anal. Chem.* **261**, 31 (1972).
82. WOIDICH, H. and PFANNHAUSER, W., Quantitative analysis of mercury in biological samples: I, Simultaneous extraction of mercury and copper as dithizonates and their separation, *Z. Lebensm. Unters. Forsch.* **149**, 1–7 (1972); *Chem. Abstr.* **77**, 86701d (1972).
83. CORVI, C., Microdetermination of total mercury in food products and cosmetics, *Mitt Geb. Lebensmittelunters. Hyg.* **63**, 134–141 (1972); *Chem. Abstr.* **77**, 16070d (1972).
84. HEIZMANN, P. and BALLSCHMITER, K., Analytical application of 1,2-diketone bisthiobenzyhydrazones, *Z. Anal. Chem.* **259**, 110–114 (1972).
85. CRESSER, M. S., A new method for the U.V. spectrophotometric determination of mercury, *Anal. Lett.* **6**, 375–380 (1973).
86. USATENKO, Y. I., SUPRUNOVICH, V. I., and KULIKOVSKAYA, Z. B., Use of 8-mercaptoquinoline for the photometric determination of mercury, *Khim. Tekhnol. Kharkov* 1971, 117–121; *Chem. Abstr.* **77**, 83219s (1972).
87. ANISIMOVA, L. G., Triphenylmethane dye reagents for photometric determination of group II elements of the periodic system, *Sb. Nauch. Soobshch. Dagestan. Univ. Kafedra Khim* 90–94 (1971); *Chem. Abstr.* **77**, 83219s (1972).
88. KANKE, M., INOUE, Y., WATANABE, H., and NAKAMURA, S., Spectrophotometric determination of mercury by using Ruhemann's purple, *Bunseki Kagaku* **21**, 622–626 (1972); *Chem. Abstr.* **77**, 69702n (1972).
89. TANDON, S. N. and GUPTA, C. B., Spectrophotometric determination of mercury(II) with 1,5-diphenylcarbazide, *Indian J. Chem.* **10**, 543–544 (1972); *Chem. Abstr.* **77**, 159827m (1972).
90. KE, P. J. and THIBERT, R. J., Kinetic method for the determination of mercury at the nanogram level using an iodide-catalyzed arsenite-cerium reaction, *Mikrochim. Acta* 1972, 768–783.
91. KE, P. J. and THIBERT, R. J., Anticatalytic microdetermination of mercury using cerium–arsenite reaction, *Mikrochim. Acta* 1973, 15–24.
92. KE, P. J. and THIBERT, R. J., Kinetic microdetermination of mercury in natural waters and biological materials, *Mikrochim. Acta* 1973, 417–427.
93. WEISZ, H. and LUDWIG, H., A continuous kinetic-catalytic method of analysis using a flow-through cell, *Anal. chim. Acta* **62**, 125–135 (1972).
94. DATTA, K. and DAS, J., Determination of mercury(II) from the rate of catalyzed substitution of α,α'-dipyridyl for cyanide in potassium ferrocyanide, *Indian J. Chem.* **10**, 116–118 (1972); *Chem. Abstr.* **77**, 13687u (1972).
95. LINDEN, W. E. VAN DER, BEIJER, S., and DEN BOEF, G., Photometric titrations of mercury(II) based on UV-absorption of mercury(II)-complexes, *Mikrochim. Acta* 1972, 334–338.
96. LINDEN, W. E. VAN DER and DIEKER, J., Amperometric titration of mercury(II) with EDTA, DTPA and TRIEN in the ppm-range, *Z. Anal. Chem.* **264**, 353–355 (1973).
97. HOLZBECHER, J. and RYAN, D. E., The fluorimetric determination of mercury, *Anal. chim. Acta* **64**, 311–315 (1972).
98. TANDON, S. N., SRIVASTAVA, P. K., and JOSHI, S. R., 2-Amino-4-methylthiazole as a reagent for the gravimetric determination and extraction of mercury(II), *Anal. chim. Acta* **59**, 311–315 (1972).
99. KULIKOVSKAYA, Z. B., SUPRUNOVICH, V. I., and USATENKO, Y. I., Gravimetric determination of mercury in the form of diiodobis (8-quinolinethiol)-mercury(II), *Khim. Tekhnol. Kharkov*, 1971, 121–125; *Chem. Abstr.* **77**, 28544u (1972).
100. BUSEV, A. I. and EUSIKOV, V. V., Phenyl-substituted dithiolethiones as analytical reagents: gravimetric determination of mercury(II), *Vestn. Mosk. Univ. Khim.* **13**, 81–85 (1972); *Chem. Abstr.* **77**, 28591g (1972).
101. JONES, P. and NICKLESS, G., The estimation of inorganic mercury (at low concentration) by gas chromatography, *J. Chromatog.* **76**, 285–289 (1973).
102. WESTÖÖ, G., Determination of methylmercury salts in various kinds of biological materials, *Acta chem. scand.* **22**, 2277 (1968).
103. FISHBEIN, L., Chromatographic and biological aspects of inorganic mercury, *Chromatogr. Rev.* **15**, 195–238 (1971).
104. MCNERNEY, J. J., BUSECK, P. R., and HANSON, R. C., Mercury detection by means of thin gold films, *Science* **178**, 611–612 (1972).
105. VASSOS, B. H., HIRSCH, R. F., and LETTERMAN, H., X-ray microdetermination of chromium, cobalt, copper, mercury, nickel, and zinc in water using electrochemical preconcentration, *Anal. Chem.* **45**, 792–794 (1973).
106. MIYAI, Y. and MURAO, Y., Determination of mercury in brine purification mud by X-ray fluorescence spectrometry, *Bunseki Kagaku* **21**, 608–613 (1972); *Chem. Abstr.* **77**, 109099e (1972).

A Review of the Status of Total Mercury Analysis (W. Dickinson Burrows)

DISCUSSION *by* JAMES H. FINGER and TOM B. BENNETT
Environmental Protection Agency, Region IV, Surveillance and Analysis Division, Chemicals Services Branch, Athens, Georgia 30601

INTRODUCTION

The Southeast Region of the United States has seven river basins contaminated by mercury discharges from chlor-alkali plants. Several basins are severely polluted to the extent that the human consumption of fish is either limited or prohibited. A total of 106 potential industrial users of mercury were investigated by the United States Environmental Protection Agency (USEPA) during late 1970 and early 1971. Of 30 industries found to be using mercury in their processes, 10 chlor-alkali plants were found to be the principal users and dischargers of mercury into the aquatic environment.[1]

Investigations and monitoring were jointly undertaken by States and Federal Agencies. This involved total mercury analysis of industrial effluents, receiving stream waters, sediments, and fish tissues. The results showed varying degrees of mercury contamination in areas of these principal waste dischargers.

Since the regulation of industrial discharge and the regulation of fish consumption were based on total mercury levels, the evaluation of mercury analytical procedures was necessary. Our evaluation was made through national round robin and regional split-sample studies in addition to EPA performance quality control programs.

DISCUSSION

Total mercury rather than methylmercury analysis was selected because the US Food and Drug Administration Standard on edible foods was 0.5 μg/g (ppm) total mercury. Also, since the transformation of mercury into various mercurial species and the ultimate fate of mercury are not clearly defined, it was decided that environmental samples should be monitored for total mercury.

Various chemistry principles can be used for total mercury determination. These include neutron activation,[2] spectrophotometry,[3] atomic absorption spectrophotometry,[4,5] X-ray fluorescence,[6] and spark source mass spectrometry.[7] Many techniques of sample preparation prior to analysis are used and vary from no preparation to complete combustion or destruction of the sample.

After a literature search of methodology and several months of experimentation with the wire amalgamation–flameless atomic absorption we decided that flameless atomic absorption detection preceded by wet digestion was the technique of choice for the Chemical Services Branch. This decision was based on the following:

(a) Flameless atomic absorption spectrophotometry is very sensitive and trace amounts of mercury can be detected.
(b) Atomic absorption instruments were available in our laboratory and in most other analytical chemistry laboratories.
(c) Wet digestion is rapid and does provide for the preparation of large numbers of environmental samples for analysis. It requires minimum manpower and laboratory equipment.

Industrial waste and waters analysis

Hatch and Ott's method,[5] in which the sample is digested at room temperature in an acid–permanganate medium, was investigated for the analysis of waters. The mercury(II) is reduced with stannous sulfate to the elemental state and detected by the flameless atomic absorption system. Precision and recovery of mercuric chloride were good, but the recovery of some organomercurials was poor. Recovery of methyl mercuric chloride was less than 5%. The digestion procedure was modified to include the addition of potassium persulfate and an elevation of temperature to 95°C to obtain complete recovery of organomercurials.

After experimentation with this procedure, a joint round-robin test was conducted by the EPA and the American Society for Testing and Materials (ASTM).[10–12] Table 1 is a summary of the test data

TABLE 1
ASTM–EPA method study of total mercury in water

	Sample 1		Sample 2		Sample 3		Sample 4	
	Distilled water	Natural water by difference	Distilled water	Natural water by difference	Distilled water	Natural water by difference	Distilled water	Natural water by difference
True value (μg/l)	0.21	0.21	0.27	0.27	0.51	0.51	0.60	0.60
Mean recovery (μg/l)	0.418	0.349	0.450	0.414	0.653	0.674	0.744	0.709
Accuracy as % relative error (bias)	98.9	66.4	66.5	53.3	28.1	32.1	24.0	18.1
Standard deviation (μg/l)	0.279	0.276	0.325	0.279	0.376	0.541	0.466	0.390
Relative deviation (%)	66.9	78.9	72.2	67.5	57.5	80.3	62.6	55.0
Range (μg/l)	1.60	1.27	1.80	1.20	2.26	3.99	2.97	2.30
	Sample 5		Sample 6		Sample 7		Sample 8	
	Distilled water	Natural water by difference	Distilled water	Natural water by difference	Distilled water	Natural water by difference	Distilled water	Natural water by difference
True value (μg/l)	3.4	3.4	4.0	4.1	8.8	8.8	9.6	9.6
Mean recovery (μg/l)	3.40	3.41	4.26	3.81	8.48	8.77	9.38	9.10
Accuracy as % relative error	0.03	0.34	3.85	−7.11	−3.7	−0.4	−2.3	−5.2
Standard deviation (μg/l)	1.29	1.49	1.42	1.12	2.60	3.69	3.24	3.57
Relative deviation (%)	37.9	43.7	33.2	29.3	30.7	42.1	34.5	39.2
Range (μg/l)	11.1	12.4	10.4	6.3	19.4	27.5	24.4	25.8

which shows the following precision for natural water:[10]

$$S_T = 0.3855X + 0.1071,$$

where S_T is the overall precision and X the determined concentration of mercury (μg/l).

Goulden[13] automated the Hatch and Ott procedure, but since our data showed poor recovery of methylmercury compounds with this procedure, we modified it to include a persulfate digestion.[14] Table 2 contains precision and recovery data for the modified automated method and the data show complete recovery of organomercurials.

Fish tissue analyses

As a flameless atomic absorption method was used for water samples, a compatible method for fish tissues was examined. Uthe's Method[15] involves digestion of tissue in sulfuric and nitric acids at 58°C, oxidation with potassium permanganate at room temperature, reduction of mercury to the elemental state, and detection by flameless atomic absorption spectrophotometry.

This procedure is highly sensitive. It has a 0.04 μg/g minimum detection limit using a 0.5 g sample. Duplicate data generated by the Chemical Services Branch in the range of 0.0–1.50 μg/g, using 312 observations, had an average standard deviation of approximately 20%. These data are shown in Table 3. Replicate analyses of a shad muscle tissue had a standard deviation of 0.054 at the 0.90 μg/g level or roughly 6%.

As a check on accuracy, eight tissues were analyzed both by our laboratory and by neutron activation analysis in another laboratory. Results are shown in Table 4. Samples D and F and samples G and J are duplicates, unknown to the neutron activation laboratory. Based on these data, the two methods were assumed comparable and accurate.

In 1972 a round robin test involving 64 laboratories was conducted. All data along with the method of analysis used are listed in Table 5. In addition to total mercury, the samples were analyzed for methylmercury in 10 of these laboratories and the data are presented in Table 6. A statistical summary of all data is presented in Table 7. The standard deviation for total mercury using all the data ranged from 28–40% of the means for three tissue samples. In laboratories using only the flameless atomic absorption method, the standard deviation ranged from 26% to 34% of the means for the same three tissues.

Since different analytical techniques were used

Discussion

TABLE 2
Automated method[14]
Precision data

Theoretical concentration (μgHg/l)	Mean (μgHg/l)	Standard deviation (μgHg/l)
0.5	0.420	0.04
1.0	1.00	0.07
2.0	1.92	0.09
5.0	4.85	0.20
10.0	10.03	0.40
20.0	19.88	0.84

Recoveries of organic mercurials (at the 10 μg/l concentration).

	Percent recovery	
Compound	Distilled water	Natural water
Methyl mercuric chloride	101	105
Methyldicyandiamide (Panogen)	125	117
Merthiolate	98	95
Phenyl mercuric nitrate	97	90
MEMMI[a]	101	96
Diphenyl mercury	98	106
Phenyl mercuric benzoate	100	93
EMMI[b]	101	95
Phenyl mercuric hydroxide	99	93
Phenyl mercuric acetate	92	87

[a]MEMMI, N - methylmercuri - 1,2,3,6 - tetrahydro - 3,6endo - methano - 3,4,5,6,7,7 - hexachlorophthalimide.
[b]EMMI, N-ethylmercuri - 1,2,3,6tetrohydro - 3,6endo - methano - 3,4,5,6,7,7 - hexachlorophthalimide.

TABLE 3
Duplicate analyses, fish tissue, flameless atomic absorption

Range (μg/g)	No. of observations	Standard deviation (μg/g)
0.00–0.25	95	0.039
0.26–0.75	146	0.069
0.76–1.50	64	0.144
>1.50	7	0.349

to analyze these tissue samples, and the means of the results of the flameless method and of all other techniques are comparable, we feel that the method is accurate.

Sediment analyses

We also needed a total mercury method for sediments compatible with the method for waters and tissues. After experimentation with several digestion procedures followed by flameless atomic absorption detection, the EPA proposed two alternate digestion procedures:[16]

(a) an aqua-regia–potassium permanganate digestion at 95°C; and
(b) a sulfuric–nitric acid–permanganate digestion with autoclaving at 121°C and 15 psi.

We conducted a round robin test using three environmental sediment samples. Table 8 is a list of the data from all labs. Table 9 is a statistical summary of all these data in addition to the data from those laboratories using only the EPA procedure (aqua regia digestion). Only four laboratories

TABLE 4
Comparison of flameless atomic absorption and neutron activation analyses

Tissue No.	Dish No.	Flameless atomic absorption (μg/g)	Neutron activation analyses (μg/g)	
			First value	Second value
647	A	0.87	0.90	0.88
644	B	0.69	0.84	0.80
665	C	1.74	2.20	1.80
609	D	1.82	1.60	1.60
674	E	0.55	0.46	0.38
609	F	1.82	1.60	1.70
774	G	2.22	2.80	3.40
668	H	0.72	0.84	0.66
632	I	0.38	0.48	0.41
774	J	2.22	2.8	2.8

TABLE 5
Mercury in fish round robin, 1972. Total mercury, µg/g

Lab code No.	Sample number 72C1222	72C1223	72C1224	Method
10	2.3, 2.3, 2.5, 2.0, 2.3	7.4, 6.4, 5.9, 5.8, 6.8	8.0, 7.8, 7.5, 8.6, 7.0	EPA, 1972, CVAA, auto-detection
12	2.21, 2.40	6.29, 6.61	8.01, 8.01	H_2SO_4–HNO_3 at 80°C, $KMnO_4$, CVAA
12A	2.94, 2.18	5.64, 5.51	6.44, 6.54	Pyrolysis
12B	2.34, 2.34	6.02, 7.47	8.61, 7.98	H_2SO_4–HNO_3 at 270°C, $KMnO_4$, CVAA
13	2.17, 2.31, 2.20, 2.35	5.79, 5.57, 5.24, 5.65	6.58, 6.68, 6.40, 7.07	Combustion, amalgamation, CVAA
14	1.65	4.01	7.87	Combustion, CVAA
15	1.8 ± 0.3	7.6 ± 0.4	9.1 ± 0.3	Uthe, CVAA (Gilbert and Hume, in press)
17	2.05 ± 0.10	6.05 ± 0.16	7.61 ± 0.36	EPA, 1972, CVAA
19	2.31	6.62	8.19	H_2SO_4, $KMnO_4$, boil, CVAA
22	1.9, 2.0	5.8, 5.8	7.2, 7.0	Reflux, HNO_3, H_2SO_4, $KMnO_4$, CVAA
24	6.58, 6.80	3.50, 4.40	11.48, 16.23	EPA, 1972, CVAA (reported contamination)
26	2.36, 2.43	6.50, 6.14	7.70, 8.17	EPA, 1972, CVAA
27	1.31, 1.52	5.00, 5.12	6.96, 6.97	EPA, 1972, CVAA
29	1.90, 1.82	4.20, 4.04	5.46, 5.54	Pyrolysis, CVAA
30	2.19, 2.18	5.82, 5.85	7.62, 7.70	EPA, 1972, CVAA
34	0.50, 0.50	2.70, 3.30	2.80, 2.50	EPA, 1972, CVAA
35	3.31, 3.34	9.81, 9.80	12.34, 12.48	Combustion, CVAA
37	0.87, 0.47, 0.93	2.56, 2.25, 2.81	4.56, 5.52, 5.28	H_2SO_4, HNO_3, $KMnO_4$, boil, CVAA
38	2.00, 2.00	10.00, 9.90	11.00, 11.20	EPA, 1972, CVAA
39	6.07, 5.42	6.53, 6.98	10.29, 9.67	H_2SO_4, HNO_3, $KMnO_4$ at room temp., CVAA, auto-detection
42	1.78, 1.58	6.13, 5.91	7.00, 6.83	EPA, 1972, CVAA
43	2.53, 2.73	7.20, 7.01	8.73, 8.94	HNO_3, $HClO_4$, reflux, CVAA
46	2.41, 2.67	7.25, 8.25	9.10, 9.85	NAA–destructive
47	2.14, 2.07	6.99, 6.78	7.93, 8.09	EPA, 1972, CVAA
48	1.71, 1.62	5.22, 5.43	6.31, 6.58	EPA, 1972, CVAA
55	2.23, 2.21	5.73, 5.93	7.49, 7.50	EPA, 1972, CVAA
56	2.26, 1.87, 2.28	6.52, 6.13, 6.42	7.49, 7.77, 7.91	EPA, 1972, CVAA
57	0.28, 0.19	3.16, 3.23	4.14, 4.02	EPA, 1972, CVAA
59	2.16, 2.28	5.92, 5.72	7.24, 7.04	H_2SO_4, HNO_3, $KMnO_4$, $K_2S_2O_8$ at 95°C, CVAA
61	1.92, 2.46, 1.12, 1.53, 1.44, 1.11	5.71, 4.74, 5.89, 3.69, 6.63, 4.80	8.49, 9.87, 8.07, 8.40, 4.25, 6.57	EPA, 1972, CVAA

Discussion

TABLE 5 (*continued*)

Lab code No.	72C1222	Sample number 72C1223	72C1224	Method
62	2.03, 2.40	6.17, 6.72	8.44, 8.82	EPA, 1972, CVAA
67	1.80, 1.46	6.60, 6.05	7.54, 7.72	—
70	2.24, 2.29	5.56, 5.48	—	—
72	2.15, 2.25	6.25, 6.25	6.40, 6.25	JAOAC **55**, 96–100 (1972)
74	2.12, 2.01, 1.96	6.22, 5.89, 5.89	7.96, 7.86, 7.68	Uthe and Armstrong, *AA Newsletter* **10**, 101 (1971)
75	2.12, 2.04	5.61, 5.26	7.11, 7.15	EPA, 1972, CVAA, auto-detection
77	2.20, 2.36	6.39, 6.20	8.10, 7.90	Munns and Holland, *JAOAC* **54**, 202–5 (1971)
79	2.17, 2.10, 1.95	5.97, 6.39, 6.09	8.25, 7.02, 7.24	EPA, 1972, CVAA
81	2.9, 2.9, 2.8	7.9, 6.9, 7.6	8.6, 11.4, 9.5	EPA, 1972, CVAA
83	2.47, 2.34	3.82, 4.56	6.10, 6.23	HNO_3–$HClO_4$ at 75°C, CVAA
84	2.12, 2.12	6.32, 6.46	7.57, 7.63	EPA, 1972, CVAA
86	2.67, 2.84	7.80, 8.16	9.49, 9.49	EPA, 1972, CVAA
87	2.00, 1.90	5.80, 6.40	7.30, 7.50	EPA, 1972 at 80°C, CVAA
87A	—	5.60, 5.60	7.10, 7.30	EPA, 1972 at 58°C, CVAA
90	2.67, 2.54, 2.45	6.54, 7.05, 6.56	8.18, 7.91, 8.45	H_2SO_4, HNO_3, V_2O_5 at 120°C, CVAA
91	1.80, 1.81	4.85, 5.32	—	H_2SO_4, HNO_3, $K_2S_2O_8$ at 90°C, CVAA
92	2.14, 2.00	5.34, 5.16	6.43, 6.21	Combustion
95	1.1, 1.6	3.5, 3.3	4.8, 5.5	EPA, 1972, CVAA
128	1.74, 1.60	4.24, 4.50	6.65, 6.65	EPA, 1972, CVAA
134	1.12, 1.23	2.68, 3.26	4.44, 4.41	—
144	1.30, 1.24	4.32, 4.39	5.24, 5.21	NAA–nondestructive
154	1.88, 2.14	6.25, 7.70, 6.01, 6.23	6.87, 6.93	H_2SO_4, HNO_3, $KMnO_4$, CVAA
155	1.41, 1.67	3.92, 3.47	4.59, 3.88	EPA, 1972, CVAA
158	1.21, 1.34	4.54, 4.63	4.52, 4.56	H_2SO_4, HNO_3, $KMnO_4$, $K_2S_2O_8$ at 98°C, CVA
160	2.00, 1.80	5.50, 5.90	6.20, 6.70	
163	2.79, 3.12	6.60, 7.44	9.30, 9.95	EPA, 1972, CVAA
166	1.77, 1.85, 1.65	4.80, 4.90, 4.85	5.75, 5.70, 5.65	Combustion, CVAA
168	1.9, 1.9	6.6, 5.9	7.7, 6.9	EPA, 1972, CVAA
169	1.38, 1.25	3.57, 3.37	4.41, 4.52	H_2SO_4, HNO_3, $HClO_4$, CVAA
170	1.51, 1.40	4.23, 4.68	5.54, 5.59	EPA, 1972, CVAA
173	0.72	2.16, 2.03	2.44, 2.46	Uthe, 1970, CVAA
174	5.25, 4.86	11.77, 9.47	13.82, 14.37	H_2SO_4, solid $KMnO_4$, CVAA
175	1.45, 1.37	4.43, 4.45	5.79, 5.92	H_2SO_4, H_2O_2, $KMnO_4$ at 60°C, CVAA
176A	2.4 ± 0.6	7.3 ± 0.7	7.6 ± 0.7	X-ray fluorescence

CVAA, Cold vapor atomic absorption.

TABLE 6
Mercury in fish round robin, 1972, methyl mercury (μg/g)

Lab code no.	Sample number 72C1222	72C1223	72C1224	Method
12	1.64, 1.64	4.40, 4.05	5.38, 5.37	Mercury complex broken by cupric sulfate and sodium Br. ME Hg Br in toluene and then into thiosulfate complex broken by cupric bromide and ME Hg Br into Benzene—GC
34	0.22, 0.23	0.60, 0.70	1.00, 0.70	EPA, Prov., 1972
47	1.99, 2.04	5.69, 5.88	6.18, 6.44	EPA, Prov., 1972
57	2.05	5.33	6.44	Technical Report 27, EWRE, Vanderbilt University, pp. 35–38.
70	2.24, 1.73	6.02, 5.76	6.70, 7.38	—
73	2.12, 1.93	5.29, 5.59	8.00, 7.64	Uthe et al., JAOAC 583–589, 1972
92	1.62, 1.64	4.80, 4.78	5.94, 6.11	JOAC 55, 583 (1972)
130	2.1, 2.0	6.4, 6.0	6.6, 6.6	Uthe et al., 1972, "Rapid semi-micro method for determination of methyl mercury in fish"
175	1.85, 1.94	6.67, 6.67	6.69, 6.86	EPA, Prov., 1972
176	2.18, 2.15	5.75, 5.78	7.06, 6.74, 6.90, 6.70	HCL, Benzene, GC with plasma discharge detector

TABLE 7
Statistical summary of mercury in fish round robin, 1972 (μg/g)

Sample No.	Mean	Min.	Max.	Standard deviation	Number of analyses
A. Total mercury, all laboratories and all methods					
72C1222	2.06	0.19	6.07	0.83	134
72C1223	5.75	2.03	11.77	1.62	139
72C1224	7.23	2.44	14.37	2.02	133
B. Total mercury, laboratories using the EPA Method (April 1972)					
72C1222	1.90	0.19	3.12	0.64	61
72C1223	5.75	2.68	10.00	1.52	63
72C1224	7.28	2.50	11.40	1.87	63
C. Methylmercury, all laboratories and all methods					
72C1222	1.75	0.22	2.24	0.57	19
72C1223	5.06	0.60	6.67	1.70	19
72C1224	6.07	0.70	8.00	1.85	21

Discussion

TABLE 8
Sediment round robin, mercury, 1972

Lab. code No.	Sediment sample No. 72C5643	72C5644	72C5645	Method
12	106, 105, 104	49, 42, 42	<0.01, <0.01, <0.01,	Aqua regia, $KMnO_4$, boil, CVAA
	102, 99, 105	51, 44, 41	0.01, <0.01, 0.01	Acid, $KMnO_4$ $(NH_4)_2$ S_2O_8, 80°C, CVAA
	—	—	0.042	Pyrolysis
13	117, 105, 99.0, 108	44.0, 44.9, 44.0, 48.8	0.0653, 0.0665	Combustion, cold trap, CVAA
14	136.8, 128.0	52.9, 51.9, 55.9	0.090, 0.094	Direct combustion, flameless AA
15	127, 124	52, 54	0.054, 0.051	Acid digestion and reflux, CVAA
16	101, 100	38, 42	<0.1, <0.1	Reflux with HNO_3, CVAA
17	160, 130	57, 49	<0.1, <0.1	Aqua regia at room temp., HgS, spectrographic analysis
18	124, 121	45, 46	0.090, 0.11	EPA, PROV., aqua regia at 95°C, CVAA
19	123, 126	52.3, 52.3	0.10, 0.10	EPA, Prov., aqua regia at 95°C, CVAA
25	99.6 ± 5.7	42.5 ± 3.0	0.072 ± 0.012	HCl–$KMnO_4$, boil, CVAA
26	112, 119	41, 45	0.05, 0.05	HNO_3–$KMnO_4$, boil, CVAA
27	107.1, 107.1, 108.0, 111.0	51.5, 52.5, 53.4, 53.4	0.23, 0.22	EPA, prov., aqua regia at 200°F, CVAA
28	104, 109, 106	36.8, 36.9, 37.0	0.09, 0.07, 0.08	EPA, prov., aqua regia, steam bath, CVAA
29	119.1, 120.8	48.9, 48.4	0.11, 0.08	H_2SO_4–HNO_3–$KMnO_4$, CVAA
30	118, 118, 117, 117	53, 48, 51, 51	0.20, 0.21, 0.16	EPA, prov., aqua regia at 95°C, CVAA
32	177, 182, 188	73, 69, 70	0.56, 0.42	EPA, prov., aqua regia at 95°C, CVAA
33	90.0, 100	35.0, 30.0	0.05, 2.2	EPA, prov., aqua regia at 95°C, CVAA
34	100.7, 94.3	40.7, 39.9	—	Combustion at 950°C, Hg trapped in ICl, CVAA
36	104.66, 101.27, 104.10	40.27, 41.59, 40.50	0.06, 0.06, 0.06	EPA, prov., aqua regia at 95°C, CVAA
	102.26, 108.35, 97.23	43.54, 40.72, 39.09	0.08, 0.08, 0.02	H_2SO_4–HNO_3–$KMnO_4$, boil, CVAA
37	104.23, 110.26	33.37, 33.37	1.00, 1.005	EPA, prov., aqua regia at 95°C, CVAA
38	113, 108	47.51	0.088, 0.84	CVAA
41	96.25, 100.00	45.50, 44.20	1.84, 2.04	EPA, prov., aqua regia at 95°C, CVAA
42	113, 110	42.9, 43.9	0.089, 0.056, 0.065, 0.072, 0.101, 0.070	$HClO_4$–$K_2Cr_2O_7$ at 200°C, CVAA
45	100, 101	38.6, 38.7	<1, <1	Non-destructive, NAA
45B	135, 115	47.0, 46.6	0.050, 0.060	NAA, irradiated, leached with acid, counted HgS ppt

TABLE 8 (continued)

Lab. code No.	Sediment sample No.			Method
	72C5643	72C5644	72C5645	
45B	143, 128	49, 43.2	—	NAA
46	130, 130	51, 51	0.08, 0.08	NaCl, H_2SO_4, HNO_3 digestion, CVAA
47	108, 112	39.2, 40.6	0.45, 0.48	H_2SO_4–HNO_3 at 60°C, $KMnO_4$, CVAA
51	120, 120, 116	49, 48, 48	0.12, 0.17, 0.10	EPA, prov., acid, $KMnO_4$, autoclave, CVAA
	110, 112, 122	46, 45, 41	0.13, 0.10, 0.15	EPA, prov., aqua regia at 95°C, CVAA
52	88.5, 92.5, 87.5	36.0, 38.0, 36.0	0.06, 0.08, 0.08	EPA, prov., aqua regia at 95°C, CVAA
	84.0, 96.5	39.6, 39.6	0.07, 0.07	H_2SO_4–HNO_3 at 60°C, CVAA
53	104	28	0.5	Combustion, trap Hg in $KMnO_4$, CVAA
54	112, 108	44.5, 45.6	0.06, 0.06	EPA, prov., aqua regia at 95°C, CVAA
56	117, 118, 121	44.7, 42.7, 46.3	0.046, 0.053	EPA, prov., aqua regia at 95°C, CVAA
57	162, 159	66.0, 65.0	1.02, 1.04	EPA, prov., acid, $KMnO_4$, autoclave digestion, CVAA
	164, 171	55.4, 58.2	—	EPA, prov., aqua regia, digestion at 95°C, CVAA
58	91, 120	29, 55	1.48, 0.00	EPA, prov., aqua regia, steam bath, CVAA
	68.7, 69.5	24.0, 21.2, 23.5	0.06	—
66	94.4, 96.4	46.3, 46.7	0.07, 0.08	H_2SO_4–HNO_3–$KMnO_4$, CVAA
67	115.6, 115.9	50.6, 49.9	0.12, 0.11	EPA, prov., aqua regia, CVAA
70	119, 119	51.2, 49.5	0.30, 0.31	Combustion Hg to HCl to Ag, then CVAA
71	72, 66	35, 31	0.2, 0.2	HNO_3 digestion Hg on CdS, CVAA
72	60.6, 60.8	23.4, 21.4	0.39, 0.35	Aqua regia, $KMnO_4$, steam bath, CVAA
	—	—	0.75, 0.46	Aqua regia, boiled, CVAA
73	103, 102, 103, 108, 96, 105	40.3, 41.6, 40.6, 37.1, 35.6, 31.7	0.08, 0.09, 0.08, 0.07, 0.07, 0.07,	Aqua regia at 70°C, CVAA
74	115, 111	49, 45	<0.1, <0.1	EPA, prov., aqua regia at 95°C, CVAA
	107, 105	36, 34	<0.1, <0.1	Munns and Holland, JAOAC **54**, 202–5 (1971)
76	137, 124	47, 49	0.09, 0.08	Aqua regia, CVAA
82	92.4, 97.1	42.6, 39.7	0.096, 0.105	Acid, $KMnO_4$, persulfate at 95°C, CVAA
83	>100	47.7, 46.2	0.063, 0.056, 0.059	CVAA ES and T, 6, p. 274, 1972
85	91.50, 94.60	49.50, 52.90	0.27, 0.33	HNO_3–H_2SO_4 vanadium pentoxide at 120°C, CVAA

TABLE 8 (continued)

Lab. code No.	Sediment sample No.			Method
	72C5643	72C5644	72C5645	
86	103.5, 109.3, 106.6	40.8, 42.7, 41.8	0.0608, 0.0656, 0.0593	CVAA, Acid, $K_2S_2O_8$, digestion at 90°C
88	106	44	0.080	H_2SO_4–HNO_3, CVAA, Ag screen*
	104	41	0.085	Aqua regia, CVAA, Ag screen*
89	83, 79, 80	31, 30, 28	0.13, 0.12	EPA, prov., aqua regia at 95°C, CVAA
91	108, 107	50.8, 50.8	0.103, 0.090	HNO_3–H_2SO_4, room temp., digestion, CVAA
92	77.6, 75.0, 91.3	37.7, 33.9	0.059, 0.064	EPA, prov., aqua regia at 95°C, CVAA
	83.1, 83.4	30.2, 31.4	0.040, 0.032	EPA, prov., acid, $KMnO_4$, autoclave, digestion, CVAA

*For 72C5643 and 72C5644 only; 72C5645 method of USCS research paper 800B, pp. 8151–8155, 1972.

TABLE 9
Statistical summary of mercury in sediment round robin, 1972 ($\mu g/g$)

Sample No.	Mean	Min.	Max.	Standard deviation	No. of analyses
A. Total mercury, all laboratories and all methods					
72C5643	109.55	60.6	188	21.50	135
72C5644	43.91	21.2	73	9.13	138
72C5645	0.22	0	2.2	0.38	114
B. Total mercury, laboratories using the USEPA Method (April 1972) aqua regia digestion					
72C5643	110.50	75	188	29.27	52
72C5644	44.99	28	73	10.16	51
72C5645	0.338	0	2.2	0.566	41

used the EPA autoclave procedure, therefore no statistical treatment of the data was attempted. Comparison of the means of all techniques with those of the flameless atomic absorption method at high levels (44–110 $\mu g/g$) indicates that the EPA method is acceptable as the means are comparable for most samples. However, at the low level (0–0.5 $\mu g/g$) there is approximately 30% difference in the means. A more accurate procedure should be investigated. For the two high level samples, the standard deviation for all methods is approximately 20% of the mean. However, for the low level sample, the standard deviation is not acceptable for an analytical method. Further research and development of methodology for low level analysis of sediments are needed.

REFERENCES

1. ZELLER, H. D. and FINGER, J. H., Investigations of Mercury Pollution in the Aquatic Environment of the Southeastern United States, US Environmental Protection Agency, Region IV, Atlanta, Georgia.
2. ROTTSCHAFER, J. M., JONES, J. D., and MARK, JR., H. B., A simple rapid method for the determination of trace mercury in fish via neutron activation analysis. (Prepublication copy, September 1970, University of Michigan.)
3. Official Methods of Analysis of Association of Official Agricultural Chemists, 10th edn., pp. 275–277, 1965.
4. PERKIN ELMER CORP., Norwalk, Conn., Analytical Methods for Atomic Absorption Spectrophotometry, March 1973.
5. HATCH, W. R. and OTT, W. L., Determination of submicrogram quantities of mercury by atomic ab-

sorption spectrophotometry, *Analyt. Chem.*, **40** (14) 2085–2087, December 1968.
6. GIAUQUE, R. D., GOULDING, F. S., JAKLEIRC, J. M., and PEHL, R. H., Trace element determination with semiconductor detector X-ray spectrometers, *Analyt. Chem.*, **45** (4) 671, April 1973.
7. CARTER, J., Determination of ppb concentrations of mercury using a spark source mass spectrometer sample changer, *Analyt. Lett.*, **4**, (6) 351–55 (1971).
8. BRANDENBERGER, H. and BADER, H., The determination of nanogram levels of mercury in solution by a flameless atomic absorption technique, *Atom. Absorp. Newsl.* **6**, (5) 101–103, Sept.–Oct. 1967.
9. WHEAT, J. A., *Determination of Mercury in Radioactive Samples by Flameless Atomic Absorption Spectrophotometry*, AEC Research and Development Report, E. I. Dupont de Nemours & Co., Savannah River Laboratory, August 1968.
10. *ASTM–USEPA Method Study of Total Mercury in Water*, Environmental Protection Agency, Methods Development and Quality Assurance Research Laboratory, Cincinnati, Ohio, December 1972.
11. *Standard Method of Test for Total Mercury in Water*, D3223-73, 1973 Annual Book of ASTM Standards, Part 23, November 1973, American Society for Testing and Materials, 1916 Race Street, Philadelphia, Pa. 19103.
12. KOPP, J. F., LONGBOTTOM, M. C., and LOBRING, L. B., Cold vapor method for determining mercury, *J. Am. Wat. Wks. Ass.* **64**, 20 (1972).
13. GOULDEN, P. D. and AFGHAN, B. K., An automated method for determining mercury in water, *Adv. in Automated Analysis, 1970*, Medicad Inc., Tarrytown, NY, vol. 2, 1971.
14. BENNETT, T. B., MCDANIEL, W. H., and HEMPHILL, R. N., Automated flameless atomic absorption procedure for mercury in water, *Adv. in Automated Analysis, 1972*, Medicad Inc., Tarrytown, NY, vol. 8, 1973.
15. UTHE, J. F., ARMSTRONG, F. A. J., and STAINTON, M. P., Mercury determination in fish samples by wet digestion and flameless atomic absorption spectrophotometry, *J. Fish. Res. Bd. Can.*, **27**, 805 (1970).
16. *Mercury-in-Sediment*, Environmental Protection Agency, Methods Development and Quality Assurance Laboratory, Cincinnati, Ohio, April 1972.

ANALYTICAL TECHNIQUES FOR HEAVY METALS OTHER THAN MERCURY

Herbert A. Laitinen[†]

Professor of Chemistry, University of Illinois at Urbana-Champaign, Urbana, Illinois 61801

INTRODUCTION

The purpose of this paper is to consider the factors entering into the choice of analytical techniques for heavy metals in environmental samples. Mercury is excluded because it is being covered in another paper.

In undertaking to compare various analytical techniques, it is necessary to consider not only the attributes of the measurement techniques, but also the nature of the problem to be solved or the questions to be answered by analysis. We shall first, therefore, take up some general considerations in environmental analysis, then briefly review a number of possible techniques for heavy metals in relation to the needs of environmental analysis, and, finally, draw some conclusions as to the factors to be considered in the choice of analytical methodology and the need for future effort.

GENERAL CONSIDERATIONS

The first point to be settled is the type of information to be provided by analysis. Which components of the environment are involved—the air, water, soil, plant, or animal materials? What type of information is needed? Are we interested in describing the distribution of one or more elements, the rate of flow of these elements, or are we interested in more detailed information such as the chemical form of the element, its distribution throughout the sample on a microscopic scale? If we are interested in a materials balance type of study, then a special set of requirements stressing accuracy is imposed. Thus, whether the element is present in smaller or larger concentrations, the accuracy requirement is determined by the total flow of that element in the various parts of the system. As an example, the heavy metal content of water moving through the environment may be of the order of micrograms per liter, which is normally considered in the range of ultratrace analysis. Yet the total volume of water involved may be so large that it contains a significant amount of metal, which means that an unusually accurate determination may be required to achieve the desired materials balance.

Important factors in the choice of analytical methods are the requirements for sensitivity, precision, and accuracy. The term sensitivity as applicable to a method or instrument is often loosely used to mean the amount of instrument response per unit of weight or concentration. A far more meaningful term is detection limit, corresponding to the minimum weight or concentration that can be detected with a stated confidence, say 50 or 90%. This is done by designating the weight or concentration giving a signal a stated factor higher than random noise (say $2s$, $3s$, or $4s$).

It is important to decide upon realistic goals for precision and accuracy and not demand higher standards than are justified by the research plan. The demands for precision and accuracy should be relaxed, if possible, as the levels of contaminant decrease.

We shall not concern ourselves with a detailed discussion of precision and accuracy except to remind ourselves that the standard deviation is a measure only of random errors. The existence of bias or determinant errors, and therefore the absolute accuracy of an analytical method, can be a difficult and time-consuming operation. The accepted procedure is to analyze field samples, i.e., samples that have not been spiked artificially, by two or more methods based on different principles. If concordant results are obtained by two methods, one involving chemical destruction of the sample and the other one requiring no sample treatment, then the results can be regarded as reliable. This is the procedure used by the National Bureau of Standards in certifying a trace metal content of their standard reference materials (SRMs). The existence of several SRMs, notably orchard leaves, tomato leaves, and bovine liver, represents an important advance during the last few years. However, a great many types of samples of environmental interest are not yet available in the form of SRMs. For example, soils are an important environmental component, yet no SRMs are available for trace metals, and their

---

[†] *Present address:* Dept. of Chemistry, University of Florida, Gainesville, Florida 32611.

preparation would pose difficulties because of the wide variety of types of soil. Another problem is that the trace element of interest may or may not be present at "normal" levels in the particular SRMs available. Therefore, for the foreseeable future, there seems to be no alternative to cooperative interlaboratory test programs involving the exchange of samples.

At this point it may be useful to remind ourselves of the difference between trace analysis and microanalysis, the former being addressed to low concentrations and the latter to small samples. Some analytical techniques are inherently trace techniques in the sense that they are applicable directly to liquid solutions or solids at low concentrations. The terms trace analysis and ultratrace analysis are sometimes used to indicate concentrations in the range of micrograms per gram or micrograms per kilogram, respectively. In passing, it should be noted that the terms ppm (parts per million) or ppb (parts per billion) are to be discouraged, especially in view of the British usage of ppM (parts per milliard) and the British usage of billion to represent 10^{12} instead of 10^9.

Having decided upon the type of information that is desired, the next step is to devise a sampling procedure consistent with the research plan and the analytical methods that are available. If we are interested in traditional analytical information, i.e., the average elemental composition, then the sampling procedure is designed in such a way as to provide the average composition of the material sampled. If, on the other hand, special information, such as the distribution of elements on a microscopic scale or within individual particles, is desired, then clearly the sampling procedure must reflect the experimental design as well as be compatible with the analytical technique. Sometimes the sampling operation is at least in part combined with the measurement, as, for example, in the use of electron microprobe analysis for microscopic distribution.

In sampling the atmosphere for particulates, we often wish to have information about particle size distribution because of its importance in relation to the retention of particles in the lungs. If the same samples are to be used for elemental analysis, then serious questions can arise as to the validity with which the collected fractions actually reflect the airborne particles. When filter media are used to collect particulates from the air, questions as to the completeness of collection and possible contamination from trace elements in the filter must be answered. Finally, the importance of chemical species as opposed to elements is dramatized by the need for distinguishing asbestos fibers from other silicates commonly present in airborne dusts.

The next step in the analytical operation following the sampling is the preparation of the sample for analysis.[1] Here again there is close interdependence between the procedures involved in sample preparation and the final analytical measurement. In general, storage of the sample is required. Proper precautions must be taken to avoid contamination of the sample, losses from the sample, or deterioration of sample integrity. In the determination of heavy metals, the principal dangers are contamination from such sources as airborne dusts or residues on container walls, and losses due to adsorption on container walls. Both contamination and loss can also occur in preliminary steps such as grinding, ashing, and dissolution of samples. Whenever reagents are involved, the problem of contamination from impurities must be considered. Losses of trace elements during ashing depend not only on the element to be determined, but also on the nature of the matrix. For example, metal chlorides are usually more votatile than oxides, and fluxes used in ashing operations may act to increase or decrease volatility losses depending upon the chemical forms involved. Wet ashing is often preferred for biological samples, but at the cost of possible contamination due to impurities and inconvenience in handling samples of larger size. Special procedures can be used, such as the so-called low temperature ashing in which the temperature is only 100–200°, and the oxidation is effected through an oxygen atmosphere excited by radio frequency energy. Unfortunately, this method is relatively slow, and therefore restricted to small samples. A final step in the sample preparation stage is that of separation, which may involve procedures such as distillation or volatilization, liquid–liquid extraction, ion exchange, or high temperature techniques such as combustion. As a general rule, separation steps, with their inherent dangers of contamination and loss, are to be avoided whenever possible in trace analysis. However, in certain instances, e.g. in neutron activation analysis, the separation of interfering species need not be accompanied by dangers of contamination. In other instances, as in isotope dilution analysis, a quantitative recovery is not required. In still other situations, the separation step is closely tied with the measurement step. Thus, in mass spectrometry, separation not only by elements but even by isotopes is an inherent part of the measurement.

Next we turn to the type of response given by the analytical method. Some methods respond equally to all chemical forms of an element, as in the case of neutron activation analysis. Other methods respond to the total elemental composition, but in a manner dependent upon the species present or the matrix in which the element is contained. As an example, spark source mass spectrometry responds to practically every element in practically every chemical form, but in a manner highly dependent upon the chemical form and matrix. Thus, biological samples must be destroyed to avoid a multiplicity of carbon-containing species in the response. Atomic absorption spectrometry responds to the atomic population in the flame or atmosphere in which the determination is made, but in turn is often sensitive to the matrix or the chemical form in which the element is present. Electrochemical methods are occasionally useful in distinguishing chemical forms, but, of course, are restricted to solutions, which may or may not reflect the chemistry of the original sample. The sensitivity or insensitivity of an analytical method to chemical species may be an advantage or a disadvantage, depending upon the problem to be solved.

This brings up a point that I believe should be particularly stressed, namely that the analytical procedure should be matched to the precise question that is to be answered. Until the question is framed exactly, a rational choice of method cannot be made. As an example, several different types of questions might be asked about heavy metals in natural waters. The question might relate to the motion of heavy metals through the environment, and in this case the natural water is defined as the water itself and the suspended sediments that move with it. Once valid samples representing water thus defined are taken in the field, it matters little whether some of the suspended solids settle out or what arbitrary separation might be made in the solution and particulate phases, just so long as the total sample integrity is maintained and all of it analyzed. On the other hand, the question might be related to the amount of metal in true solution and not on suspended matter. In this case, it is impossible to make a valid phase separation that would distinguish these forms because filtration or centrifugation would inevitably leave small amounts of colloidal matter in suspension. Therefore, a method truly responding to ionic metal, such as an electrochemical technique, would be required. Still another question might be to distinguish between free and complexed metal ion in true solution. Here a method responding to metal ion activity, such as an ion selective electrode, might give the desired answer.

Let us now discuss briefly the special needs of a trace analytical facility. The need for clean facilities goes up directly in proportion to the sensitivity requirements of the methods. Special provisions for minimizing dust in the air, either in the form of clean rooms or more localized clean hood areas, are highly desirable if not essential. In certain cases, protection of the samples from contamination from the analyst himself must be provided. Clothing, hair, breath, and skin are all potential sources of contamination. Cross contamination within a given laboratory must be avoided, especially if different levels of composition are being handled simultaneously.

The need for reagents of unusual purity should be stressed. One example of a round robin experiment in Great Britain was once cited to me by Dr. R. A. Chalmers of the University of Aberdeen. The results of the testing program were gratifyingly close in agreement until it was realized that all of the participants had used identically the same procedure with reagents from the same source, and that the blank contribution from the reagents was considerably higher than from the sample.

The demands for specially trained personnel with good technique are strongly dependent upon the analytical method. Not only the sophisticated modern techniques, such as spark source mass spectroscopy, but even, and perhaps especially, the classical methods give reliable results only in the hands of good analysts. The time-honored dithizone spectrophotometric method for metals is a sensitive, selective, and accurate technique in the hands of a skilled and experienced analyst, but subject to gross errors in incompetent hands.

ATTRIBUTES OF ANALYTICAL METHODS

It is not our purpose here to review the physical or chemical principles underlying various methods, but rather to compare a variety of methods from the viewpoint of the ultimate user of the results. We shall adopt the rather unusual classification of lumping together all methods applicable to solutions and then similarly cover a group of methods applicable to solid samples. The reason is primarily that for solution techniques, sensitivities are expressed in concentration units, whereas for solid techniques, the detection limits are usually expressed in terms of absolute quantities. The two methods of expressing sensitivity are, of course, related through the weight of sample used, and the volume of solution needed for a single analysis. It seems more meaningful, however, to compare various methods in this way because a method generally has physical limitations as to the size of the sample necessary or useful in a test.

Solution techniques

Chemical methods based on kinetics of reactions in solution are capable of great sensitivity, often below 10^{-6} M. Under proper conditions they can be specific or highly selective, with precision and accuracy of plus or minus 10% relative or better. One class of methods of this type is represented by the catalysis of redox reactions by heavy metals.[2] Another class is represented by the catalysis of a chelate exchange reaction which follows a chain mechanism.[3] These methods are relatively undemanding of equipment unless unusually rapid kinetics are being followed, e.g. with the stopped-flow method. The methods can be automated to make a series of observations on each set of solutions to increase the precision of measurement. Catalytic methods are demanding of operator skill and experience and purity of reagents. Prior separations, with their attendant difficulties to guard against sample contamination and loss, are often required.

Spectrophotometric methods fall into two types. In the first type, a color-producing reagent is added, together with any necessary masking or buffering reagents, and the intensity of color read on a spectrophotometer. Such methods can be made simple and inexpensive, but often lack the required sensitivity. A sample calculation will serve to illustrate this point. Suppose that it is desired to make measurements at concentrations of 10^{-6} M using a 1 cm cell and measuring absorbance change of 0.01 unit. The molar absorptivity required is 10,000, a value which is reached by only the more sensitive reagents. The second general procedure is to use a reagent forming a chelate complex that can be extracted into an organic solvent. A familiar example is diphenylthiocarbazone (dithizone), which is used in the form of a chloroform solution into which the metal complex is extracted. Selectivity is achieved by regulation of pH and use of masking reagents. Preliminary separations are often made by extracting interfering ions or the desired element under appropriate conditions. Sandell *et al.*[4] state that one to two micrograms of lead in 10 ml samples can be determined with an accuracy of around 2%, using a 5:1 ratio of aqueous to chloroform phase.

Electrochemical techniques represent an important class of solution techniques for trace analysis.[5] The methods most applicable to trace analysis are pulse polarography, differential linear sweep voltammetry, and anodic stripping voltammetry (ASV), which will be considered in turn. Classical polarography is generally considered to have a detection limit of 10^{-6} M and an operating range for quantitative work for 10^{-5} to 10^{-3} M. The main limiting factor to sensitivity is the residual current which is largely due to double layer charging of the electrode. This can be minimized by the use of pulse polarography, in which the charging current is allowed to decay away before the measurement is made. Sensitivities of the order of 10–100 times that of classical polarography can be achieved in favorable cases, such as lead and cadmium.[6] In the differential linear scan voltammetry method, the charging current is largely eliminated by using a dummy cell as a reference. Use of blank reagents in the reference solution permits a direct subtraction of blank. At the National Bureau of Standards, metals at concentrations down to a few micrograms per kilogram have been determined in nearly a hundred standard reference materials by this method,[7] and comparisons have been made with other analytical methods. The standard deviations were of the order of 2–10% for concentrations of the order of micrograms per gram in the solid sample, usually using samples of the order of 0.1–1 g. Many other types of environmental samples have been analyzed by this method.

The most commonly used electrochemical technique for environmental samples is ASV using thin-layer mercury electrodes. As ordinarily applied, this method involves the deposition of a constant fraction of the desired metal at controlled potential, after which, a linear scan of potential in the anodic direction causes the appearance of peaks in the current time curve corresponding to the stripping of the metal from the amalgam. This method is being routinely used for analysis of water in the range of 1–10 µg of lead, cadmium, and zinc per liter without prior concentration steps. Electrochemical methods are applicable to relatively few metals, being most appropriate for those metals reversibly plated and dissolved as amalgams. In our laboratory, electrochemical methods are used for water and for other samples where sensitivity is insufficient by atomic absorption spectrometry, or where information about complexation is desired. These techniques are demanding of operator skill, and, of course, attention must be paid to impurities in the necessary reagents. If a solid sample is reduced to a solution, the reagent blanks are no more serious than for other methods because the reagents serve as supporting electrolytes. A special problem associated with ASV is the possible formation of intermetallic compounds in the amalgam, which may affect the stripping characteristics. This problem is less severe at lower concentrations, and can sometimes be minimized by controlling the deposition potential. Another attribute of electrochemical methods is their sensitivity to chemical species, which can be an advantage or a disadvantage depending upon the problem being attacked. It has been found for example that the ASV method responds differently to metals such as lead and copper in natural waters whether the water is acidified or not. This has been attributed to the formation of stable complexes in the natural water which are broken up upon acidification. In principle then, it is possible to determine labile and bound metal in such waters.[8] It appears that a good deal more research is necessary before such discrimination can be made with confidence. For one thing, although the finding was associated with relatively high molecular weight organic substances in the natural waters thus studied, it would seem that inorganic materials such as colloidal clays would contribute appreciable metal binding in certain waters. Furthermore, depending upon the "time constant" of the method and the kinetics of release of the metal from the complex, a greater or lesser fraction of the bond metal should respond. In this respect, the time scale of pulse polarography, about 50 ms, is much faster than that for linear sweep voltammetry (about 1 s) or ASV (several minutes). The fact that ASV did not respond to total metal, however, indicates that in the samples reported thus far, the dynamics of ligand exchange must be extremely slow.

It might seem that the ultimate in species determination would be to measure single metal ion activities through ion selective electrodes. While this is true in principle, it does not follow that environmental impact is proportional to solution activity because chemically bound forms can be rendered biologically active through natural processes.[9] Another difficulty is that the response of most ion selective electrodes begins to decrease seriously at the levels of greatest environmental concern (less than 10^{-5} M). Other problems, such as the possible interference of surfactants present in natural samples have not yet been adequately explored. Electrochemical techniques as a class are relatively inexpensive in terms of equipment, but demanding in operator skills and purity of reagents.

Another solution technique, which should be classified more as a micro method than as a trace method, is the ring oven technique originally introduced by Herbert Weisz in Vienna[10] and applied to environmental samples especially by Professor P. W. West in this country.[11] It is essentially an adaptation of qualitative tests on filter paper. The method has been applied to the separation of up to 26 cations, which can be separated and estimated semiquantitatively. Simple inexpensive apparatus and small volumes of solution are used. The analysis of airborne particulate materials is worthy of note, considering the speed, simplicity, and scope of application for field analysis. At the present time, however, few direct comparisons have been made with other analytical techniques.

Probably, more environmental samples are being analyzed for heavy metals by atomic absorption spectrometry (AAS) than by any other analytical technique. The method as generally run involves the reduction of the sample to a solution, either through dry-ashing or wet-ashing techniques. In either case, reagent blanks and losses due to volatility must be considered. For some samples, notably soils and fiber glass filters, strong acid extraction rather than complete sample dissolution is adequate. The sensitivity of conventional AAS varies from one element to another. In a recent listing of flame AAS sensitivities,[12] the detection limits in micrograms per milliliter ranged from 0.005 for cadmium to 2 for titanium. Several types of flameless AA techniques use tantalum or carbon boats or rods. Small volumes of sample solutions (10 μl or less) are subjected to an increasing range of temperatures during which drying, pyrolysis, and atomization occur. Improvement of sensitivity over conventional AAS by two to three orders of magnitude is claimed. For the same list of elements, detection limits ranging from 7×10^{-12} g for cadmium to 4×10^{-9} g for titanium were listed. If these are converted to solution sensitivities assuming 10 μl samples, then the sensitivites range from 0.0007 μg/ml for cadmium to 0.4 μg/ml for titanium. This represents an increase of five–tenfold over the conventional solution technique. Flame emission spectroscopy is complementary to atomic absorption spectroscopy in the sense that absorption represents excitation from ground state atoms, whereas emission represents the reverse. For a given element it is qualitatively clear that there is an optimum temperature at which the atomic population is at a maximum and another temperature at which the excited atoms are at a maximum and beyond which ionized atoms begin to take over. Flame methods which correspond to relatively low excitation temperatures are therefore best suited for elements of greatest volatility and lowest ionization potentials. Atomic fluorescence spectroscopy represents a special modification in which the excitation is not purely thermal but arises from absorption of incident radiation. It is beyond the scope of this discussion to cover all the excitation methods used in emission spectroscopy ranging from flames through plasmas to d.c.-arc and a.c.-spark methods. Suffice it to say that the arc and spark methods, which were a generation or two ago considered to be the most sensitive and general trace analytical methods, have been largely supplanted by AAS and in some cases flame emission and atomic fluorescence because of their better sensitivities. However, it is worth pointing out that classical emission spectroscopy still has a place in surveying the elemental composition of unknowns, especially when there is no prior knowledge of composition. For such screening, the method is often applied directly to solid samples, whereas for the most accurate quantitative work solutions are customarily used.

Solids techniques

Neutron activation analysis represents the most important of a group of methods classified as nuclear activation techniques. Dr. W. Wayne Meinke, Chief of the Analytical Chemistry Division of the National Bureau of Standards, uses the term activation spectrometry for procedures that involve only nuclear activation followed by detection of ratiation by gamma-ray spectrometry. He confines the term activation analysis to procedures which involve not only nuclear activation and detection but radiochemical separations. In activation spectrometry, excited atomic nuclei are created by any of several methods including neutron activation, proton activation, or heavy particle bombardment. By using an energy discriminating detector, and by taking advantage of the varying decay times of the activated nuclei, discrimination among various atoms is possible. Meinke points out[13] that in most practical samples, activation spectrometry begins to fail at concentration levels between 1 and 0.1 μg/g. To go below these limits it is necessary to perform chemical separations. The important distinction between activation analysis techniques and other trace analytical methods in this respect is that by minimizing handling operations before radiation, and by performing the separations on the irradiated sample, it is possible to eliminate the very important problem of contribution of the trace element from the "blanks" in the reagents. Likewise contaminations from the surroundings, including the analyst himself, need not be of concern. Of course, losses of the activated material and the relatively short decay times of some of the elements of concern need to be considered in making the separations.

The main disadvantages are the investment in facilities, cost of irradiation, and the slowness of the analysis, especially for elements yielding nuclei of long half life. To gain the necessary sensitivity, the activation time must be correspondingly lengthened. Also, a "cooling" period is needed to help discriminate against interference by short-lived components. A common problem in environmental samples is encountered with sodium which undergoes neutron activation to yield Na^{24}. To avoid the interference, three approaches are possible. First, sodium may be removed from the sample prior to irradiation but with the hazard of contamination as previously mentioned. Second, Na^{24} activity may be allowed to decay, but this procedure limits the analysis to long-lived isotopes. Third, a post-irradiation separation of Na^{24} is often a method of choice. For certain elements of environmental concern, notably lead and cadmium, the sensitivity of neutron activation analysis is poor. Other activation techniques, notably photon and heavy particle activation, are useful in filling such gaps.

In summary, the neutron activation analysis technique is best suited to multi-element screening where time, cost, and detailed chemical information are not at a premium.

Three types of mass spectrometry are important for environmental analysis.[14] First, there is isotope dilution mass spectrometry, in which a sample is spiked with a known addition of an enriched isotope, and the dilution of that isotope as measured by mass spectrometry is used as a measure of the element in the sample. This method is important because it comes closest to being an absolute method that in principle requires no calibration. It does require that the sample be reduced to the same chemical form as the added isotope so that proper dilution actually takes place. The method is capable of extremely high precision and sensitivity, but, of course, due precautions must be taken to avoid contamination. The demands on instrument cost and operator training are high, and the analyses are expensive. Essentially, this method is usually regarded as a referee method for special purposes.

The second mass spectrometry approach is spark source mass spectrometry (SSMS) which involves high-energy excitation of solid samples followed by a double focusing mass spectrometric separation of ionic fragments from the sample. The double focusing feature is required not for high resolution separations, but to allow for the wide energy distribution as well as mass distribution of the particles. Isotopic as well as elemental composition is revealed. Usually the readout is via photographic plates; thus multi-element analyses are performed without prior knowledge of composition. The method is capable of extremely high sensitivity, especially in terms of total mass, because of the small sample size involved. It is of extremely wide applicability, some 70 elements lying in its scope. Furthermore, the presence of unsuspected elements can be revealed. The method is inherently a trace method, because its accuracy, which is usually 20–30% but which can approach 10% with adequate standards, is essentially the same at all concentration levels.

Among the limitations of the SSMS are the limited accuracy, the high cost of equipment, and need for specially trained personnel. Interpretation of the results requires careful attention to matrix effects. In general, solid samples are required, so the method is awkward for dilute solutions. Biological materials need prior destruction, because of the formation of complex carbonaceous species. Careful calibration with standard reference materials is required for quantitative work, although some semiquantitative estimates of composition can be made without calibration. In summary, SSMS is a powerful trace multi-element technique giving elemental and rough isotopic composition data of great sensitivity but limited accuracy on small solid samples. It yields no information about chemical species.

The third form of mass spectrometry is ion probe mass spectrometry,[15] which involves the use of a narrow bombarding beam to scan the sample. Ionic fragments are resolved into individual isotopic components by means of double focusing mass spectrometry. The important feature of the ion probe is its applicability to individual small particles or to scan heterogeneous samples for elemental composition. It offers three-dimensional resolution through its ability to "wear" through the sample under continuous probing. Associations of elements by multiple scans give valuable information about species. Relatively few ion probe mass spectrometers are in operation, and therefore this is not yet a routine analytical tool. It appears likely to become increasingly important in the years to come, not only for determination of elemental composition and distribution but also as an isotope ratio instrument intermediate in accuracy between the large thermal ionization mass spectrometer and the spark source mass spectrometer.

Several types of analytical techniques culminate in x-ray emission from the elements in a sample. In each case, the emitted x-rays are characteristic of the elements and not dependent upon the chemical form. However, the nature of the matrix affects the emission of background radiation as well as specific radiation. In addition, the absorption of the emitted x-rays is determined not only by the chemical composition but also the geometry (thickness) of the sample. The various methods are distinguished by the nature of the excitation process.

The x-ray fluorescence(XRF) method involves excitation by an x-ray source. Its limitations are a lack of sensitivity, dependence on sample thickness and matrix effects, and limitation to the heavier elements. In special cases, for example, trace metals in thin biological specimens, where the matrix elements are light, good results have been reported in the range of micrograms per gram.

The electron microprobe, either as a freestanding instrument or as a feature of the scanning electron microscope, is essentially a micro rather than a trace analysis technique. It is capable of resolution of an area of the order of 1 square μm, with a penetration depth of the electron beam of a few micrometers. Thus, it is an elegant technique for analysis of individual particles or for spatial resolution of surface composition. It has been successfully used where the average composition is in the range of micrograms per gram but the localized concentration is considerably higher.

Recently, heavy particle excitation for x-ray emission has received increased attention. Some form of particle acceleration, such as the van de Graaff accelerator or cyclotron, is used to generate a beam of heavy particles such as protons or alpha particles. The emitted x-rays are analyzed with an x-ray spectrometer. The advantages to be gained are greatly increased sensitivity and scope of applicability. Matrix effects, especially due to absorption of the less energetic x-rays, represent a serious problem. The principal limitation, however, is the necessity for specialized equipment, coupling a heavy-particle accelerator with an x-ray spectrometer.

Although emission spectroscopy was considered earlier under solution techniques, it should be mentioned once again as a solids technique. Where the utmost sensitivity is not required, a direct-reading spectrograph can give results acceptable for many purposes for repetitive analyses involving one to twelve components. Modern developments such as time-resolved emission spectroscopy can greatly increase sensitivity and accuracy for a given element, but with increased cost of equipment and requirement for skilled personel.

The technique of gas chromatography has not

been widely applied to metals analysis, but nevertheless deserves mention. Whether the sample be liquid or solid, the essential requirement is for the metal to be transformed in a sufficiently volatile species to be chromatographed. This has been accomplished, for example, for chromium and beryllium in picogram amounts using organic complexing agents.[16] Another technique that has received virtually no application is the conversion of trace metals into fluorides, which are then separated and detected by the use of special GC columns.[17]

CONCLUSIONS

In choosing the analytical technique for the determination of heavy metals in environmental samples, there are two general factors that need to be evaluated and matched. The first concerns the nature of the problem: an exact formulation of the question to be answered, the development of a research plan defining the sampling strategy and the number of samples and their variety, the elements to be determined, the type of information needed, the levels of accuracy and precision required.

The second factor concerns the attributes of the methods available, the sensitivity, precision, accuracy, the time required and the type of information furnished need to be considered. The relative costs of different techniques involve not only the cost of the final determination but the cost of sample preparation, the amortized cost of permanent equipment, the cost of personnel, which, of course, varies with the needed training, the cost of consumable supplies and reagents. With regard to speed, both the amount of time required in the laboratory, which determines the sample throughput per day, and also the total elapsed time from the beginning to the end of the analysis are to be considered. It is essentially the matching of these two factors that is the role of the professional analytical chemist.

I shall conclude by listing the most obvious needs for the immediate future as far as the analysis of environmental components for heavy metals is concerned. The first such need is for a greater variety of standard reference materials of established composition. This is a task of awesome proportions which never will be fully done. Perhaps a practical compromise would be to seek a limited number of standard reference materials representative of the sample matrices to be encountered in the environmental work. The second general need is for more critical intermethods and interlaboratory comparisons for precision and accuracy. These comparisons can be carried out not only on standard reference materials as certified by the National Bureau of Standards but also on a localized and informal basis for materials for which standard reference materials do not now, and perhaps will not, exist. The third need is for improved methodology. There are many questions that we cannot yet satisfactorily answer because of the lack of analytical techniques. These especially relate to more detailed information as to chemical species and distribution on a finer scale of resolution. Finally, there is need for increasing the sensitivity, accuracy, specificity, reliability, and speed of analysis while decreasing the cost and need for highly skilled or specialized personnel. For the foreseeable future, it appears that the thrust will be towards increased sophistication of methodology for better answers for more difficult problems. It is important to recognize that an investment in improving analytical methodology is necessary in planning new environmental research programs.

REFERENCES

1. TOLG, G., *Talanta* **19**, 1489 (1972).
2. LAITINEN, H. A., *Chemical Analysis*, p. 460, McGraw-Hill, 1960.
3. MARGERUM, D. W. and STEINHAUS, R. K., *Analyt. Chem.* **37**, 222 (1965).
4. MATHRE, O. B. and SANDELL, E. B., *Talanta* **11**, 295 (1964).
5. LAITINEN, H. A., in *Trace Characterization, Chemical and Physical*, pp. 75–107. (W. W. Meinke and B. F. Scribner, eds.), NBS Monograph 100, 1967.
6. FLATO, J. B., *Analyt. Chem.* **44** (11) 75A (1972).
7. MAIENTHAL, E. J., *NBS TN* **273**, 8 (1965); **403**, 22 (1966); **425**, 23 (1967); **455**, 19 (1968); **505**, 17 (1969); **545**, 41 (1970); **583**, 39 (1973).
8. BENDER, M. E., MATSON, W. R. and JORDAN, R., On the significance of metal complexing agents in secondary sewage affluents, *Environ. Sci. Technol.* **4** (6) 520–521 (1970).
9. GADDE, R. R. and LAITINEN, H. A., *Environ. Lett.* **5** (2) 91 (1973).
10. WEISZ, H., *Mikrochim. Acta* 1954, 140.
11. WEST, P. W., *APCA Jl* **16**, 601 (1966).
12. HWANG, J. Y., *Analyt. Chem.* **44** (14) 20A (1972).
13. MEINKE, W. W., *NBS Techn. News Bull.* **57**, 108 (1973).
14. MORRISON, G. H., *Trace Analysis—Physical Methods*, Interscience, 1965.
15. EVANS, C. A. JR., *Analyt. Chem.* **44** (13) 67A (1972).
16. WOLF, W. R., TAYLOR, M. L., HUGHES, B. M., TIERNAN, T. O. and SIEVERS, R. E., *Analyt. Chem.* **46**, 616 (1972).
17. JUVET, R. S. and FISHER, R. L., *Analyt. Chem.* **37**, 1752 (1965).

Analytical Techniques for Heavy Metals other than Mercury (H. A. Laitinen)

DISCUSSION by R. K. SKOGERBOE
Department of Chemistry, Colorado State University, Fort Collins, Colorado 80521

Professor Laitinen has presented a very lucid discussion of the criteria that enter into the selection of an analytical technique for the analysis of heavy metals of environmental concern. I agree completely with the criteria he has listed as primary in the choice of the analytical technique. If one were to change the emphasis of his discussion in any way, perhaps it would be to stress that the selection of an analysis technique must be based on a compromise. This compromise is reached after careful consideration of the factors he listed recognizing that optimization of the choice in favor of some single factor may be in direct conflict with the optimization of another factor. If I were to take issue with any aspect of Professor Laitinen's discussion, it would be with his evaluation of the attributes of some of the analytical techniques. We are, after all, human, and our judgements necessarily reflect our own personal preferences or prejudices. I am suggesting, therefore, that I would assign different levels to the attributes of some analytical techniques and I am admitting *a priori* that these disagreements may originate from the fact that Professor Laitinen has stepped on some of my own personal preferences and prejudices. Before proceeding to our minor differences of opinion, however, I should like to discuss some observations regarding the selection and utilization of analytical techniques in environmental laboratories.

If one were to survey the country to determine what analytical techniques are most widely used for the determination of heavy metals of environmental interest, it would be determined that the list of techniques is restricted to a few. Those techniques enjoying the most widespread use would include: colorimetry, atomic absorption spectrophotometry, flame spectrometry, and a few electrochemical techniques particularly anodic stripping voltammetry. This is not especially surprising when it is recognized that these techniques enjoy some mutual advantages. They are, after all, adequate for many of the required determinations; the necessary instrumentation is relatively simple to use; the equipment acquisition and maintenance costs are generally low in comparison with some of the more sophisticated (and therefore complex?) techniques; and most instrument manufacturers have well-staffed applications laboratories willing to show us how to use these techniques or to trouble-shoot problems for us. Surely these are valuable practical attributes that should be considered in selection of an analytical technique. Moreover, the economic thread that runs through this list of attributes serves (perhaps too often) as an essential ingredient in our deliberations. Because these techniques are so widely used, it seems appropriate to discuss each of them in more detail.

COLORIMETRY

This technique has stood the test of time and will continue in advantageous use for decades. Heavy metals are most often determined by the formation of an intensely colored metal complex such as the dithizonates. The absorption bands measured are typically broad so spectral interferences from other sample constituents are relatively common. This problem may be alleviated by the use of masking agents, regulation of pH, and/or separation by solvent extraction or ion exchange methods. The separation steps may be used advantageously to preconcentrate trace constituents, allowing measurements not possible otherwise. Although the colorimetric technique itself is generally simple to use, the methodologies required to eliminate interferences and/or obtain adequate preconcentration add to the overall complexity and contribute to the time–cost commitment. In the hands of a careful and patient analyst, these methods can be used to obtain reliable results; for someone less cautious or patient the results may be disastrous. This latter factor should be carefully considered in making a decision to use a colorimetric method.

ATOMIC ABSORPTION AND EMISSION SPECTOPHOTOMETRY

The fact that this technique enjoys very extensive use can hardly be disputed. According to a recent survey,[1] more than 40 elements can be determined at solution concentrations between 0.0001 and 0.1 μg/ml. The use of the parallel technique, flame emission spectrometry, expands this list to nearly 60 elements. Clearly the availability of these two techniques (both of which may rely on the same instrumentation), makes it possible to carry out analyses for a wide variety of heavy metals at concentration levels of environmental interest. Table 1 summarizes the heavy metal detection limits that can be realized through the use of these two techniques and permits a comparison of their relative sensitivities. It should be noted that for those analysis problems requiring higher sensitivity, extraction of the trace constituent into an organic solvent can frequently be used to obtain preconcentration and/or sensitivity enhancement. Because of the specificity of these two techniques, separation of elements from each other to avoid interferences is not usually required. This permits the use of simultaneous extraction of several elements from each sample through the use of broad spectrum complexing agents such as the dithiocarbamates.[2] Because additional extraction operations to increase the separation selectivity are not required, the process is subject to less potential error. This gives atomic absorption and emission measurements which do not require separation a very distinct edge over comparable measurements by those colorimetric methods which require selective separations.

The literature on flame emission indicates that chemical and physical interferences exist for the technique. Much of the early literature on atomic absorption insisted that the technique should not be subject to interferences, but this expectation has been shown fallacious in several instances. Certainly the effects which change the degree of atomization in a flame affect both emission and absorption. Fassel et al.[3] have shown that spectral interference due to absorption line overlap occurs for several elements. In a more recent paper[4] it was also shown that interferences due to the lateral diffusion of species in a flame can also be prominent. Perhaps the most serious interference problems which have been recognized for AA are those due to molecular absorption and light scattering.[5-7] To compensate for these interferences, it is advisable to make a dual-beam measurement with a continuous light source to correct for the scattering and/or molecular absorption.[5,8-12] In the author's experience, nearly half of all environmental samples are subject to some degree of molecular absorption or scattering effects for those elements determined at wavelengths below approximately 2800 Å. Thus, such corrections are essential for many analyses.

While these techniques can be used for numerous determinations, adequate sensitivity is often lacking for those elements present at low concentrations or for those cases in which the amount of sample available is limited. This has led to the development of nonflame methods of atomization.[11-13] In these systems, the sample (solution or possibly solid) is placed in a device which has a particular heating cycle. The first stage of the cycle goes up to a temperature of about 110°C to dry the sample; it is then increased to a temperature sufficient to vaporize organic matter that may be

TABLE 1
Detection limits for the determination of heavy metals by flame emission and atomic absorption spectrophotometry[1]

Element	Detection limits (μg/ml)	
	Emission[a]	Absorption[b]
Ag	0.02	0.005
Al	0.005	0.1
As		0.2
Au	0.5	0.02
Ba	0.001	0.05
Bi		0.05
Cd	2	0.005
Co	0.05	0.005
Cr	0.005	0.005
Cu	0.01	0.005
Fe	0.05	0.005
Hg	0.5	0.5
Mg	0.005	0.0003
Mn	0.005	0.002
Mo	0.1	0.03
Ni	0.03	0.005
Pb	0.2	0.03
Se		0.5
Sn	0.5	0.06
Sr	0.0001	0.01
Ti	0.2	0.1
Tl	0.02	0.2
V	0.01	0.02
Zn		0.002

[a] For nitrous oxide-acetylene flame.
[b] For nitrous oxide or air-acetylene flames depending on best choice.

present; finally, it is stepped to a level that may be in excess of 2000°C to vaporize and atomize the analytical species of interest. The purpose is to produce a high concentration of atoms per unit time contained in a relatively confined space. Because the magnitude of the atomic absorption is determined by the density of atoms in the optical path, an increase in sensitivity is observed. Such systems have proven useful for analyses at very low concentration levels and have served as means for solving difficult trace analysis problems. Consequently, they are valuable complements to flame analysis systems.

It should be emphasized that these systems are subject to both molecular absorption and scattering effects;[11,13,14] background correction appears to be essential for most analyses based on the nonflame approach. Moreover, at least one recent paper[15] indicates that the rate and the extent of atomization is dependent on the compound form in which the metal exists in the sample. In short, one would be well advised to utilize these nonflame atomization methods with caution lest interferences cause erroneous results.

ANODIC STRIPPING VOLTAMMETRY

This technique is rapidly emerging as one favorable for the determination of heavy metals—particularly cadmium, copper, lead, and zinc.[16-19] In the most advantageous mode of operation, the trace metals in an electrolyte solution are electrolyzed into a thin film of mercury on a graphite electrode. Following this step, which concentrates the trace metals in the thin film, the potential is scanned to strip the metals out of the film. By measuring the analytical current at the respective potentials for each element, quantitation can be accomplished. The sensitivity can be enhanced by using a pulsed stripping procedure.[16-19] Although it has been stated that the addition of supporting electrolyte is neither required nor desirable,[16,17] it has been shown that this is not true.[18] Used properly, ASV can provide analytical information at exceedingly low concentrations for a number of trace elements.[17] Several elements can be determined in each plating–stripping cycle, so the method offers the desirable multi-element analysis capability. It must be emphasized that certain elements tend to form intermetallic compounds, e.g. ZnCu, GaCu, and AuCu, in the mercury thin film. Because such formations can lead to analytical errors, precautions must be taken to circumvent or correct for their occurrence. We may expect to see increased adoption of this useful technique.

If we consider these widely used techniques, it is apparent that a high percentage of the analytical problems can be solved with them. In general, the techniques are reliable. Errors which occur are largely due to the misuse of the methods by the analysts. The burden, then, is that of insuring that the analysts are fully aware of the capabilities and limitations of the techniques they use.

It was previously implied that the widespread adoption of these techniques has been based on economic reasons. Often, this has been true but I should like to point out that these may have been false economic considerations. With the exception of ASV, these techniques are largely single-element in nature, i.e., one element is determined at a time. To determine 10 elements by atomic absorption, for example, a time commitment of 1–2 h per sample would be a reasonable estimate if separation procedures are not used in the method. In other words, about 5–8 samples can be processed for 10 elements per man day. Some of the multi-element analysis techniques permit much more rapid processing of the same set of samples. Direct reading emission spectrometry, for example, can be used to determine 10 elements per sample in 10–15 min. This means that about 30 samples can be processed per man day. One may conclude that, although a direct reading emission spectrometer may cost three times the price of an atomic absorption unit, it may offer a factor of 3–4 improvement in sample throughput. If an analyst's time is worth as much as that of an electrician or a plumber, the instrument cost differences are made up in a few months.

Emission spectroscopy is one technique where I must take issue with Professor Laitinen's evaluation of attributes. He has stated that emission spectrometry (ES) has been replaced by atomic absorption and other flame techniques because they offer better sensitivity. An evaluation of the detection capabilities for ES shows that this is true only in a limited number of instances.[20-22] Emission methods can be used to determine a wide variety of elements in solution samples at levels below the upper limits imposed by the USPHS for drinking water. Direct analyses of solids (e.g. filterable solids from water) can also be conveniently carried out.

X-ray fluorescence (XRF) is another useful multi-element analysis technique that may offer an economic advantage in terms of analysis time when several elements are of interest. Professor Laitinen has indicated that XRF lacks in sensitivity, shows a dependence on sample thickness, and is subject to matrix effects. The sensitivity statement may be subject to question. Gilfrich[23] has determined submicrogram amounts of several elements as have Campbell et al.(24-26) The latter group has used ion exchange resin-loaded paper disks as media for removing trace quantities of elements from water solutions for XRF analysis. Under these collection conditions, the sample thicknesses are uniform, matrix effects are not generally important, and the capabilities of XRF are essentially optimized. In effect, the problems cited by Dr. Laitinen are minimized, so one may expect to see increased utilization of this technique.

Of the various multi-element analysis techniques available, it would seem that ES and XRF offer the advantages outlined above and the fact that they are less expensive than neutron activation or spark source mass spectrometry. While it can often be shown that they also show an economic advantage over the single element techniques if multi-element analyses are required, it should be pointed out that both techniques generally require a higher level of analyst proficiency. Given that proficiency, their use should be recommended.

To conclude, I should state that I agree enthusiastically with Professor Laitinen's list of analytical needs for the future. Progress in environmental research will continue to be paced by progress in the development of analytical methods and techniques. Certainly one may expect that increasingly sophisticated methods are forthcoming. While we are waiting, however, we should evaluate our current status regarding those techniques currently available. Are we using them to the fullest extent of their capabilities? I think not! Are we properly matching the technique to the problem? The answer is very often—No! Can we change our usage patterns to cut costs and/or increase our capabilities for characterizing environmental systems? In many cases, the answer is Yes! Adoption of multi-element analysis approaches may serve as just one example of how this can be accomplished.

REFERENCES

1. PICKETT, E. E. and KOIRTYOHANN, S. R., Analyt. Chem. **41** (14), 28A (1969).
2. KOIRTYOHANN, S. R. and WEN, J. W., Analyt. Chem. **45**, 1986 (1973).
3. FASSEL, V. A., RASMUSON, J. O., and COWLEY, T. G., Spectrochim. Acta **23B**, 579 (1968).
4. WEST, A. C., FASSEL, V. A., and KNISELEY, R. N., Analyt. Chem. **45**, 1586 (1973).
5. KOIRTYOHANN, S. R. and PICKETT, E. E., Analyt. Chem. **37**, 601 (1965).
6. KOIRTYOHANN, S. R. and PICKETT, E. E., Analyt. Chem. **38**, 585 (1966).
7. KOIRTYOHANN, S. R. and PICKETT, E. E., Analyt. Chem. **38**, 1087 (1966).
8. SPRAGUE, S. and SLAVIN, W., Atom. Absorp. Newsl. **5**, 10 (1966).
9. KAHN, H. L., Atom. Absorp. Newsl. **7**, 40 (1968).
10. BARNETT, W. B. and KAHN, H. L., Atom. Absorp. Newsl. **8**, 21 (1969).
11. KAHN, H. L. and MANNING, D. C., Am. Lab. **6**, 51 (1972).
12. DICK, D. L., URTAMO, S. J., LICHTE, F. E., and SKOGERBOE, R. K., Appl. Spectrosc. **27**, 467 (1973).
13. AMOS, M. D., Am. Lab. **2** (8) 33 (1970).
14. HWANG, J. Y., ULLUCCI, P. A., and MOKELER, C. J., Analyt. Chem. **45**, 795 (1973).
15. MATOUSEK, J. P. and BRODIE, K. G., Analyt. Chem. **45**, 1606 (1973).
16. SIEGERMAN, H. and O'DOM, G., Am. Lab. **4**, 59 (1972).
17. FLATO, J. B., Analyt. Chem. **44** (11) 75A (1972).
18. COPELAND, T. R., CHRISTIE, J. H., OSTERYOUNG, R. A., and SKOGERBOE, R. K., Analyt. Chem. **45**, 995 (1973).
19. COPELAND, T. R., CHRISTIE, J. H., OSTERYOUNG, R. A., and SKOGERBOE, R. K., Analyt. Chem. **45**, 2171 (1973).
20. DEKALB, E. L., KNISELEY, R. N., and FASSEL, V. A., Ann. NY Acad. Sci. **137**, 235 (1966).
21. DICKENSON, G. W. and FASSEL, V. A., Analyt. Chem. **41**, 1021 (1969).
22. LICHTE, F. E. and SKOGERBOE, R. K., Analyt. Chem. **45**, 399 (1973).
23. GILFRICH, J. V., Adv. X-Ray Analysis **16**, 1, Plenum Press, New York (1973).
24. CAMPBELL, W. J., SPANO, E. F., and GREEN, T. E., Analyt. Chem. **38**, 987 (1966).
25. CAMPBELL, W. J., GREEN, T. E., and LAW, S. L., Am. Lab. **28** (June, 1970).
26. GREEN, T. E., LAW, S. L., and CAMPBELL, W. J., Analyt. Chem. **42**, 1749 (1970).

SESSION III
TRANSPORT MECHANISMS (1)

FIELD OBSERVATIONS ON THE TRANSPORT OF HEAVY METALS IN SEDIMENTS

A. J. DE GROOT

Head, Department of Chemistry, Institute for Soil Fertility, Haren (Groningen), Netherlands

and

E. ALLERSMA

Head, Department of Hydrodynamics and Morphology, Delft Hydraulics Laboratory, Delft, Netherlands

INTRODUCTION

The pollution of rivers with degradable organic wastes, persistent materials such as heavy metals and chlorinated hydrocarbons, as well as oil, is a pressing problem, in particular for the Rhine in western Europe. For the Netherlands delta this also applies, to a lesser extent, for the rivers Meuse and Scheldt. Of the substances mentioned above, the heavy metals are probably the most harmful. Compared with the natural uncontaminated environment, the heavy metals can occur in relatively high concentrations and influence the fluvial ecosystem and, after their transport to the sea, also influence the food chains in the marine environment. The harmful effects are linked to the accumulation in biological systems, even in their lowest forms of development.

For the Dutch delta, information could be obtained on the occurrence and the chemical behavior of nine heavy metals. Notwithstanding the rather low concentrations of suspended matter in river water (generally 40–80 mg/l), many metals are predominantly bound to the suspended material rather than being dissolved. It is therefore important, when considering the contamination of rivers with heavy metals, to pay detailed attention to the elements fixed onto the suspended matter.

In this study the contents and behavior of heavy metals in a number of types of suspended matter will be described. Special attention will be given to the changes which the heavy metal composition of some types of suspended matter undergoes upon the displacement of the material from the rivers through their estuaries to the open sea. The fate of heavy metals in suspended matter can only be studied efficiently, however, if the transport paths of the material from the rivers to the marine coastal areas are well known. As an introduction to the problem, some details on these displacements of material, especially referring to the Dutch delta, are given. In this paper, those solid constituents having a diameter $< 16 \mu m$ are studied. All experiments refer to freshly deposited material from the different locations.

Further, some data shall be given on the contents of a few heavy metals in filtered river and sea water. By combining the amounts of fine-grained material and water discharged by the river with the metal contents of these components, an estimate can be made of the total load of a river with heavy metals.

ORIGIN AND TRANSPORT OF SEDIMENTS

The fine-grained material (often called mud) transported along the western European coast, originates mainly from the rivers. A detailed insight into the movements of this material was obtained on the basis of the comparison of the considerably divergent Mn contents of sediments from different origins, as shown in Fig. 1, in which the mud of the River Thames was chosen as a standard.[1]

In this paper we confine ourselves to the transport behavior of the mud within the Netherlands delta (Fig. 2). This delta is mainly influenced by the Rhine, Meuse, Scheldt, and Ems. The suspended matter of the Scheldt has only a restricted sedimentation area and is deposited chiefly within the eastern part of the Western Scheldt. The mouthing area of the Western Scheldt receives its fine-grained material mainly

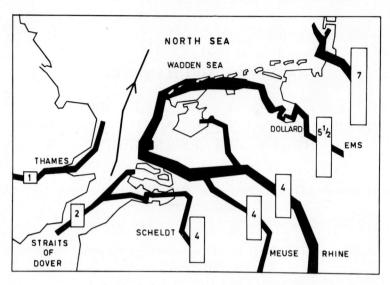

FIG. 1. Manganese contents and movements of mud in western Europe (schematic).

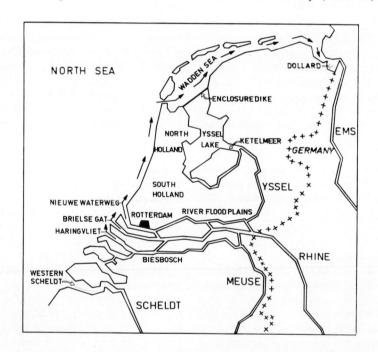

FIG. 2. Movements of sediments in the Rhine estuary, North Sea, and Wadden Sea.

from the south, coming through the Straits of Dover to this part of the Netherlands coast. The main source of mud in the Dutch delta is the Rhine, transporting yearly 3.5 Mtons of fine-grained material in 75 km^3 of water. From this material 10% is transported via the distributary IJssel to Lake IJssel. After leaving the river outlets by the Haringvliet and Nieuwe Waterweg, the Rhine material (Figs. 1 and 2), mixed with a smaller contribution of Meuse sediment, is transported mainly in a northeasterly direction in a narrow zone along the coast of the provinces of South and North Holland (average speed 0.05–0.10 m/s). Then the material reaches the western Wadden Sea from which it is transported further towards the east over the Wadden Sea flats. A part of the Rhine mud finally reaches the Dollard area. The Ems carries much less suspended matter than the Rhine, so its deposition is restricted to a part of the Dollard, especially along the German border of this area.

Although the preceding sediment transport studies were carried out according to the manganese method, based upon the property of this element to remain fixed to the suspended matter during transport in aerated water, nowadays more advanced techniques are available. Our knowledge of the behavior of metals in suspended matter is more detailed now and much progress has been achieved in the analytical techniques to determine these metals. So lanthanum, scandium, and a number of rare earths were found to behave like manganese and can be easily determined by nondestructive neutron activation analysis. Furthermore, we use, especially for sediment transport studies over shorter distances, tracer techniques (in connection with siltation problems of harbors and navigation channels). Therefore an element, which either does not occur in the sediment or only occurs in minute quantities, is fixed to the mud from the river or sea arm. After the material is marked it is returned to the water course where it mixes with the solids moving naturally. At specified points throughout the water course sediment samples are taken to determine the marking element by activation analysis. This gives an insight into the flow path of the suspended matter. As a marking element, tantalum was found to be successful in this respect.[2,3]

EXPERIMENTAL

Due to a preferred occurrence of the heavy metals in the finest grain-size fractions, linear relationships are found between the contents of the heavy metals and the fraction of particles less that 16 μm in size (expressed as a percentage of the $CaCO_3$-free mineral constituents in the oven-dry sediment) in samples from the same location. The metals are generally present as a coating around the particles, so a larger surface area per unit of weight causes a higher content of the relevant metal. In Fig. 3 these relationships are shown for a number of elements in sediments of the Ems.

These linear relationships make it possible to characterize the content of a specific metal of a whole group of co-genetic sediments by a single value, the content being obtained by extrapolation to 100% of the fraction < 16 μm.[1] These types of values will now be used in this paper for the description of the heavy metal composition at different localities.

The analysis of metals has been carried out by

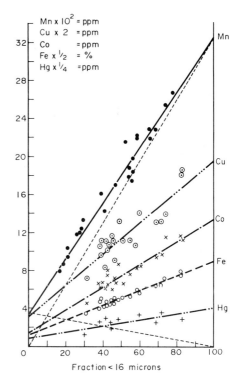

FIG. 3. Linear relationships between metal contents and percentage of fraction < 16 μm (Ems sediment).

different techniques. Originally many metals were estimated by classical spectrophotometry. This technique has now been replaced by atomic absorption for the estimation of cadmium, chromium, copper, nickel, lead, and zinc. For mercury, arsenic, and antimony, as well as for a number of elements estimated in connection with sediment transport studies activation analysis is used.[4,5]

HEAVY METALS IN RIVER DEPOSITS

Immediately following, a treatise is given on the contents of heavy metals in deposits from the Rhine, Meuse, Scheldt, and Ems. A few words will further be devoted to metals in sediments deposited on river flood plains as a consequence of high water discharge of the river. Finally, some details on the suspended matter composition of some tropical rivers will be given.

Rhine, Meuse, Scheldt, and Ems

Within western Europe the Rhine is the prototype of a river in which industrial wastes and other pollutants are drained in large quantities. From the four rivers governing the Dutch deltaic area, the Rhine is therefore predominant from a viewpoint of pollution. The Ems and its tributaries, on the contrary, flow through a sparsely populated area with only a limited amount of industry. The Meuse and the Scheldt are intermediate between the Rhine and Ems.

The composition of sediments from the freshwater part of the Rhine, Meuse, and Ems with respect to a number of heavy metals have been analyzed. For the Scheldt the deposits originated from the brackish-water tidal area of this river. The results are given in Table 1.

From Table 1 it is obvious that the sediments of the Rhine, compared with those of the Ems, have very high contents of zinc, chromium, and copper. This applies to a lesser extent to nickel. Although it has not been mentioned in Table 1, it was already found before 1970 that the contents of lead, mercury, and arsenic are also much higher in the Rhine deposits that in those of the Ems.[5] The contents of heavy metals in the Meuse sediments are also high, although the values are lower than for the Rhine. Striking are the high cadmium values for the Meuse, which are as high as those for the Rhine.

The Scheldt sediments generally have no higher contents of metals than the Ems, with the exception of chromium. We should take into account, however, that the Scheldt deposits have been taken from the brackish-water tidal area of the river. Wollast[6] found that farther upstream in the river the heavy metal contents of the mud are appreciably higher. This is in accordance with the findings to be mentioned later in this paper about the solubilization of metals from sediments in the estuarine part of a river (section on the mobilization of metals).

During the last few years attention has been focused, mainly in the daily press, on a steady increase of the heavy metal pollution of our rivers, especially of the Rhine. We were able to compare in this respect some rivers for the years 1960 and 1970. For the Rhine the results have been given in Table 2.

Except for cadmium, there is, on the average, no strong increase in the contents of the heavy metals during the past decade. The contents of arsenic, zinc, and lead decreased; copper, chromium, and nickel, on the other hand, increased. Alarming is the very strong increase of the cadmium contents of the Rhine sediments. This increase is continuing after 1970, now being more than twice as high as the mercury contents. The industrial use of cadmium seems to increase as a constituent of rubber, dyes, alloys, batteries, etc.

River flood plains

The contents of heavy metals in Rhine sediments, as mentioned in the preceding sections, refer to sediments as they are transported under normal flow conditions of the river. The suspended matter then mainly originates from the source areas of the river. After deposition we characterize this material as "original mud."

Under conditions of high water discharge the erosion of the river bed and of the shores of the river gives rise to the formation of a suspension with different sedimentation characteristics, referred to as "erosion mud." This material forms aggregates of sufficient size to easily settle again when the high flow velocities are reduced somewhat. This mainly happens where the river enters the area of the river flood plains. These flood plains consist of meadows and are occasionally flooded during wintertime. In

TABLE 1

Metal contents, expressed in ppm, in sediments from the Rhine, Meuse, Scheldt, and Ems in 1970 (extrapolated to 100% of the fraction $< 16 \mu m$)

	Rhine	Meuse	Scheldt	Ems
Zn	2900	2500	800	1100
Cr	1240	620	380	180
Cu	600	340	140	160
Ni	100	83	53	79
Pb	800	600	—	—
Cd	45	45	—	—
Hg	23	—	—	—
As	220	—	—	—
Sb	18	—	—	—

TABLE 2

Comparison of metal contents in Rhine sediments in 1960 and 1970, expressed as percentages of the amounts in 1960

	1960	1970
As	100	66
Zn	100	77
Pb	100	85
Sb	100	89
Cu	100	110
Cr	100	125
Hg	100	129
Ni	100	147
Cd	100	194

the summer season these plains are used for grazing cattle.

In 1970 samples of this erosion mud were taken from the river flood plains of the Rhine after extremely high water levels of the stream. In Table 3 the contents of a number of metals in this erosion mud are given, compared with the contents of these metals in original mud (the latter sampled under normal flow conditions of the river).

From this table it is obvious that the metal contents in erosion mud are lower than in the original Rhine mud. From these lower contents it must be concluded that these materials had less contact with pollutants, introduced into the river, than the original mud. It is further remarkable that the erosion mud shows large regional differences in the metal contents (indicated in parentheses in Table 3). At one location, however, the sediments have a uniform composition. We have not yet found the cause of the regional differences in the sediment composition of the river flood plains. It must be attributed, however, to any process of selective sedimentation.

Tropical rivers

Some attention has been paid to the natural contents of heavy metals in sediments from tropical rivers. Generally these contents are low. In Table 4 this has been demonstrated for the River Chao Phya in Thailand[4] and for the River Tji Tarum on the island of Java, Indonesia,[7] compared with the Ems.

It is obvious that the level of the metals, which act as trace metals in agriculture (zinc, copper, and cobalt), is critical. In this connection attention should be paid to the very low zinc contents of the sediments of the River Chao Phya. This river irrigates the central rice area in Thailand.

The most striking value in Table 4 is the very high natural mercury content of the sediment from the River Tji Tarum. This value is as high as that of the very polluted Rhine. The high mercury contents of this tropical river must be attributed to the volcanic deposits in the drainage area of this river.

Tropical river sediments not only deviate in their metal contents from those deposited under temperate climatic conditions, but also the behavior of the metals in the estuarine part of the river is different. Later on in this paper we will pay attention to this aspect.

DISCHARGE OF HEAVY METALS BY THE WATTER AND BY THE SUSPENDED MATTER OF THE RHINE

The discharge of a number of heavy metals by the Rhine water can be calculated from the mean water discharge (2200 m^3/s) and the mean contents of the relevant metals in the filtered river water. The latter were obtained from the Government Service for Public Health in the Netherlands. The results of these calculations, expressed as tons per year, have been given in Table 5.

In the same table the discharges of heavy metals by the suspended matter have been given. These estimates could be made by combining water discharge, concentration of suspended matter in the river water (38 mg/l of the fraction less than 16 μm) and the contents of heavy metals in the suspended matter. From the ratios metal in water/metal in sediments it is obvious that upstream in the river there is for lead, chromium, copper, arsenic and mercury, a more pronounced occurrence in the

TABLE 3
Contents of zinc, copper, chromium, and mercury, expressed in ppm, in original mud and in erosion mud from the Rhine (extrapolated to 100% of the fraction < 16 μm)

	Zn	Cu	Cr	Hg
Original mud	2900	600	1240	23
Erosion mud	1300	270	530	9
(Flood plains)	(900–1500)	(140–370)	(280–760)	(3–13)

TABLE 4
Metal contents, expressed in ppm, in sediments from the rivers Ems, Chao Phya, and Tji Tarum (extrapolated to 100% of the fraction < 16 μm)

	Zn	Cu	Cr	Co	Pb	Hg	As
Ems	1100	160	180	40	100	3	60
Chao Phya	30	50	100	20	30	—	50
Tji Tarum	80	37	40	27	—	20	7

TABLE 5
Discharge of heavy metals by the water and by the suspended matter of the Rhine

	In the water (tons/year)	Fixed to sediments (fraction < 16 μm) (tons/year)	Metal in water/ metal in sediments
Pb	695	1830	1:2.6
Cr	1250	2820	1:2.3
Cu	765	1355	1:1.8
As	375	500	1:1.3
Hg	42	53	1:1.3
Cd	125	105	1:0.8
Zn	11380	6705	1:0.6
Ni	765	235	1:0.3

suspended material. So the transport of these metals mainly takes place in a solid form.

Similar results have been found by Gibbs[8] for the Amazon and Yukon rivers, where the ratios were even more extreme.

Mobilization of metals in the tidal area of the delta

As long as a river does not undergo the influence of the sea, the metals remain fixed to the suspended matter. Downstream of the freshwater tidal area of the river, however, a number of metals are mobilized to a greater or lesser extent, going into solution into the surrounding water, partly as organo-metallic complexes.[4, 5, 9] For the Rhine these mobilization processes, measured according to the conditions in 1960, have been demonstrated in Fig. 4. Compare Fig. 2 for the relevant localities.

There are large differences among the several metals in their degree of mobilization, varying from more than 90% of the total quantity originally present in the sediment to complete immobility. The most striking elements in this respect are cadmium and mercury, followed by copper, zinc, lead, chromium, and arsenic. Mobilization at an intermediate level takes place with nickel. The elements lanthanum, scandium, samarium, and manganese are not subject to solubilization processes. The usefulness of manganese and also of lanthanum, scandium, and samarium for sediment transport studies is emphasized by these observations.

The main cause of these mobilization processes is the intensive decomposition of the organic matter in the sediments from the freshwater tidal area, especially as far as the mouth of the estuary. This organic matter also changes in its carbon/nitrogen ratios (the Biesbosch, Haringvliet, and Wadden Sea are 21, 14, and 11, respectively). In laboratory experiments, evidence has been obtained that decomposition products of the organic matter form soluble organometallic complexes with the metals from the suspended matter. The degree of mobilization depends on the stability constant of the metal under consideration with the organic ligand (the group in a molecule able to form a metal complex). For some metals (copper and mercury) the mobilization can be promoted by the possibility of forming stable complexes with both positively and negatively charged organic ligands. Finally, inorganic ions also, such as Cl^-, can play a role in the mobilization processes.[5] It is well known in this respect that mercury forms very stable complexes with Cl^- ions (e.g. $HgCl_2$ and $HgCl_4^{2-}$). The stability of these complexes can compete with several organic mercury compounds. The mobilization processes in the Rhine distributaries are so intensive that great quantities of a number of metals in the sediments from the upper reaches of the estuary are reduced to normal quantities in the lower courses of this area.

For the Ems and the Scheldt the processes occurring from the freshwater tidal area to the deposition area in the Dollard are similar to those in the Rhine, as far as the solubilization of the metals is concerned, apart from the fact that the Ems sediment does not contain an excessive amount of heavy metals.

A number of investigations have been carried out to characterize the organic compounds responsible for the mobilization of the metals. Freshly deposited sediments from the freshwater tidal regions of the Rhine and Ems were incubated with distilled water and from the dissolved organic matter the fulvic and humic acid fractions were isolated according to Kononova[10] (fulvic acid soluble in acid and alkali, humic acids insoluble in acid and soluble in alkali). A calculation based on iron and organic-matter contents of the fractions pointed out that the fulvic acid fraction is mainly responsible for the metal mobilization.

The fulvic acid fractions subsequently have been subjected to gel filtration column-chromatography on Sephadex and to functional group analysis.

With respect to the molecular weight distribution it can be reported that the fulvic acid fraction is composed of three groups: one having molecular weights less than 1000; the second, from 1000 to 10,000; and the third, larger than 10,000. Only the second group, which comprises by far the greatest part of the fulvic acid fraction, possesses metal-

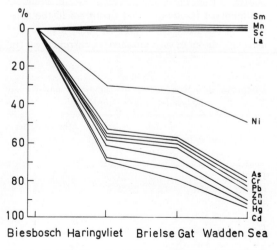

FIG. 4. Mobilization of metals in the Rhine estuary, North Sea, and Wadden Sea, expressed as percentages of the original contents (1960).

chelating properties. The major oxygen-containing functional groups in these compounds were carboxyls, phenolic hydroxyls, and carbonyls.

A comparison between the fulvic acid fractions of Ems and Rhine sediments demonstrated further that the Ems has a greater mobilizing capacity than the Rhine. In this connection it can be remarked that the fulvic acids responsible for the mobilization show some differences for the two river systems. For the Rhine these fulvic acids have a predominantly aliphatic character, while the mobilizing compounds of the Ems largely proved to consist of phenols, the latter being the result of the peaty character of the drainage basin of this river.

In a preceding section it has been reported that generally the contents of heavy metals in deposits from tropical rivers are low. Another important difference compared with the Rhine, Scheldt, and Ems is the behavior of the metals in the suspended matter on their way from the freshwater tidal region of the river to the sea. For the River Chao Phya it was found that no mobilization of the heavy metals took place. The contents of all the metals in the sediments remained constant. The cause of this immobility is the lack of organic matter in such a river. Even in the freshwater tidal region of the river, the organic contents of the sediments are very low and no significant decomposition takes place on the way to the marine area.

HEAVY METALS IN RHINE AND COASTAL WATERS

The heavy metals, separating from the suspended matter as a consequence of mobilization processes, are solubilized into the water. For the freshwater tidal area this means that the water is burdened with these metals, almost without dilution. In the brackish-water tidal area, however, dilution with salt water from the sea takes place. So the water in the mouthing area of the Rotterdam harbour contains about 75% of sea water.

At sea a further mixing with the water of the North Sea takes place. Under the influence of the tides and the wind, the sea water is constantly in motion. Through the Straits of Dover 1500 to 2000 km^3 of water enters the North Sea annually.[11] This amount of water moves at an average speed of 0.05–0.1 m/s along the eastern shores of the North Sea.

The velocity with which dissolved substances are spread throughout the coastal waters of the North Sea can be derived from some tracer experiments.[12] It was found that an injection of 1 kg of tracer was spread in 12 h over an area of about 1 km^2 with a maximum concentration of 10^{-6} kg/m^2. This concentration further diminishes with time (estimated range t^{-2} to t^{-3}), whereas the surface of the area increases proportionally.

The dilution of the river water (Rhine water, 75 km^3/year) with the sea water (the latter coming in via the Straits of Dover) will not be smaller than the ratio of their discharges 75 : 1750 = 1 : 23. After a few weeks a further mixing with the total water mass of the North Sea takes place.

The large amounts of metals added to the water in consequence of mobilization processes are not found back in the coastal water in the North Sea, as may be seen from Table 6. The strong dilution of the river water with sea water is responsible for this.

The metal contents in the North Sea water are generally much lower than in the river water (Table 6). The high copper contents in the coastal water may be the result of dumping in the investigated area.

INFLUENCE OF CIVIL ENGINEERING PROJECTS ON THE BEHAVIOR OF HEAVY METALS IN SEDIMENTS

The mobilization of heavy metals, as has been described in a previous section, is related to the existence of the tidal effects in the fluvial and marine regions, especially in the estuaries of rivers. The intensive decomposition of the organic matter in estuaries will be caused by an optimum microbial activity in these fresh- and brackish-water tidal areas.

TABLE 6
Metal contents in filtered water, expressed in µg/l

Location	Ref.	Zn	Cu	Cr	Ni	Pb	Cd	Hg
Rhine		164	11	18	11	10	1.8	0.6
North Sea (coast)		31	21	—	—	—	0.9	0.1
North Sea	13	6	6	—	—	2	0.3	0.1
Wadden Sea		30	2	—	1	2	<0.1	—
Sea water (mean value)	14	5	3	1	2	<0.1	<0.1	<0.1

Large civil engineering projects are carried out in the mouthing areas of the Rhine in connection with the well-known Delta Plan. This means that by the end of 1970 the action of the tides was stemmed in one of the most important Rhine outlets—the Haringvliet. In several respects the hydrology is disturbed as a consequence of such actions.

Such a situation has existed even longer in Lake IJssel, which was separated 40 years ago from the Dutch Wadden Sea by the construction of the Enclosure Dike. Under these circumstances, 10% of the Rhine water is flowing into Lake IJssel without passing of a tidal area.

In the following we pay some attention to the influence of the above-mentioned projects in the behavior of the heavy metals in Lake IJssel, the Delta area (main mouthing area of the Rhine), and the harbors of Rotterdam.

Lake IJssel

Since the origin of Lake IJssel (after the enclosure of the former Zuider Sea, 40 years ago) the main deposition area of the river IJssel is the Ketelmeer, directly in the vicinity of the mouth of the river. Recently, the heavy metal composition of the Rhine deposits in the Ketelmeer has been examined. The results of these investigations, compared with the composition of the original Rhine, have been given in Table 7.

From Table 7 it is obvious that the composition of the freshly deposited material (superficial layer) in the Ketelmeer and even of the underlying more consolidated sediments is very close to the Rhine sediments deposited in 1960 and 1970. There is no evidence of any mobilization process. The lead and cadmium contents in the Ketelmeer deviate somewhat from those in the Rhine river. In this connection we should be aware of the fact that the lead contents of the Rhine sediments have been further decreasing since 1970; the cadmium contents, on the other hand, still showed a considerable rise from 1970 to 1972.

Any suspended matter from the River Yssel which is not deposited in the Ketelmeer is spread out over Lake IJssel. Probably this material is settling out in the former tidal channels of the lake. Only slight amounts of the suspended matter escape through the sluices of the Enclosure Dike to the Wadden Sea.

Delta area

At the end of 1970 the Haringvliet was closed by a dam, as a part of the Delta Plan, thus stemming the action of the tides in a second important Rhine distributary.

The first consequence of this enclosure dam is that large quantities of mud, originally discharged via the Haringvliet to the North Sea, are now settling in the area between Biesbosch and Haringvliet. The experience obtained in the Ketelmeer gives rise to the expectation that the heavy metals in these deposits are also not mobilized as a consequence of the stemming of the action of the tides. At this moment a good deal of our effort is focused on the study of these sediments.

Rotterdam harbors

Originally there were two main outlets of the Rhine—the Haringvliet and the Nieuwe Waterweg (compare Fig. 2). The latter distributary flows through the Rotterdam harbor area. Since the enclosure of the Haringvliet, the main part of the Rhine water is discharged via the Waterweg. As a consequence of the great depth of the harbors, intensive siltation processes occur. The deposited material

TABLE 7
Heavy metals in Ketelmeer deposits (1972), expressed as percentages of the Rhine deposits in 1970

	Zn	Cu	Cr	Pb	Cd
Rhine, 1960	130	91	80	118	52
Rhine, 1970	100	100	100	100	100
Ketelmeer (superficial layer)	114	92	86	81	115
Ketelmeer (consolidated layer)	112	86	87	81	104

TABLE 8
Heavy metals in dredged sediments from the Rotterdam harbors (1972), expressed as percentages of the Rhine deposits in 1970

	Zn	Cu	Cr	Pb	Cd	Dredged quantities (Mton/year)
Inner harbors	74	74	70	68	79	0.4
Intermediate harbors	45	41	35	38	42	2.1
Outer habors	12	9	15	12	6	1.0

contains more, or less, heavy metals, depending on the location of the relevant harbor. In consequence of the siltation, large-scale dredging operations are necessary. A review of the quantities of dredged sediments and the contamination of these materials has been given in Table 8.

From Table 8 it is apparent that the degree of contamination depends on the measure of exposure of the relevant harbors to effects of the sea. The mud of the inner harbors must be regarded as severely contaminated, although the contents of the heavy metals are lower than farther upstream in the Rhine. The sediments from the outer harbors, on the other hand, must be regarded as uncontaminated. The low contents of heavy metals in the latter sediments are partly caused by mobilization processes of the heavy metals from the suspended matter in the river. Introduction of uncontaminated material from the sea into the mouthing area of the Nieuwe Waterweg may also play a role.

At this time an exact discrimination between materials of fluvial and marine origin in the Rotterdam harbor area is not yet possible. We are convinced, however, that the main portion of the deposited material comes from the Rhine.

For the dredged materials, deposition areas have to be found. The mud is deposited either at sea (about 10 miles outside of the harbor mouth) or on the land. The latter deposits are later used for suburban development or are used as arable land. Insofar as the deposits are contaminated with heavy metals and other pollutants (chlorinated hydrocarbons and oil) serious problems arise for the disposal of these materials. Much attention has to be paid to these problems in the near future.

PERSPECTIVES FOR FURTHER INVESTIGATIONS

Some important questions for further investigations in the near future on the occurrence and behavior of heavy metals in estuary regions and the nearby coast involve deeper insights into the mechanisms of the mobilization processes and into the sorption of metals to different kinds of suspended matter.[15] Although it is beyond the scope of this paper and we will not go into further details on this subject, we mention the severe gaps in our knowledge of the uptake and accumulation of the heavy metals in the biosphere.

Mobilization processes

The mobilization of metals, especially in the estuarine part of the river, is one of the most striking phenomena found in some estuaries of the temperate climatic region. It is doubtful, however, if these processes occur generally in estuaries around the North Sea. It is important to assemble evidence on the existence of estuaries contrasting in this respect, and to carry out intensive investigations in these estuaries.

Detailed investigations on the processes of mobilization of the metals may also throw more light on the constitution of organic and also inorganic metal compounds in the water of the various coastal regions. Also important in this respect is knowledge of the fates of trace materials after they have been solubilized (so far only iron has been studied in our investigations, and it again adheres to solid substances outside the estuarine region). These undeveloped fields of research must be considered to be of prime importance for studies of bioaccumulation of heavy metals.

Metal sorption on suspended matter

For bottom sediments, taken from the same location, a linear relationship is always found between the contents of the heavy metals and the fraction of particles less than 16 μm (Fig. 3). The granulometric composition was determined after destruction of the organic matter.

There are other factors which influence the metal contents of suspended material. In the first place, the type of the suspended matter depends on the velocity of the river currents. Under normal flow conditions of the river the main component of the suspended matter is the so-called original mud. For the Rhine this material originates mainly from its source area. It is this type of material, gathered under low tidal conditions as a freshly deposited, thin, superficial layer in estuarine and marine areas, from which the general conclusions in this publication have been drawn. But earlier it was mentioned that under conditions of high water discharge the erosion of the river bed gives rise to the formation of a suspension with different sedimentary characteristics, referred to as erosion mud. In consequence of processes of pedogenesis on the river bottom, this material forms aggregates of sufficient size to easily settle again when the high flow velocities are reduced at the place where the river enters the area of the flood plains. The latter is still upstream with respect to the freshwater tidal region.

The original mud, as it was collected from the main deposition areas in estuarine and coastal regions, has been regarded for many years as representative of the material present as a suspension in the river water under normal flow conditions.

However, it has been pointed out recently that the suspended matter contains components which do not settle out within the normal sedimentation areas. These components have a higher content of organic matter than those settling out normally. It is even possible that at least a part of these components never settles out at all.

The characteristics of the different forms of this suspended matter, apart from the granulometric composition after destruction of the organic matter, are of the highest importance. Besides differences in sedimentation characteristics, there exist appreciable differences in contents of heavy metals. For the Rhine it has been found that the metals supplied to the stream by industrial wastes adhere to the solids in decreasing contents in the following order: suspended matter, original mud (the main component settling in shallow areas), and erosion mud. The distribution of the solid substances in the environment diminishes in the same order (i.e., the erosion mud is deposited very locally on the river flood plains; a part of the suspended matter, on the other hand, moves very far). So those constituents which have been contaminated in the most severe way have the most widespread distribution in the aquatic environment.

A more detailed insight into the relations between contamination characteristics and physical properties of the solid constituents of the aquatic environment is urgently needed, especially in view of processes of bioaccumulation.

SUMMARY

In addition to their natural content of heavy metals, sediments transported by some rivers carry large amounts of these elements resulting from pollution.

The distribution of these elements between the suspended matter and the surrounding water, upstream with respect to the estuary, indicates for a number of elements a preferred adsorption to the solid phase. The composition of the fine-grained material from the rivers influencing the Netherlands delta (mainly the Rhine) is described, as well as that from some tropical river systems.

Under temperate climatic conditions the sediments on their way to the sea can undergo changes in their metal composition by mobilization of these elements as soluble metal-organic complexes. These processes lead to less-contaminated sediments in the lower courses of the delta. Special attention is paid to the influence of civil engineering projects (enclosure of river outlets) on these processes.

Finally, a treatise is given on the pollution with heavy metals of the diverse types of suspended matter. Dependent on the velocity of the river currents, different types of suspended matter are transported. These types differ in their sedimentation characteristics. It appears that those constituents, which have been contaminated in the most severe way, have the most widespread distribution in the aquatic environment.

REFERENCES

1. GROOT, A. J. DE, Origin and transport of mud in coastal waters from the Western Scheldt to the Danish Frontier, in *Developments in Sedimentology* 1, 91–103, Amsterdam, 1964.
2. GROOT, A. J. DE, ALLERSMA, E., BRUIN, M. DE, and HOUTMAN, J. P. W. Cobalt and tantalum tracers measured by activation analysis in sediment transport studies, *Isotope Hydrology*, IAEA Vienna, pp. 885–898, 1970.
3. GROOT, A. J. DE, ALLERSMA, E., BRUIN, M. DE, and HOUTMAN, J. P. W. Use of activable tracers, Technical Reports Series No. 145, *Tracer Techniques in Sediment Transport*, IAEA Vienna, pp. 151–166, 1973.
4. GROOT, A. J. DE, ZSCHUPPE, K. H., BRUIN, M. DE, HOUTMAN, J. P. W., and AMIN SINGGIH, P., Activation analysis applied to sediments from various river deltas, *Proc. 1968 International Conference on Modern Trends in Activation Analysis, Gaithersburg (USA)*, pp. 62–71, 1968.
5. GROOT, A. J. DE, GOEIJ, J. J. M. DE, and ZEGERS, C., Contents and behaviour of mercury as compared with other heavy metals in sediments from the Rhine and Ems rivers, *Geologie en Mijnbouw* 50, 393–398 (1971).
6. WOLLAST, R., Discharge of particulate pollutants in the North Sea by the Scheldt, *North Sea Science, NATO-conference (Aviemore, Scotland)*, working papers II, 9 pp., 1971.
7. HOUTMAN, J. P. W., Trace element behaviour in soil of some Indonesian sawahs and in sludge of an Indonesian river. *Delft Progress Report* 1, Ser. A, pp. 5–16 (1973).
8. GIBBS, R. J., Mechanisms of trace metal transport in rivers. *Science* 180, 71–73 (1973).
9. GROOT, A. J. DE, Mobility of trace elements in deltas, *Transactions, Commissions II and IV, International Society of Soil Science, Aberdeen*, pp. 267–279, 1966.
10. KONONOVA, M. M., *Soil Organic Matter*, Pergamon Press, New York, 58 pp., 1966.
11. FAIRBRIDGE, R. W., *The Encyclopedia of Oceanography*, New York, 1966.
12. DAM, G. C. VAN, *Dispersie van opgeloste en zwevende stoffen in zee gebracht ter hoogte van Wijk aan Zee op 3 km uit de kust*. Rijkswaterstaat (Netherlands Public Works for Roads and Waterways), Report MFA 6812, 1968.
13. ELSKENS, I., *Les métaux et les métabolites de pesticides. Programme national sur l'environnement*

physique et biologique. Project mer, Modèle mathématique, Rapport de synthèse (*Luik*), pp. 123–137, 1971.

14. RILEY, J. P. and CHESTER, R., *Introduction to Marine Chemistry*, New York, p. 465, 1971.

15. GROOT, A. J. DE, Occurrence and behavior of heavy metals in river deltas, with special reference to the Rhine and Ems rivers, in *North Sea Science*, pp. 308–325 (MIT Press), 1973.

Field Observations on the Transport of Heavy Metals in Sediments (A. J. de Groot and E. Allersman)

DISCUSSION by I. R. JONASSON
Geological Survey of Canada, Ottawa, Canada

and

M. H. TIMPERLEY
Chemistry Division, DSIR, Petone, New Zealand

INTRODUCTION

We shall begin this discussion by defining some arbitrary classifications which serve to describe some of the various geochemical regimes of a contaminated river such as the Rhine. This exercise is solely for the purpose of providing a basis for our discussion so that we can treat in turn heavy metal dispersion in what might otherwise seem to be unrelated situations.

CATCHMENT REGIME

The first regime is that of the source or catchment basin of the river, from where most of the suspended load which courses the entire river system is derived. It is here that the river collects what may be referred to as its natural load of heavy metals; terms in more popular usage which mean much the same thing, are "background load" or "baseline level." In this regime, the geology of the enclosing rock formations determines the chemical, and, to a large extent, the physical nature of both the suspended load and the coarser river bottom sediments.

If mineralization enriched in heavy metals is present in the catchment area, then it is likely that it will be reflected by enhanced metal levels in the sediments and water. So it is no coincidence that many of the studies made under these circumstances often involve geochemists in search of economic deposits of base metals.

The most extensive study of the migration and deposition of heavy metals in catchment basins which comes to mind is that of our colleagues, Allan et al.[1] who sampled silty lake sediments (<63 μm fraction) over a 90,000 km^2 area of the northern Canadian Shield. They were able to show that the major element distribution of the fine sediment fraction was very similar to that of the surrounding rocks, and that the trace metal contents showed broad regional patterns which were more intense in areas of known mineral potential. Maps of these element distribution patterns have either been published or are in preparation for more than 20 elements (see *Geol. Surv. Can., Prelim. Maps*, 9–15, 1972). In areas where unknown mineralization may be proved to be present, some sediments contained up to 500 ppm of zinc, copper, and uranium as well as high amounts of arsenic. These values surely represent a high degree of what one might term natural contamination of these arctic zone lakes and their feeder streams.

Other similar survey programs have been carried out by Arnold[2] in northern Saskatchewan and by Nickerson[3] in the Yellowknife district, NWT. This latter study is particularly interesting to environmentalists because it is possible to establish from it baseline data for the widespread As contamination found in the lakes and streams near Yellowknife itself. Arsenic in sediments from areas removed from industrial activity may reach 300 ppm compared with contamination levels of up to 300 ppm in the vicinity of mining operations and smelters.

The most recent work along these lines is being carried out by Hornbrook et al.[4] who are currently sampling lake and stream sediments across the entire province of Newfoundland, and by Cameron and Durham[5] who are looking into the problems of heavy metal dispersion in some mineralized, permafrosted areas of the Canadian Arctic barren lands.

The movement of heavy metals in streams and in snow-melt waters in the Arctic exhibits a number of peculiarities unique to permafrost terrain. For example, once the spring thaw is completely finished, little trace of base metals such as zinc, copper, or mercury, remains in the stream-bed materials. The accumulation of salts, such as sulfate and carbonate, in winter snow cover and in the active zone of the permafrost, ensures that these metals are carried away in soluble form very rapidly to the lakes where they are adsorbed by fine silts and clays. There is little organic matter involved in these transport processes except perhaps in the case of uranium dispersion. Elements such as arsenic, lead, and silver, which are physically trapped in iron oxide precipitates and crusts, survive in the stream sediments in much greater proportion and are therefore more useful in geochemical prospecting.

We have been involved over the last 3 years with fallen-snow sampling, which, in Canada at least, represents an approach to heavy metal mobilization at the first level.[6,7] There is now little doubt that metals move upward from soils to snow over the course of the winter, even in permafrosted ground, and are then washed into streams in spring. We have established that metals which move in cationic form can concentrate in fallen snow to high ppb (μg/l) levels (e.g. copper, silver, mercury, zinc, bismuth, and nickel) but to date we have found no significant quantities of anionic forming species such as molybdenum, uranium, arsenic, or antimony, even in mineralized areas where they are known to occur in quantity in the host rocks and residual soils. A short study, now almost complete, of the forms of copper in fallen snow, i.e., copper which has moved upward from the soils beneath, indicates that copper is present almost exclusively as the aquocupric ion. The dominant controlling factor appears to be snow pH, which is usually around 4, and may well be due to considerable dissolved CO_2 at 0°C.[8]

The study by Reeder et al.[9] of the hydrogeochemistry of the surface waters of the Mackenzie River drainage basin in the Northwest Territories of Canada, looks into one of the next stages of heavy metal transport in the Arctic. The waters of this system are virtually unpolluted. Although the study is not detailed, it does at least indicate that the dissolved metal loads are quite characteristic of the geological nature of the enclosing rocks. For example, the average content of uranium (0.5 μg/l) is higher than usual for such a river system and undoubtedly represents a steady contribution from the uranium-rich rocks around Great Bear Lake.

The comment by Dr. de Groot on the high mercury content of the River Tji Tarum sediment in Java (20 ppm) is very interesting to a geochemist. Experience suggests that this very high value is not so much attributable to volcanic rocks as to the mercury mineralization which is likely contained in them. In the absence of such mineralization one might expect mercury contents of volcanic rocks to reach less than 1 ppm,[10] so it seems highly probable, especially in the absence of complexing organic matter, that mercury sulfide, in the form of cinnabar, may be present in the fine silts.

A study by Dall'Aglio[11] on hydrogeochemistry in the Mt. Amiata region of Tuscany (20,000 km^2) showed that cinnabar, which is quite resistant to oxidation but very easily pulverized, can be transported in river sediments at least 20 km before background levels are again recorded. In this study, background for mercury in sediments was at 0.05 ppm with most elevated values falling into the range 0.5–1.5 ppm and with local highs of up to 50 ppm. The unfiltered water samples ranged from a low of 0.01 μg/l up to highs of 15 μg/l in the vicinity of cinnabar mineralization. Water-borne cinnabar and perhaps dissolved ionic mercury were observed to persist for 20–30 km.

Mercury mineralization contained in similar volcanic rocks is known in Java, so the conclusion that primary sulfides may be responsible for the mercury levels recorded by Dr. de Groot is not unreasonable.

CONTAMINATION REGIME

The second regime may be broadly described as the contamination regime under which may be grouped man-inspired activities involving pollution from industrial, agricultural, and urban–domestic sources. In general, metal contributions to the river system from such industrial sources far outweigh any input from natural sources other than, perhaps, from those areas where extensive mining activity may be operative.

One of our recently completed studies[12] involved a survey of the potential input of Hg from base-metal (copper, lead, silver, and zinc) mining operations in the Noranda–Matagami–Chibougamau regions of northwest Quebec, and has demonstrated that if adequate precautions are taken with containment of pyritic tailings, then very little mercury enters the river drainage systems of the area (Bell and Nottaway rivers); in fact it is suspected that less mercury is derived from mining sources than from some of the common rock types of the area, such as Precambrian pyritic shales. Thus, the mere presence of sulfide ores which contain significant levels of mercury does not necessarily lead to contamination in nearby streams or groundwaters for the simple reason that these ore bodies are relatively small sources, are quite localized in extent, and can be virtually isolated from the natural drainage.

However, aside from such considerations on the effects of mining practices, the great complexity and variety of man's industrial efforts causes relatively clean sediments entering the contamination regime of a river to rapidly acquire a pollution load of heavy metals. Contrary to the situation within the catchment regime, where certain generalizations on the chemistry of sediments and the migration behavior of trace metals can be made quite legitimately in terms of known geology and ore occurrence, the chemistry of the contamination regime will be as complex and as varied as the individual sources of heavy metals. Relationships between certain metal pairs such as zinc and cadmium, zinc and mercury, or cobalt and nickel, for example, where common natural sources may occur, can be effectively studied in the uncontaminated headwaters of a river system; but similar studies of these metals from one particular industrial source may well be quickly obscured by the superimposition, interaction, and mixing of metals from other quite diverse industrial sources.

It is interesting to note in Dr. de Groot's paper that there seem to be few apparent genetic relationships between the different metals in the sediments. Could it be that the samples taken from estuarine sediments are too far removed from their sources to reflect common origins of certain metals? And can the converse argument be applied to this question— If the industrial sources are known and some data on the chemical forms in which the metals enter the river system are available, can they be traced to the estuary in such a way that the individual migration paths can be recognized and perhaps characterized? We would be interested to learn whether Dr. de Groot's very extensive data on the Rhine system can be used in this way.

ESTUARINE REGIME

The third regime which we arbitrarily define is that which covers the distributary and estuarine zones of the river system. These zones are characterized by intense interaction between seawater and river water and by the drastic changes in the chemistry of the heavy metals and of the transporting sediments. The work described by Dr. de Groot was carried out largely in this regime.

One of the most interesting features of his work is that the physical nature of the suspensions and sediments and the types of metal contaminants entering the study area have not changed markedly in character over the 10-year study period. The muds, according to Dr. de Groot, remain quite characteristic of their Alpine sources and continue to act as host carriers of heavy metals. What has changed, however, and quite dramatically, are the relative quantities of metals such as mercury, nickel, and cadmium, which have increased nearly twofold, and arsenic and zinc which have decreased significantly. The immediate inquiry which comes to mind is whether or not there is a viable explanation in terms of the changed nature of industrial activity.

The second point to note is that again there seem to be no obvious genetic relationships for like element pairs such as cadmium and zinc; in this instance one has increased, the other has decreased. Assuming that the respective chemical behavior between source and deposition, or, if you like, sink areas in the water sediment system is similar, and this is reasonable in view of the known simple chemistry and adsorption behavior of each, then is it likely that zinc and cadmium come from quite different sources? In support of the assumption of similar chemical behavior are the respective distribution ratios for zinc and cadmium between suspended matter and water (zinc 1:0.6, and cadmium 1:0.8).

Other pairs which may also be regarded as chemically similar in terms of their simple inorganic chemistry, e.g., copper and nickel, manifest greatly different distribution ratios which are indicative of quite different chemical behavior during their mobilization and dispersion processes.

The next point, which may be somewhat irrelevant to Dr. de Groot's study, concerns the oxidation–reduction characteristics of the Rhine water and its bottom sediments. Firstly, what if any, are the changes measured in oxygen content of the river water between the sources of industrial contamination and the estuaries? Secondly, are the bottom muds at all reducing (H_2S present) in the lower reaches of the river distributaries, particularly in the sediment deposition areas behind dams or tidal barrages such as exist on Lake IJssel and at the Haringvliet dam? It is likely that if the river bottom is at all reducing, then some sulfides will be present in the muds which would firmly fix most of the heavy metals of interest (mercury, copper, zinc, etc.). Have any of the estuarine and river bottom sediments been analyzed for sulfide sulfur, especially in cores? This question is particularly interesting in view of the intense remobilization processes described by Dr. de Groot for the heavy metals as they enter areas of tidal encroachment. Any metal sulfides contained in muds will be particularly vulnerable to dissolution when they come into contact with freshly oxygenated tidal waters, especially in the presence of increased chloride ion concentration.

In our own hydrogeochemical work we have observed similarly intense mobilization processes in small lakes (e.g. Perch Lake, Eastern Ontario) when the water level drops sufficiently to leave bottom silts and muds in an oxidizing rather than a reducing environment. It was quite noticeable that samples collected from the same locations a few weeks apart had lost more than half of their original contents of zinc, copper, lead, nickel, and arsenic from the sulfide-rich muds, and more than 80% of their original sulfur and mercury contents. Carbon content stayed roughly constant at around 35%. Specifically, sulfur decreased from about 2.5% to around 0.5% and mercury dropped from about 250 to 50 ppb. The other metals were originally present in amounts ranging from 4 to 150 ppm depending on the metal in question.

Some further work on these and some samples from other lakes[13] has attempted to determine the forms of binding of metals such as copper, iron, and zinc in reducing, organic-rich muds and gels. One of the main conclusions drawn was that the accumulation of copper was mainly by sulfide precipitation whereas organic complexing was more important to the binding of zinc. One might expect cadmium to behave like zinc, but that arsenic and mercury would be largely bound as sulfides. Preliminary work on these metals suggests that this is so; the controlling factor seems to be direct competition between sulfide and organic ligands for the metal, so some predictive value is possible on the basis of sulfide solubilities.

The point we wish to make from all of this discussion is that if sulfides are present in bottom muds, or for that matter, in what Dr. de Groot refers to as "erosion muds," then they will be rapidly oxidized and solubilized on contact with aerated sea-water or upon deposition on flood plains either naturally or by dredging. It could well be that the lower metal contents of flood plain muds from the Rhine are due in part to oxidative remobilization processes.

We are intrigued by the vigorous and extensive mobilization of As reported in the Rhine estuary. This element is not noted for its ability to form stable fulvates or humates (which means that we have not seen any data on stability constants), but it seems to be mobilized as effectively as copper or zinc. Could it be that again sulfides are involved which become oxidized to soluble arsenite, arsenate or their thio-analogues on contact with oxygenated tidal waters?

CONCLUDING REMARKS

We have attempted to draw upon, for discussion purposes, our own research in hydrogeochemistry and upon that of our colleagues in the Geological Survey of Canada. Most of what we have commented on is very recent work, much of it still in the pre-publication stage. We have also attempted to extract from Dr. de Groot's excellent work on the Rhine delta some related aspects, which in our view require further elaboration and which, perhaps, may indicate the need for further research. The overall objective is, of course, to spark a lively discussion of heavy metals in sediments and their respective pathways of migration.

REFERENCES

1. ALLAN, R. J., CAMERON, E. M., and DURHAM, C. C., Reconnaissance geochemistry using lake sediments of a 36,000 square-mile area of the northwestern Canadian Shield, *Geol. Surv. Can. Pap.* 72–50, 70 pp., 1972.
2. ARNOLD, R. G., The concentrations of metals in lake waters and sediments of some Precambrian lakes in the Flin Flon and La Ronge areas, *Geol. Div. Sask. Res. Council Circ.* 4, 30 pp., 1970.
3. NICKERSON, D., An account of a lake sediment geochemical survey conducted over certain volcanic belts within the Slave structural province of the Northwest Territories during 1972, Dept. Indian Affairs and Northern Development, Canada, *Geol. Surv. Can. Open File Rep.* 129, 24 pp., 1972.
4. HORNBROOK, E. H. W., DAVENPORT, P. H., and GRANT, D. R., Regional and detailed geochemical exploration studies in glaciated terrain in Newfoundland; to be published by the Province of Newfoundland and Labrador, Dept. Mines and Energy Publications.
5. CAMERON, E. M. and DURHAM, C. C., Follow-up investigations on the Bear-Slave Operation, *Geol. Surv. Can. Pap.* 74-1A, "Report of activities", pp. 53–60, 1974.
6. JONASSON, I. R. and ALLAN, R. J., Snow: A sampling medium in hydrogeochemical prospecting in temperate and permafrost regions, *Geochemical Exploration—1972*, Proc. 4th Int. Geochem. Explor. Symp. (London), Publ. 1973, Inst. Mining Met., London (Jones, M. J., ed.), pp. 161–176, 1972.
7. JONASSON, I. R., Migration of trace metals in snow, *Nature, Lond.* 241 (5390), 447–448 (1973).
8. TIMPERLEY, M. H. and JONASSON, I.R., Chemical forms of copper in snow, (paper in preparation, 1974).
9. REEDER, S. W., HITCHON, B., and LEVINSON, A. A., Hydrogeochemistry of the surface waters of the Mackenzie River drainage basin, Canada; I. Factors controlling inorganic composition, *Geochem. cosmochim. Acta* 36, 825–865 (1972).
10. JONASSON, I. R. and BOYLE, R. W., Geochemistry of mercury and origins of natural contamination of the environment, *Can. Mining Met. Bull.* 65 (717) (1972).
11. DALL'AGLIO, M., Comparison between hydrogeochemical and stream sediment methods in prospecting for mercury, *Geochemical Exploration*,

Proc. 3rd Int. Geochem. Explor. Symp. (Toronto), CIM Spec. Vol. 11, Publ. 1971 (Boyle, R. W. and McGerrigle, J. I., eds.), pp. 126–131, 1970.

12. MACLATCHY, J. E. and JONASSON, I. R., The relationship between mercury occurrence and mining activity in the Nottaway and Rupert river basins of Northwestern Quebec, Geol. Surv. Can. Paper Ser., in press, 1974.

13. TIMPERLEY, M. H. and ALLAN, R. J., The formation and detection of metal dispersion haloes in organic lake sediments, *J. Geochem. Explor.* 3, 167–190 (1974).

Discussion (I. R. Jonasson and M. H. Timperley)

SOME REMARKS by A. J. DE GROOT

With regard to the excellent discussion of my paper by Jonasson and Timperley I will comment upon their views on the mobilization of metals. The relatively low content of heavy metals in erosion muds may well be caused by the accumulation of sulfides in the bottom sediments and subsequent loss of metals as a consequence of reoxidation when these sediments have been taken up by the stream.

This theory, however, does not apply to the mobilization processes in the estuarine part of the river. All samples taken in these areas were freshly deposited materials with redox potentials generally between 400 and 500 mV (pH 7–8). Only in a few cases could these sediments be expected to have been moved from former deposition areas with reducing conditions. This could be checked with the help of manganese analyses in these sediments. A former deposition under reduced conditions would give rise to losses of this element (reduction of insoluble manganese oxides to Mn^{2+}-ions and subsequent exchange by ions of the surrounding water).

The view of Jonasson and Timperley on the accumulation of sulfides under reduced conditions may give rise to serious problems when dredging is undertaken. Huge quantities of reduced sediments from the Rotterdam harbor are deposited at sea, as has been mentioned previously. The consequences of introducing these reduced sediments into aerated water with concomitant release of heavy metals are now under serious consideration.

METABOLIC CYCLES FOR TOXIC ELEMENTS IN THE ENVIRONMENT

A STUDY OF KINETICS AND MECHANISM†

J. M. WOOD

Director, Freshwater Biological Research Institute, University of Minnesota, College of Biological Sciences, St. Paul, Minnesota

The evolution of metabolic pathways to cope with the naturally occurring toxic compounds, both for their synthesis and degradation, provides a balance for the natural levels of such compounds. The dynamic aspects of this problem cannot be overemphasized, because for the metabolism of toxic compounds one must be aware of the circumstances which lead to an increase in the *rate* of synthesis or release of toxic metabolites compared to the rate of their degradation or removal. The danger associated with naturally occurring toxic metabolites largely depends on their distribution in nature, which may or may not place them in a situation where they can pose serious public health problems. The toxic elements are undoubtedly the most difficult to study because background levels are always present in the environment, and a small perturbation on an ecosystem could bring about a rapid increase in the rate of synthesis of these compounds.

From an environmental pollution standpoint, the elements can be generally considered as (a) noncritical, (b) toxic and relatively accessible, and (c) toxic but very insoluble or very rare. Using these broad definitions potential pollution candidates can be defined as presented in Table 1.

The movement of toxic elements under natural conditions depends on the geocycle which causes natural Bio-leaks (e.g. volcano eruptions, etc.). However, man's activities do provide new sources for toxic elements which can dramatically influence the mobility and metabolism of these elements. The metabolic versatility of microorganisms plays a significant role in the mobility of toxic elements. Therefore it is extremely important to understand in detail the kinetics and mechanisms employed for interconversions of toxic elements, and how physical parameters such as pH, temperature, light, etc. influence the kinetics of these potentially dangerous reactions.

---

†This research was supported by Grants USPH AM 12599 and NSF GB 26593X.

TABLE 1
Classification of elements according to their toxicity

Designation[a]	Elements
Noncritical elements	Na, K, Mg, Ca, H, O, N, C, P, Fe, S, Cl, Br,[b] F,[b] Li, Rb, Sr, Ba, Al, Si
Very toxic and relatively accessible	Be, Co, Ni, Cu, Zn, Sn, As, Se, Te Pd, Ag, Cd, Pt, Au, Hg, Tl, Pb, Sb, Bi
Toxic but very insoluble or very rare	Ti, Hf, Zr, W, Nb, Ta, Re, Ga, La, Os, Rh, Ir, Ru

[a] Many elements are omitted because they cannot be designated in this way.

[b] Some may argue with this designation, but we do add F^- to drinking water.

For inorganic complexes it should be recognized that microorganisms have evolved enzymes capable of changing the oxidation state of elements. For example, methanogenic bacteria can reduce As^{+5} to As^{-3}, as shown in the 8 electron reduction of arsenate to dimethylarsine.[1] When an element is introduced into a microbial environment we have to assume that each valence state for that element is made available for metabolic interconversions. With mercury we must consider the following disproportionation at all times:

$$Hg_2^{2+} \rightleftharpoons Hg^{2+} + Hg^{\circ}$$

Vaporization of mercury metal (Hg°) shifts the equilibrium to the right, but return of Hg° from the atmosphere to the earth's crust shifts the equilibrium to the left. Interconversions between these three inorganic forms of mercury can be catalyzed by microorganisms. Aerobes can solubilize Hg^{2+} from HgS (solubility product 10^{-53} M) by oxidizing the sulfide through sulfite to sulfate.[2] Once Hg^{2+} has been solubilized it can be reduced to Hg° by an enzyme which is present in a number of bacteria, and this enzyme has been shown to require reduced pyridine nucleotide as a coenzyme for catalysis:[3]

$$Hg^{2+} + NADH + H^+ \rightleftharpoons Hg° + 2H^+ + NAD^+$$

The conversion of Hg^{2+} to $Hg°$ can be regarded as a detoxification mechanism because the $Hg°$ has sufficient vapor pressure to be lost from the aqueous environment into the vapor phase.

Some bacteria employ a second detoxification mechanism by converting Hg^{2+} to methylmercury and dimethylmercury. A survey of the methylating agents which are available for methyl-transfer reactions in biological systems reveals that there are three coenzymes which are known to be involved in this reaction: (1) S-adenosylmethionine, (2) N^5-methyltetrahydrofolate derivatives, and (3) methylcorrinoid derivatives. S-adenosylmethionine and N^5-methyltetrahydrofolate derivatives are not capable of transferring methyl groups to Hg^{2+} because for both of these coenzymes the methyl group is transferred as CH_3^+. Therefore, the only *known* methylating agents capable of methyl transfer to Hg^{2+} are methylcorrinoids because they are capable of transferring methyl groups as CH_3^-.[4-7]

carbonium ion donor such as S-adenosylmethionine, e.g.:

$$Hg^{2+} + NADH + H^+ \rightarrow Hg° + NAD^+ + 2H^+$$
$$Hg° + SAM \rightarrow CH_3Hg^+ + SAH$$
$$SAM = S\text{-adenosylmethionine}$$
$$SAH = S\text{-adenosylhomocysteine}$$

Recent studies have elucidated the details of the mechanisms for the synthesis of both methylmercury and dimethylmercury by the B_{12} catalyzed reaction.[7, 11, 12]

In 1968 two groups[4, 13] demonstrated the synthesis of methylmercury in biological systems. Jensen and Jernelov showed the synthesis of methylmercury in biologically active sludge.[13] Wood *et al.*[4] showed that methyl cobalamin was implicated in both the enzymatic and nonenzymatic synthesis of methylmercury and dimethylmercury.

When methyl-B_{12} is mixed with an excess of mercuric diacetate in acetate buffer at pH 5.0, then a rapid color change from red to yellow is observed followed by a slower reappearance of the red color. The spectrum of the yellow intermediate (λ_{max} 475 nm) is similar to protonated "base off" methyl-B_{12}. These spectral changes are illustrated in

Bz = 5,6 dimethylbenzimidazole, but this can be displaced by other bases. In fact, in the B_{12} enzymes, bases in the protein can increase the electron density on CH_3, which facilitates its displacement as CH_3^-.[8] Landner[9] has shown that *Neurospora crassa* is capable of the synthesis of methylmercury from mercuric ions. It is generally believed that *Neurospora* does not contain B_{12} enzymes,[10] and so some alternative mechanism for methylmercury synthesis must be present in this genus. Landner[9] found evidence for the involvement of the methionine biosynthetic pathway and he suggests the formation of a homocysteine–Hg^{2+} complex as a prerequisite for methyl-transfer to give homocysteine–Hg–CH_3. If this mechanism is correct then a yet undiscovered methyl-transfer enzyme must be present which is capable of transferring CH_3^-, and this carbanion transfer must be B_{12} independent. Alternatively, in *Neurospora* a tightly coupled series of enzymes may be present, probably in the cell membrane, which reduce Hg^{2+} to $Hg°$ followed by CH_3^+ transfer to the metal by a

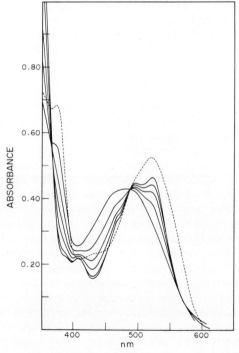

FIG. 1. Changes in the visible spectrum of methyl-B_{12} (6.95×10^{-5} M) upon the addition of $Hg(OAc)_2$ (15.0×10^{-5} M). Methyl-B_{12} (····); solid lines are scans in ascending order at 525 nm at 5 sec and 2, 4, 6, 8 min after mixing.

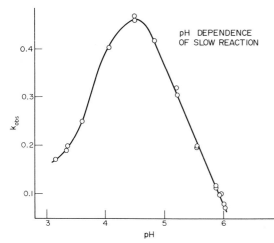

FIG. 6. The pH dependence of the observed rate constant for the demethylation reactions (reactions 2 and 3).

converts Hg^{2+} to Hg_2^{2+}. The NMR spectrum presented in Fig. 7c shows that no mercuric species is coordinated to benzimidazole at the end of the reaction since the C_{20} methyl resonance at -0.55 ppm indicates that benzimidazole is coordinated to the cobalt and the remainder of the spectrum is identical to that of aquo-B_{12} at the same pH.[11,14]

One point of minor importance is the fact that methylmercury itself reacts with methyl B_{12} to give dimethylmercury as the product. Therefore, reactions 1–3 theoretically could be written for methylmercury as well as for mercuric acetate. However, the reaction with methylmercury is several orders of magnitude slower than the reaction with mercuric acetate, and, therefore, this reaction is of minor significance in the kinetic experiments under our reaction conditions.

These experiments contrast sharply with those published by Imura et al.[15] who claimed that dimethylmercury is the first product of this reaction, and that dimethylmercury then reacts with excess mercuric ions to disproportionate to give monomethylmercury.

The choice of mercuric diacetate as the attacking agent in the study of this reaction was made as this is a suitably soluble mercury salt, but it also has the advantage that it could be similar to possible attacking agents in the bacteria from sediments.[16] Weak acids, such as acetic, propionic, and butyric, are common fermentation products of these bacteria. There is no mercuric salt which is free of all complexation and dissociation equilibria and the necessity of working in a buffered medium to control pH and ionic strength is unavoidable.

In this kinetic study we have shown that displacement of 5,6-dimethylbenzimidazole from the coordination sphere of the cobalt atom in methyl-B_{12} strengthens the Co–C bond to make electrophilic displacement by Hg^{2+} difficult. However, electrophilic attack by Hg^{2+} on methyl-B_{12} molecules which have not formed a "base off" complex with Hg^{2+} (or H^+) causes the equilibrium K_1 to shift to the left to provide a fresh source of substrate for the rapid synthesis of methylmercury. Furthermore, we have shown the methyl-B_{12} and $Hg(OAc)_2$ dependence of k_{obs} and the pH rate profile in terms of a mechanism involving electrophilic attack by $Hg(OAc)_2$ on the Co–C bond. The equilibrium K_1 and the demethylation reactions (reactions 2 and 3) are first order in Hg^{2+} and in methyl-B_{12}.

Recent studies on the B_{12}-enzymes which are involved in methyl-transfer reactions show that thiol groups (i.e. cysteine residues) are important in facilitating methyl-transfer.[8] In a number of cases, the reaction of ϕ-chloromercuribenzoate (PCMB) with B_{12}-apoenzymes blocks B_{12}-binding so that active holoenzymes are not formed. We have shown that sulfur can coordinate to the cobalt in place of benzimidazole, to increase the electron density on the methyl carbon, and decrease the stability of the cobalt–carbon bond.[8] These observations explain why electrophilic attack by metals such as Hg^{2+} can be quite rapid in the enzymes. The sulfur coordinated to the cobalt atom is in a protected environment to the extent that it cannot react with the Hg^{2+}, e.g.

$$\begin{array}{c}CH_3\\|\\Co^{+3}\\|\\Bz\end{array} + R-SH \rightleftharpoons \begin{array}{c}CH_3\\|\\Co^{+3}\\|\\S\\|\\Bz\quad R\end{array} + Hg^{2+} \longrightarrow \begin{array}{c}CH_3Hg^+\\\\Co^{+2}\\\uparrow\\Bz\end{array} + RS^- + H^+$$

R = B_{12}-apoenzyme.

The pH optimum for methyl-transfer to Hg^{2+} to give methylmercury either under laboratory conditions or in the sediments from Lake St. Clair is 4.5.[16] At more basic pH conditions methylmercury disproportionates to give dimethylmercury and Hg^{2+}. Dimethylmercury is volatile and once in the atmos-

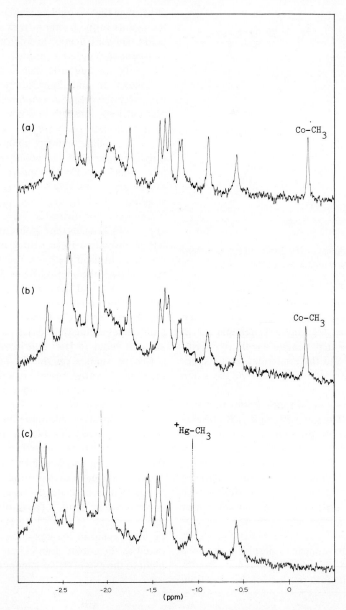

FIG. 7. 220 MHz NMR spectra of the high field portion of the methyl-B_{12} spectrum in D_2O: (a) methyl-B_{12}, (b) methyl-B_{12} + Hg(OAc)$_2$ + ascorbic acid, (c) methyl-B_{12} + Hg(OAc)$_2$. The products observed in spectrum (c) are aquo-B_{12} and methylmercuric acetate. The large resonance at -2.1 ppm is the methyl resonance of acetate.

phere it is photolysed by UV light to give methane plus ethane and Hg°, i.e.

$$(CH_3)_2Hg \xrightarrow{h\nu} Hg° + 2CH_3^{\cdot}$$

Methylradicals can abstract hydrogen atoms to give methane, or couple to give ethane.

In addition to those microorganisms which can synthesize methylmercury there are those which can detoxify their environment by reducing methylmercury to Hg° plus methane. This detoxification reaction converts the neurotoxin methylmercury to the less toxic and more volatile Hg°.[3]

From the mercury example it is clear that biological cycles exist for toxic elements. This mercury cycle is summarized in Fig. 8. These interconversions of mercury set up a dynamic system which

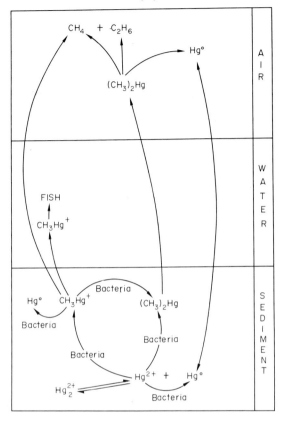

FIG. 8. The mercury cycle.

leads to a steady state concentration of methylmercury in sediments. These steady state concentrations of methylmercury need not reflect the rate of synthesis of methylmercury, and so determination of the concentration of methylmercury in sediments is in this case a meaningless exercise. The parameters which are most useful to assess a particular mercury pollution situation are the concentration of *total* mercury in the sediments, and the *rate of methylmercury* uptake in fish.

From our experience with methylmercury we can predict which other heavy metals can be transformed in a similar way. For example, using the same approach it can be predicted that tin, platinum, gold, and thallium will be methylated in the environment, but lead, cadmium, or zinc will not be methylated, not only because the latter alkyl-metals are not stable in aqueous systems, but also mechanistically the methyl group cannot be transferred to these elements. The above predictions have proved to be correct.[17] Cadmium does form coordination complexes with a variety of ligands in biological systems, but no metabolic interconversions have ever be demonstrated for this element. The heavy metals which are methylated should be watched closely. The replacement of lead compounds by tin compounds as gasoline additives should be of particular concern.

In addition to the biological cycle for the synthesis of methylmercury, under special circumstances, chemical synthesis of methylmercury can upset the balance of the cycle. DeSimone has shown that water soluble methylsilicon compounds react with Hg^{2+} to give methylmercury.[18]

$$(CH_3)_3\text{—}Si\text{—}(CH_2)_2\ SO_2\ Na + Hg^{2+} + 2\ OAc^- \rightarrow$$
$$(CH_3)_2\text{—}Si\text{—}(CH_2)_2\ SO_3\ Na + [CH_3\ Hg^+][OAc^-]$$
$$\underset{OAc}{|}$$

Also, Jernelöv has shown that contaminating methyl groups in ethyl-lead compounds lead to the chemical synthesis of methylmercury from Hg^{2+}.[19] Furthermore, the chemical synthesis of methylmercury as a fungicide does contribute to upset the balance of the mercury cycle. When considering the synthesis of neurotoxins such as methylmercury one should always recognize that these are natural biosynthetic processes which are influenced by a variety of parameters. For example, the rate of synthesis of methylmercury depends on the concentration of available Hg^{2+}, the microbial population, the pH, temperature, redox potential, and the synergistic or antagonistic effects of other metabolic or chemical processes.

The metalloids

By looking at the Periodic Table together with a current knowledge of the properties of the toxic elements one can predict how most of these toxic elements should behave in the environment. Arsenic compounds are reduced and methylated by anaerobes to give dimethylarsine and trimethylarsine as volatile products of extreme toxicity.[1,20] Fortunately, these arsines are readily oxidized in the aerobic environment to give products less toxic such as cacodylic acid. However, cacodylic acid has been shown to be an intermediate in the synthesis of dimethylarsine from arsenic salts. Therefore, a natural cycle exists for arsenic just as with mercury (Fig. 9). Alkylarsenic compounds have been found to accumulate in shell fish analyzed in Norway.

On the basis of our understanding of arsenic chemistry, similar metabolic reactions are predicted and found to occur for selenium, tellurium, and sulfur.[1]

Clearly, with such a dynamic system of cycles for

The Arsenic cycle

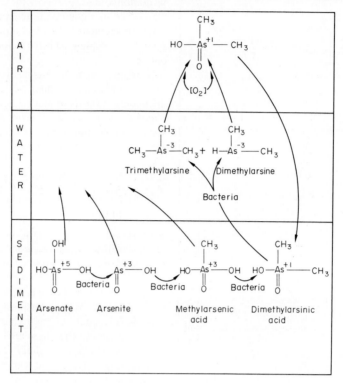

FIG. 9. The arsenic cycle.

the toxic elements, small perturbations of these cycles can change the equilibria involved to change the concentrations of toxic intermediates. Great care must be taken when deciding which species to monitor in the environment, and neglect of these biochemical transformations can often make the development of models for the flow of chemicals through the environment a futile exercise.

REFERENCES

1. McBride, B. C. and Wolfe, R. S., *Biochemistry* **10**, 4312 (1971).
2. Jensen, S. and Jernelöv, A., *Mercury Contamination in Man and His Environment*, Technical Reports No. 137, International Atomic Energy Commission, Vienna, chap. 4, 43–47, 1972.
3. Wood, J. M., *Science* **183**, 1049–1052 (1974).
4. Wood, J. M., Kennedy, F. S., and Rosen, C. G., *Nature* **220**, 173 (1968).
5. Hill, H. A. O., Pratt, J. M., Ridsdale, S., Williams, F. R., and Williams, R. J. P., *Chem. Commun.* **6**, 341 (1970).
6. Bertilsson, L. and Neujahr, H. Y., *Biochemistry* **10**, 2805 (1971).
7. DeSimone, R. E., Penley, M. W., Charbonean, L., Smith, S. G., Wood, J. M., Hill, H. A. O., Pratt, J. M., Ridsdale, S., and Williams, R. J. P., *Biochim. biophys. Acta* **273** (2), 265 (1973).
8. Law, P. Y. and Wood, J. M., *J. Am. Chem. Soc.* **95** (3), 914 (1973).
9. Landner, L., *Nature* **230**, 452 (1971).
10. Selhub, J., Burton, E., and Sakami, W., *Fedn. Proc.* **28**, 352 (1969).
11. Wood, J. M. and Brown, D. G., *Structure and Bonding*, **11**, 47–105, 1972.
12. Wood, J. M., Penley, M. W., and DeSimone, R. E., *Mercury Contamination of Man and His Environment*, Technical Reports No. 137, International Atomic Energy Commission, Vienna, chap. 5, pp. 49–65, 1972.
13. Jensen, S. and Jernelöv, A. *Nordforsk* **14**, 3–6 (1968).
14. Penley, M. W., Brown, D. G., and Wood, J. M., *Biochemistry* **9**, 4302 (1970).
15. Imura, N., Sakegawa, S., Pan, E., Nagaro, K., Kim, J. Y., Kwan, T., and Ukita, T., *Science* **172**, 1248 (1971).
16. Wood, J. M. and Lien, E. L., Report to the Government of Ontario, 1971.
17. Agnes, Y., Hill, H. A. O., Pratt, J. M., Ridsdale, S. C., Kennedy, F. S., and Williams, R. J. P., *Biochim. biophys. Acta* **252**, 207 (1971).
18. DeSimone, R. E., *Chem. Commun.* **13**, 780 (1972).
19. Jernelöv, A., Report to the Government of Ontario, 1971.
20. Challenger, F., *Chem. Rev.* **36**, 315 (1945).

Metabolic Cycles for Toxic Elements in the Environment: A Study of Kinetics and Mechanism (J. M. Wood)

DISCUSSION by J. J. BISOGNI JR. and A. W. LAWRENCE
Department of Environmental Engineering, Cornell University, Ithaca, NY

The author is to be commended for an excellent paper which both summarized and amplified the important research of Wood and his colleagues on metabolic conversion of toxic elements. Their research has contributed significantly to understanding metabolic conversion of heavy metals in aquatic systems. The nature and complexity of Wood's research, however, has dictated that the experiments be conducted with cell-free systems. It is the intent of this discussion to suggest that Wood's mechanistic and kinetic explanations for the methylation of mercury can be extended to *in vivo* microbial systems.

In this paper and in previous work[1,2] Wood has presented a cobalamin dependent kinetic model for the methylation of mercuric ion. The model is based on the postulate that there are several reactions which control the rate of methylation of mercury. In particular, there is a fast reaction, reaction 1, and several rate-limiting slow reactions of which reaction 2 predominates:

$$\begin{bmatrix} CH_3 \\ | \\ Co^{+3} \\ \uparrow \\ Bz \end{bmatrix} + Hg^{+2} \rightleftharpoons \begin{bmatrix} CH_3 \\ | \\ Co^{+3} \\ \\ BzHg^{+2} \end{bmatrix} \quad K_1 \text{ (reaction 1)}$$

$$\begin{bmatrix} CH_3 \\ | \\ Co^{+3} \\ \uparrow \\ Bz \end{bmatrix} + Hg^{+2} \xrightarrow[k_2]{H_2O} \begin{bmatrix} H\,O\,H \\ \downarrow \\ Co^{+3} \\ \uparrow \\ Bz \end{bmatrix} + CH_3Hg^+ \quad \text{(reaction 2)}$$

The overall rate of transfer of methyl group to mercuric ion via reaction 2 can be expressed as follows:

$$\text{Rate of methylation} = k_2 \begin{bmatrix} CH_3 \\ | \\ Co \\ \uparrow \\ Bz \end{bmatrix} [Hg^{+2}] \quad (1)$$

De Simone et al.[1] normalized this methylation rate by dividing the overall reaction rate by total methylcobalamin concentration, $T_{CH_3-B_{12}}$, where

$$T_{CH_3-B_{12}} = \begin{bmatrix} CH_3 \\ | \\ Co \\ \uparrow \\ Bz \end{bmatrix} + \begin{bmatrix} CH_3 \\ | \\ Co \\ \\ BzHg^{+2} \end{bmatrix} \quad (2)$$

The resulting reaction rate was denoted as k_{obs}:

$$k_{obs} = \frac{\text{rate of methylation}}{T_{CH_3-B_{12}}}$$

$$= \frac{k_2[Hg^{+2}]}{\begin{bmatrix} CH_3 \\ | \\ Co \\ \\ BzHg^{+2} \end{bmatrix} / \begin{bmatrix} CH_3 \\ | \\ Co \\ \\ Bz \end{bmatrix}} \quad (3)$$

After substituting the equilibrium relationship (reaction 1) into eqn. (3), De Simone et al. obtained an equation essentially identical to the following:

$$k_{obs} = \frac{k_2[Hg^{+2}]}{1 + K_1[Hg^{+2}]} \quad (4)$$

This model was verified by a series of kinetic experiments using cell-free methylcorrinoid preparations and mercury in the form of mercuric diacetate. However, in the development of eqns. (1) through (4), shown here, free mercuric ion is indicated rather than the mercuric diacetate as used by De Simone et al. The purpose for specifying uncomplexed mercuric ion is to make the equations more useful in natural aquatic systems. More specifically, when the physiological pH range is considered, hydroxides and oxides, rather than acetate, are more likely to limit the availability of mercuric ion (in aerobic systems).

Wood, in this paper, and Wood et al.,[2] stated that the significance of studying methylcobalamin–mercury reactions is that methylcorrinoids are common in natural aquatic sediments and hence are implicated in the methylation of mercury in natural systems. In a natural system Wood states that synthesis of methylmercury will depend on factors such as microbial population, availability of Hg^{+2}, pH, redox potential, temperature, and synergistic or antagonistic effects of other metabolic or chemical processes. The kinetic model, as represented by eqn. (4), was not intended to incorporate these parameters and consequently cannot be used, in this form, to predict methylmercury synthesis rates in natural systems. Therefore, it is most appropriate to investigate the degree to which the mechanistic and kinetic models of Wood are applicable to the regime of in vivo microbial systems. Bisogni and Lawrence,[3] guided by the methylation mechanisms proposed by Wood, have developed and experimentally verified a kinetic model which they feel can be applied in natural microbial systems. This model describes the rate of microbial synthesis of methylmercury in terms of micrograms of methylmercury produced per day per gram of volatile biomass.

This synthesis rate is denoted NSMR and is described as follows:

$$NSMR = \gamma \beta^n (Hg_{total})^n \qquad (5)$$

where NSMR is the net specific methylation rate,

$$\frac{\mu g \text{ of } (CH_3)_2Hg \text{ or } CH_3Hg^+ \text{ as } Hg}{\text{gram of VSS-day}},$$

γ the coefficient determined by the microbial growth rate of the system; β the ratio of available mercuric ions to total inorganic mercury; n the pseudo-order of the reaction which depends mainly on the redox potential; Hg_{total} the concentration of total inorganic mercury (mg/l); VSS the concentration of volatile suspended solids (a measure of the biomass) (g/l).

The kinetic parameter NSMR is not unlike the parameter k_{obs}. In fact, NSMR and k_{obs} can be shown to be generically equivalent. To demonstrate this equivalency, appropriate conversion factors must be employed, i.e., to convert μg of CH_3Hg^+ to moles CH_3Hg^+; to convert days to seconds; and, finally, to convert grams of volatile biomass to moles of total methylcobalamin. While the kinetic parameters (NSMR and k_{obs}) possess equivalent units, it is quite apparent that the kinetic expressions by which they are described are similar only in their dependency on Hg^{+2} concentration. With appropriate manipulation it is possible to compare Wood's kinetic data with the in vivo kinetic data represented by the NSMR model of the discussors. The majority of Wood's data was collected at pH conditions and mercury concentrations where the NSMR model is not valid. Figures 4 and 6 (in the subject paper), however, allow Wood's kinetic data to be extrapolated to physiological pH and nonlethal mercury concentrations, where the NSMR model is most suitable.

It is critical that the proper NSMR coefficients be selected so as to closely match the conditions of Wood's cell-free studies. Wood's studies can be considered nonenzymic aerobic synthesis reactions. Wood et al.[2] speculated that similar (nonenzymic) processes occur during aerobic microbial synthesis of methylmercury. Of the cases studied by Bisogni and Lawrence,[3] the one that most closely matched Wood's experimental condition, was a high metabolic activity, aerobic microbial system. The NSMR coefficients for this particular system are shown in eqn. (6).

$$NSMR = 7(Hg^{+2})^{0.30} \qquad (6)$$

To effect the desired comparison of k_{obs} and NSMR, appropriate conversion factors were employed. Only the conversion of grams of volatile biomass to moles of methylcobalamin offered any difficulty in this regard. Direct measurements of methylcobalamin fraction of biomass were not made in the studies of Bisogni and Lawrence.[3] It was necessary to rely on information presented in the literature to estimate the methylcobalamin content of the microbial mass used in determining the coefficients of eqn. (6).

In the studies of the discussors, the microbial mass was a heterogeneous culture of microorganisms similar to activated sludge. Hoover et al.[4] and Knivett[5] have shown that such a culture can be expected to have a total B_{12} content of about 10 μg (expressed as cyanocobalamin) per gram of dry weight of biomass. This is equivalent to about 12.5 μg B_{12} (cyanocobalamin) per gram of volatile biomass. An extensive search of the literature by the discussors yielded little information concerning methylcobalamin fraction of total B_{12} in microorganisms. Problems encountered in reviewing the literature to determine microbial methylcobalamin content included: variability among organisms; the evidence that methylcobalamin is a transient metabolic intermediate; and the relatively small

TABLE 1

A comparison of observed cell-free (k_{obs}) and in vivo microbial (k'_{obs}) mercury methylation rates

Hg^{+2} concentration (mg/l)	NSMR[a] (μg/g day)	$k'_{obs} \times D$	k'_{obs} (s^{-1}) $D = 0.10$	k'_{obs} (s^{-1}) $D = 0.05$	k'_{obs} (s^{-1}) $D = 0.01$	k_{obs}[b] (s^{-1})
400	42.5	267×10^{-6}	267×10^{-5}	534×10^{-5}	267×10^{-4}	390×10^{-5}
800	51.8	326×10^{-6}	326×10^{-5}	652×10^{-5}	326×10^{-4}	550×10^{-5}
1000	55.4	348×10^{-6}	348×10^{-5}	696×10^{-5}	348×10^{-4}	578×10^{-5}
2000	68.5	433×10^{-6}	433×10^{-5}	866×10^{-5}	433×10^{-4}	680×10^{-5}

[a] $NSMR = 7(Hg^{+2})^{0.30}$.
[b] Data taken from Fig. 4 and extrapolated to neutral pH using Fig. 6 of discussed paper.

fraction of total B_{12} which is methyl B_{12}.[6–9] To account for these uncertainties, a factor D was introduced into the analysis presented here to allow for some latitude in determining the microbial methyl B_{12} fraction of total B_{12}. The factor D is defined as the ratio of methyl B_{12} to total B_{12}.

By using the proper numerical conversion factors and the factor D, eqn. (7) was developed to represent the relationship between k_{obs} and NSMR for a mixed culture of aerobic microorganisms. The symbol k'_{obs} is used to distinguish the methylation rate calculated using NSMR from the k_{obs} (at neutral pH) observed in Wood's kinetic experiment.

$$k'_{obs} = \frac{(6.3 \times 10^{-6})(NSMR)}{D} \qquad (7)$$

Table 1 compares the k_{obs} of Wood's experiment with the k'_{obs} calculated from eqn. (7) for a range of D factors.

This comparison shows that the cell free kinetic studies of Wood are in reasonably close agreement with methylation rates predicted for an *in vivo* microbial system over a range of reasonable D values. Given the vastly different reaction environments involved, the degree of agreement shown in Table 1 should be reassuring and gratifying to both groups of researchers.

The anaerobic enzymatic methylation of mercury as described by Wood *et al.*[2] has also been modeled by Bisogni and Lawrence[3] using the microbial NSMR model with appropriate coefficients. Unfortunately there is no cell-free study with which to compare these results.

In conclusion, the discussers feel that the cell-free kinetic mechanisms and models presented by Wood provide a useful basis for understanding the mode of microbially mediated methylation of some heavy metals, in particular mercury. In addition, these mechanisms can be used, as demonstrated above, to aid in constructing kinetic methylation models which are applicable to natural microbial systems.

REFERENCES

1. DE SIMONE, R. E., PENLEY, M. W., CHARBONNEAU, L., SMITH, S. G., WOOD, J. M., HILL, H. A. O., PRATT, J. M., RIDSDALE, S., and WILLIAMS, R. J. P., *Biochem. biophys. Acta* **273** (2), 265 (1973).
2. WOOD, J. M., PENLEY, M. W., and DE SIMONE, R. E., *Mercury Contamination of Man and His Environment*, Technical Report No. 137, International Atomic Energy Commission, Vienna, chap. 5, 49–65 (1972).
3. BISOGNI, J. J. and LAWRENCE, A. W., *Kinetics of Microbially Mediated Methylation of Mercury in Aerobic and Anaerobic Aquatic Environments*, Technical Rpt. No. 63, Cornell University Water Resources and Marine Sciences Center, Ithaca, New York, 1973.
4. HOOVER, S. R., JASEWICZ, L. B., and PORGES, N., *Science* **114**, 213 (1951).
5. KNIVETT, V. A., *Prog. Ind. Microbiol.* **2**, 29 (1960).
6. FOSTER, M. A., DILWORTH, M. I., and WOODS, D. D., *Nature* **201**, 39 (1964).
7. LEZIUS, A. G. and BARKER, H. A., *Biochem.* **4**, 511 (1965).
8. TOOHEY, J. I., PERLMAN, D., and BARKER, H. A., *J. Biol. Chem.* **236**, 2119 (1961).
9. VOLCANI, B. E., TOOHEY, J. I., and BARKER, H. A., *Archs. Biochem. Biophys.* **92**, 381 (1961).

SORPTION PHENOMENON IN THE ORGANICS OF BOTTOM SEDIMENTS

ROBERT S. REIMERS
Battelle, Columbus Laboratories, Columbus, Ohio

PETER A. KRENKEL
Director of Environmental Planning Division, Tennessee Valley Authority, Chattanooga, Tennessee 37401

and

MARGARET EAGLE and GREGORY TRAGITT
Vanderbilt University, Nashville, Tennessee

INTRODUCTION

From recent incidents of mercury poisoning it is obvious that investigations of the transfer of mercury in the natural environment are mandatory. High concentrations of mercury were found in fish in both Japan and Sweden, and because of these high concentrations, a study of waters and sediments in both areas was instigated. The investigators observed high levels of mercury in the sediments in each region of high mercury contamination even though the primary sources are different.[1,2]

Investigations of mercury in sediments have indicated an association between mercury content and organic content in sediments, the observation being first noted by Stock and Cucuel[3] in the 1930s. Warren et al.[4] noted high mercury concentrations in sediments containing high organic concentrations and/or clay contents, and Andersson[1] observed significant mercury in decayed sludge.

In the late 1960s, Kennedy et al.[5] reported on 132 sediment analyses from Lake Michigan and found a correlation between mercury content, organic content, and sulfur content; however, the less than 2 μm size clay fraction showed little or no correlation. It was assumed that the base level of mercury was 0.03–0.06 ppm on a dry weight basis, the highest concentration observed being 0.88 ppm with most values being considerably less. Klein and Goldberg[6] observed higher than normal mercury concentrations in oceanic sediments near the Los Angeles sewage outfall. Obviously, the organic content was also high compared to that of a normal oceanic sediment. Applequist et al.[7] made similar observations in the sediments of New Haven Harbor.

In contrast, Aidin'yan et al.[8] found the lowest content of mercury to be in the upper portions of the humic acid horizon and the highest content to be in the mineral portion, with contained iron and manganese. Also, Matsumura et al.[9] stated that the inorganic component of sediments binds mercury rather strongly and does not yield mercury to organic complexing agents.

OBJECTIVES OF THE INVESTIGATIONS

The major objectives of these studies were as follows: to observe the sorption characteristics of sediment organics; to determine the effects of varying pH, chloride concentration, and mercury species, compounds, and concentrations on the process; to convert all available mercury sorption data to a common base in order that comparative sorption efficiencies can be made; and to determine the potential for inhibition of mercury methylation in the sediment organics by adsorption. It is hoped that an outgrowth of this work will be the development of a predictive model for the fate of trace-metal contaminants in the environment, taking into account the physical–chemical problems of sediment organic sorption and the hydraulics of its residence time in the sediment.

METHODS AND MATERIALS

Sorption tests

The adsorption capacitance tests were run on a batchwise basis at a constant temperature of 25°C and with concentrations of 1 ppm mercuric chloride or 1 ppm methylmercury chloride or hydroxide as mercury. The chloride ion concentration and the pH were set at approximately 0, 100, or 10,000 ppm, and 5, 7, or 9 pH units, respectively, and the experiments were conducted for a period of 4 h. Subsequent to the filtration of the sediments, the desorption tests were run for 24 h under the same conditions, but without the mercuric chloride or the methylmercury chloride. The ability of sediments of known composition to adsorb or desorb methylmercury chloride or mercuric chloride was thus ascertained. If the sediments illustrated significant adsorption, then batch isothermal tests (everything being held constant except for the mercury solution concentrations) were conducted.

Since the mercuric species have been shown by Hem[10] to be very stable in aqueous solutions and methylmercury is also stable in semiaerobic and alkaline sediments, these compounds were chosen for this study. In the mercuric chloride adsorption studies, the pHs of 5, 7, and 9, and chloride concentrations of 0, 100, and 10,000 ppm were used because they are common in nature. Three inorganic forms of mercury, $HgCl_4^=$, $HgCl_2$, and $HgO \cdot H_2O$ are predominant under various pH conditions and chloride concentrations. The pH values and chloride concentrations used will result in the methylmercury being either in the form of methylmercuric chloride or methylmercuric hydroxide, as demonstrated by Fig. 1.

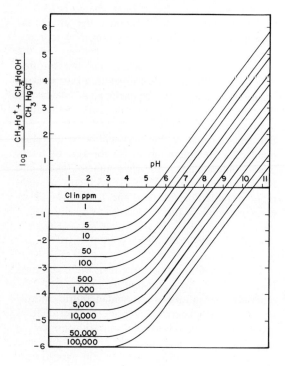

FIG. 1. Effect of pH & chlorides on formation of methylmercury chloride.

The mercury solutions were 1 ppm mercuric chloride or methylmercuric chloride in distilled water; the various chloride concentrations were made using sodium chloride; and the pH of solutions was adjusted by adding the needed quantity of nitric acid or sodium hydroxide, avoiding the addition of buffers which could either complex, precipitate, or change the oxidation state of the mercury. Deionized distilled water used for standard solutions and distilled water used for the sorption studies were tested for chlorides, chlorine, pH, calcium hardness, and total hardness. Since the chloride concentration was varied five orders of magnitude, the chloride content was analyzed by specific conductance using a conductivity bridge. The chloride concentrations were obtained by analyzing all samples and standards at 25°C and comparing the specific conductance of the sample to a log–log plot of chlorides in ppm versus specific conductance. The redox potential readings were checked by comparison with standard curves obtained by preparing solutions of known potential.

The temperature in all the sorption runs was held at 25°C and the flasks were shaken sufficiently to cause complete mixing. Initial and final analyses were performed on each run for mercury (inorganic or methyl) and pH, but the chloride content and redox potential were only measured in the initial solutions. Since there was a potential problem with mercury adsorbing onto the surfaces of glassware, it was washed in aqua regia and was then rinsed with large quantities of distilled water so that the acid residue would be removed.

Sediment matrix

Because of the lack of data concerning metal sorption phenomena, an important aspect of this study was the creation of a standard system that would enable the comparison of the transfer of various heavy metal contaminants in the natural environment. For studies of the effect(s) of organics on sediment adsorption, three specific organic functional groups were investigated: the carboxyl group, the amine group, and the mercaptan group, along with dodecane which contains no functional group. Each of the organic chemicals incorporated long chain alkanes, which are insoluble in water, and had a single functional group. The compounds utilized were stearic acid (R–COOH), octadecylamine (R–NH$_2$), and dodecanethiol (R–SH). By placing these three organic materials in ether, in which they are soluble, it was possible to evaporate the ether and place the organics on sands in the 70–100 mesh range. Prior to use, the sand was washed with distilled water and then heated at 300°F until it was completely dry. Twenty percent organics by weight were added to the sand because sediments containing high organics have been observed to be 15–20% organic material by weight.[11]

The synthetic sediments of the sand–organic mixtures contained 10–17% COD. Only the alkylamine–sand mixture had any noticeable organic nitrogen, and the sulfhydryl alkyl–sand mixture seemed to have a high content of organic sulfur. Illite was the only constituent possessing significant sulfur content, which was 0.356% total sulfur, probably in an oxygenated form. None of the sands or organic–sand mixtures were found to change the pH of unbuffered solutions of pHs of 5, 7, and 9 more than 0.6 of a unit.

Analytical techniques

Prior to analysis for total mercury, all samples were acidified with nitric acid in order to stabilize the mercury. Nitric acid was used because it does not form complexes with mercury as does hydrochloric acid, nor does it allow mercury to volatilize as is the case with sulfuric acid. Since digestion of the samples was not necessary, only the last steps of the Hatch and Ott method[12] were used.

If the sample was in liquid form, it was centrifuged, the liquid supernatant was acidified with nitric acid, the inorganic mercury was reduced to elemental mercury with stannous chloride, the elemental mercury was stripped from the water by aeration with compressed air, and the stripped mercury was analyzed by flameless atomic absorption spectroscopy. If the sample was a sediment, a 5–10 g sample was combusted at 1000°C with oxygen and the combusted elemental mercury was trapped in potassium permanganate. The reduced permanganate was reduced further with hydroxylamine to solubilize the coprecipitated mercury. Stannous chloride was added to reduce the mercury to elemental mercury which was stripped with compressed air and analyzed by flameless atomic absorption spectroscopy.

Liquid samples were analyzed for methylmercury using the last steps of the modified Westöö procedure,[13] eliminating digestion steps. Solid samples were mixed with sulfuric acid, potassium bromide, and benzene for 3 h, and the benzene extract was then analyzed, also according to the modified Westöö procedure.

The chemical oxygen demand (COD) and the total organic nitrogen were measured in the sediments in order to substantiate the organic content and the purity of the inorganic constituents. The methodology utilized was taken from Ballinger and McKee[14] and Standard Methods.[15] Sulfur, being a critical constituent of the sediments, was also determined. The procedure, of the three tested, which yielded the most accurate sulfur results, utilized a sodium peroxide oxidation step.[16]

Calculations

The data obtained by experimentation or in the literature were converted to comparable terms in the form of instability partition coefficients. The instability partition coefficient K_{INS} may be defined as a measure of the degree to which mercury will desorb from an adsorbent. It is expressed in the equation below:

$$K_{INS} = \frac{\text{ppm of mercury in solution (amount of mercury in the water)}}{\text{ppm of mercury in the adsorbent (amount of mercury in the adsorbent)}}.$$

With data from the literature and these investigations, Freundlich isotherms were calculated for possible utilization in determining the engineering applicability of a promising sorbent. The isotherms were arranged so that the concentrations for the adsorbents were in ppm of mercury as in natural sediments and the aqueous equilibrium solutions were in ppb of mercury as in natural water. The Freundlich isotherm equation utilized is illustrated below:

$$\text{ppm of mercury adsorbed by a sorbent} = kC^{1/n},$$

where C equals the aqueous equilibrium concentration in ppb of mercury.

RESULTS AND DISCUSSION

Mercuric compounds with organic reagents

The effect of specific functional groups on the adsorption of mercuric chloride has been investigated by Reimers and Krenkel[17] and Feick et al.[18] Kinetic data of Reimers and Krenkel[17] demonstrated that the dodecane thiol and sand mixture and the octadecylamine and sand mixture adsorbed all of the mercuric chloride or mercuric hydroxide within a period of 10 min. However, the stearic acid and sand mixture attained a maximum capacity in approximately 1 h. The descending order of uptake rates of those materials for mercuric chloride in μg of mercury compound as mercury per gram of

TABLE 1
Instability partition coefficient for long-chained alkyls with mercuric chloride

Adsorbent	Time	Mercury concentration (ppm) Dry sediment	Water	Inst. part. coef. $K = \dfrac{[Hg^{++}]H_2O}{[Hg^{++}]_{sed}}$	pH	Cl⁻	DO
\multicolumn{8}{c}{Effect of initial concentration and dissolved oxygen}							
\multicolumn{8}{c}{Data from Feick et al.,[18] each sample contains 100 g of Georgia kaolin}							
5 g CaCO$_3$	7 days	314.0	11.5	3.7×10^{-2}	7.4		7.5
1 g CaCO$_3$ and 1 ml NDM[a]	7 days	378.0	0.00003	7.9×10^{-8}	7.5		
1 g CaCO$_3$ (low O$_2$)	7 days	378.0	0.00002	5.3×10^{-8}	7.7		3.0
5 g CaCO and 1 ml NDM[a]	7 days	1000.0	0.00002	2.0×10^{-8}	6.8		11.5
1 ml of NDM[a]	7 days	321.0	0.05	1.6×10^{-4}	5.1		10.0
1 ml of NDM[a]	7 days	1000.0	0.00015	1.5×10^{-7}	4.4		12.0
1 ml of NDM[a] (aged 36 days)	7 days	300.0	0.0154	5.2×10^{-5}	5.0		15.0
\multicolumn{8}{c}{Effect of varying dissolved oxygen, reducing agents and strengths of mercury solutions.}							
\multicolumn{8}{c}{Data from Feick et al.[18]}							
10% NDM,[a] 0.01% Armac-T	7 days	307.0	0.0005	1.6×10^{-6}	6.4		7.1
10% NDM,[a] 0.01% and sands	7 days	108.0	<0.0004	$<3.7 \times 10^{-7}$	6.2		6.1
10% (5 g CaCO$_3$ NDM)[a]	7 days	112.0	0.0004	3.6×10^{-7}	8.5		4.9
10% NDM[a] (5 g F, low O$_2$)	7 days	125.0	0.0004	3.2×10^{-7}	7.2		5.0
10% NDM[a] (5 g F, low O$_2$)	7 days	93.0	0.0016	1.7×10^{-5}	7.3		1.1
10% NDM[a] (5 g F, low O$_2$)	7 days	93.0	0.012	1.3×10^{-4}	6.8		1.2
10% NDM[a] (aged)	7 days	92.0	0.0094	1.0×10^{-4}	6.9		1.1
\multicolumn{8}{c}{Effect of different alkyl thiol compounds with various reducing agents.}							
\multicolumn{8}{c}{Data from Feick et al.;[18] each test contains 100 g of Georgia kaolin}							
1 g CaCO$_3$ and 1 ml MTM[a]	7 days	378.0	<0.00002	$<5.3 \times 10^{-8}$	7.6		
1 g CaCO$_3$ and 1 ml THM[a]	8 days	378.0	<0.00002	$<5.3 \times 10^{-8}$	7.6		5.0
1 ml DDD[a]	7 days	300.0	6.2	1.7×10^{-2}	5.4		6.2
1 ml DDD[a] and 5 g Zn dust	7 days	300.0	0.0114	3.8×10^{-5}	5.4		0.8
1 ml DDD[a] and 5 g Fe powder	7 days	300.0	0.0254	8.5×10^{-5}	7.6		0.8
1 ml DDD[a] and 100 ml sawdust extr.	7 days	300.0	3.8	1.3×10^{-2}	6.0		6.4
1 g CaCO$_3$ and 1 ml NDM[a]	7 days	378.0	0.00003	7.9×10^{-8}	7.5		
\multicolumn{8}{c}{Effect of CaCl and NaCl on the adsorption of HgCl$_2$ by NDM.}							
\multicolumn{8}{c}{Data from Feick et al.;[18] each sample contains 100 g of Georgia kaolin}							
5 g CaCO$_3$ and 1 ml NDM[a]	7 days	1000.0	0.00002	2.0×10^{-8}	6.8		11.5
5 g CaCO$_3$ and 1 ml NDM[a]	7 days	1000.0	0.00024	2.4×10^{-7}	7.2	21,000	12.0
5 g CaCO$_3$ and 1 ml NDM[a]	7 days	300.0	0.0025	8.3×10^{-6}	5.4	165 g CaCl$_2$	2.0
5 g CaCO$_3$ and 1 ml NDM[a]	7 days	300.0	0.00006	2.0×10^{-7}	7.2	21,000	8.5
\multicolumn{8}{c}{Effect of alkanes on the adsorption of HgCl$_2$[17]}							
20% ND[a] and sand	4 h	10.5	0.120	1.1×10^{-2}	7	~0	
20% ND[a] and sand	4 h	13.9	0.090	6.7×10^{-3}	7	~100	
20% ND[a] and sand	4 h	5.9	0.655	1.1×10^{-1}	7	~10,000	
\multicolumn{8}{c}{Effect of chlorides, pH, and mercury strength on the adsorption of HgCl$_2$ by carboxyl groups[17]}							
Sand with 20% STA[a]	4 h	19.5	0.030	1.5×10^{-3}	5	24.5	
Sand with 20% STA[a]	4 h	12.8	0.100	7.8×10^{-3}	7	9.7	
Sand with 20% STA[a]	4 h	11.8	0.140	1.2×10^{-3}	9	11.2	

TABLE 1 (continued)

Adsorbent	Time	Mercury concentration (ppm) Dry sediment	Water	Inst. part. coef. $K = \frac{[Hg^{++}]H_2O}{[Hg^{++}]_{sed}}$	pH	Cl⁻	DO
Sand with 20% STA[a]	4 h	6.7	0.775	1.2×10^{-2}	5	117.0	
Sand with 20% STA[a]	4 h	10.7	0.057	5.3×10^{-3}	7	130.0	
Sand with 20% STA[a]	4 h	13.8	0.123	9.6×10^{-3}	9	122.0	
Sand with 20% STA[a]	4 h	5.4	0.775	1.4×10^{-1}	5	10,060	
Sand with 20% STA[a]	4 h	6.6	0.515	7.8×10^{-2}	7	10,020	
Sand with 20% STA[a]	4 h	4.7	0.520	1.1×10^{-1}	9	10,404	
Sand with 20% STA[a]	4 h	0.4	0.005	1.2×10^{-2}	7	~0	
Sand with 20% STA[a]	4 h	161.5	2.400	1.5×10^{-2}		~0	
Effect of chlorides, pH, and mercury strength on the adsorption by amine groups[17]							
Sand with 20% ODA[a]	4 h	18.8	0.073	3.9×10^{-3}	5	24.5	
Sand with 20% ODA[a]	4 h	13.1	0.078	6.0×10^{-3}	7	9.7	
Sand with 20% ODA[a]	4 h	13.3	0.046	3.4×10^{-3}	9	11.2	
Sand with 20% ODA[a]	4 h	17.8	0.081	4.5×10^{-3}	5	177.0	
Sand with 20% ODA[a]	4 h	11.6			7	130.0	
Sand with 20% ODA[a]	4 h	15.2	0.074	4.8×10^{-3}	9	122.0	
Sand with 20% ODA[a]	4 h	18.4			5	10,060	
Sand with 20% ODA[a]	4 h	15.0	0.014	9.4×10^{-4}	7	10,020	
Sand with 20% ODA[a]	4 h	13.3	0.004	3.0×10^{-4}	9	10,040	
Sand with 20% ODA[a]	4 h	1.1	0.016	1.5×10^{-2}	8	~0	
Sand with 20% ODA[a]	4 h	128.0	0.600	4.7×10^{-3}	7	~0	
Sand with 20% ODA[a]	4 h	199.8	1.35	6.7×10^{-3}	7	~0	
Effect of varying concentrations of mercuric chloride on the adsorption by mercaptans[17]							
Sand with 20% NDM	4 h	16.7	0.003	1.8×10^{-4}	7		
Sand with 20% NDM	4 h	367.9	0.003	7.9×10^{-6}	7		
Sand with 20% NDM	4 h	1768.0	0.005	2.8×10^{-6}	7		
Sand with 20% NDM	4 h	7568.0	0.003	3.9×10^{-7}	7		
Sand with 20% NDM	4 h	19,854.0	0.005	2.5×10^{-7}	7		
Sand with 20% NDM	4 h	34,912.0	7.150	7.0×10^{-4}	7		

[a] Organic compounds.
NDM, n-dodecyl mercaptan.
MTM, mixed tertiary mercaptan.
THM, t-hexadecyl mercaptan.
DDD, di-t-dodecyl disulfide.
ODA, n-octadecylamine.
STA, stearic acid.
ND, n-dodecane.

sediment per minute was:

Mercaptans (R–S–H) > Amines (R–NH₂) > Carboxyls (COOH)
     (8550)   (10.5)   and Alkanes (R–H)
                                                  (7.3)

These data were obtained using solutions containing 10 ppm mercuric chloride (except the mercaptans which adsorbed up to 10 ppm of mercury in 10 min), a pH of 7, and no chloride additions. In addition, the rates were assumed to be of zero order as a simple approximation.

The same descending order as shown for the uptake rates of these materials was observed for the capacitance isotherms of mercuric chloride in a solution with a pH of 7 and no added chlorides. The dodecane and sand mixture followed the same capacitance pattern as the stearic acid and sand mixture inasmuch as the same capacitance range and approximately the same instability partition coefficients were observed at a pH of 7 and varied chloride (0–10,000 ppm). These results imply that the inorganic mercury reacts with noncharged carbon and not with the carboxyl group. The instability partition coefficients were found to be approximately the same for the dodecane thiol and sand mixture by Feick et al.[18] and in this study.

The desorption of mercuric chloride from the organics investigated was found to be negligible using leaching times of 24 h and previously described analytical techniques.[17]

The effect of chlorides on the various functional

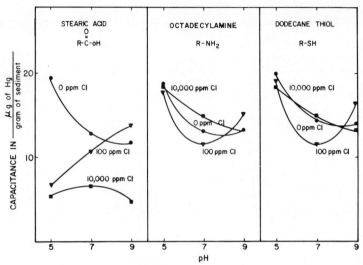

FIG. 2. Organic component capacitance for $HgCl_2$, various [Cl^-], at 25°C, $C_0 = 1$ ppm. (After Reimers and Krenkel.[17])

groups observed in the sediment was also studied by Reimers and Krenkel. Figure 2, which was taken from their work, reveals that the carboxyl group's capacity for adsorption of mercury is affected by salinity, while the effect of salinity on the adsorption of mercury by sulfhydryl and amine groups is minimal. In each case, the experiments were conducted with a mixture of the specific organic and sand in water. Feick et al.[18] observed the same lack of effect with mercaptans as did Reimers and Krenkel. The equilibrium results found by Reimers and Krenkel for mercuric chloride adsorption onto the various organic functional groups are shown in Table 1. With stearic acid, the instability partition coefficients increased by two orders of magnitude with an increase in chloride concentration. The instability partition coefficients for stearic acid and sand mixtures are seen to be in the same range as those found for dodecane and sand mixtures in aqueous solutions at varied chlorides (from 0 to 10,000 ppm) and a pH of 7. The results of these tests imply that the water molecules may have a higher affinity for the carboxyl group than the mercuric hydroxide or mercuric chloride does and that the inorganic mercury actually adsorbs physically to the noncharged organic carbons atoms. This physical adsorption is similar to organic solvent extractions where the nonpolar compounds, such as mercuric chloride, have a greater affinity for nonpolar organics than for water.

As shown in Fig. 2, there was no apparent trend of pH effects on the adsorption by the carboxyl group. The greatest adsorption by other groups was found at a pH of 5, and lower capacitances were observed at a pH of 9 and 7, respectively. These observations imply that these materials tend to adsorb (in descending order): mercuric chloride, mercuric hydroxide, a mixture of mercuric chloride and mercuric hydroxide, and mercuric tetrachloride complex.

Feick et al.[18] looked at the effects of dissolved oxygen, aging of thiol-containing mixtures, and the type of carbon to which the mercaptan is bound. Varying the dissolved oxygen content appeared to have no effect on the instability partition coefficient for the adsorption of mercuric chloride onto dodecane thiol and sand. However, when iron was added to aid in reduction, the instability partition coefficients were observed to increase in value indicating a decrease in the capacitance of the dodecane thiol and sand. One milliliter of dodecane thiol was also mixed with 100 g of clay, and the mixture was aged in air for 2 months. After 2 months of aging, the instability partition coefficient was 5.2×10^{-5} rather than 1.5×10^{-7}, a change of almost three orders of magnitude. Even though this a relatively small instability partition coefficient, these oxidized sediments are still fairly capable scavengers of mercury. Feick et al. found no change in the capacity or partition coefficient for mercuric chloride by changing the type of carbon to which the sulfhydryl group is attached.

Changes in the initial concentration of mercury in the sediment did not cause a change in the magnitude of the instability partition coefficient for the octadecylamine and sand mixture and the stearic acid and sand mixture; the average partition coefficients for each being 9.1×10^{-3} and 1.3×10^{-2}, respectively. On the other hand, the mercaptan group demonstrated instability partition coefficients of 2.0×10^{-4} for sediment containing from 0 to 100 ppm of mercury, 1.0×10^{-6} for sediment containing from 100 to 2000 ppm of mercury, and 7.0×10^{-7} for sediments containing from 2000 to 20,000 ppm of mercury. At concentrations greater than 20,000 ppm of mercury, the instability partition coefficients began to increase as illustrated in Table 1.

Mercuric compounds with natural sediments

A review of the literature concerned with the adsorption of mercuric chloride by organically enriched sediments reveals a paucity of quantitative data. Perusal of numerous field investigations showed few efforts resulting in quantification of the adsorption of mercuric chloride or mercuric oxide by naturally organically rich sediments.[1, 18, 22]

A comparison of uptake capacities for various natural organically enriched soils is given in Table 2, from which the materials can be arranged in decreasing order of uptake as follows:

Peat > ashland sediment
 > oceanic humic acids > coniferous soil.
Swedish humic acids > deciduous soils
 > grass soil > pine mull.

Feick et al.,[18] Strohal and Huljev,[21] and de Groot et al.[22] have investigated the effects of changing chloride concentrations on the adsorption of mercuric chloride by natural sediments with a high organic content. Data from Feick et al. indicate a rise in the instability partition coefficient when the chloride content was increased, using peat as the adsorbent. Furthermore, if higher initial concentrations of mercuric chloride were used, the instability partition coefficient increased even more. It is interesting to note that when calcium chloride was used as a source of chlorides instead of sodium chloride, a larger instability coefficient resulted.

Strohal and Huljev observed this same increase with increased chloride content for the adsorption of mercuric chloride by humic acids when similar tests were run with distilled water and oceanic water. They found that when the sodium chloride content was increased by a factor of 27 in the mercury-containing water, the instability partition coefficient increased by a factor of approximately 20, and when that increased sodium chloride content was doubled, only a small increase in the instability partition coefficient resulted. The increase in the instability partition coefficient due to chlorides caused approximately a 1% loss in mercury. These observations substantiate investigations of de Groot demonstrating that mercury concentrations in sediments are reduced with an increase in salinity.

The effects of pH on the adsorption of mercuric chloride by the organics contained in natural sediments have been studied by Trost[19] and Andersson.[1] Andersson observed optimal adsorption of mercuric chloride at low pH, as illustrated in Fig. 3. Andersson also observed an increase in the solubility of the mercury-humate complex with an increase in pH, mercury concentration, and humus sediment. He postulated that the decrease in the mercury adsorptive capacity at a higher pH was not caused by desorption mechanisms alone, but was influenced by the hydrolysis of the soluble humate-mercury complex. Reimers and Krenkel[17] found that mercuric chloride tended to interact with the inert carbon of the long chain fatty acid and not with the carboxyl group. Therefore, as the pH becomes greater than 5, the organic-bonded mercury of the mercury-humate complex will become water soluble as the carboxyl groups become charged (the pK_a of organic acids ranges between 3 and 6). The desorbed inorganic mercury may, therefore, be a humate-mercury complex instead of mercuric chloride or mercuric hydroxide.

Trost observed that humic acids tend to complex cinnabar (HgS).[19] The solubility of cinnabar was increased by 30 times at a pH of 5 and even more at pHs of 6 and 7 by the addition of 850 ppm of humic acid expressed as carbon. An increase in the solubility of mercuric oxide and mercuric chloride by a

TABLE 2
The strength of various natural organic soils and sediments

Soil or sediment	pH	Inst. part. coef. $K = \dfrac{[Hg^{++}]_{H_2O}}{[Hg^{++}]_{sed}}$	Ref.
Acton peat	5.0	$(50 - 1.4) \times 10^{-8}$	18
Ashland sediment	6.0	$(2 - 5) \times 10^{-5}$	18
Oceanic humic acids	8.3	4.8×10^{-4} (1 day) 7.6×10^{-5} (44 days)	21
Pine A (coniferous soils)	6.0	1.3×10^{-3}	19
Swedish humic acids	6.0	2.0×10^{-3}	1
Aspen A (deciduous soils)	6.0	3.8×10^{-3}	19
Grass A (grass soils)	6.0	5.0×10^{-3}	19
Peat moss[a]	8.3	5.0×10^{-3}	20
Pine mull	6.0	6.7×10^{-3}	19

[a] Analysis performed in 1956 before recent improvements in analysis.

The smaller the instability partition coefficient, the greater the affinity of the soil for mercury.

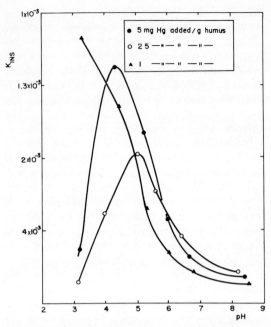

FIG. 3. The influence of pH on the distribution of mercury ions in the system organic soil and water. (Modified from Andersson.[1])

factor of 3 and 5, respectively, was also noted. Trost also found that at high pHs mercuric carbonate will begin to precipitate and thus begin to remove some of the soluble mercury complexes. It may be concluded from the studies of Andersson and Trost that the solubility of mercuric chloride or mercuric hydroxide will increase with an increase in soluble humic acid, pH and humic acid, or pH up to 8.3. If the chloride content rises to the vicinity of that of sea water, or approximately 18,000 ppm,[23] then the inorganic mercury may not precipitate with the carbonate because it will either be in the mercuric oxide form or in the tetrachloride complex form.

Feick et al.[18] investigated the effect of various environmental abiotic parameters such as dissolved oxygen, aging in the presence of air, contact time, varied initial mercuric chloride strengths, and the possibilities of leaching mercury from the sediments on the adsorption of mercuric chloride on peat and ashland sediments. They observed that peat had both a high sulfide content and a high biochemical oxygen demand, which implies that the peat should have an excellent mercury binding capacity and should be either anoxic or reducing. Dissolved oxygen had a significant effect on the instability partition coefficient, it being increased by an order of magnitude with the addition of 300 ml of air. With small additions of dissolved oxygen, the instability partition coefficient was increased by a factor of 2. With respect to the effect of aging of the peat in the presence of air for 2 months, the instability partition coefficient was increased by 2 orders of magnitude, which indicates that dredged peat or peat which is left as spoil could leach mercury back into the

TABLE 3
The effect of abiotic parameters on natural sediments' uptake of inorganic mercury (after Feick[18])

Description	Time	[Hg^{++}] in ppm Dry sediment	Water	Inst. part. coef. $K = \dfrac{[Hg_{++}] H_2O}{[Hg]_{sed}}$	pH	DO
		Effect of time and concentration				
Acton peat	7 days	476.0	<0.0002	<5.3 × 10^{-7}	5.3	0.0
Acton peat	4 days	1430.0	<0.0002	<1.4 × 10^{-7}	4.9	0.0
Acton peat	7 days	1430.0	0.00002	1.4 × 10^{-8}	5.2	0.0
Acton peat	7 days	2610.0	0.0044	1.7 × 10^{-6}	5.1	0.0
Acton peat	7 days	8000.0	0.192	2.4 × 10^{-5}	4.7	0.0
		Effect of dissolved oxygen and aging				
Acton peat	7 days	2610.0	<0.0044	1.7 × 10^{-6}	5.1	0.0
Acton peat (low O$_2$)	7 days	2610.0	0.0258	9.7 × 10^{-6}	5.1	0.0
Acton peat (300 ml of air)	7 days	2610.0	0.074	2.8 × 10^{-5}	4.8	1.0
Acton peat (300 ml of air)	7 days	2610.0	0.062	2.3 × 10^{-5}	5.0	1.0
Acton peat	7 days	1470.0	<0.00002	<1.4 × 10^{-8}	5.2	0.0
Acton peat (aged 2 months)	7 days	1335.0	0.0031	2.3 × 10^{-6}	4.8	0.6
Acton peat	7 days	1470.0	<0.00002	1.4 × 10^{-8}	5.2	0.0
Acton peat	7 days	1320.0	8.8	6.7 × 10^{-3}	4.1	0.6
Acton peat	7 days	1365.0	0.044	3.2 × 10^{-3}	5.4	0.4
Acton peat	7 days	890.0	0.09	1.0 × 10^{-4}	5.8	0.5

environment. As would be expected, the instability partition coefficients were found to decrease with an increased contact time as shown in Table 3. As was the case with dodecane thiol and sand, an optimum sediment mercury concentration occurred before the instability partition coefficient increased with increased initial mercury concentration. This concentration appeared to be approximately 2000 ppm of mercury at a pH of 5.

The possibility of leaching mercuric chloride from the sediments with additives was studied, and even with cysteine hydrochloride, the most effective additive, 99% of the mercury was still bound at equilibrium with the sediment and an equal weight of water solution. The other two additives, thiourea and sodium thiosulfate, were found to be even less effective because of the colloidal properties observed with peat.

Based on the preceding results, several comments can be made. The peat and ashland sediments used by Feick et al.[18] contained elemental sulfur or organic sulfur. The ashland sediment obviously contained significant organics and sulfur because hydrogen sulfide resulted upon acidification. Because both of these sediments contain sulfur, low instability partition coefficients indicating high mercury uptake would be expected and were obtained. The oceanic humic acids used by Strohal and Huljev[21] could have contained sulfur because of their low instability partition coefficients, but because of the desorption work, these authors postulated that their humic acids contained no sulfur. Because of the larger instability partition coefficients obtained with other natural organic sediments, it is assumed that these organics had only nitrogen- and oxygen-containing organic functional groups. This relationship becomes quite important when evaluating the cycling of organic and inorganic mercury in aquatic ecosystems. The Swedish humus samples had a high instability partition coefficient, indicating a significantly lower binding capacity for mercury. The American sediments investigated by Feick et al. had low instability partition coefficients and high binding capacities for mercury. It is obvious that the organics in the sediments must be identified before a proper understanding of the mercury transport by field sediments can be understood.

Methylmercury with organic reagents

Reimers and Krenkel observed significant adsorption of methylmercury only with the thiol or mercaptan functional group.[17] The rate of uptake was observed to be approximately 116.8 g of methylmercury as mercury per gram of sediment per minute, assuming a zero order reaction rate. As was the case with inorganic mercury, desorption of methylmercury was not observed over a 24 h period. This significant adsorption only by thiols and mercaptans could explain the incomplete adsorption of mercury by all films, fibers, and organic sediments that lacked sulfhydryl groups, e.g., nylon, polyvinyl chloride, etc., as reported in recent literature.[24-26] Clarkson[27] substantiated this phenomenon in his attempts to cure "minamata disease" for humans, inasmuch as methylmercury adsorption occurred only with ion exchange resins containing mercaptans.

The order of stability for methylmercury derivatives follows the sequence:

	S	>	RSH^-	>	CN^-	>	OH^-	>	$R-NH_2$	>	C_6H_5O	>	$RCOO^-$
$\log K_s$	45.5		16.5		14.1		9.5		8.8		6.5		3.6

From the stability constants, it can be seen that only sulfides, mercaptans, and cyanides form complexes more stable than methylmercury hydroxide in aqueous solutions. Chlorides or pH do not greatly affect the adsorption capacity of the mercaptan–sand mixture for methylmercury, although the instability partition coefficient decreased approximately one order of magnitude for 100 ppm of chlorides and two orders of magnitude for 10,000 ppm chlorides. These results are illustrated in Table 4.

Feick et al.[18] tested four parameters that could affect the adsorption of methylmercury by mercaptans. They found a decrease in the instability partition coefficient by 2 orders of magnitude when the dissolved oxygen was decreased by a factor of 8. This lowered the equilibrium concentrations of methylmercury from 4 ppm as mercury to approximately 0.1 ppm. No effect on capacitance or the instability partition coefficient was noted when reducing agents such as ferrous iron were added. The use of clay instead of sand decreased the magnitude of the partition coefficient by 2, probably due to the surface area differences. These variables are shown in Table 4. With respect to variation of initial methylmercury chloride concentration, Feick et al. observed no effective change, the instability partition coefficients ranging from 8×10^{-2} to 4×10^{-2}.

TABLE 4
Instability partition coefficients for methylmercury chloride adsorption onto long-chained thiols at 24–25°C

Description	Time	[Hg^{++}] in ppm Dry sediment	Water	Inst. part. coef. $K = \frac{[Hg^{++}]_{H_2O}}{[Hg^{++}]_{sed}}$	pH	Cl$^-$	DO
Effect of varied initial methylmercury concentrations[18]							
NDM, armac-T, sand	7 days	100.0	8.0	8.0×10^{-2}	6.3	—	6.0
NDM, armac-T, sand	7 days	30.0	1.21	4.0×10^{-2}	6.4	—	5.7
NDM, armac-T, sand	7 days	19.0	0.65	3.4×10^{-2}	6.4	—	4.9
NDM, armac-T, sand	7 days	106.0	6.35	6.0×10^{-2}	6.7	—	9.4
Effect of dissolved oxygen and reducing agents on adsorption[18]							
NDM, armac-T, sand (5 g CaCO$_3$)	7 days	96.5	4.25	4.4×10^{-2}	6.9	—	7.8
NDM, armac-T, sand (low O$_2$1)	7 days	119.0	0.02	1.7×10^{-4}	7.0	—	1.2
NDM, armac-T, sand (5 g Fe + low O$_2$)	7 days	126.0	0.11	8.7×10^{-4}	7.6	—	1.2
NDM, armac-T, sand (5 g Fe + low O$_2$)	7 days	132.0	0.35	2.7×10^{-3}	8.4	—	1.2
NDM, armac-T, sand (old batch)	7 days	111.0	0.10	9.0×10^{-4}	7.2	—	1.1
Effect of utilization of clays instead of sand and armac-T[18]							
NDM, kaolin (5 g CaCO$_3$)	7 days	300.0	0.24	8.0×10^{-4}	7.0	—	9.1
NDM, sand	4 h	11.1	0.007	6.3×10^{-4}	5	11.0	—
NDM, sand	4 h	10.6	0.007	6.6×10^{-4}	7	0.9	—
NDM, sand	4 h	12.2	0.0003	2.5×10^{-5}	9	0.7	—
NDM, sand	4 h	11.3	0.0009	8.0×10^{-5}	5	142.0	—
NDM, sand	4 h	10.8	0.008	7.4×10^{-4}	7	110.0	—
NDM, sand	4 h	11.5	0.001	8.7×10^{-5}	9	112.0	—
NDM, sand	4 h	10.9	0.0001	9.5×10^{-6}	5	10,000.0	—
NDM, sand	4 h	10.5	0.001	9.2×10^{-6}	7	10,000.0	—
NDM, sand	4 h	11.0	0.0001	9.1×10^{-6}	9	10,000.0	—
Effects of varied initial mercury concentrations[17]							
NDM, sand	4 h	1.6	0.052	3.2×10^{-2}	7	—	—
NDM, sand	4 h	7.0	0.609	8.7×10^{-2}	7	—	—
NDM, sand	4 h	49.7	0.665	1.3×10^{-2}	7	—	—
NDM, sand	4 h	582.3	0.803	1.4×10^{-3}	7	—	—
NDM, sand	4 h	1,168.2	0.779	6.7×10^{-4}	7	—	—

TABLE 5
Partition coefficients for methylmercuric chloride with Acton peat (after Feick et al.[18])

Description	Time	Mercury conc. (ppm) Dry sediment	Water	$K = \frac{[Hg^{++}]_{H_2O}}{[Hg^{++}]_{sed}}$	pH	DO
Acton peat (aged 2 months)	7 days	2860	6.5	2.3×10^{-3}	5.3	0.4
Acton peat (aged 2 months)	7 days	143	0.048	3.4×10^{-4}	5.4	0.2
Acton peat (fresh)	7 days	2630	2.76	1.0×10^{-3}	5.1	0.4
Acton peat (fresh)	7 days	1470	1.0	6.8×10^{-4}	5.2	0.2

The changes noted in the partition coefficients with differing dissolved oxygen concentrations are significant, inasmuch as sediments could leach methylmercury to the environment in the presence of air or aerobic waters. Thus, the dredging of mercury-laden sediments where the sediment can change from an anaerobic condition to an aerobic state may add unwanted mercury to the environment, no matter what the form of the mercury originally adsorbed to the sediment.

Methylmercury with natural sediments

The only natural organically enriched sediment found in the literature that was investigated for its adsorptive properties for methylmercury chloride was peat, as described by Feick et al.[18] No changes in the adsorptive capacity or instability partition coefficients were observed with aging in the presence of air for 2 months. However, a significant increase in the instability partition coefficient and a decrease in the capacitance for methylmercury were noted when the initial mercury concentrations were increased by one order of magnitude, which, in turn, increased the sediment concentrations by an order of magnitude, as noted in Table 5. As was discussed previously, of the natural, organically enriched sediments studies, only peat and the ashland sediments demonstrated any traces of sulfur. Therefore, only these two sediments would have had the ability to bind methylmercury. Since Swedish humus, coniferous, deciduous, and grass soils and oceanic humic acids did not demonstrate the presence of sulfur by their ability to bind inorganic mercury, they would not bind methylmercury.

This becomes important when considering the ratio of methylmercury to total mercury for Swedish, American, and Japanese fish. In the United States and Japan, the sediments demonstrated significant sulfur contents near the outfalls where mercury pollution occurred. Thus, according to this investigation, the microbially converted methylmercury would be retained in the sediments and result in the fish having less methylmercury than in Sweden, where the methylmercury could more easily go into the water and thence be adsorbed by the fish. In Canada and Sweden, both of which are located in northern climates, the sediments apparently do not contain significant sulfur-containing organics and are located in lakes with great depths. Therefore, organic mercury would be the only form transported in the sediment because almost all natural sediments tend to bind the various forms of inorganic mercury.

CONCLUSIONS

Using principles obtained from the chemistry of mercury, the literature on mercury sorption on organics and the results of the sorption experimentation described earlier, an attempt has been made to analyze and compare the sorption characteristics of mercury in synthetic and natural organic sediments. From the previous discussion, it may be concluded that:

(1) All forms of organic matter adsorb inorganic mercury, but only sulfur-containing organics have a capacity greater than that of clays, sands, and synthetic media as shown in Fig. 4. These results agree with the basic chemistry of mercury.

(2) Even though methylmercury apparently has a

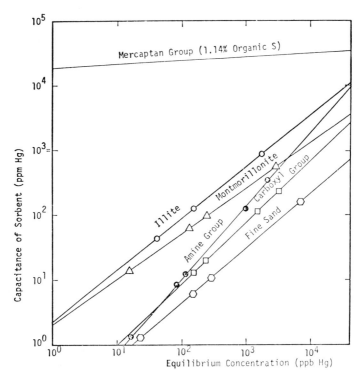

FIG. 4. Freundlich isotherms of various materials for the adsorption of $HgCl_2$ at 25°C and pH = 7.

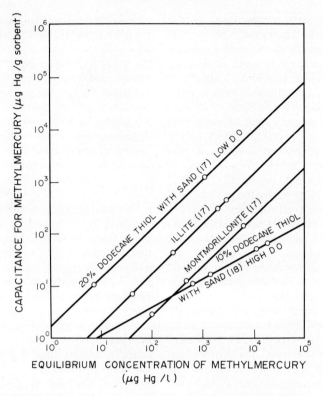

FIG. 5. Isotherms of methylmercury adsorption by sediment components at 25°C and pH = 7.

lesser attraction than mercuric chloride to the adsorbents tested, all sediments or scavengers containing sulfur possess a high affinity for methylmercury. Other organics and organic functional groups do not appear to have an affinity for methylmercury.

(3) The degree of binding of methylmercury by natural sediments depends on the composition and the environmental conditions of the organic sediments. Under anaerobic conditions, sulfur-containing sediments have a high binding capacity for methylmercury; however, in aerobic conditions, the capacitance of these sediments is reduced below that of the three-dimensional clays as shown in Fig. 5.

(4) The work of Widman and Epstein[26] and Bongers et al.[24] has demonstrated that polyethylene–nylon films, sands, and gravel inhibited the aqueous transfer of mercury to fish in aquarium studies. Since these materials have a low capacity for inorganic mercury and almost none for methylmercury, the inhibition mechanism must not be sorption or binding but stabilization of the colloidal sediment by reduction of sediment agitation.

(5) At a constant pH, the only synthetic organic sediments which were affected by chlorides were the alkanes and organic acids. They reduced in their capacitance for mercuric chloride with increased chlorides.

(6) The effects of pH on the attraction of mercuric chloride to the organic sediments studied were varied. The amines and the mercaptans tended to adsorb more at pHs of 5 and 9 while the alkanes and alkyl acids appeared to be unaffected by pH changes. As the pH was increased, the humate organics did not release mercury; however, they became water soluble, thus releasing a soluble mercury-humate complex with methylmercury. Adsorption of methylmercury by sulfur-containing organic sediments was also unaffected by pH fluctuations.

(7) When inorganic or organic mercury was adsorbed by sediment organics, the desorption was observed to be negligible.

REFERENCES

1. ANDERSSON, A., Mercury in soil, *Grundforbattring* **20**, 95 (1968), Dept. Sec. State Transl. Bureau No. 5433, 1970.
2. UI, J., Mercury pollution of sea and fresh water, its

accumulation into water biomass, *Rev. Inter. Oceanogr. Med.* **22** and **23** (1971).

3. STOCK, A. and CUCUEL, F., Die Quantitative Bestiminung Kleinster Quecksilbermengen, *Naturwissenschaften* **22**, 270 (1934).
4. WARREN, H. V., DELAVAULT, R. E., and BARAKSO, J., Some observations on the geochemistry of mercury as applied to prospecting, *Econ. Geol.* **61**, 1010 (1966).
5. KENNEDY, E. J., RUCH, R. R., and SHIMP, N. F., Distribution of mercury in unconsolidated sediments from southern Lake Michigan, studies of Lake Michigan bottom sediments No. 7, *Illinois Geol. Survey Environmental Geology Note*, 1971.
6. KLEIN, D. H. and GOLDBERG, E. D., Mercury in the marine environment, *Environ. Sci. Technol.* **4**, 765 (1970).
7. APPLEQUIST, M. D., KUTZ, A., and TUREKIAN, K. K., Distribution of mercury in the sediments of New Haven (Conn.) Harbor, *Environ. Sci. Technol.* **6**, 1123 (1972).
8. AIDIN'YAN, N. KH., TROITSKII, A. J., and BELAUSKAYA, G. A., Distribution of mercury in the soils of the USSR and Vietnam, *Geokhim.* No. 7, 654 (1964).
9. MATSUMURA, F., GOTOH, Y., and BOUSH, G. M., Factors influencing the transformation of mercury in river sediment, *Bull. of Environ. Contam. and Toxicol.* **8** (1972).
10. HEM, J. D., Chemical behavior of mercury in aqueous media, *Mercury in the Environment*, Geological Survey Prof. Paper 713, 19 (1970).
11. REIMERS, R. S., A stable carbon isotope study of a marine bay and a domestic waste treatment plant, Thesis, University of Texas, Austin (1968).
12. HATCH, W. R. and OTT, W. L., Determination of sub-microgram quantities of mercury by atomic adsorption spectrophotometry, *Analyt. Chem.* **40**, 2085 (1968).
13. WESTÖÖ, G., *Determination of Methylmercury by Gas Chromatography*, Swedish Water and Air Pollution Research Laboratory, Stockholm, Sweden, August 1970.
14. BALLINGER, D. G. and MCKEE, G. D., Chemical characterizations of bottom sediments, *J. Wat. Pollut. Control Fed.* **43**, 216 (1971).
15. JARVAS, M. J., GREENBERG, A. E., HONK, R. D., and RAND, M. C. (eds.), *Standard Methods for the Examination of Water and Wastewater*, 13th edn., American Public Health Association, February, 1970.
16. CHAPMAN, H. D. and PRATT, P. F., *Methods of Analysis for Soils, Plants and Waters*, Division of Agricultural Sciences, University of California at Riverside, California, 185, 1961.
17. REIMERS, R. S. and KRENKEL, P. A., The kinetics of mercury adsorption and desorption in sediments, *J. Water Poll. Control Fed.* **46**, (2) 352 (1974).
18. FEICK, G., JOHANSON, E. E., and YEAPLE, D. S., *Control of Mercury Contamination in Fresh Water Sediments*, Environmental Protection Technology Series EPA-R2-72-077, October, 1972.
19. TROST, P. B., Effects of humic-acid-type organics on secondary dispersion of mercury, unpublished PhD thesis, Colorado School of Mines, 1, 1970.
20. KRAUSKOPF, K. B., Factors controlling the concentrations of thirteen rare metals in sea water, *Geochim. cosmochim. Acta* **9**, (1 & 2) 1 (1956).
21. STROHAL, P. and HULJEV, D., Investigations of mercury pollutant interaction with humic acids by means of radiotracers, *Proc. Symposium on Nuclear Techniques in Environmental Pollution*, Int. Atomic Energy Agency, Vienna, February 1971.
22. DE GROOT, A. J., DE GOEIJ, J. J. M., and ZEGERS, C., Contents and behavior of mercury as compared with other heavy metals in sediments from the rivers Rhine and Ems, *Geologie Mijnb.* **50**, 393 (1971).
23. LYMAN, J. and FLEMING, F. H., Composition of sea water, *J. Marine Res.* **3**, 134 (1940).
24. BONGERS, L. H. and KHATTAK, N. M., *Sand and Gravel Overlay for Control of Mercury in Sediment*, Water Pollution Control Series 16080 HTE, January, 1972.
25. TRATNYEK, J. P., *Waste Wool as a Scavenger for Mercury Pollution in Waters*, Water Pollution Control Research Series 16080 HAB, April, 1972.
26. WIDMAN, M. A., and EPSTEIN, M. M., *Polymer Film Overlay System for Mercury Contaminated Sludge—Phase I*, Water Pollution Control Series 16080 HTZ, May, 1972.
27. CLARKSON, T. W., *ICES Invitational Symposium*, Chapel Hill, North Carolina, September 26, 1972.
28. FRIEDMAN, M., HARRISON, C. S., WARD, W. H., and LUNDGREN, H. P., *Sorption Behavior of Mercuric and Methylmercuric Salts on Wool*, Division of Water, Air, and Waste Chemistry Meeting, ACS, Los Angeles, California, 109, March 29–April 2, 1971.
29. FRIEDMAN, M. and WEISS, A. C., Mercury uptake by selected agricultural products and by-products, *Environ. Sci. Technol.* **6**, 457 (1972).
30. MCKAVENEY, J. P., FASSINGER, W. P. and STIVERS, D. A., Removal of heavy metals from water and brine silicon alloys, *Environ. Sci. Technol.* **6**, (13) 1109 (1972).

Sorption Phenomenon in Organics of Bottom Sediments
(Reimers et al.)

DISCUSSION by E. A. JENNE†
U.S. Geological Survey, Menlo Park, California

INTRODUCTION

The preservation and improvement of our environment require that we develop an understanding of the processes whereby toxic substances are sorbed by sediments and transported in fluvial systems. Reimers et al. at this conference have expressed the hope that studies of the type they have carried out will facilitate the development of "... a predictive model for the fate of trace metal contaminants in the environment...." Their study involved laboratory simulation using certain specific organic compounds deposited onto the surfaces of river sand. Therefore it is appropriate to inquire in this discussion as to the normal organic carbon content of sediments and the dominant reactive groups of these organics for comparison with the laboratory experiments. The significance of organic matter relative to other trace element sinks in sediments and aspects of the collection and presentation of mercury sorption data are also assessed. Finally, the additional information required to model the fluvial transport of mercury at the level of detail required by the use of sorption isotherms of mercury onto particulate organics is considered.

Additional experimental data were obtained for and are presented in this discussion. The additional chemical characterizations were made on portions of the same materials used by Reimers et al. These data have been supplemented by other published information to aid in the interpretation of the findings of Reimers et al.

CONCENTRATION, COMPOSITION, AND REACTIVE GROUPS OF NATURAL ORGANICS

To the extent that it is an important sink for mercury, the quantity of particulate organic matter in sediments and its temporal and spatial homogeneity may be of significance to deterministic (mechanistic) modeling of trace-element sorption and transport in surface waters, and perhaps ground waters as well. Reimers et al. have added organic coatings to relatively clean Tennessee river sand in the amount of 20% (w/w). This amounts to 15.2, 16.2, and 14.2% organic carbon, respectively, for stearic acid, n-octadecylamine, and n-dodecyl mercaptan. Reported organic carbon values from some published reports for bed sediments range from 0.3 to 8.6% in rivers, 1.5 to 13.8% for harbors and bays, 0.1 to 3.0% in the continental shelf, and in deeper marine sediments from 0.08% in Pacific Ocean and red clays to 11.5% in clayey muds of the Black Sea. In the suspended sediment of major rivers, reported organic carbon contents range from 2.3 to 9.0%, excepting a colored Florida water which contained 35%.[1] The organic carbon content of sediments of a given water body may vary by an order of magnitude, owing in large part to variations in particle size distribution.

Thus, the organic carbon content of most sediments is less than half of the value used by Reimers et al.

Soluble organics occur in significant quantities in ground waters as well as in surface and interstitial waters and sometimes exceed the quantity in transport in the particulate phase where suspended sediment concentrations are low. Dissolved organic carbon commonly occurs to the extent of 3–10 mg/l in natural surface waters with upper values of 7.7 mg/l for the larger streams of the northwestern United States and up to 27–29 mg/l for lakes.[2,3] Interstitial waters of sediments of eutrophic lakes may contain much higher levels. The interstitial water of Herman Lake, South Dakota, was reported to contain 209 mg/l of organic carbon.[4] Concentrations of organic carbon as high as 500 mg/l have recently been found in the interstitial waters of sediment in Tampa Bay, Florida.[1] In contrast, surface water of the open ocean contains 0.5–5 mg/l of dissolved organic carbon,[5,6] and the interstitial waters of marine sediments (Pacific and Okhotsk Sea clayey sediments) contain from 10 to 19 mg/l.[7]

Soluble organics are of much greater importance in the geochemical cycling of mercury and other

---

†Approved for publication by the Director, US Geological Survey.

trace elements than has been generally recognized. In studies of the sorption of trace elements by earth materials containing organics, it is necessary either to determine the quantity of the trace element present in a soluble organic complex or demonstrate that it is of little importance. This is vital because the sorption characteristics of trace elements complexed by soluble organics are much different than those of the uncomplexed or even of most inorganic ligand complexes. This source of error was not evaluated by Reimers et al. or by Feick et al.[8] whose data were utilized in the paper under discussion.

Reimers et al. have used organics containing reactive carboxyl (stearic acid), amine (n-octadecylamine), and mercaptan (n-dodecyl mercaptan) groups to represent the reactive groups of particulate organics of bottom sediments. The exact nature of the bonds between natural organics and trace elements is not very clear; different trace elements may in fact be preferentially bonded by different reactive groups. Thus, fulvic acids are said to complex more iron per unit carbon than do humic acids.[9,10] Although the carbon and sulfur content of fulvic acids do not differ greatly from humic acids, fulvic acid (from a podzolic soil) contained approximately twice the quantity of acidic functional groups (primarily carboxyl, phenolic, and enolic hydroxyl groups) as did the companion humic acid.[11]

Imide and amide nitrogen groups and amino acids are potential trace-element bonding sites. However, it appears that carboxylic and aromatic hydroxy acid sites dominate the soluble organic fraction of fluvial systems although the possibility that this is an artifact of the extractive techniques used must be borne in mind. The concentration of organic carbon is not a good index of the concentration of available complexing sites, but in many surface waters major fractions of the dissolved trace elements may be organically complexed.

TRACE-ELEMENT SINKS

The general presumption that organic matter is the predominant sink for mercury in freshwater sediments results from the known ability of plankton, peat, etc., to sorb mercury as well as from the positive correlations frequently found between mercury and organic carbon contents in sediments. However, there are several lines of evidence which indicate that organic matter is not the only important trace element sink in sediments. A number of recent studies suggest that additional trace-element sinks are the hydrous oxides of iron and manganese. This was generally the situation for first transition series metals in soils and sediments,[12] and stream sediments,[13] cobalt in a lake,[14] and in the ocean beyond the continental shelf excepting anoxic regions.[15] In addition, Reimers et al. obtained mercury sorption values for selected reference clays which are higher than most of the values obtained with sands to which they had added 20% (w/w) organic matter as long chain alkanes (one of which contained a mercaptan group) (Fig. 1). These authors also found nearly as much mercury sorbed by the uncoated sand itself as by the organic coatings.

Organic carbon and organic sulfur, as well as oxidic iron and manganese, were determined on splits of 80–120 mesh Tennessee river sand and the reference clays used by Reimers et al. (Table 1). These data suggest (Fig. 2) that the oxidic iron present as an impurity in the reference clays may account for the differences in their sorption capacities.

Another sink which may be significant in aquatic systems is the precipitation of mercury as the sulfide or, more likely, coprecipitation with ferrous sulfide in anaerobic sediments. Several recent studies show that mercury is concentrated in anoxic sediments.[16-18] Since there is a tendency for anaerobic conditions to coincide with high organic-matter levels, it is difficult to separate the effect of these two factors.

Oxidic iron and manganese were determined on samples from Bellingham Bay, Washington, provided by Mike Bothner and Roy Carpenter of the

TABLE 1
Mercury sorption by and chemical characterization of samples used by Reimers et al. (this Conf. Proc.)

Earth material	Mercury[a] sorbed (mg/kg)	Free Oxides		Organic	
		Fe (mg/kg)	Mn	Cn (%)	S
Kaolinite	30	1108	1.6	0.04[b](0.14)[c]	NA[d]
Montmorillonite	77	3032	492	0.83[b](0.21)[c]	0.034
Illite	167	5407	243	0.79[b](0.11)[c]	0.46
Sand[d]	10(13)[e]	1815	91	0.04[b](0.1)[b]	0.012

[a]Initial mercury concentrations ranged from 810 to 950 µg/l, values averaged for all pH values at 100 mg/l of chloride.
[b]Organic carbon by combustion.
[c]Chemical oxygen demand measurements of Reimers et al.
[d]NA = not analyzed for.
[e]70–100 mesh was used for sorption studies whereas 80–120 mesh were used for physicochemical characterization.
[f]Average if one low value omitted.

Discussion

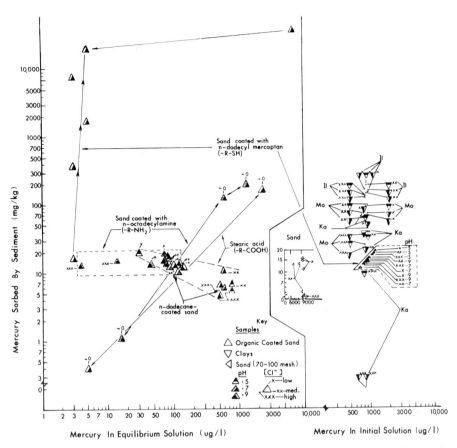

FIG. 1. Quantity of mercury sorbed by various earth materials and sand coated with specific organics as a function of mercury concentration. (After Reimers *et al.*, this Conference Proceedings.)

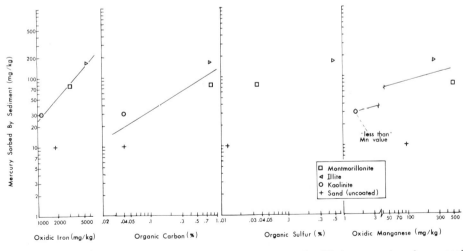

FIG. 2. Concentration of mercury sorbed versus concentrations of oxidic iron, organic carbon, organic sulfur, and oxidic manganese in the 70–100 mesh sand and the reference clays used by Reimers *et al.* (this Conference Proceedings). (Mercury values are averaged across pH levels for ≈ 100 mg/l of chloride concentration.)

University of Washington. Mercury concentrations were found to correlate fairly well with total carbon, organic carbon, and oxidic iron but not with oxidic manganese.[19] The mercury–total carbon relation results from the very minor amounts of inorganic carbon as compared to organic carbon present in the samples analyzed. Similar analyses were performed on samples from southern Lake Michigan which were provided by Joyce (Kennedy) Frost and Neil Shimp of the Illinois Geological Survey. A general correspondence between mercury concentrations and clay (minus 2 μm) content was found in the 53 samples examined although the points scatter considerably. This implies that mercury is, indeed, concentrated in the clay-size fraction, but not in a uniform manner. Therefore, the chemical parameters were normalized to a 100% clay basis. Positive linear relationships were thus found which were similar for mercury versus oxidic iron, organic carbon, and oxidic manganese.[19] However, there also appeared to be a positive correlation between organic matter and oxidic iron and between oxidic iron and oxidic manganese. Therefore, these parameters may not be independent variables.

DATA COLLECTION AND PRESENTATION
Partitioning coefficients

The main justification for publishing environmental biogeochemical data in general, and sorption–desorption data in particular, in the archival literature is that the information contributes toward either conceptual modeling (i.e. the understanding of reactions and processes) or mathematical modeling of biogeohydrochemical systems. Sorption–desorption data must be so collected as to be susceptible to mathematical description in order to be of more than local usefulness. Most such studies utilize either adsorption isotherms, exchange constants (including stability constants for organics), or distribution coefficients (also known as partition coefficients, instability coefficients, etc.).

Nearly all of the experimental data in the paper by Reimers et al. has been collected into a single plot (Fig. 1). It is apparent that most of the data (i.e., that in the dashed "boxes") show little effect of particular reactive groups with which the sands were coated or of chloride values (≈ 10, ≈ 100, $\approx 10,000$ mg/l). Indeed, the mercury sorption values for the n-dodecane (no reactive group) coated sand and the uncoated sand itself all fall in the same range (4–25 mgHg/kg sample) as the organic-coated sand although the concentration of mercury in the equilibrium solution varied by more than two orders of magnitude. This is a very surprising result. In view of the lack of data on analytical precision or of any error estimates, it seems possible that these differences among the sorption values simply represent experimental error.

It will be noted (Fig. 1, right-hand side) that for the n-dodecyl mercaptan-coated sand, the amount of mercury sorbed increased slightly with an increase in *initial* mercury concentration. This suggests the beginning of development of a Freundlich sorption isotherm wherein the solution concentration of mercury is of greater importance than chloride concentration or pH. In a later set of experiments with $Cl \approx 0$ and $pH = 7$, Reimers et al. obtained the expected sorption isotherm development (Fig. 1, left-hand side) with indications of a maximum value for mercury sorption for the stearic acid and n-octadecylamine. Because the sorption capacity of the substrate influences extrapolation of the results, the data should be described by the Langmuir equation rather than the Freundlich equation.

Percent sorbed

Whenever the amount of mercury or other trace element sorbed by the solid is in excess of 90% of either the quantity or concentration added, the experiments should be repeated using higher initial concentrations. This is because a sorption curve must approach the 100% line in an asymptotic manner, regardless of the remaining sorption capacity of the solid. Thus, plots of percent sorbed versus time may give a *false* appearance of equilibrium. Likewise, when investigating the kintics of sorption reactions it is important that the maximum amount sorbed be less than 90%. Another reason that sorption parameters calculated from sorption data in the plus-90%-sorbed range are of questionable value is that analytical measurements become substantially less accurate and precise as the decrease in the total quantity remaining in solution approaches two orders of magnitude of that initially added. Frequently, differences between successive analyses in the low concentration range are subject to large errors.

pH

A very critical parameter in most sorption studies is pH. It should always be carefully measured in equilibrium solutions, even if not controlled. Reimers et al. found the pH to vary by 1 pH unit from the stated initial value. When the pH is unstable, sorption values should be plotted as a function of measured hydrogen ion activity and the amount sorbed at any particular pH can be read from these plots for further plotting or calculations.[21]

The combination of a high percentage of mercury sorbed from solution and uncontrolled pH variations may account for much of the apparent experimental error.

Chemical oxygen demand measurements

The chemical oxygen demand measurement as an estimate of organic carbon content of the reference clays and Tennessee river sand gives values that differ significantly from those obtained by dry combustion of splits of the same samples (Table 2). The value for the kaolinite is overestimated and the values for the other clays are markedly underestimated by the chemical oxidation demand method.

TABLE 2

Comparison of dry combustion and chemical oxygen demand estimates of organic carbon in reference clays and river sand

Sample	Organic carbon	
	Dry combustion (Huffman Laboratories, Inc., Wheatridge, Col.) (%)	Wet combustion (Chemical Oxygen demand) (Reimers et al.) (%)
Kaolinite	0.04	0.14
Illite	0.79	0.11
Montmorillonite	0.83	0.21
Sand (70–100 mesh)	0.04	
Sand (80–120 mesh)		0.1

Techniques

The accuracy of laboratory experimentation with mercury,[22] silver,[23] and probably all trace elements can be improved by using a given set of containers which have been pre-equilibrated for a given concentration range, especially for concentrations of 1 μg/l or less.

Mercury sorption studies also require mass balance measurements to insure that significant quantities are not being lost to containers or by vaporization.

Since high organic contents of sediment are known to interfere with mercury measurement by gold foil retention,[24] it would have been desirable for Reimers et al. to demonstrate that this was not a source of error in their study.

MODELING

Many papers currently being published in the field of environmental chemistry are partially justified on the basis of their contribution to simulation modeling of aquatic systems. It is, therefore, timely to consider the potential significance of such studies to simulation modeling.

The development of a simulation model may require input from the disciplines of hydrology, physics, chemistry, biology, engineering, and meteorology. Clearly, the profitable inclusion of chemical information such as the speciation of solute mercury and sorption isotherms of mercury by particulate organics (Reimers et al.) requires that information be introduced from the other disciplines which may have a similar magnitude of effect on the reliability of the model. For example, this may necessitate the incorporation of data on the rate of biological uptake of specific chemical species as a function of solute concentration and time, the extent of sediment interchange between the bed and the overlying fluid, the concentration of other sorption sinks in the sediment, etc. This is not to suggest that it is incumbent upon the environmental chemist to develop the entire model but rather that he needs to place his planned contribution in perspective. He must be cognizant of the equivalent level of detail or complexity of the other parts of the model which the introduction of a given detail of chemical information will call for.

SUMMARY

Quantitative estimates of trace element transport in fluvial systems requires the development of simulation (mathematical) models. However, the first step is the development of conceptual models which qualitatively relate the sinks, sources, pathways, and processes. In turn, this requires detailed studies of the important equilibrium and kinetic chemical parameters which control the reactions of a trace element such as mercury in the fluvial environment. Particulate organic carbon, oxidic iron, oxidic manganese, and sulfides are the predominant sinks for mercury in sediments. Thus, studies such as the one of Reimers et al. are valuable to the extent that they produce additional new insight into the relevant reactions, provide equilibrium or kinetic values which are needed for simulation modeling or that they systemize available information. The contribution of Reimers et al. lies primarily in their systemization of their own and the previously available information on the sorption potential of particulate organics for dissolved mercury.

Acknowledgements—It is a pleasure to acknowledge the analytical assistance of James W. Ball and the bibliographic and plotting assistance of J. M. Burchard; thanks are due to Margaret Eagle for supplying parts of the samples used by Reimers *et al.*

REFERENCES

1. MALCOLM, R. L., written communication, November 1973.
2. NELSON, K. H. and LYSYJ, IHOR, Organic content of southwest and Pacific coast municipal waters, *Environ. Sci. Technol.* **2**(1) 61–62 (1968).
3. MALCOLM, R. L., oral communication, November 1973.
4. GAHLER, A. R., Sediment-water nutrient interchange, *Proc. Eutrophication-Biostimulation Assessment Workshop* (Middlebrooks, E. J., Maloney, T. E., Powers, C. F., and Kaack, L. M., eds.), pp. 243–257, 1969.
5. BADER, R. G., HOOD, D. W., and SMITH, J. B., Recovery of dissolved organic matter in sea-water and organic sorption by particulate material, *Geochim. cosmochim. Acta* **19**, 236–243 (1960).
6. MENZEL, D. W., The distribution of dissolved organic carbon in the Western Indian Ocean, *Deep-Sea Res.* **11**, 757–765 (1964).
7. STARIKOVA, N. D., Organic matter of the liquid phase of marine muds, *Int. Oceanographic Congress Preprints*, pp. 980–981, 1959.
8. FEICK, G., JOHANSON, E. E., and YEAPLE, D. S., *Control of mercury contamination in fresh water sediments*, Environ. Prot. Tech. Ser., EPA-R2-72-077, 1972.
9. D'YAKONOVA, K. V., Iron-humus complexes and their role in plant nutrition, *Sov. Soil Sci.* **7**, 692–698 (1962).
10. HENRY, A. S., The course of behavior of applied zinc to soil containing different levels of freshly applied organic matter, PhD thesis, Utah State Univ., Logan, Utah, 1966.
11. MALCOLM, R. L., Mobile soil organic matter and its interactions with clay minerals and sesquioxides, PhD thesis, Univ. North Carolina, Raleigh, NC, 1964.
12. JENNE, E. A., Controls on Mn, Fe, Co, Ni, Cu, and Zn concentrations in soils and water: The significant role of hydrous Mn and Fe oxides, in *Trace Inorganics in Water*, Amer. Chem. Soc., Adv. in Chemistry Series No. 73, pp. 337–387, 1968.
13. ROSE, A. W., DAHLBERG, E. C., and KEITH, M. L., A multiple regression technique for adjusting background values in stream sediment geochemistry, *Econ. Geol.* **65**, 156–165 (1970).
14. MARSHALL, J. S. and LE ROY, J. H., Iron, manganese, cobalt and zinc cycles in a South Carolina reservoir, pp. 465–473, in Nelson, S. J. (ed.), *Proc. Third Nat. Symposium in Radioecology, May 10–12, 1972*, Conf. 710501-P7 (Available from NTIS, Springfield, Va.).
15. CARVAJAL, M. C. and LANDERGREN, STURE, 4, Marine sedimentation processes—the interrelationships of manganese, cobalt and nickel, *Stockholm Contributions in Geology* **XVIII**(4) 99–122 (1969).
16. BURTON, J. D. and LEATHERLAND, T. M., Mercury in a coastal marine environment, *Nature* **231**, 440–441 (1971).
17. THOMAS, R. L., The distribution of mercury in the sediments of Lake Ontario, *Can. J. Earth Sci.* **9**(6) 636–661 (1972).
18. VERNET, J.-P. and THOMAS, R. L., The occurrence and distribution of mercury in the sediments of the Petit Lac (Western Lake Geneva), *Eclog. geol. Helv.* **65**(2) 307–316 (1972).
19. JENNE, E. A., Unpublished data.
20. REIMERS, R. S. and KRENKEL, P. A., Personal communication, November 1973.
21. ANDERSON, B. J., JENNE, E. A., and CHAO, T. T., The sorption of silver by poorly crystallized manganese oxides, *Geochim. cosmochim. Acta* **37**, 611–622 (1973).
22. ROSAIN, R. M. and WAI, C. M., The rate of loss of mercury from aqueous solution when stored in various containers, *Analytica chim. Acta* **65**, 279–284 (1973).
23. CHAMBERS, C. W., PROCTOR, C. M., and KABLER, P. W., Bacterial effect of low concentrations of silver, *J. Am. Wat. Ws. Ass.* **54**, 208–216 (1962).
24. WEISSBERG, B. G., Determination of mercury in soils by flameless atomic absorption spectrometry, *Econ. Geol.* **66**, 1942–1947 (1971).

SESSION IV

TRANSPORT MECHANISMS (2)

ROLE OF HYDROUS METAL OXIDES IN THE TRANSPORT OF HEAVY METALS IN THE ENVIRONMENT

G. Fred Lee

Institute for Environmental Sciences, University of Texas, Dallas, Texas

INTRODUCTION

The presence of heavy metals in natural water systems has in some instances caused significant ecosystem degradation. In the past the concern for heavy metals in the environment generally focused on milligram per liter concentrations of these contaminants. Studies conducted during the past few years have shown that while concentrations of many heavy metals approaching milligram per liter levels will cause acute lethal toxicity; microgram per liter levels will cause significant adverse sublethal toxicity such as impairment of an organism's reproductive capacity. Water quality standards for many heavy metals in many states are still at the milligram per liter concentrations. Within a few years, these will be reduced by a hundred- to a thousandfold in order to protect aquatic ecosystems and man from excessive exposure to heavy metals. As water quality standards are reduced to lower levels, a much better understanding of the aqueous environmental chemistry of heavy metal contaminants must be available in order to properly evaluate the significance of a heavy metal discharge to the environment.

One of the first steps in any systematic study of the aquatic chemistry of a potential contaminant is an elucidation of the principal modes of transport and reservoirs for the contaminant. For heavy metal contaminants such as copper, cadmium, lead, zinc, nickel, mercury, iron, and manganese, the hydrous oxides of aluminum, iron, and manganese may play dominant roles in determining reservoirs and modes of transport of these metals. This paper discusses the potential significance of hydrous metal oxides in the environmental chemistry of heavy metal contaminants in natural water systems.

The ability of hydrous metal oxides to interact with heavy metals has been known for many years. One of the methods used to concentrate heavy metals from aquatic systems prior to analysis involves the co-precipitation of these metals with ferric or aluminum salts. Kolthoff and Sandell[1] in their classical text on *Quantitative Inorganic Analysis* discuss the co-precipitation of heavy metals with ferric, manganese and aluminum salts.

Sandell[2] in his book on colorimetric metal analysis presents an extensive discussion of the use of hydrous metal oxides for the recovery of heavy trace metals from dilute aqueous solutions. The advent of new, more sensitive analytical techniques, such as atomic absorption spectrophotometry and improved complexing agents, has virtually eliminated the use of hydrous metal oxides as co-precipitation agents as a means of improving sensitivity of analytical procedures. Renewed interest was developed in this technique in the 1940s and 1950s as a means of recovery of radioisotopes from natural water systems. Overman and Clark[3] in their text on radioisotope techniques discuss the factors influencing the recovery of radioisotopes from environmental samples by precipitation techniques.

CHEMISTRY OF HYDROUS METAL OXIDES

In order to understand the role that hydrous metal oxides may play in the environmental chemistry of heavy metal contaminants, it is essential to have some knowledge of the environmental chemistry of hydrous metal oxides. Stumm and Morgan[4] in their text on *Aquatic Chemistry* provide an introduction to the general topic of the aqueous environmental chemistry of hydrous metal oxides. This work should be consulted for further information on this topic. They also briefly discuss the potential importance of hydrous metal oxides in the transport and aqueous environmental chemistry of various heavy metals such as cobalt, iron, nickel, and zinc in natural water systems.

Parks[5] summarizes the factors controlling the sign and magnitude of surface charge of oxides and mineral oxides, especially hydrous metal oxides. He notes that the metal oxides exhibit ion exchange properties and the ion exchange capacity of the simple oxides arises from the existence of a pH dependent surface charge. According to Parks, in acid solutions, the surface charge is positive and therefore the hydrous metal oxides act as anion exchangers. In basic solutions, the surface charge is negative and they are cation exchange particles. In neutral solutions, the surface charge is mixed, both plus and minus and the particles show a limited capacity for exchange of both cations and anions. Parks also discusses the hydroxo complexes in solutions of a precipitate from hydrous metal oxides. Much of the discussion on the solution characteristics of hydroxo complexes is directly applicable to the complexes on the surfaces of hydrous metal oxide coatings on various types of particles as well as on discrete hydrous metal oxide particles.

Parks[5] notes that the charge on hydrous metal oxides is instrumental in determining the state of dispersion, rheology, and the extent to which the solids act as ion exchangers for sorption sites. He further notes that it is possible that these materials could play important roles in the concentration of trace elements in natural water systems.

Morgan[6] reported that the oxidation of manganese(II) by dissolved oxygen yielded a stoichiometric $MnO_{1.9}$ only under highly alkaline conditions. Oxidation under other conditions led to considerable adsorption of manganese(II) from solution.

Hem[7] presented an extensive discussion of the environmental chemistry of manganese in natural waters. His work, as well as that of Stumm and Morgan[4] and Delfino and Lee,[8] should be consulted for further information on the environmental behavior of manganese(II) and manganese hydrous oxide. In general, it has been found that under reducing conditions (absence of dissolved oxygen) the manganese dioxide is reduced to manganese(II). Under oxidizing conditions, however, both manganese(II) and manganese(IV) are present. Although thermodynamically unstable, manganese(II) can exist in the presence of dissolved oxygen because of the slow rates of reaction between the two species under acid-neutral or slightly alkaline pH conditions.

Stumm and Lee[9,10] studied and reviewed the aqueous environmental chemistry of iron. As contrasted to manganese, under oxidizing conditions such as in the presence of dissolved oxygen, ferric iron is the only species found in slightly acid to alkaline pH range. Ferrous iron is stable in the presence of dissolved oxygen under strongly acid conditions such as would occur in extreme cases of acid mine drainage. In addition to precipitating as a hydroxide, ferrous iron can precipitate as a carbonate under conditions of moderate to high carbonate alkalinity. Ferric iron precipitates as a hydroxide, generally in an amorphous form.

Langmuir and Whittemore[11] discuss the characteristics of hydrous ferric oxide precipitates in natural water systems. Ferric iron tends to form complexes with natural water organics. Subsequent sections of this paper deal with the interactions of organics and ferric iron with respect to their potential significance in iron transport.

The aqueous environmental chemistry of aluminum is somewhat simpler than that of manganese and iron due to the single oxidation state involved. There have been numerous studies on the various characteristics of aluminum hydroxide precipitates. Hem et al.[12] have reviewed this work and discussed their studies which demonstrate a reaction between aluminum hydroxide and dissolved silica to form an alumino silicate mineral with a clay-like structure. Schenk and Weber[13] and Porter and Weber[14] found that ferric iron interacts with silica to form soluble complexes. Elderfield and Hem[15] have investigated the characteristics of aluminum hydroxide in natural waters and have found that there are $Al(OH)_3$ particles present in the solution under conditions where aluminum hydroxide should be soluble. These particles are formed on aging aluminum hydroxide. Elderfield and Hem propose that these materials represent an aged polymeric aluminum hydroxide complex which is in the process of forming gibbsite. Hahn and Stumm[16] studied the adsorption of aluminum on a silica dispersion. They reported that the destabilization of silica dispersions results from specific adsorption of positively charged hydroxo aluminum complexes onto the negatively charged colloid surface causing a decrease and ultimately a reversal of the sign of the particle's surface potential.

There are a number of practical applications of the use of the sorption ability of hydrous metal oxides in water and waste-water treatment. No attempt will be made to review the literature in this area. However, such a review would show that the use of iron or aluminum in domestic and industrial water treatment for the purpose of coagulation is often accompanied by the removal of significant amounts of trace metals, organic contaminants and other chemical species from the water. It is reasonable to propose, based on the previous studies, that the normal water treatment practice of coagulation tends to reduce the concentrations of these materials that enter domestic and industrial process or drinking water. Additional evidence for the potential significance of hydrous metal oxides of iron and aluminum in removing trace contaminants from natural waters is provided by the studies that were done during the 1950s on the removal of radioisotopes in water and wastewater treatment processes.

SOURCES OF HYDROUS METAL OXIDES FOR NATURAL WATER SYSTEMS

Hydrous metal oxides can arise from a variety of sources including the weathering of various mineral species. They enter natural water systems from both surface and groundwater. Generally in a groundwater system they would occur in the reduced oxidation states such as manganese(II) and iron(II). Upon contact with the natural water which contains oxygen, they oxidize to the hydrous metal oxide. The relative rates of oxidation of iron and manganese have been studied in detail. It has been reported by Stumm and Lee[10] and Morgan and Stumm[17] that while iron is readily oxidized by dissolved oxygen to the ferric form in the alkaline-neutral to slightly acid pH range, manganese on the other hand requires a much higher pH for equivalent rates of oxidation. A considerable part of the manganese oxidation may take place at the surface of particles such as calcite where there is a microzone of higher pH. Also, the manganese oxidation may be mediated to a considerable extent by microorganisms.

In lakes with anoxic sediments which have reducing conditions, it is generally found that both iron and manganese would tend to migrate in the sediments through the interstitial waters until they come in contact with oxygen, where a precipitation of the hydrous metal oxide should occur. Generally, the precipitation of iron would occur first. In lakes with anoxic hypolimnia, considerable concentrations of iron and manganese in their reduced state do build up in the water column below the thermocline. As a result of thermocline erosion, generally due to high-intensity wind stress, there could be a continual production of hydrous metal oxides which would become part of the epilimnion. Stauffer and Lee[18] have studied this mode of transport for phosphorus in Lake Mendota as well as other Wisconsin lakes. It is the most significant source of phosphorus during the summer. This source is one of the dominant controls of the blue-green algal blooms in the lake throughout the summer period.

Since the hypolimnion often contains higher concentrations of iron and manganese in their reduced forms, thermocline erosion and leakage of hypolimnetic waters at the thermocline sediment interface may be important sources of freshly precipitated hydrous metal oxides in the surface waters of lakes.

PREVIOUS STUDIES ON SIGNIFICANCE OF HYDROUS METAL OXIDES IN NATURAL WATER SYSTEMS

There have been numerous studies which point to the potential significance of hydrous metal oxides in influencing chemical contaminants in the environment. Jenne[19] has proposed that the hydrous metal oxides of manganese and iron are the principal control mechanisms for cobalt, nickel, copper, and zinc in soils and freshwater sediments. He states that the common occurrence of these oxides as coatings allows them to exert a chemical activity far in excess of their total concentrations. He further indicates that the uptake or release of these metals from these oxides is a function of such factors as increased metal ion concentration, the concentrations of other heavy metals, pH, and the amount and type of organic and inorganic complex formers in solution. Jenne claims that the information available on the factors that control copper, zinc, nickel, and cobalt in natural waters suggests that the organic matter, clays, carbonates, and precipitation as discrete oxides or hydroxides cannot explain the aqueous environmental chemistry of these elements. According to Jenne, this explanation must include, as one of the dominant factors, the environmental behavior of the hydrous oxides of iron and manganese. The primary basis for Jenne's remarks is the literature on the behavior of these metals in soil systems. It is certainly reasonable to extend this behavior to the aquatic sediment systems since, in some respects, they are somewhat like some soils. There are significant differences, however, between sediments and soils that must be considered in any specific location and care must be exercised in extrapolating soil chemistry studies to the area of aquatic chemistry of sediments.

Morgan and Stumm[17] have presented a general review of the role of hydrous metal oxides in limnological transformations. Other studies in this area include the work of Shimomura et al.,[20] who found that mercuric ions in the presence of chloride could be adsorbed onto ferric hydroxide. This adsorption was independent of pH but was dependent on the chloride, bromide and iodide content.

Lockwood and Chen[21] reported that mercury(II) adsorption by manganese oxides was rapid, taking place in a few minutes, when added to aged suspensions of MnO_2 at low ionic strength. The addition of sodium chloride at 0.6 M repressed adsorption below pH 9 but not above pH 10. The addition of sodium perchlorate at 0.6 M decreased the rates of adsorption by an order of magnitude. They propose that the uncharged $Hg(OH)_2$ is the adsorbed species. They concluded that MnO_2 may be an important mercury scavenger in fresh and brackish waters.

Krauskopf[22] discussed the factors controlling the concentrations of Zn, Cu, Pb, Bi, Cd, Ni, Co, Hg, Ag, Cr, Mo, W, and V in seawater. He noted that one of the principal mechanisms for controlling the concentrations of these various elements in seawater was the adsorption on hydrated ferric oxide or manganese dioxide.

Schindler[23] discusses the heterogeneous equilibria for hydrous metal oxides in seawater and specifically notes the potential importance of these oxides in the seawater systems. Slowey et al.[24] noted that the distribution of copper, manganese, and zinc was related to the distribution of ferric hydroxide in these waters.

Gibbs[25] has demonstrated the potential importance of metallic coatings of hydrous metal oxides on river-borne sediments as one of the major phases responsible for the transport of transition metals in natural water systems. He notes that these metallic coatings are mainly ferric hydroxide.

Posselt et al.[26] studied the sorption of metal ions and tensioactive organic substances found on hydrous manganese dioxide. The metal ions investigated included Ag^+, Ba^{2+}, Mg^{2+}, Mn^{2+}, Nd^{3+}, Sr^{2+}, anionic, nonionic and cationic surface active agents.

They found that the rates of sorption of all the metals studied were rapid, with equilibrium attained within a matter of several minutes. The cationic organic solute was also rapid. However, neither the anionic or nonionic organic solute sorbed to any significant extent. An exchange-type mechanism appears to be the principal mode of metal ion sorption. The equilibrium distributions between the sorbate and sorbent fit a Langmuir sorption equation with equilibrium capacities in the range of 0.1–0.3 mole/mole of manganese dioxide.

Murray et al.[27] reported that group one and two cations are strongly sorbed on manganese dioxide. They found that the sorption was independent of small pH changes at high concentrations of the cations and was highly pH dependent at low concentrations of the cations. They propose that in dilute solutions the adsorption occurs as counter ions in the diffuse double air, while in the high concentration the sorption occurs within the manganite lattice. They also reported that Ni^{2+}, Cu^{2+} and especially Co^{2+} exhibited marked specific adsorption on the MnO_2.

Hingston et al.[28] studied the adsorption of selenite on goethite and found that the specific adsorption increased the pH of the suspension and the negative charge of the oxide surface. They proposed that the mechanism of adsorption involved the release of water molecules from the surface when selenite ion is adsorbed.

The above discussion is not meant to represent a comprehensive review of the previous studies on potential significance of hydrous metal oxides on the environmental chemistry of various contaminants. Instead, it is designed to be illustrative of the types of studies that have been done in this area. It is clear from the literature that hydrous metal oxides could play a very significant role in a wide variety of chemical contaminants in natural water systems. Certainly any discussion on the transport and cycling of heavy metals in natural waters would be incomplete if it did not consider the role of hydrous metal oxides. It is somewhat surprising to find that in the fall of 1972, a conference was held that was sponsored by the US Environmental Protection Agency, National Science Foundation and Battelle-Columbus Laboratories, concerned with the cycling and control of metals in the environments. Examination of the proceedings of this conference[29] points to the fact that none of the authors chose to discuss the potential significance of hydrous metal oxides as a mode of transport of heavy metals in natural water systems. It is felt that this is a significant deficiency of the 1972 conference since it is certainly improper to discuss heavy metal contaminant cycling without considering the potential role of hydrous metal oxides.

FACTORS AFFECTING THE HYDROUS METAL OXIDE METAL TRANSPORT

Some of the factors that could be significant in influencing the role that hydrous metal oxides play in metal ion transport are discussed below.

The review by Jenne[19] should be consulted for further information on the potential role of various factors in controlling the transport of heavy metals in environmental systems, especially for references on factors controlling heavy metals in sediments and soils.

Age of precipitates

The age of a hydrous metal oxide precipitate could play a significant role in the ability of these precipitates to interact with heavy metals and other chemical contaminants. Morgan and Stumm[17] summarized the work of several investigators (also see Morgan and Stumm[30]) on the sorption characteristics of manganese dioxide for various metallic species. They found that freshly precipitated manganese dioxide has a very significant sorption capacity for heavy metals. They also found that this sorption capacity had a marked pH dependence in the neutral to slightly alkaline pH range with increasing sorption with increasing pH. However, as pointed out by Lee (see Morgan and Stumm[17]) this high sorption capacity and marked pH dependence may only be applicable to freshly precipitated hydrous metal oxides. The aging of the precipitate would likely reduce sorption capacity as a result of molecular rearrangements which improve the crystallinity of the precipitate. In addition, the sorption capacity of hydrous metal oxides may be changed significantly with age due to the sorption of other materials in solution. Of particular concern is the role of natural water organics on the sorption process. This will be discussed further in a subsequent section.

Of potential significance in the role of heavy metal oxides on metal ion transport is the fact that the degree of interaction between the heavy metals and hydrous metal oxides is likely to be dependent on whether the heavy metal was present at the time of formation of the hydrous metal oxide precipitate or coating. It has been known for some time that much greater incorporation of chemicals into ferric hydroxide precipitates occurs when the precipitation takes place in the presence of the contaminant. For example, Malhotra et al.[31] found that almost twice as much phosphate was incorporated into a ferric hydroxide floc if the formation of the floc took place in the presence of the phosphate as compared to addition of the preformed floc to a phosphate solution.

If it is found that freshly precipitated hydrous metal oxides tend to have higher sorption capacities, then the regions where hydrous metal oxide formation is occurring could be areas where they would have their greatest influence on heavy metal transport in natural water systems. Particular attention should be given to boundary areas between anoxic and oxic systems such as the thermocline region in eutrophic lakes which have anoxic hypolimnia; selected trenches and fjords in the oceans and coastal zone; the boundary of anoxic sediments and the oxidized sediment, soil, or overlying water; and the point where anoxic groundwaters enter surface waters or are discharged to the surface through springs. The other boundary condition which could be important in influencing the role of hydrous metal oxides on heavy metal contaminants is the situation where there is neutralization of strongly acidic waters, such as the neutralization of acid mine drainage. At the point of neutralization, two phenomena could be occurring. One of these is the precipitation of the oxidized forms of iron and aluminum as the pH of the solution is raised. The other is the enhanced rates of oxidation of reduced forms of iron and manganese which take place at higher pH values. Morgan and Stumm[17] have summarized the work in this area and have noted that the rate of oxidation of both iron and manganese increases by a factor of 100 for each pH unit increase in pH. A practical example of this phenomenon is found in the work of Theobold et al.[32] in their studies on the junction of Deer Creek with Snake River in Colorado. The junction of these two rivers results in the precipitation of large amounts of iron and aluminum hydrous oxides arising from the neutralization of acidic waters. These studies have shown that the precipitates contain large amounts of various metallic species. Not only do the hydrous metal oxides exert a significant influence on the heavy metals, but also the heavy metals such as copper may have an influence on the hydrous metal oxides. For example, Stumm and Lee[10] have found that the presence of copper greatly catalyzes the oxidation of ferrous sulfate by dissolved oxygen. The copper in turn would tend to interact with the ferric hydroxide formed under oxidations which take place in neutral to slightly alkaline conditions. A similar type of catalysis has been noted for the oxidation of manganese during water treatment as reported by Jenne.[19]

In order to better understand the role of hydrous metal oxides in the chemistry of contaminants in natural water systems, there is a need for research on the factors influencing the sorption characteristics of hydrous metal oxides as a function of age of the precipitate, sorption on the performed precipitates, and the role of organics and other chemical constituents on sorption.

Role of organics

Jenne[19] has presented a review of the factors that could influence metal ion transport in soils and to a lesser extent in aquatic systems. In considering the aqueous environmental chemistry of metals in natural waters, the potential role of organic matter must be considered. Organics could play a dominant role in the transport of metals in natural water systems from several points of view. One of these would be the physical transport of particulate organics in which the metal and the organics would become associated either in the form of insoluble complexes or peptized colloidal species. Shapiro[33] found that extracts of natural water organics tended to influence the size of ferric hydrous oxide precipitates where a dominant size fraction occurred in the 0.1–0.45 μm size range in the presence of natural organic matter. In its absence, the size fraction was larger than this and the bulk of the precipitated ferric species could be removed on 0.45 μm pore size membrane filters. In this case, it appears that the organics played a dominant role in peptization of the iron species. This in turn could be of significance in interaction between the hydrous iron oxide and other organic and inorganic species in solution.

The work of Hall and Lee[34] has shown that natural water organic matter obtained under conditions which probably most closely simulate the material in the natural water systems is in true solution, provided the iron content of the solution is small. As ferric iron is added to the system, the organics and the iron become associated to form larger particles and tend to become colloidal in character. This observation probably explains the wide discrepancy among various investigators on the nature of natural water coloring matter; where some have claimed that it is colloidal, while others claim it is in true solution. Based on the Hall and Lee work, it appears that whether it is true solution or not depends directly on the amounts of iron present in the system.

In the South Atlantic and Gulf Coast areas of the United States, many of the rivers entering the estuaries contain large amounts of coloring matter which are derived from the leaching of forests and marshes. In general, the studies in these areas have shown that the coloring matter tends to be in a precipitated form. Such a situation would likely be of significance to filter-feeding organisms such as oysters and certain crustaceans as a result of the fact that as part of their filter feeding process, they may pick up iron organic-color precipitates. Therefore, not only must there be an understanding of the occurrence and mode of transport of the hydrous metal oxides as it may affect the transport of heavy metals such as zinc, copper, nickel, cadmium, etc.; information must also be available on the role of natural organics in influencing the behavior of hydrous metal oxides. Natural organics could influence how various heavy metals interact with organics and with hydrous metal oxides. This in turn would influence how various types of organisms could obtain excessive exposure to potentially hazardous chemicals.

Recent studies by Hall[35] on the toxicity of zinc to algae demonstrate the potential significance of the interactions between Fe^{3+}, complexing agents and trace heavy metals in affecting algal growth. Using standard AAP algal media, Hall found that the toxicity of zinc to *Microcystis* in batch culture was a function of the order of addition of iron, EDTA and zinc to the culture medium. A different toxicity of zinc was found if the EDTA was added to the culture medium after the iron rather than before it. This pattern is probably the result of the fact that under the pH conditions that exist in the culture medium, hydrous ferric oxide would tend to be formed. The EDTA added to the culture medium is not a sufficiently strong complexing agent to prevent $Fe(OH)_3$ precipitation and incorporation of zinc into the precipitate. Also, the zinc would tend to form complexes with the EDTA which may tend to cause it to be nontoxic to algae if the zinc-EDTA complex behaves like the copper-EDTA complex. The interactions between iron, complexing agents and heavy metals could be very important in determining the transport and toxicity of these elements to aquatic organisms.

Filter-feeding organisms tend to be highly selective in the particle size of the food that they take in. It is possible that heavy metals could interact with hydrous metal oxides whose particle size would be influenced by natural water organics in such a manner as to make the potential contaminants more or less available than would occur in the absence of natural organics.

It is generally believed that the primary role of natural organics in influencing the transport of heavy metals is complexation. Often complexation is invoked as a means of explaining a lack of behavior according to simple mass action relationships. Frequently, what could be explained as the formation of soluble complexes could also be explained as reactions of hydrous metal oxides in many situations. Sometimes investigators utilize an increase in free metal ion concentration with decreasing pH as an indication of complexation. Actually, the same kind of behavior would be expected for metal ions associated with hydrous metal oxides of the colloidal size range. Any systematic study of the metal ion complexes of natural water systems must include efforts to determine whether the metal ion transport is due to colloidal hydrous metal oxides. It is clear that additional studies must be done in this area before the role of soluble organic complexes in influencing metal ion transport versus that of hydrous oxide transport in natural water systems can be understood. For example, Sanchez and Lee[36] studied the factors controlling copper in Lake Monona, Madison, Wisconsin. This lake has received 1.5 million lb of copper sulfate for algae control over the past 50 years. The copper precipitates in the system and has become incorporated in the sediments. The purpose of the Sanchez and Lee study was to ascertain whether or not there is any evidence for soluble organic complexes of copper influencing the amount of copper in solution. Based on these studies, they concluded that soluble organic complexes play a very minor role in the chemistry of copper in this lake. The concentrations found could be readily explained by either the basic carbonates in an aerated system or the sulfides in an anoxic system. For further discussion of the possible role of organics in influencing the transport of metals in natural water systems, consult the review by Saxby[37] on the role of metal organic complexes in geochemical cycles.

Barber[38] has recently published the results of some studies on the role of organic complexing agents on controlling algae growth in natural waters. He notes that in some natural waters there are considerable amounts of toxicity to algae which can be eliminated by the addition of complexing agent. Similar types of results may occur whenever there are large amounts of hydrous metal oxides in the area since the metal ions which are responsible for the toxicity may be removed from solution by sorption onto the hydrous metal oxide. Kharkar et al.[39] have shown that many of the transition metals are not readily desorbable from iron hydroxide precipitates. Once the co-precipitation occurs, the possibility of removing these metals under natural water conditions is quite poor. Therefore, the interaction between heavy metals and hydrous metal oxides under aerobic conditions could represent a more or less permanent sink for potentially hazardous heavy metals. However, iron and manganese hydrous oxides dissolve under anoxic conditions which could potentially lead to a significant amount of metal contaminant release to the water. Normally, however, a buildup of heavy metals under anoxic conditions does not occur due to the fact that under these conditions there is concomitant production of hydrogen sulfide which forms highly insoluble compounds with many of the heavy metals. This was the situation that Sanchez and Lee[36] found for copper in Lake Monona. About the only aquatic environment where it would be expected that the reduction of hydrous metal oxides could result in significant metal contaminant release, excluding iron and manganese, is an environment where sulfide production is severely limited by the amount of sulfate present in the water. It is estimated by the author that such an environment would have a sulfate concentration of less than 1–2 mg/l.

Interaction of hydrous iron oxide and phosphate

There is substantial literature on the interactions between iron and aluminum hydrous oxides and phosphate in natural water systems. No attempt will be made in this paper to review this literature except to point out that iron and aluminum hydroxide can effectively remove phosphate from natural water systems. The interaction between these species is used as a basis for advanced waste treatment methods for phosphate removal from domestic waste-waters. More recently aluminum has been added to lakes to remove phosphate in a lake-wide treatment program. Dramatic results have been attained both in Sweden and in the United States using this kind of treatment. While the alleged benefits of adding alum to lakes are for phosphate removal, it cannot be ascertained without special studies whether the reduced algal growth is due to the aluminum hydroxide removal of trace metals which are necessary for algal growth. The significance of this kind of situation has been demonstrated in some unpublished work by the author in which an investigation was made on the potential benefits that might be derived by reducing the phosphorus content of various lakes by alum addition. Samples of the lake water were taken to the laboratory and various amounts of aluminum sulfate were added to the water. Algal growth that occurred with or without the addition of the alum was noted in the standard laboratory bioassay technique. In order to check whether or not it was the removal of phosphate that was the key to influencing algal growth, the samples that have been treated with alum received phosphate in an amount equal to that originally present and an algal assay was again conducted. In some samples, it was found that very poor growth occurred when the phosphorus was added back to the samples. It was reasoned that this poor growth was due to the alum floc removing trace metals to an extent that they then became limiting for algal growth in the media. One of the methods that could be used to correct this situation is to add a trace metal supplement to the sample which had received the alum treatment. When this was done, then reasonably consistent growth occurred which was more or less proportional to the amount of growth that was achieved without the addition of alum, indicating that phosphorus was one of the key limiting elements in the samples for the algal population present.

HYDROUS OXIDE COATINGS

Clay minerals and some other mineral species have significant cation exchange capacity. It is sometimes stated that they could play a dominant role in the transport of heavy metals. However, it is doubtful that the cation exchange capacity of layer silicates, such as clay minerals, plays a significant role in heavy metal transport for several reasons. First, the cation exchange capacity represents a small part of the sorption capacity of the natural water particulate matter for cations. Studies by Fruh and Lee[40] have shown, for the uptake of cesium on Lake Mendota sediments and on vermiculite, that cation exchange capacity represents a small part of the total sorption capacity for this species. It was found that a large part of the sorption could be removed by repeated washing with low-metal-content water. Another factor to consider is that competing for the cation exchange sites with the heavy metals of interest are the bulk metal species such as calcium, magnesium, and sodium which occur at concentrations many-fold over the heavy metals. Since, in general, cation exchange reactions have distribution coefficients of approximately the same order of magnitude for the various metallic species, it would be expected that calcium and magnesium would be the dominant ions occupying the cation exchange sites with very few of them being covered by metal ions of the heavy metal type. Jenne[19] has noted that there is little relationship between the cation exchange capacity of the soils and the fixation of the heavy metals in the soils.

While it is generally found that the distribution coefficients for the uptake of various cations are approximately the same for the sorption by clay minerals, Morgan and Stumm[30] found that the distribution coefficient for heavy metals or freshly precipitated manganese dioxide was greater than for alkaline or alkaline earth metals. Therefore, there could be a preferential sorption of heavy metals on hydrous metal oxides even in the presence of large amounts of other cations.

It should be noted that when considering the sorption capacity of mineral fragments for heavy metal species, consideration must be given to the possibility of hydrous metal oxide coatings on the surface of these particles which would in turn play a dominant role in the chemistry of heavy metals. For example, studies by Plumb and Lee[41] have shown that taconite tailings derived from iron ore mining in the Mesabi Range in northern Minnesota tend to show significant sorption capacity for various metal ions such as copper, zinc, cadmium and phosphates. This capacity is manifested even though the tailings, which are composed primarily of quartz and of an iron–magnesium silicate (cummingtonite), were found to be a fraction of 1% soluble under Lake Superior conditions. Even though there was release of magnesium and silicates to a solution, there was no release of the trace amounts of copper, zinc, and nickel present in the silicate lattices. Actually, there was uptake of these species by the mineral fragments. This behavior can be explained by the fact that the iron which was released from the taconite particles would precipitate on the surface as a ferric hydroxide and would tend to remove phosphate and the heavy metals. A significant part of this removal was likely to be associated with surface coatings of the hydrous metal oxides on the surface of the mineral fragments.

A similar type of situation could develop for several types of natural water precipitates. Of particular importance would be that of the calcium carbonate species such as calcite, argonite and dolomite. These species in natural waters would tend to have a microzone of higher pH surrounding the particle than the bulk solution. This microzone of higher pH could readily promote the oxidation and precipitation on the surface of the calcium carbonate species. This hydrous metal oxide on the surface would tend to sorb various metal species from solution as has been found for the pure metal oxides investigated by Morgan and Stumm.[17]

Plumb and Lee[41] found that drying the taconite tailings at 105°C for 1 h markedly changed the initial release of copper. Under these conditions, it was found that the copper present in the tailings was initially partially released. After a period of time, however, this copper was adsorbed to a significant extent back onto the tailings. It is reasoned that this drying step markedly changed the surface character of the hydrous iron oxide which inhibited its initial copper sorption. The resorption after extended periods of time on the dried tailings is related to the diffusion of fresh iron from the cummingtonite particle lattice to the surface where copper sorption could occur once again. This change in the surface structure of the taconite tailings was readily demonstrated by the amount of ammonium acetate leachable copper. Plumb and Lee found that the amount of the ammonium acetate leachable copper from taconite tailings was quite small under conditions where tailings had never been dried. However, drying caused a significant increase in the ammonium acetate-leachable copper. This further points to a change in the surface chemistry of the tailings most probably related to the hydrous metal oxide on the surface of the particle such as cummingtonite.

It is important to emphasize that the control of heavy metals by mineral fragments with hydrous oxide coatings may actually be a tertiary or possibly a quaternary system where organic matter in the form of colloidal compounds or dissolved species or combinations of both may actually be involved. Few studies have been done on tertiary systems of this type involving heavy metals. Wang et al.[42] have conducted some studies on tertiary systems involving clay minerals, organics and pesticides. It was found that the sorption of pesticides on clays was both enhanced and retarded with the presence or absence of certain types of organic compounds. In one case, a certain type of organic would enhance the sorption of parathion on montmorillonite, while another organic would inhibit parathion sorption on montmorillonite.

A somewhat analogous situation developed when nitrilotriacetic acid was added to a solution which had been equilibrated with Lake Monona sediments. As noted above, these sediments contained large amounts of copper as a result of the fact that copper has been added to the lake for algae control. The copper was found in an aerated system to be controlled by the basic carbonate. Upon addition of NTA to the solution in the presence of sediments, it was found that the copper decreased in concentration rather than increased. Since the concentrations of NTA used were in the order of a milligram per liter, there should have been significant complex formed between the copper in NTA. Therefore, it was postulated that this soluble complex was strongly sorbed by the hydrous ferric oxide from the oxidation of the iron sulfides present in the sediments. In other words, the hydrous metal oxides tended to cause NTA to remove copper from solution rather than making it available for interactions with aquatic organisms.

Hem and Skougstad[43] conducted a study on the incorporation of copper into ferric hydroxide floc. They found that significant copper incorporation occurred in the acid to neutral pH range. Under alkaline conditions, it was not possible to distinguish between incorporation in the floc and the direct precipitation of copper hydroxide.

Iron and manganese nodules

One of the most pronounced examples of the sorption capacity of hydrous metal oxides for trace metals is found in the manganese nodules from the oceans (see Goldberg[44]). Numerous studies have shown that these nodules contain large amounts of trace metals. The concentration of some trace metals in the nodules is sufficient to cause serious consideration of nodule mining for recovery of heavy metals. While the exact mechanism of incorporation is not known, it is likely to involve a sorption of the metal ions on the hydrous metal oxides.

Peterson[45] found, in a study of iron and manganese encrustations on rocks taken from northern Wisconsin lakes, that these encrustations were deficient in heavy metals compared to the surrounding sediments. The reason that the marine nodules tend to concentrate heavy metals and the fresh water nodules tend to be deficient in heavy metals is not known at this time.

SUMMARY AND CONCLUSIONS

Jenne has proposed that the hydrous metal oxides of manganese and iron are nearly ubiquitous in soils and sediments, both as partial coatings on other minerals and as discrete oxide particles. He proposes that these oxides act as a sink for heavy metals. The available evidence discussed in this paper strongly supports Jenne's proposal. It has been found that the uptake and release of heavy metals is influenced by the pH of the solution and by the presence of organic and inorganic complexes. Further, it is noted that one should not judge the potential role of heavy metals in influencing the aqueous environmental chemistry of copper, zinc, nickel, cadmium, etc. based on the concentration of iron and manganese in the solution. Actually, as noted by Jenne, often hydrous metal oxide sorption activity is far in excess of what would be predicted, based on their concentrations as a result of their occurring on the surfaces of various types of mineral and detrital particles. This situation is further complicated by the fact that the reactions between the hydrous metal oxides and the heavy metals are often nonstoichiometric. Also, many of these reactions are not reversible. This makes simple solution equilibria and the simple chemical kinetic approach essentially nonapplicable to the systems. The system is further complicated by the fact that the hydrous metal oxides would not be expected to show constant characteristics as a function of the age of the oxide. A freshly precipitated oxide, such as ferric hydroxide or manganese dioxide, would likely have markedly different sorption characteristics than aged oxides or hydroxides. Further, in the presence of natural waters, it is possible that the natural water organic matter could play a significant role in uptake and release of heavy metals on hydrous oxide coatings or discrete particles.

It is evident from the discussion that while there is no doubt that hydrous metal oxides are important sinks and modes of transport for heavy metals in the environment, the quantitative magnitude of this role is not known for a variety of natural water conditions. It is clear that, as greater emphasis is placed on the control of heavy metals in the environment by water pollution control regulatory agencies, a much better understanding of the interactions between heavy metals and hydrous metal oxides must be available in order to affect technically sound and economically feasible control programs.

Acknowledgement—This paper was supported by the Institute for Environmental Sciences, University of Texas at Dallas.

REFERENCES

1. KOLTHOFF, I. M. and SANDELL, E. B., *Textbook of Quantitative Inorganic Analysis*, 3rd edn., Macmillan, New York, 1953.
2. SANDELL, E. B., *Colorimetric Determination of Traces of Metals*, 3rd edn., Interscience, New York, 1959.
3. OVERMAN, R. T. and CLARK, H. M., *Radioisotopes Techniques*, McGraw-Hill, New York, 1960.
4. STUMM, W. and MORGAN, J. J., *Aquatic Chemistry*, Wiley, New York, 1970.
5. PARKS, G. A., Aqueous surface chemistry of oxides and complex oxides minerals, in *Equilibrium Concepts in Natural Water Systems*, Advances in Chemistry Series, Vol. 67, pp. 121–160, American Chemical Society, Washington DC, 1967.
6. MORGAN, J. J., Chemical thermodynamics, in *Equilibrium Concepts in Natural Water Systems*, Advances in Chemistry Series, Vol. 67, pp. 1–29, American Chemical Society, Washington DC, 1967.
7. HEM, J. D., *Chemical Equilibria and Rates of Manganese Oxidation*, US Geol. Survey Water-Supply Paper 1667-A, A1–A64, 1963.
8. DELFINO, J. J. and LEE, G. F., Chemistry of manganese in Lake Mendota, Wisconsin, *Environ. Sci. Technol.* **2**, 1094–1100 (1968).
9. STUMM, W. and LEE, G. F., The chemistry of aqueous iron, *Schweizerische Zeitschruft fur Hydrology* **XXI**, 295–319 (1960).
10. STUMM, W. and LEE, G. F., Oxygenation of ferrous iron, *Ind. Eng. Chem.* **53**, 143–146 (1961).
11. LANGMUIR, D. and WHITTEMORE, D. O., Variations in stability of precipitated ferric oxyhydroxides, in *Nonequilibrium Systems in Natural Water Chemistry*, Advances in Chemistry Series, Vol. 106, pp. 209–234 (1971).
12. HEM, J. D., ROBERSON, C. E., LIND, C. J., and POLZER, W. L., *Interactions of Aluminum with Aqueous Silica at 25°C*, US Geol. Survey Water-Supply Paper 1827-E, 1973.
13. SCHENK, J. and WEBER, W. J., Chemical interactions of dissolved silica with iron II and iron III, *J. Am. Wat. Wks. Ass.* **60**, 199–212 (1968).
14. PORTER, R. A. and WEBER, W. J., Interaction of silicic acid with iron III and uranyl ions in dilute aqueous solutions, *J. Inorganic and Nucl. Chem.* **33**, 2443–2449 (1971).
15. ELDERFIELD, H. and HEM, J. D., The development of crystalline structure in aluminum hydroxide polymorphs on aging, *Min. Magazine* **39**, 89–96 (1973).
16. HAHN, H. H. and STUMM, W., Coagulation by Al(III)—the role of adsorption of hydrolyzed aluminum in the kinetics of coagulation, in *Adsorption from Aqueous Solution*, Advances in Chemistry Series, Vol. 79, pp. 91–111, American Chemical Society, Washington DC, 1968.
17. MORGAN, J. and STUMM, W. with discussion by LEE, G. F. Role of multivalent hydrous metal oxides on limnological transformations, *Proceedings 2nd International Conference of Water Pollution Research*, Pergamon Press, 1965.
18. STAUFFER, R. and LEE, G. F., The role of the thermocline in controlling blue–green algal blooms, to appear in the *Proceedings of Eutrophication Process Modeling Workshop, Utah State University* (September 1973).
19. JENNE, E. A., Controls on Mn, Fe, Co, Ni, Cu, and Zn concentrations in soils and water: the significant role of hydrous Mn and Fe oxides, Advances in Chemistry Series, Vol. 73, *Trace Inorganics in Water*, pp. 337–387, 1968.
20. SHIMOMURA, S., NISHIHARA, Y., FUKUMOTO, Y., and TANASE, Y., Adsorption of mercuric ion on ferric hydroxide, *J. of Hygienic Chem.* **15**, 84–89 (1969).
21. LOCKWOOD, R. A. and CHEN, KY, Adsorption of Hg(II) by hydrous manganese oxides, *Environ. Sci. Technol.* **11**, 1028–1034 (1973).
22. KRAUSKOPF, K. B., Factors controlling the concentrations of thirteen rare metals in seawater, *Geochim. cosmochim. Acta* **9**, 1–32 (1956).
23. SCHINDLER, P. W., Heterogeneous equilibria involving oxides, hydroxides, carbonates and hydroxide carbonates, in *Equilibrium Concepts in Natural Water Systems*, Advances in Chemistry Series, Vol. 67, pp. 196–221, American Chemical Society, Washington DC, 1967.
24. SLOWEY, J. F., HEDGES, D. H., ISBELL, E. R., and HOOD, D. W., Distribution studies of Cu, Mn, and Zn in the Gulf of Mexico, *Geophys. Union* (A) **46**, 548 (1965).
25. GIBBS, R. J. Mechanisms of trace metal transport in rivers, *Science* **180**, 71 (1973).
26. POSSELT, H. S., ANDERSON, F. J., and WEBER, W. J., Cation sorption on colloidal hydrous manganese dioxide, *Environ. Sci. Technol.* **2**, 1087–1093 (1968).
27. MURRAY, D. J., HEALY, T. W., and FUERSTENAU, D. W., The adsorption of aqueous metal on colloidal hydrous manganese oxide, in *Adsorption from Aqueous Solu-*

tion, *Advances in Chemistry Series*, Vol. 79, pp. 74–81, American Chemical Society, Washington DC, 1968.
28. HINGSTON, F. J., POSNER, A. N., and QUIRK, J. P., Adsorption of selenite by goethite, in *Adsorption from Aqueous Solution*, Advances in Chemistry Series, Vol. 79, pp. 82–90, American Chemical Society, Washington DC, 1968.
29. ANONYMOUS, Cycling and control of metals, *Proceedings of Environmental Research Center*, US Environmental Protection Agency, Cincinnati, Ohio, 1973.
30. MORGAN, J. and STUMM, W., Colloid-chemical properties of manganese dioxide, *J. Coll. Sci.* **19**, 347–369 (1964).
31. MALHOTRA, S. K., LEE, G. F., and ROHLICH, G. A., Nutrient removal from secondary effluent by alum flocculation and lime precipitation, *J. Air Wat. Pollut.* **8**, 487–500 (1964).
32. THEOBOLD, P. K., JR., LAKIN, H. W., and HAWKINS, D. B., The precipitation of aluminum, iron, and manganese at the junction of Deer Creek with the Snake River in Summit County, Colo., *Geochim. cosmochim. Acta* **27**, 121–132 (1963).
33. SHAPIRO, J., The effect of yellow organic acids on iron and other metals in water, *J. Am. Wat. Wks. Ass.* **56**, 1062–1082 (1964).
34. HALL, K. and LEE, G. F., Molecular size and spectral characterization of organic matter in a meromictic lake, *Wat. Res.* **8**, 239–251 (1974).
35. HALL, R. H., personal communication, Procter and Gamble Corporation, Cincinnati, Ohio, 1973.
36. SANCHEZ, I. and LEE, G. F., Sorption of copper on Lake Monona sediments—effect of NTA on copper release from sediments, *Wat. Res.* **7**, 587–593 (1973).
37. SAXBY, J. D., Metal–organic chemistry of the geochemical cycle, *Revs. Pure Appl. Chem.* **19**, 131–150 (1969).
38. BARBER, R. T., Organic ligands and phytoplankton growth in nutrient-rich seawater, in *Trace Metals and Metal–Organic Interactions in Natural Waters*, pp. 321–338, Ann Arbor Science Publications, 1973.
39. KHARKAR, D. P., TUREKIAN, K. K., and BERTINE, K. K., Stream supply of dissolved silver, molybdenum, antimony, selenium, chromium, cobalt, rubidium, and cesium to the oceans, *Geochim. cosmochim. Acta* **32** (3) 285 (1968).
40. FRUH, E. G. and LEE, G. F., unpublished results, University of Wisconsin.
41. PLUMB, R. H. and LEE, G. F., *Effect of Taconite Tailings on Water Quality in Lake Superior*, report to Reserve Mining Company, University of Wisconsin, Madison, 1973.
42. WANG, W. C., LEE, G. F., and SPYRIDAKIS, D., Adsorption of parathion in a multi-component solution, *Wat. Res.* **6**, 1219–1228 (1972).
43. HEM, J. D. and SKOUGSTAD, M. W., *Co-precipitation effects in solution containing ferrous, ferric and cupric ions*, US Geol. Survey Water-Supply Paper 1459-E, 95–110 (1960).
44. GOLDBERG, E. D., The oceans as a chemical system, in *The Sea*, vol. 2, *Composition of Seawater, Comparative and Descriptive Oceanography* (Hill, ed.), 1963.
45. PETERSON, J., Aqueous environmental chemistry of lead, PhD thesis, University of Wisconsin, Madison, 1973.

Role of Hydrous Metal Oxides in the Transport of Heavy Metals in the Environment (G. F. Lee)

DISCUSSION by J. D. Hem
US Geological Survey, Menlo Park, California 94025

SIGNIFICANCE AND CALCULATION OF REDOX POTENTIALS

The effectiveness of hydrous iron oxide as an adsorbing agent for heavy metals in aqueous systems is described in many published papers. The mechanisms whereby the adsorption occurs may include several kinds of processes, and a clear understanding of which ones are most effective in a given situation is difficult to achieve by studying existing literature on the subject.

In this discussion, principles and techniques of equilibrium chemistry are applied to several types of chemical reactions involving the amorphous freshly precipitated form of ferric hydroxide. This material participates reversibly in oxidation–reduction and pH controlled precipitation–dissolution reactions with certain other metals in ways that can substantially influence their solubility at equilibrium. The following is a theoretical treatment using thermodynamic data from published sources and the usual limitations to such an approach apply. One assumes here an equilibrium condition and ignores such effects as temperature and pressure deviations from standard conditions and the effects of unconsidered solution complexes. Various kinds of Eh–pH diagrams used separately and together can help to clarify the chemical reactions that are involved. Some method, in any event, is needed to simplify the simultaneous consideration of many variables. Although the approach used has limitations, it is, nevertheless, strikingly evident that chemical reactions of a rather well-defined nature may actually be responsible for setting limits on solubilities of certain metal ions in systems where rather ill-defined chemical sorption processes have often been cited as the controlling factors.

The redox potential (Eh) of a chemical system is a measure of the intensity of oxidizing or reducing conditions within the system. The Eh in a solution is somewhat analogous to the activity of H^+ in that one represents in effect the activity of electrons and the other the activity of protons. The thermodynamic function generally used for relating solution concentrations and standard electrochemical potentials of redox couples, or half-reactions, is the Nernst equation. In conventions of sign used by the writer in numerous earlier publications, the relationship has the form

$$Eh = E° + \frac{0.0592}{n} \log \frac{\text{(oxidized species)}}{\text{(reduced species)}}$$

when the redox couple is written as a reduction. Reacting species concentrations are in molar activities. The standard free energy change in the redox couple (ΔG_R) is related to the standard potential $E°$ by the relation

$$E° = \frac{-\Delta G_R}{23.06n},$$

where n is the number of electrons shown in the redox half-reaction. Both these expressions yield a value of $E°$ or Eh in volts applicable at 25°C and 1 atm pressure. Increasing positive values of Eh indicate progressively more oxidizing conditions.

A variation of this method of calculating and expressing redox potentials was advocated by Sillén.[1] A redox potential expressed as the log of the activity of electrons (pE) is related to Eh by the expression

$$pE = \frac{Eh}{0.0592}$$

at standard conditions (25°C, 1 atm pressure).

The behavior of iron is more nearly ideal in oxidation–reduction reactions than that of most of the other metals. Redox reactions involving iron generally are rapid at the pH levels of interest in natural systems and the applicability of equilibrium redox models to the behavior of iron in natural ground water systems (for example, ref. 2) has been demonstrated. Similar models also appear useful in studies of bottom muds in lakes and other surface water bodies.[3] However, the measurement of Eh in dilute aqueous systems poses difficulties, as described in detail by Stumm and Morgan.[4]

THE Eh–pH DIAGRAM

To the extent that the iron species in an aqueous system are in a state of equilibrium, they constitute a useful measure of the Eh of the system, especially that near the interface between solid iron species and iron-bearing solutions where the equilibrium is established. The practical utilization of this concept is facilitated by suitable Eh–pH graphs.

Figure 1 is an Eh–pH diagram showing the dominant dissolved iron species in a simple Fe–H$_2$O system. Three oxidation states are involved, Fe^{+2} and Fe^{+3}, both occurring in three forms, and ferrate, an Fe^{+6} species which is dominant only in strongly oxidizing solutions. This diagram identifies the solute form of iron to be expected at any pH and Eh in the absence of other ligands that might form complexes with iron, but gives no solubility information. In some of the stability fields the solubility of iron is very low.

If the diagrams are to be useful for studying natural systems, other components must be included. The most important elements in controlling metal ion solubilities in natural systems commonly are sulfur and carbon. For present purposes, it is convenient to specify a constant total activity of each of these elements, amounting to $10^{-4.00}$ moles/l of sulfur, about 10 mg/l as SO_4^{-2}, and $10^{-3.00}$ moles/l of carbon, equivalent to 61 mg/l of HCO_3^-.

Figure 2 is a stability-field and solubility diagram for the solids in this system. The four solids considered are ferric and ferrous hydroxide, siderite (FeCO$_3$), and pyrite (FeS$_2$). The more stable oxides hematite and magnetite are not included in the solid species considered because the available evidence suggests they do not display reversible equilibrium solubility in systems of the type being modeled. Dashed lines drawn in the stability fields show the activity of the predominant dissolved iron species, as indicated in Fig. 1, in equilibrium with the designated solid.

The value used for the free energy of formation of Fe(OH)$_3$c was arbitrarily selected as the least negative (highest solubility) of those published in standard references. This value (-166.0 kcal mole^{-1}) fits laboratory experiments of the writer. A somewhat lower solubility has been observed in the field,[2] but values shown in Fig. 2 probably are reasonable for most natural water systems.

In a system where the stable solid is ferric hydroxide, the concentration of dissolved ferrous iron, if it can be measured or estimated, and the observed pH can be used with Fig. 2 to estimate the Eh. Within the domains of Fe^{+2} and FeOH$^+$ the total dissolved iron should be nearly all ferrous, and in more oxidizing systems above pH 5 the dissolved ferric species and total dissolved iron should be less

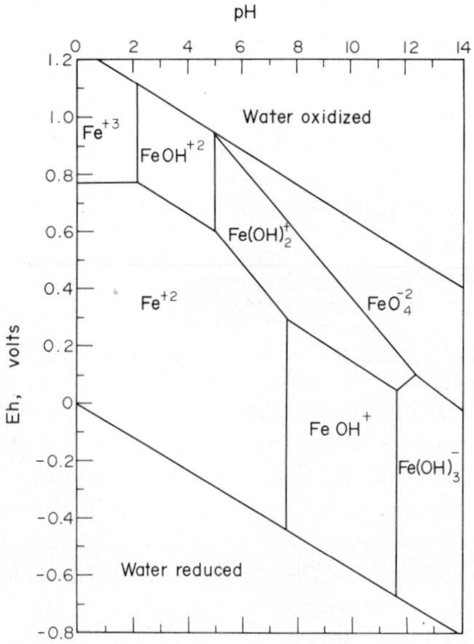

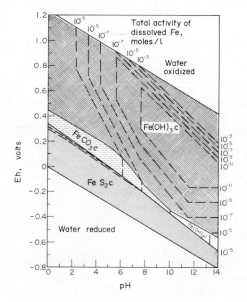

FIG. 1. Eh–pH diagram showing areas of dominance of seven solute species of iron at equilibrium at 25°C and 1 atm. System Fe–H$_2$O–C–S. Total dissolved C = $10^{-3.00}$ moles/l; total dissolved S = $10^{-4.00}$ moles/l.

FIG. 2. Eh–pH diagram showing fields of stability of solids and total equilibrium activity of dissolved iron at 25°C and 1 atm. System Fe–H$_2$O–C–S. Total dissolved C = $10^{-3.00}$ moles/l; total dissolved S = $10^{-4.00}$ moles/l.

than 1 μg/l. For some waters the total dissolved iron value can be considered equivalent to the ferrous iron concentration, but this is not always a reliable indicator. Some waters, for example, contain colloidal ferric hydroxide or other ferric species that can pass through filters and appear as part of the dissolved iron. Concentrations and chemical behavior of dissolved iron, as well as that of other metals, may also be influenced by organic complexing agents. If, however, an Eh can be estimated, even approximately, it can in turn be used to predict the stable solid form and to calculate the solubility of other metals, if Eh of the system is a significant control.

Certain metals, such as copper, tend to have a lower solubility in moderately reducing than in more oxidized environments. Figures 3 and 4 are Eh–pH diagrams for the Cu–CO$_2$–S–H$_2$O systems. It is evident that from the Cu–CO$_2$–S–H$_2$O stability diagram (Fig. 4) alone one could not predict the solubility of copper in a particular water at pH 7 without an Eh value. If the dissolved-iron content and pH are known, however, an estimate of Eh can be made from Fig. 2 and used in Fig. 4 to give an equilibrium solubility for copper.

The solubility of copper in a mildly reducing environment is very much less than it would be in an oxidizing one (Fig. 4) because cuprous oxides and especially metallic copper, the reduced solids, are less soluble than the cupric species. A somewhat similar behavior is probably characteristic of other metals that are stable in the uncombined metallic form in water.

REDUCTION OF OTHER METALS BY FERROUS IRON

Chemical equilibria used to calculate values of Eh are not complete reactions. For a reduction reaction to take place there must be a source for the electrons shown in the half-reaction that represents the reduction process. This source will generally be another half-reaction involving an oxidation. In other words, when a chemical reduction of one species occurs it is accompanied by the oxidation of an equivalent quantity of some other species. Using the Eh–pH diagrams for iron and copper in series is equivalent to coupling the reaction for copper reduction to that of ferrous iron oxidation, giving

$$Cu^+ + Fe^{+2} + 3H_2O = Fe(OH)_3c + Cu^0c + 3H^+$$

From chemical thermodynamic data available in the literature,[6] an equilibrium constant for this reaction can be calculated. It is then possible to write

$$\frac{[H]^3}{[Fe^{+2}][Cu^+]} = 10^{-8.04}$$

In a solution at pH 7, containing about 500 μg/l Fe (which would be essentially all ferrous at this pH)

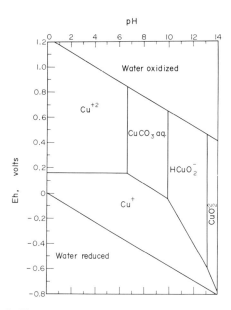

FIG. 3. Eh–pH diagram showing areas of dominance of five solute species of copper at equilibrium at 25°C and 1 atm. System Cu–H$_2$O–C–S. Total dissolved C = 10$^{-3.00}$ moles/l; total dissolved S = 10$^{-4.00}$ moles/l.

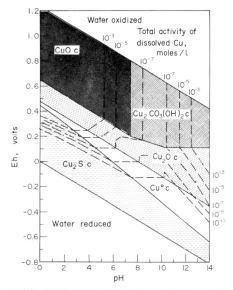

FIG. 4. Eh–pH diagram showing fields of stability of solids and total equilibrium activity of dissolved copper at 25°C and 1 atm. System Cu–H$_2$O–C–S. Total dissolved C = 10$^{-3.00}$ moles/l; total dissolved S = 10$^{-4.00}$ moles/l.

the activity of copper would be about 10^{-8} M or less than 1 µg/l. The Eh in this system would be near zero v, a value commonly encountered in ground water, and compatible with Fe^{+2}–$Fe(OH)_3c$ equilibria. This implies depletion of oxygen, or at least suggests it is not controlling the Eh at the site of reaction.

Electrochemical reactions that involve solid species occur at surfaces, and it is the surface redox potential of the dominant iron solid species that is of primary significance. Measurement of solution Eh with platinum and calomel electrodes in systems containing iron have generally yielded values corresponding to those one might calculate for the ferrous ion-ferric hydroxide solid equilibrium. Doyle[3] observed an actual deposition of a film of ferric hydroxide on the platinum electrode and attributed the measured potential to this material. It seems reasonable to suppose that when ferric hydroxide surfaces are present, they will have a characteristic redox potential that will tend to control the behavior of many metal ions.

An investigation of the coprecipitation of copper with iron[7] gave results in general agreement with this theoretical model, although the methods utilized then for determining dissolved copper were not sensitive enough to test the calculations rigorously.

Many authors have noted the increase in the rate of oxidation of ferrous iron in the presence of dissolved copper. The oxidation rate of Fe^{+2} in aerated water is pH dependent and very slow at low pH. It appears likely that the energy barrier for the electron transfer from one metal ion to another is lower than that involved in the rather complex mechanism of oxidation by dissolved oxygen. Details of the mechanism are not well understood, however. The role of copper in this mechanism does imply that the oxidation of ferrous iron to ferric hydroxide will act as an effective scavenger for copper and in most ground water systems copper will be extensively influenced by reactions involving iron.

It also has been noted that manganese oxides could be alternately precipitated with ferric hydroxide by altering the ratio of dissolved Fe^{+2} to Mn^{+2}. However, the precipitation of manganese oxide requires a very low Fe^{+2} activity.[8] Conditions of this sort may occur in lakes where concretions of iron and manganese oxides develop on the bottom.

FORMATION OF MULTIPLE OXIDES

There are other types of interaction between other metal species and iron which may be significant. Instead of producing a mixed solid, a compound of both metals with a definite chemical composition can be produced in some instances. Chromium, for example, may enter water as chromate anions, through industrial waste disposal. One reaction that might occur is reduction to Cr^{+3}, followed by precipitation as $Cr(OH)_3$, having a low solubility:

$$CrO_4^{-2} + 3Fe^{+2} + 7H_2O$$
$$= 3Fe(OH)_3c + Cr(OH)_2^+ + 3H^+$$

and

$$Cr(OH)_2^+ + H_2O = Cr(OH)_3c + H^+$$

Or, perhaps more likely, some of the ferrous iron would combine with the reduced chromate, to form a stable chromite:

$$2Cr(OH)_2^+ + Fe^{+2} = FeCr_2O_4c + 4H^+$$

The equilibrium constant for this reaction is $10^{-1.06}$ and from the expression

$$\frac{[H]^4}{[Cr(OH)_2^+]^2[Fe^{+2}]} = 10^{-1.06}$$

it can be shown that at pH 7.00 with $[Fe^{+2}] = 500$ µg/l, the solubility of chromium would be about 10^{-11} M (0.0005 µg/l), an amount far below detection by almost any analytical procedure. This solid is less soluble than $Cr(OH)_3$.

From Fig. 2 it appears that even if a rather strongly oxidizing condition prevailed so that dissolved iron is 10^{-9} M (0.06 µg/l) the solubility of chromium would still be below 1 µg/l.

These estimates entail the assumption that iron is present in sufficient excess to predominate over the other metals. Stoichiometric effects could become important, however, if the concentration of iron were to be as low as those of the "trace metals."

Ferrites, compounds in which the companion metal is combined with the FeO_2^- ion, are reported by Wagman et al.[6] to have very low free energies of formation and could be important in decreasing the solubility of metals in the presence of ferrous-ferric species. Among ferrites of potential significance in natural systems are those of nickel and cobalt, both of which tend to be associated with ferric hydroxide in lake-bed and shallow marine deposits.[9]

CONCLUSIONS

It appears possible to explain the scavenging effect of ferric hydroxide for dissolved ions of certain other metals by means of oxidation–reduction equilibria at the ferric hydroxide surface. The formation of the ferric hydroxide may itself serve as an electron source, and in some instances new compounds incorporating both metals may be formed. The chemistry of these reactions can be understood more readily with the aid of Eh–pH diagrams, as the systems involved contain many variables that need to be evaluated simultaneously. Computer programs for modeling these systems are available.[10]

REFERENCES

1. SILLÉN, L. G., Master variables and activity scales, in *Equilibrium Concepts in Natural Water Chemistry*, Advances in Chemistry Series, vol. 67, p. 52, 1967.
2. BARNES, I. and CLARKE, F. E., *Chemical Properties of Ground Water and their Corrosion and Encrustation Effects on Wells*, U.S. Geol. Survey Prof. Paper 498-D, 1969.
3. DOYLE, R. W., The origin of the ferrous ion-ferric oxide Nernst potential in environments containing dissolved ferrous iron, *Am. Jr. Sci.* **266**, 840–859 (1968).
4. STUMM, W. and MORGAN, J. J., *Aquatic Chemistry*, Wiley-Interscience, New York, 1970.
5. HEM, J. D., Equilibrium chemistry of iron in ground water, in *Principles and Applications of Water Chemistry*, (Faust, S. D. and Hunter, J. V., eds.), Wiley, New York, pp. 625–643, 1965.
6. WAGMAN, D. D., EVANS, W. H., PARKER, V. B., HALOW, I., BAILEY, S. M., and SCHUMM, R. H., *Selected Values of Chemical Thermodynamic Properties—Tables for Elements 35 through 53 in the Standard Order of Arrangement*, NBS Tech. Note 270-4, National Bur. of Standards, Washington DC, p. 88, 1969.
7. HEM, J. D. and SKOUGSTAD, M. W., *Coprecipitation Effects in Solutions Containing Ferrous, Ferric and Cupric Ions*, US Geol. Survey Water-Supply Paper 1459-E, pp. 95–110, 1960.
8. HEM, J. D., *Deposition and Solution of Manganese Oxides*, US Geol. Survey Water-Supply Paper 1667-B, 1964.
9. MANHEIM, F. T., Manganese-iron accumulations in the shallow marine environment, in *Symposium on Marine Geochemistry*, pp. 217–276, Occ. Publ. No. 13-1965, Narragansett Marine Laboratory, Univ. of Rhode Island, 1965.
10. KHARAKA, Y. K. and BARNES, I., *SOLMNEQ: Solution-mineral Equilibrium Computations*, US Geol. Survey Computer Contribution PB 215-899, 1973.

THE ACCUMULATION AND EXCRETION OF HEAVY METALS IN ORGANISMS

Jorma K. Miettinen

Department of Radiochemistry, University of Helsinki, Helsinki, Finland

INTRODUCTION

By the term heavy metals we usually understand, at least in the biological context, iron and metals denser than it. Some of these metals, e.g., iron, manganese, copper, cobalt and zinc, are essential to many organisms, while others are either nonessential, harmful, or outright toxic. However, the concept of "essentiality" is under constant review, and "harmfulness" and "toxicity" vary widely with organism and concentration. All heavy metals are potentially harmful to most organisms at some level of exposure and absorption.

Three heavy metals which are toxic to most organisms at the lowest concentrations and which are probably never beneficial to living things, are cadmium, mercury, and lead. Due to human activity, their presence in the environment has increased in some areas to levels which threaten the health of aquatic and terrestrial organisms, man included. Especially dangerous situations may arise due to heavy metal compounds, such as methylmercury, which are efficiently enriched from one trophic level to another along the food chains.

Information regarding the accumulation and excretion of these three heavy metals and those of their compounds able to persist in the environment has therefore become highly desirable.

TABLE 1
Some toxic heavy metals and compounds thereof occurring in aquatic food chains and the corresponding critical organs in man

Compound	Critical organ
Cadmium (Cd^{2+})	Kidney
	Liver
Mercury (Hg^{2+})	Kidney
Methylmercury (CH_3Hg^+)	CNS[a]
Lead (Pb^{2+})	Hematopoetic system
	CNS and PNS[b]
	Kidney

[a] CNS, central nervous system.
[b] PNS, peripheral nervous system.

This review is concerned mainly with mechanisms and rates of absorption and excretion, with particular reference to biological half-times and accumulation. The primary emphasis is on accumulation in man, although other organisms, especially fish and shellfish which are important constituents of food chains leading to man, are also briefly treated. Only the most recent results and viewpoints are discussed, without full coverage of the earlier literature. In this respect, reference is made to the recent report of an international Task Group on Metal Accumulation[1] which reviews the literature extensively.

RELEVANT COMPOUNDS, CORRESPONDING CRITICAL ORGANS IN MAN, AND SCHEME OF ABSORPTION (BY INGESTION) AND OF ELIMINATION

The compounds concerned and their critical organs in man are presented in Table 1. By critical organ we understand that organ in which the first undesirable functional changes occur under increasing intake of the toxicant. This is not necessarily the same organ in which the highest accumulation takes place. Sensitivity varies widely between individuals, and the critical concentrations usually form a Gaussian distribution within a population (Fig. 1). The variability may be so great as to make one organ critical for one individual while a different organ can be so for another individual. This variability depends upon individual differences in retention and elimination rate, possibly even in the capacity to detoxify the toxic metal. It is known, for example, that mammals and even fish contain an adaptive protein, metallothionein, which can bind large amounts of mercury, cadmium, and zinc.

Absorption via ingestion is the only route by which heavy metals enter man from the aquatic milieu. In addition to this, however, man obtains heavy metals and their compounds from atmospheric pollution via inhalation. This source has to be reckoned with when the effects of these metals upon man's health are studied, since some occupational, industrial, and even urban environments may contri-

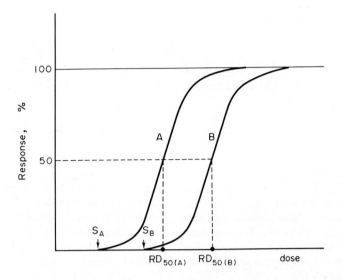

FIG. 1. Dose–response curves obtained for two toxic substances, A and B, within a population. The most sensitive individuals (S) respond to a much lower dose than do 50% of the population (RD_{50}). The curves represent integrals of Gaussian distribution functions.

bute significant amounts compared with the intake from food. Metals are to some extent absorbed directly from the lungs, but a portion of these in particulate form is transferred by mucociliary clearance into the gastrointestinal tract and may be absorbed there. In this paper, however, only intake via ingestion is considered.

In Fig. 2 a model for the absorption and elimination of metals is presented. The major source of mercury, cadmium, and lead is food, and, especially for mercury and cadmium intake by man, aquatic food chains play an important role. In seafood, cadmium, lead, and ionic mercury exist primarily as bivalent cations, more or less loosely bound to proteins. Their availability to man, equal, whether from proteinate or from ionic solution, depends upon nutritional factors. A deficiency of dietary calcium increases the absorption of lead and cadmium,[2, 3] while a deficiency of dietary iron increases the absorption of lead.[4] In children, and in pregnant and lactating women, the absorption rates of lead and cadmium may be doubled due to their great demand for calcium. A low intake of protein and vitamin D also increases cadmium absorption.[5] A low protein diet may increase lead absorption as well.[6]

Mercury is often present in seafood in the form of its organometallic compound, methylmercury. In Scandinavia, about 95% of the total mercury in fish is usually in this form.[7] Elsewhere, lower percentages have been reported, e.g., <50% in certain marine fish (blue marlin) of the Hawaii area.[8]

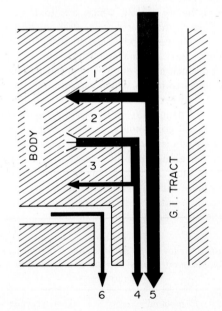

FIG. 2. A model for the absorption and elimination of heavy metals via the gastrointestinal tract: 1, absorption; 2, gross excretion; 3, reabsorption; 4, net excretion; 5, unabsorbed; 6, excretion via urine; $1+5$, intake; $4+5+6$, elimination.

Selenium in the diet may influence the absorption of methylmercury[9] and nutritional status may influence its retention, but these factors have not as yet been sufficiently well investigated.

In *marine animals* the direct uptake of metals from water via the gills (fish) or the sieve tubes (molluscs) is possible, although this factor probably does not play a major role except in heavily polluted areas. Methylmercury and the metal ions are all readily absorbed by plankton and bacteria, thus being mostly particle-bound in the sea water and able to be taken up only through food chains.

ABSORPTION AND ELIMINATION OF CADMIUM IN MAN

One method of studying the absorption and retention of cadmium in man is the "balance study", i.e. measurement of intake and excretion in a number of subjects over a long period of time. Alternatively, an attempt may be made to estimate the approximate cadmium intake of individuals for whom kidney and liver analyses have been performed *post mortem*. Both of these methods are extremely tedious and require a large number of subjects, since the intake of cadmium varies daily, weekly, seasonally, and perhaps even over longer periods for most individuals. The bulk of the human cadmium burden is usually obtained by utilizing certain foods, e.g. kidney, liver, some cereals or seafood, which the majority consume only occasionally. As the retained amount is very small compared with the amounts taken in and excreted, its reliable quantitative determination would require high accuracy of the analytical method. This is why cadmium balance studies have not, thus far, resulted in a reliable figure regarding the percentage of retention.

Use of the *radioisotope* ^{115m}Cd *and the whole-body counting technique*, however, provides for the precise, direct determination of the retained fraction. Unfortunately, the physical half-life of ^{115m}Cd (43 days), although highly useful for the determination of retention, is too short to permit determination of the biological half-time of the slowly excreted portion of the element, which appears to be quite long, of the order of years.

A study on cadmium metabolism in man has been recently reported[10,11] in which a single oral dose of labeled, mostly protein-bound cadmium, was administered to each of five voluntary male subjects ranging in age from 19 to 50 years. The labeled dose was carefully prepared to resemble normal food. The radioisotope and a solution of stable carrier were mixed with homogenized calf kidney cortex and kept under refrigeration overnight. Analysis demonstrated that approximately 80% of the cadmium became protein-bound. Portions of the homogenate containing 100 μg Cd and 4.8–6.1 μCi ^{115m}Cd were mixed with dressing and served to the volunteers as a sandwich.

Whole-body counting, as well as the analysis of feces and urine samples which were collected quantitatively during the first days following the administration of the dose, revealed that approximately 75% of the ingested ^{115m}Cd activity was *eliminated within 3–5 days* and 94% within 10–15 *days*. This elimination occurred primarily via the feces with only a minor portion being excreted in urine (Fig. 3). The elimination equation was biexponential (Fig. 4), the biological half-time of the fast component being about 2.5 days. The mean percentage of the slow component in the five individuals was 5.9 ± 0.5% of the dose. The biological half-time of the slow component could not be determined due to the low percentage of 0.935 MeV

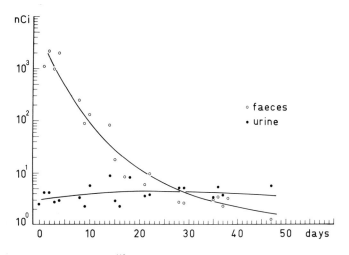

FIG. 3. Daily excretion of ^{115m}Cd in feces and urine (average of five volunteers).

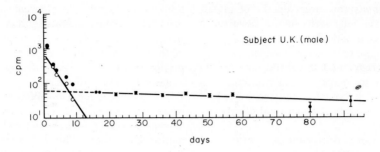

FIG 4. Whole-body retention by an individual subject of ^{115m}Cd resulting from a single oral dose of cadmium.

gamma rays emitted by ^{115m}Cd in its decay and its short physical half-life (43 days). Some idea regarding this component is obtainable from autopsy data.

Such data regarding cadmium concentration in the human kidney at 10 year age intervals were published by Schroeder and Balassa.[12] Their results reveal an almost linear increase until approximately 40 years of age and a similar decrease after 50 years. Data collected from differing populations were compared by Friberg et al.[5] According to this study, the maximum concentration of cadmium is approached in most cases at the age of 30–40 years, while the half-saturation value based upon the maximum appears to occur at about 18 years (12–24 years). The accumulation curve once again appears to be nearly linear.

A biological half-time of 18 years for cadmium would correspond to an average daily retention of about 1.6 µg, thus indicating an average daily intake of 26 µg (assuming 6% retention). These figures are feasible for the first 18 years of age if the adult intake is 50 µgCd/day.[5] The linearity of the accumulation curve may be due to increased Cd intake with advancing age (as a result of, for example, increased food consumption, greater use of kidney and liver, and the onset of smoking at 14–18 years). In light of these evaluations, a biological half-time of approximately 15–20 years appears likely for the slow component of cadmium excretion.

Cadmium circulates in blood primarily bound to red cells. It is evidently bound partly to hemoglobin and partly to metallothionein.[5]

Regarding the transport of cadmium, the plasma fraction is probably more important than that of the red cells.

ABSORPTION AND ELIMINATION OF CADMIUM IN FISH AND MOLLUSCS

The retention and elimination of cadmium in rainbow trout was studied utilizing food labeled with ^{109}Cd^{2+} and whole-body counting the live fish. The results (Fig. 5) demonstrated that the bulk of the ^{109}Cd activity was rapidly eliminated, the excretion following a power, rather than an exponential function. After 42 days, about 1% of the administered dose was retained in the whole-body, the elimination rate then being quite slow.[13] In nature, the following concentration factors have been measured:[14] fish 10^2, crustacea 10^3, mollusc 10^3–10^4 (10^5), plankton 10^4, and seaweed 10^2–10^3.

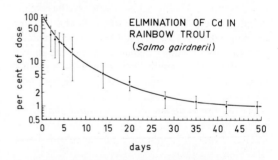

FIG. 5. Elimination of cadmium from rainbow trout determined by whole-body counting. Points indicate average retention by six fish; bars indicate one standard deviation.

The main vehicles of cadmium intake by man from marine biota are molluscan shellfish and the brown flesh of the crab, which may contain high levels of cadmium in polluted areas. Pelagic fish are low in cadmium as are coastal fish, even in polluted areas.[15] This suggests that there is no important food chain effect, since high concentrations would then be found in predatory and aged fish. It should be mentioned here that one important terrestrial source of cadmium is kidney of aged horse, elk, and reindeer. Kidney of aged horse may contain 50–200 ppm Cd, and one meal consisting of such kidney would correspond to a whole year's normal intake.

METABOLISM OF INORGANIC MERCURY AND METHYLMERCURY IN MAN AND ITS IMPLICATIONS

Two studies have been published on the absorption and elimination of methylmercury in man. Åberg et al.[16] administered *ionic* methylmercury to three male volunteers and obtained an absorption of 95% and a half-time of 70–74 days. Miettinen et al.[17] gave *protein-bound* methylmercury to fifteen volunteers and obtained nearly the same retention (94%) and half-time (76 ± 3 days) (Table 2). We also studied inorganic mercury in the same way,[18] administering protein-bound and ionic inorganic mercury to 10 volunteers and obtaining a retention of 15% and half-time of 42 ± 3 days (Fig. 6). No difference between the two forms was observed. All of these studies were performed utilizing ^{203}Hg-labeled doses and the whole-body counting technique. Both inorganic mercury and methylmercury follow, in animals, a bi-exponential elimination equation, although the fast component is not always easily measurable, being too short. It may be assumed that the fast component primarily reflects the blood and plasma level, while the slow component reflects the intracellular level in flesh, liver, and other tissues.

The ratio of the amount of mercury in the red blood cells to that in the plasma is 10 for methylmercury and 0.4 for inorganic mercury; thus, it is a simple matter to calculate the content of each in cases when both compounds are present simultaneously. However, the half-times of both compounds in the blood are considerably shorter than in the whole body. Following the administration of a single dose, the half-time of methylmercury in the red blood cells was 50 ± 7 days, while that of inorganic mercury was 16 days. If the elapsed time since the beginning of mixed intoxication is known, then the content of mercury in the body may be calculated from the blood levels, at least in cases of intoxication due to a single dose. Chronic cases require other calculations, which may be difficult if both inorganic mercury and methylmercury are present in the diet.

It is quite reasonable to assume that different mercury compounds may be present in different populations. Figure 7 presents data compiled by Suzuki.[19] These data can be interpreted as meaning that Swedish fish consumers contain practically pure methylmercury (ratio of RBC/plasma = 10), while the Japanese-Americans in Honolulu would contain nearly pure inorganic mercury (ratio of RBC/plasma = 0.5) and the Japanese in Tokyo 1/4 methylmercury and 3/4 inorganic mercury. Both rice and the Pacific fish, the prime sources of dietary mercury in the Pacific islands, seem to contain an

TABLE 2
Biological half-time of ^{203}Hg activity in the whole body after oral administration of protein-bound $CH_3^{203}Hg$ (Error marked as one standard deviation of the mean)

Volunteer	$T_{1/2}$ biol (days)	Volunteer	$T_{1/2}$ biol (days)
♀		♂	
AE	52	RE	72
EH	69	AH	88
LJ	73	PK	87
AN	88	JK	74
KR	87	MM	78
LU	56	JM	70
		VM	78
		RP	93
		HT	74
Mean	71 ± 6	Mean	79 ± 3
Mean of whole group 76 ± 3			

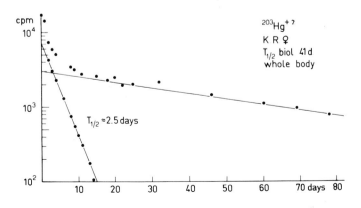

FIG. 6. Typical retention curve of $^{203}Hg^{2+}$ in the whole body.

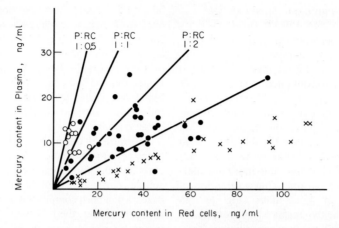

FIG. 7. Mercury content in plasma versus red blood cells of some populations, according to Suzuki.[19] ○ Japanese-Americans in Honolulu. ● Japanese in Tokyo and surroundings. × Swedes consuming fish up to three times per week.

important proportion of their mercury in the inorganic form (at least the blue marlin), thus explaining the results of Fig. 7. In the blood, most mercury is eliminated by the plasma fraction.

BIOLOGICAL HALF-TIME OF MERCURY COMPOUNDS IN AQUATIC ANIMALS

Several studies[20-24] have been performed in which labeled mercury compounds (Hg^{2+} or methyl-, ethyl-, propyl- or phenyl-Hg) have been introduced into aquatic animals by means of food, intramuscular injection or direct uptake from the surrounding water, and the absorption, distribution, and rate of elimination have been studied by whole-body counting of the individual labeled animal. The effects of the administration route, the concentration of the compound in the animal, and the ambient water temperature upon the rate of elimination have also been studied.

Briefly summarized, the excretion rate varies greatly among the different species, but usually follows a bi-exponential equation. The biological half-time at an ambient water temperature of the fast component usually varies from 1 to 10 days, while the slow one ranges from 200 to 1200 days. The excretion rate at 15°C is twice as fast as that at 4°C. Fresh, brackish, and salt water species were found to have broadly similar rates of excretion. When the concentration of methylmercury in fish increases, the biological half-time of mercury decreases: in rainbow trout, dosed with Me^{203}Hg, a concentration of 0.4 mgHg/kg is eliminated with a half-time of 340 days, while a concentration of 4 mgHg/kg is eliminated with a half-time of 170 days. Ionic and protein-bound methylmercury have similar rates of elimination when orally administered, but methylmercury absorbed by rainbow trout directly from aquarium water via the gills is eliminated faster (half-time: 268 days) than methylmercury administered orally as, for instance, a proteinate (half-time: 326 days). The half-times of inorganic Hg and phenyl- and methyl-Hg in mussel (*Pseudanodonta complanata*) were 23, 43, and (depending on age) 100–400 days, respectively, while the half-times of methyl-, ethyl-, and propyl-Hg in rainbow trout were 346, 119, and 233 days, respectively. Thus, of all these compounds, methylmercury has the slowest rate of elimination.

Both injected and orally administered methylmercury have approximately the same rate of elimination in fish, while inorganic mercury, when injected intramuscularly, is eliminated several times more slowly than when dosed in food. Methylmercury administered orally to the seal has a very long biological half-time (about 500 days). The biological half-times of the slow component of ^{203}Hg elimination for ionic and protein-bound methylmercury in three species of fish are displayed in Table 3.

ABSORPTION AND ELIMINATION OF LEAD IN MAN

The gastrointestinal absorption of lead in man is quite variable, between 1 and 16%,[25,26] and is age-dependent, being highest in the young. In blood, lead is bound to the red cells. The plasma-to-red cell ratio of lead is the smallest for any metal (0.01), yet lead passes into the brain more easily than cadmium, although it does not accumulate there.[27] More lead concentrates in cortical gray matter and basal gang-

TABLE 3
Biological half-time of the slow component of excretion for ionic and protein-bound methylmercury in three species of fish: flounder, pike, and eel. Gulf of Finland, Tvärminne Zoological Station (salinity 6%)

Fish	Route of administration and form of methylmercury	$T_{1/2}$ biol (days)	No. of fish
Flounder	*per os*, proteinate	780 ± 120	9
(*Pleuronectes*	*per os*, ionic	700 ± 50	7
flesus)	intramuscular, ionic	(1200 ± 400)	10
Pike	*per os*, proteinate	750 ± 50	5
(*Esox lucius*)	*per os*, ionic	640 ± 120	2
	intramuscular, ionic	780 ± 80	3
Eel			
(*Anguilla*	*per os*, proteinate	910 ± 40	4
vulgaris)	*per os*, ionic	1030 ± 70	4
	intramuscular, ionic	1030 ± 80	2

lia than in cortical white matter in cases of lead poisoning.[28] Organo-lead compounds penetrate the brain quite easily, especially the monovalent three-ethyl lead, which the organism readily forms from tetraethyl lead. In this respect, they resemble methylmercury. The cells evidently are not adequately protected against monovalent metal compounds which mimic alkalies. The gastrointestinal excretion of lead is low, of the order of one or a few percent.[26] In the kidney, a tubular reabsorption mechanism exists for lead in acutely exposed subjects with high blood levels (> 100 μg Pb per 100 ml) and probably also in cases of normal lead burdens.[29, 30] Details of the tubular mechanism are still unknown since the low level of lead in plasma has thus far made analysis difficult. Modern methods, especially the graphite oven technique for atomic absorption spectrophotometry, have substantially increased the possibility of analyzing plasma lead in normal persons.

Lead excretion via milk correlates with blood levels and may reach high values (0.3 ppm) in cows exposed to lead.[31]

The primary difficulty in studies of lead absorption and retention is its strong affinity to bone. The half-time of lead in human bone is not known, but the ICRP uses 10 years as the first approximation.[32] There exist two or three components with different mobilities, and excretion follows more a power function than an exponential one. The liver is another site for the storage of lead. This partition of lead between the liver and bone as well as its normally very low plasma level render it difficult to evaluate the relationships between lead concentration in blood, the critical organs and urine under normal exposure, but much data exists regarding current exposure.[33] Lead in blood and urine reflect current exposure well. The blood level is considered to be more reliable, since metabolic factors may influence renal handling of lead.

Lead metabolism in the mussel was recently studied by Kauranen and Järvenpää[34] utilizing ^{210}Pb. The biological half-time for the soft parts of *Mytilus edulis* was 300 days; for the shell, non-measurable (long). In two other marine organisms, *Mesidotae entomon* and *Harmatoe* sp., the values were 170 and 50 days, respectively. A short component of a few days exists in all these species.

CONCLUSION

Mechanisms detailing the absorption and excretion of heavy metals in animals and man are still only rather vaguely known. In the case of methylmercury, however, a reliable mathematical model exists. Although common features exist among the various metals which permit the drawing of general schemes such as that illustrated in Fig. 2, the differences among the metals are more marked. In particular, the mechanisms involved in renal excretion are not completely known. The use of radioactive or stable isotopes and new analytical methods make experimental studies easier. Better epidemiological data are necessary in order to be able to evaluate the significance of the parameters of general environmental exposure, which is chronic and at low level.

REFERENCES

1. TASK GROUP ON METAL ACCUMULATION, Accumulation of toxic metals with special reference to their absorption, excretion and biological half-times. *Environ. Physiol. Biochem.* **3**, 65–107 (1973).
2. SIX, K. M. and GOYER, R. A., Experimental enhancement of lead toxicity by low dietary calcium. *J. Lab. Clin. Med.* **76**, 933–943 (1970).
3. KOBAYASHI, J., NAKAHURA, H., and HASEGAWA, T., Accumulation of cadmium in organs of mice fed on cadmium polluted rice, *Jap. J. Hyg.* **26**, 401–407 (1971) (in Japanese, with English summary).
4. SIX, K. M. and GOYER, R. A., The influence of iron deficiency on tissue content and toxicity of ingested lead in the rat, *J. Lab. Clin. Med.* **79**, 128–138 (1972).
5. FRIBERG, L., PISCATOR, M., and NORDBERG, G., *Cadmium in the Environment*, Chemical Rubber Co., Cleveland, Ohio (1972).
6. MILEV, N., SATTLER, E.-L., and MENDEN, E., Aufnahme und Einlagerung von Blei im Korper unter verschiedenen Ernahrungsbedingungen, *Med. Ernahr*, **11**, 29–32 (1970).

7. WESTÖÖ, G., Methylmercury compounds in animal foods, *Chemical Fallout Current Research on Persistent Pesticides* (Miller, M. W. and Berg, G. G., eds.), pp. 75–93, C. C. Thomas, Springfield, Illinois, 1969.
8. RIVERS, J. B., PEARSON, J. E., and SCHULTZ, C. D., Total and organic mercury in marine fish, *Bull. Environ. Contam. Toxicol.* **8**, 257–266 (1972).
9. GANTHER, H. E., GOUDIE, C., SUNDE, M. L., KOPECKY, M. J., WAGNER, P., OH, S.-H., and HOEKSTRA, W. G., Selenium: relation to decreased toxicity of methylmercury added to diets containing tuna, *Science* **175**, 1122–1124 (1972).
10. RAHOLA, T., AARAN, R.-K., and MIETTINEN, J. K., Half-time studies of mercury and cadmium by whole-body counting, *IAEA Symposium on the Assessment of Radioactive Organ and Body Burdens, Stockholm, Sweden, November 22–26, 1971*, pp. 553–562, IAEA-SM-150/13, 1972.
11. RAHOLA, T., AARAN, R.-K., and MIETTINEN, J. K., Retention and elimination of cadmium-115m in man, *Second European Congress on Radiation Protection, Budapest, Hungary, May 3–5, 1972*, pp. 213–218.
12. SCHROEDER, H. A., BALASSA, J. J., BRATTLEBORO, V., and HOGENCAMP, J. C., Abnormal trace metals in man: cadmium, *J. Chron. Dis.* **14** (2) 236–258 (1961).
13. JAAKKOLA, T., TAKAHASHI, H., SOININEN, R., RISSANEN, K., and MIETTINEN, J. K., Cadmium content of sea water, bottom sediment and fish, and its elimination rate in fish., *Radiotracer Studies of Chemical Residues in Food and Agriculture*, 69–75, IAEA-PI-469/7, IAEA, Vienna, 1972.
14. PRESTON, A., Cadmium in the marine environment of the United Kingdom, *Mar. Pollut. Bull.* **4**, 105–107 (1973).
15. JAAKKOLA, T., TAKAHASHI, H., and MIETTINEN, J. K., Cadmium content in sea water, bottom sediment, fish, lichen and elk in Finland, *Environmental Quality and Safety, Global Aspects of Chemistry, Toxicology and Technology as Applied to the Environment*, vol. II, pp. 230–237 (Coulston, F. and Korte, F., eds.), Academic Press, New York, 1973.
16. ÅBERG, B., EKMAN, L., FALK, R., GREITZ, U., PERSSON, G., and SNIHS, J. O., Metabolism of methylmercury (^{203}Hg) compounds in man: excretion and distribution, *Arch. Environ. Hlth.* **19**, 478–484 (1969).
17. MIETTINEN, J. K., RAHOLA, T., HATTULA, T., RISSANEN, K., and TILLANDER, M., Elimination of ^{203}Hg-methylmercury in man, *Ann. Clin. Res.* **3**, 116–122 (1971).
18. RAHOLA, T., HATTULA, T., KOROLAINEN, A., and MIETTINEN, J. K., Elimination of free and protein-bound ionic mercury (^{203}Hg^{2+}) in man, *Ann. Clin. Res.* **5**, 214–229 (1973).
19. SUZUKI, T., personal communication, 1970.
20. MIETTINEN, J. K., TILLANDER, M., RISSANEN, K., MIETTINEN, V., and OHMOMO, Y., Distribution and excretion rate of phenyl- and methylmercury nitrate in fish, mussels, molluscs and crayfish, *9th Japan Conference on Radioisotopes, Tokyo, May 1969*, paper B/(11)-17, pp. 474–478, 1969.
21. MIETTINEN, J. K., HEYRAUD, M., and KĚCKES, S., Mercury as hydrospheric pollutant. II, Biological half-time of methylmercury in four Mediterranean species: a fish, a crab and two molluscs, *Marine Pollution and Sea Life* (Ruivo, M., ed.), pp. 295–298, FAO Fishing News, London, 1972.
22. JÄRVENPÄÄ, T., TILLANDER, M., and MIETTINEN, J. K., Methylmercury: half-time of elimination in flounder, pike and eel, *Suomen Kemistilehti* **B43**, 439–442 (1970).
23. TILLANDER, M., MIETTINEN, J. K., and KOIVISTO, I., Excretion rate of methylmercury in the seal, *Pusa hispida*, *Marine Pollution and Sea Life* (Ruivo, M., ed.), pp. 303–306, FAO Fishing News, London, 1972.
24. RUOHTULA, M. and MIETTINEN, J. K., Retention and excretion of ^{203}Hg-labelled methylmercury in rainbow trout, *Oikos*, **24** (in press) (1974).
25. KEHOE, R., The metabolism of lead in man in health and disease, *J. Roy. Inst. Publ. Hlth.* **24**, 81–97, 101–120a, 129–143, 177–203 (1961).
26. HURSH, J. B. and SUOMELA, J., Absorption of ^{212}Pb from the gastrointestinal tract of man, *Acta Radiol.* **7**, 108–120 (1967).
27. SCHROEDER, H. A. and TIPTON, L., The human body burden of lead, *Arch. Environ. Hlth.* **17**, 965–978 (1968).
28. KLEIN, M., NAMER, R., HARPUR, E., and ROBIN, R., Earthenware containers as a source of fatal lead poisoning, *New Engl. J. Med.* **283**, 669 (1970).
29. VOSTAL, J., Mechanisms of lead excretion, *Biochem. Pharmacol.* **2**, 207 (1963).
30. VOSTAL, J., General mechanisms of the renal and biliary excretion of toxic metals, *Proceedings of the XVIIth International Congress on Occupational Health* (in press) 1972.
31. HAMMOND, P. B. and ARONSON, A. L., Lead poisoning in cattle and horses in the vicinity of a smelter, *Ann. NY Acad. Sci.* **111**, 596–611 (1964).
32. ICRP, 2, *Recommendations of the International Commission on Radiological Protection*, report of Committee II on Permissible Dose for Internal Radiation, ICRP Publication 2, Pergamon Press, London, 1959.
33. HERNBERG, S., The value of lead analyses in blood and urine as indices of body burden, accumulation in critical organs and exposure, *Proceedings of the XVIIth International Congress on Occupational Health* (in press) 1972.
34. KAURANEN, P. and JÄRVENPÄÄ, T., Biological half-times of ^{210}Po and ^{210}Pb in some marine organisms, *Department of Radiochemistry, University of Helsinki, Annual Report, May 1972*.

The Accumulation and Excretion of Heavy Metals in Organisms
(J. K. Miettinen)

DISCUSSION by ROBERT A. GOYER[†]
University of North Carolina, Chapel Hill, North Carolina

I want to compliment Dr. Miettinen on the thoroughness of his review. I shall only emphasize a few points he has presented and attempt to show how the various transport forms of heavy metals influence toxicity, particularly at the cellular level. Cellular effects are best studied in those cells involved in the transport of metals during absorption or excretion. The kidney is probably the ideal organ for these types of studies. General principles regarding renal excretion are similar for mercury, cadmium, and lead.

All three are cleared from the plasma by glomerular filtration. It is thought that all three metals may also be excreted into tubular lumen by transtubular flow—that is from pericapillary blood—across the renal tubular lining cell into the lumen. Thirdly, all are reabsorbed to some degree from the luminal surface of the cell.

Information regarding the relative roles of these transport pathways for different dose levels and chemical forms of each metal is not available and is badly needed.

I will comment first on the cellular effects of mercury in the kidney and later on cadmium and lead.

MERCURY

Dr. Miettinen talked about differences in biological half-life and retention of inorganic mercury versus CH_3Hg. The metabolism of these two forms of mercury by the kidney may also differ. Injection of a large dose of inorganic mercury to rats may produce necrosis or death of renal tubular lining cells in 24 h.[1] The form of mercury which is presented to these cells is not known but because of the rapidity of the changes it is most likely that these effects are the result of the interaction of inorganic mercury or ionic mercury with cellular membranes.

The pathological changes following the administration of methylmercury or by injection are considerably less dramatic. There is some swelling of mitochondria and increase in number of granular appearing lysosomes. Fowler[2] at NIEHS has recently shown by using X-ray microanalysis that mercury becomes concentrated in lysosomes.

These studies are consistent with the observation of Norseth and Brendeform,[3] who found, by cell fraction methods using enzyme markers, that high concentrations of mercury were present in liver lysosomes. It appears, therefore, that mercury, administered as methylmercury, becomes sequestered in lysosomes where it may be less toxic to cellular function than if it were in the free ionic or inorganic form. Some of the difference in severity and form of reaction may be due to dose and rate of mercury administration. The studies of Piotrowski et al.[4] suggest that most of the mercury which accumulates in the rat kidney 3 weeks after injection of mercuric chloride is bound to a small protein with a molecular weight of about 10,000 and chromatographically identical to metallothionein. This is the metal-binding protein previously shown to be inducible by cadmium. Piotrowski found that about 60–75% of the total mercury in the kidney was bound to metallothionein. This work suggests that the mercury bound to metallothionein may also be sequestered by lysosomes. It is known that small molecular weight proteins, cleared from plasma by glomerular filtration are resorbed by renal tubular lining cells and latter catabolized by lysosomal enzymes. The importance of this sequence in the metabolism and excretion of methylmercury is unknown but the studies of Piotrowski, Fowler, and Norseth, looked at collectively, suggest that the lysosomal sequestration of mercury may influence not only the rate of mercury excretion but the intracellular binding or tolerance to mercury.

CADMIUM

Determination of cadmium half-life in man, for reasons reviewed by Dr. Miettinen, is a rather difficult task. There seem to be at least two compartments: one with a fast half-life of just a couple of days ($2\frac{1}{2}$), and a second compartment with a half-life of 15–20 years. In terms of adverse effects from exposure to environmental pollution with cadmium, the slow component probably has the greater signifi-

---

[†] Present address: University Western Ontario, London, Ontario.

cance. Dr. Nomiyama reviewed the metabolism and renal effects of cadmium yesterday morning. I want to review only a couple of points. Experimental animals, and man, accumulate orally administered cadmium in the liver initially. Here it is believed that cadmium induces metallothionein synthesis.[5] With continued exposure to cadmium, renal cadmium content progressively increases. Most of the renal cadmium is in the cortex and is probably in the proximal renal tubule. It has been suggested that cadmium induced metallothionein synthesis has an overall protective effect by reducing the cellular toxicity of cadmium. This hypothesis has some direct experimental support. Cadmium salts injected into mice produce severe testicular necrosis. However, if an equivalent dose of cadmium bound to metallothionein is given to mice, no testicular necrosis occurs, so that the binding of cadmium to metallothionein seems to protect against testicular necrosis.[16] This protective effect of metallothionein, however, is somewhat paradoxical. That is, binding of cadmium to metallothionein does protect against testicular necrosis, but metallothionein-bound cadmium produces severe necrosis of renal tubular lining cells that does not occur with injection of nonprotein bound cadmium salts. It appears that most of the cadmium sequestered in the tubular lining cells is not excreted until cadmium levels reach a critical level. It is thought that metallothionein-bound cadmium, like protein-bound mercury, may be resorbed by proximal tubular lining cells and sequestered into lysosomes. As mentioned by Dr. Nomiyama, Vigiliani[7] has suggested that enzymatic catabolism of protein in lysosomes is impaired by the cadmium, but when cadmium content reaches a critical level, cellular necrosis occurs accompanied by renal tubular dysfunction, proteinuria and increased urinary excretion of cadmium.

All of these studies and speculations indicate that much is known about the intracellular metabolism and transport of cadmium, particularly in terms of protein binding, but the details of the subcellular pathology and sites of intracellular deposition of cadmium have not yet been fully demonstrated. Some of the difficulties in understanding the renal metabolism of cadmium may be avoided if the process is regarded as dynamic with differences regarding dose and acuteness or chronicity.

LEAD

I would like now to make a few comments regarding the transport and metabolism of lead. The biologic half-life of lead, like cadmium, is very long.

The distribution of lead in the body is at least as complex as that for other metals, and for descriptive purposes should be considered to have three compartments (Table 1). Compartment 1 may be regarded as freely diffusible lead in blood and soft tissue. Lead in this compartment must be bound to micro-ligands and may correspond to the rapidly exchangeable pool which has been recently estimated by Rabinowitz et al.[8] to have a half-life in the adult of about 27 days. Compartment 2 contains the slowly exchangeable pools in soft tissues and probably corresponds to nondiffusible protein complexes which, in cases of lead toxicity, can be viewed in cells of some tissues as morphologically discernible inclusion bodies. The half-life of this compartment has not been estimated but, as in total soft tissue, content of lead must be quite variable from one tissue to another. In a moment I shall comment further on some aspects of soft tissue distribution of lead, in particular nondiffusible lead–protein complexes or inclusion body formation. The third compartment is that portion of body lead which is relatively fixed in bone. This component has an extremely long half-life, probably greater than 20 years, and for some portions of this compartment may be as long as 90 years.[9] In the adult, as much as 90% of whole-body lead may be in this compartment. Although this compartment is a potential contributor to compartments 1 and 2, most of the active metabolism of lead, particularly as it affects the pathogenesis of adverse health effects, is related to soft tissue lead. I should also point out that a small portion of bone lead, for kinetic purposes, must be regarded as being identical to soft tissue lead, but this must be a small fraction of total bone lead.[10]

I would like to turn now to the distribution of soft-tissue lead, particularly in the kidney, an organ where lead is readily transported into and out of tubular lining cells. Like cadmium, most lead in the kidney is located in the cortex in the area corresponding to the proximal tubular lining cells, and in lead poisoning, most of the lead in the proximal tubular lining cells is concentrated in intranuclear inclusion bodies. These bodies are discernible by

TABLE 1
Kinetic distribution of whole body lead

Compartment	Half-life
1. Blood, soft tissues (diffusible)	27 days[8]
2. Soft tissues (nondiffusible) Bone (easily diffusible)	Variable
3. Bone (fixed pool)	20 years[9]

light and electron microscopy in nuclei of renal tubular lining cells of rats fed lead experimentally, and consist of a dense central core and an outer fibrillary margin.[11] Lead-induced inclusion bodies have been shown by a number of techniques to be composed of a lead–protein complex.[12–15] About 90% of the lead in the kidney of lead–toxic rats is found in these lead–protein complexes.[16] This has suggested to us that the inclusion bodies represent a nondiffusible depot of lead in the cell. Accumulation of lead in this manner results in a relatively small increment of lead in other portions of the cell such as the endoplasmic reticulum and mitochondria. The formation of these bodies, therefore, is important in the cells of those organs which must metabolize or transport lead.[17] Inclusion bodies are common in hepatic parenchymal cells[18] as well as renal tubular cells following excessive exposure to lead, and they have also been induced experimentally in osteoclasts of bone[19] and glial cells of the brain near the surface of the cerebral cortex.[17] Their formation seems to be a universal phenomenon in most living things. They have been found in many species of mammals,[16, 18–21] in birds,[22, 23] and in moss leaf cells.[24]

Although the inclusion bodies are almost always found in nuclei, they are sometimes found in the cytoplasm.[21] Whether inclusions are formed only in the cytoplasm and then move across the nuclear membrane, or whether they may form in either site, is not known.

Studies regarding the nature of the protein forming inclusions have shown that the protein is virtually insoluble except in denaturants like 6 M urea or sodium deoxycholate.

When the inclusion bodies are incubated with a chelator such as EDTA, the lead is completely removed and the structure of the protein is reduced to an amorphous mass. These studies suggest that lead–protein binding may contribute to a tertiary structure or aggregation of protein units to form the inclusion bodies. Amino acid analysis of hydrolysates of the protein show that it is rich in acidic amino acids and contains about 4% cystine. The availability of many carboxyl groups as well as sulfhydryl groups provides potential sites for lead binding.[25]

In summary, studies to date suggest that the intracellular distribution of metals is not uniform but is concentrated within certain organelles. Mercury, and perhaps cadmium, are, in common with several other metals such as iron and copper, concentrated in lysosomes, probably bound to low molecular weight proteins. Lead on the other hand, in common with bismuth and perhaps other metals, forms discrete intracellular complexes with acidic proteins.

The formation of these protein complexes or intracellular "sinks" must certainly influence the study of dose–response relationships.

REFERENCES

1. GRITZKA, T. L. and TRUMP, B. F., Renal tubular lesions caused by mercuric chloride, *Am. J. Path.* **52**, 1225–1250 (1968).
2. FOWLER, B. A., unpublished observations, 1973.
3. NORSETH, T. and BRENDEFORD, M., Intracellular distribution of inorganic and organic mercury in rat liver after exposure to methyl mercury salts, *Biochem. Pharmacol.* **20**, 1101–1107 (1971).
4. PIOTROWSKI, J. K., TROJANOWSKA, B., WISNIEWSKA-KRYPL, J. M., and BOLANOWSKA, W., Further investigations on binding and release of mercury in the rat, in *Mercury, Mercurials, and Mercaptans*, MILLER, M. W. and CLARKSON, T. W. (eds.), C. C. Thomas, Springfield, Ill., 1973.
5. NORDBERG, G. F., PISCATOR, M., and LIND, B., Distribution of cadmium among protein fractions of mouse liver, *Acta pharmac. tox.* **29**, 456–470 (1971).
6. NORDBERG, G. F., Cadmium metabolism and toxicity, *Environ. Physiol. Biochem.* **2**, 7–36 (1972).
7. VIGLIANI, E. C., The biopathology of cadmium, *Am. Ind. Hyg. Ass. J.* **30**, 329–340 (1969).
8. RABINOWITZ, M. B., WETHERILL, G. W., and KOPPLE, J. D., Lead metabolism in the normal human: stable isotope studies. *Science* **182**, 725–727 (1973).
9. BARLTROP, D., Significance of lead in bone, *Clin. Toxic. Bull.* **3**, 63–73 (1973).
10. HAMMOND, P. B., The effects of chelating agents on the tissue distribution and excretion of lead, *Toxic. Appl. Pharmac.* **18**, 296–310 (1971).
11. GOYER, R. A., Lead and the kidney, *Current Topics in Pathol.* **55**, 147–176 (1972).
12. DALLENBACH, F. D., Uptake of radioactive lead by tubular epithelium of the kidney, *Verh. Dt. Ges. Path.* **49**, 179–185 (1965).
13. GOYER, R. A., MAY, P., CATES, M. M., and KRIGMAN, M. R., Lead and protein content of isolated inclusion bodies from kidneys of lead poisoned rats, *Lab. Invest.* **22**, 245 (1970).
14. CARROLL, K. G., SPINELLI, F. R., and GOYER, R. A., Electron probe micro-analyzer localization of lead in kidney tissues of poisoned rats, *Nature, Lond.* **227**, 1056 (1970).
15. HORN, J., Isolation and examination of inclusion bodies of the rat kidney after chronic lead poisoning, *Virchows Arch. Abt. B (Zellpath.)* **6**, 313 (1970).
16. GOYER, R. A., LEONARD, D. L., MOORE, J. F., RHYNE, B., and KRIGMAN, M. R., Lead dosage and the role of the intranuclear inclusion body, *Arch. Environ. Hlth.* **20**, 705 (1970).
17. GOYER, R. A. and RHYNE, B. C., Pathological Effects

of Lead, in *International Review of Experimental Pathology*, RICHTER, G. W. and EPSTEIN, M. A. (eds.), Vol. 12, pp. 1–77, 1973.
18. BLACKMAN, S. S., JR., Intranuclear inclusion bodies in kidney and liver caused by lead poisoning, *Bull. Johns Hopkins Hosp.* **58**, 384 (1936).
19. HSU, F. S., KROOK, L., SHIVELY, J. N., DUNCAN, J. R., and POND, W. G., Lead inclusion bodies in osteroclasts, *Science* **181**, 447 (1973).
20. HASS, G. M., BROWN, D. V. L., EISENSTEIN, R., and HERMANS, A., Relations between lead poisoning in rabbit and man, *Am. J. Path.* **45**, 691 (1964).
21. STOWE, H. D., GOYER, R. A., and KRIGMAN, M. R., Experimental oral lead toxicity in young dogs, *Arch. Path.* **95**, 106 (1973).
22. LOCKE, L. N., BAGLEY, G. E., and IRBY, H. D., Acid-fast intra-nuclear inclusion bodies in the kidneys of mallards fed lead shot, *Bull. Wild. Dis. Ass.* **2**, 127 (1966).
23. SIMPSON, C. F., DAMRON, B. L., and HARMS, R. H., Abnormalities of erythrocytes and renal tubules of chicks poisoned with lead, *Am. J. Vet. Res.* **31**, 515 (1970).
24. SKAAR, H., OPHUS, E., and GULLVÄG, B. M., Lead accumulation within nuclei of moss leaf cells, *Nature, Lond.* **241**, 215 (1973).
25. MOORE, J. F., GOYER, R. A., and WILSON, M., Lead induced inclusion bodies: solubility, amino acid content and relationship to residual acidic nuclear proteins, *Lab. Invest.* **29**, 488–494 (1973).

TRANSPORT AND BIOLOGICAL EFFECTS OF MOLYBDENUM IN THE ENVIRONMENT†

WILLARD R. CHAPPELL
Department of Physics and Astrophysics,
and The Molybdenum Project,
University of Colorado, Boulder, Colorado 80302

INTRODUCTION

Molybdenum is a trace element which occurs widely in nature. It plays a key role in agriculture because it is a micronutrient for both plants and animals (there are areas where plant growth is severely affected because of a lack of available molybdenum in the soil) and because it can be accumulated by plants to a level which is toxic to livestock.[1-3] Its chief industrial use is in various alloy steels. Other important uses are as a catalyst in petroleum processing and as a lubricant.[4,5]

Because of the known toxic effects on livestock, the Federal Water Quality Committee suggested an irrigation water standard of 5 µg/l for continuous use on all soils and 50 µg/l for intermittent use on fine-textured soils.[6] There are many streams in the United States which contain molybdenum at levels above 50 µg/l. Many of these streams are associated with potential or known industrial sources of molybdenum, and in some streams the molybdenum concentrations are as high as 1 mg/l. In addition, airborne molybdenum has been known to cause problems with livestock in the areas surrounding some molybdenum and uranium mills.

Our purpose here is to summarize the work done on molybdenum as it relates to its transport and biological effects in the environment. We discuss the distribution of molybdenum in the environment, the release of molybdenum into the environment, the aqueous transport of molybdenum, the transport of molybdenum from irrigation water to plants, and the biological role and effects of molybdenum. The emphasis is on the work done by the Molybdenum Project at the University of Colorado and Colorado State University.

---

†Supported by N.S.F. (RANN) Grant GI-34814X and GI-34814X1.

DISTRIBUTION OF MOLYBDENUM IN NATURE

The average abundance of molybdenum in the earth's crust is $10^{-4}\%$ or one part per million (ppm).[7] The distribution in individual rock types is quite variable. In igneous rocks it is present at concentrations between 1 and 6 ppm. In black shale it can vary from 1 to 300 ppm, while in phosphorites it varies from 5 to 100 ppm.[7]

The commercial deposits contain 200 ppm molybdenum or more. The lower concentrations are generally mined as a byproduct of copper mining. The world's largest molybdenum mine is located at Climax, Colorado, and produces approximately 60% of the molybdenum mined in the United States and 40% of that mined in the Western world. Smaller primary molybdenum operations are located at Questa, New Mexico, and Urad, Colorado. The production of molybdenum as a byproduct of copper mining is growing in importance. Presently, it is estimated that one-quarter of the US production and one-half of the world production is as a coproduct or byproduct of copper production.[5] The current world production is estimated at about 200 million lb per year with domestic production approximately two-thirds of that figure.[4,5] Molybdenum is one of the few metals of which the US is a net exporter. Approximately one-half, or about 50 million pounds per year, of the US production is exported.[5] The other countries which are significant producers of molybdenum are Canada, USSR, China, and Chile.

The domestic use of molybdenum is approximately:[4,5]

70% for use in high speed, tool, stainless, and low alloy construction steels.
20% for use in special alloys and casting.
10% for use as molybdenum metal, pigments, catalysts, fertilizer, lubricants, and other chemicals.

King et al.[5] point out that the breakdown in use indicates an important role in an industrially sophisticated nation. Thus, we would expect to see an increase in use as developing nations increase their need for automobiles, industrial equipment, refineries, and building construction materials.

The total molybdenum concentration in soils can vary quite widely, but most soils contain between 0.6 and 3.5 ppm.[8] The average value in the soils of the United States has been reported by the USGS to be less than 3 ppm (the limit of detection in this particular study).[9] Others have reported an average of 1–2 ppm over a wide range of soils.[10] Soils can vary quite widely in their total molybdenum content, however, with soils near ore grade deposits of molybdenum containing hundreds of parts per million.[11]

The concentration of molybdenum in plants varies quite widely. Levels as low as 0.01 ppm in lemon leaves to as high as 300–400 ppm in legumes have been reported.[3] In general, legumes will concentrate molybdenum more than other plants. Clover and alfalfa tend to be among the best concentrators of molybdenum. It is important to note at this point that if the molybdenum level is too low, e.g. 0.1 ppm in red clover, the plant is deficient in molybdenum and growth is less than optimal. When the plant molybdenum reaches 5–15 ppm the plant can become toxic to livestock.[1]

The level of molybdenum in stream sediments can vary quite appreciably. Runnells of our group has measured values ranging from 2 ppm to 400 ppm.[11] Similar results were obtained by Webb et al.[12] who have proposed a background level of less than 2 ppm in stream sediment.[13] This appears to be substantiated by the work of Runnells.[11,14]

The median concentration of molybdenum in surface waters of North America has been reported by Hem as 0.35 mg/l.[15] In the same study the median level of public water supplies was reported to be 1.4 µg/l.

Sources of anomalous amounts of molybdenum in soils, plants, and water may be either natural (e.g. weathering of molybdeniferous rocks) or industrial. A relatively large area (1,000,000 acres) in central England has been shown to be potentially harmful to livestock (at a subclinical level).[13] This work involved a large scale regional geochemical reconnaissance effort. Webb et al.[12] used stream sediments collected with a sampling density of one sample per square mile and animal tests to delineate the suspect area. In this case the bedrock (a black marine shale, averaging 11 ppm molybdenum) was the original source and normal soil forming processes were the release mechanism.

RELEASE INTO THE ENVIRONMENT

Clearly molybdenum mining and milling is an obvious potential source of molybdenum release into the environment. We have been investigating the drainage system adjoining the mine at Climax in order to gain an understanding of how much molybdenum is released and transported from the area. We have also made a brief survey of an area downstream from the molybdenum mine at Questa, New Mexico.

A major problem that one encounters in this work is the separation of the amount due to man's disturbance and the amount due to natural weathering mechanisms. In an area where there is an extensive exposed deposit this separation can be quite difficult.

In an effort to accomplish this separation, Runnells of our project and his students have studied four sites quite extensively[11,14] The study areas are shown in Fig. 1. The study areas are:

(1) *Ten-mile Creek.* This stream drains the area which includes the Climax mine and mill and flows into Dillon Reservoir. We think of this area as a mineralized and "disturbed" area. We will see that there are complications in this definition.

(2) *Mt. Aetna.* This is an area which includes an exposed, well-mineralized zone of MoS_2. This area has not been mined, however, and we think of it as a mineralized, "undisturbed" deposit.

(3) *The Upper Blue River.* This is an area which is poorly mineralized and is relatively "undisturbed." It is on the other side of a narrow range from Ten-mile Creek.

(4) *The Lower Blue River.* This is an agricultural valley which receives water from both Tenmile, the Upper Blue, and an additional stream, the Snake River, after these waters have been mixed in Dillon Reservoir (Fig. 1).

Runnells and his students have sampled water, plants, soils, and sediments from these areas. These results are given in Table 1.

We note a considerable difference between water and sediment values between Ten-mile and Mt. Aetna and similarly between the Upper and the Lower Blue River. It appears that there is a considerable release of molybdenum in the Climax area. It is difficult to separate, however, the natural component in sediments from the disturbed component.

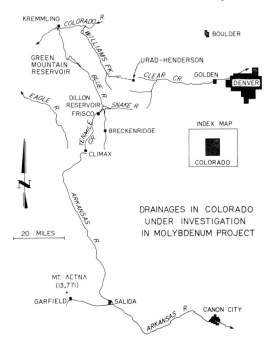

FIG. 1. Drainages in Colorado under investigation in molybdenum project. (Reproduced by courtesy of the Curators of the University of Missouri, Columbia, Mo.)

The comparison of the Mt. Aetna and Climax deposits suffers because extensive glaciation of the Climax deposit occurred whereas there was only minor glacial erosion at Mt. Aetna. Thus nature, as well as man, can "disturb" an area. In addition, the size of the exposed deposits at the two locations differs considerably.

The contribution to stream concentration due to the glacial debris and the effect of the different sizes of the deposits is not yet understood. Runnells is now carrying out laboratory experiments to try to determine how much molybdenum one would expect in waters under varying conditions. This is being done by simulating the Eh and pH conditions found in natural waters and observing the approach to equilibrium and final concentration of molybdenum derived from various natural solid materials.

The work by Runnells and his students also indicates that stream sediments seem to be a more reliable indicator than plants of molybdeniferous areas. There are no reliable indicator plants that we have found in the study areas. The large fluctuations of plant molybdenum levels found in adjacent areas are a reflection of the fact that not all of the molybdenum in a soil is available to the plant. It is well known that the amount of available molybdenum depends very sensitively on soil parameters such as pH. In a calcareous soil much of the molybdenum is labile, while in an acidic soil the molybdenum is mostly unavailable. In addition, other environmental factors such as climate can significantly affect plant uptake.

The mechanisms which determine the availability of molybdenum are not yet understood, but many believe that the concentration of molybdenum in soil water and other surface waters is largely controlled by the adsorption of $HMoO_4^-$ and MoO_4^{2-} on ferric oxyhydroxides.[7] Greater concentrations in the soil water seem to occur under alkaline conditions than under acid conditions.

The effect of pH is seen in Runnells' data (Table 1). The plants collected from Mt. Aetna, where the soil has an average content of 24 ppm and a pH of 5.5–5.7, have a molybdenum content of about 1 ppm or less. On the other hand, the plants collected from the more alkaline soils (pH 5.9–7.9) in the Lower Blue River Valley have molybdenum contents ranging up to 60 ppm despite the fact that the average content of molybdenum in the soils in this area is much lower than on Mt. Aetna (Table 1).

There are several other, perhaps more important sources of environmental molybdenum than molybdenum mining. These include the processing of molybdenum ore, the use of molybdenum products, the mining and milling of some uranium and copper ores, and the burning of fossil fuels.

In recent years we have encountered three reports of molybdenum toxicity in cattle which were associated with uranium mining and/or milling. One case in North Dakota involved a farm near a plant where a uraniferous lignite coal was ashed to upgrade the uranium content.[16] The resulting ash contained 3200 ppm molybdenum. Ash released from the plant gave rise to elevated molybdenum levels in the soils (8–50 ppm) and hay (10 ppm). Because of the relatively high levels of molybdenum in the soil, ingestion of soil may have been a significant source of molybdenum intake as well as ingestion of forage.

Another case of molybdenosis involving uranium mining and milling occurred in Karnes County, Texas, on a ranch containing several abandoned uranium mines and where the AEC had stockpiled uranium while processing mills were being constructed.[17] Analysis showed that the grasses contained from 15 to 45 ppm molybdenum and from 2 to 25 ppm copper, the soils contained from 4 to 8 ppm molybdenum, and the surface waters contained from 130 to 170 µg/l molybdenum. The authors noted that had not the pH of the soils been so low,

TABLE 1

Average concentrations of Mo and pH values in materials of Ten-mile Valley below Climax mine and mill, stream which drains molybdenum deposit on Mt. Aetna, and Blue River above and below Dillon Reservoir. Summers 1971 and 1972 [a]

Material	Ten-mile Valley 1971 Mo	pH	Ten-mile Valley 1972 Mo	pH	Mt. Aetna 1971 Mo	pH	Mt. Aetna 1972 Mo	pH
Water (ppb)	Hundreds to thousands of parts per billion, depending on season and releases from mill upstream. Average pH, 8.1 (2)		Average pH 8.3 (2)		0.5 ppb (13) (X-ray fluorescence of residue)	6.3 (10)	Not detectable by colorimetric procedure on three samples.	
Sediment (ppm)	240 ppm (6)		590 ppm (3)		17 ppm (5)		22 ppm (8)	
Soils (ppm)	25 ppm (5) Range of slurry pH values −5.8 (5)		Upper 2 in., 110 ppm (5) Deeper than 2 in., 41 ppm (5) Range of slurry pH of surface layer 5.6–6.3 (4), with one value of 8.3 Range of slurry pH of deeper soils 5.6–6.3 (4) with one value of 8.3		24 ppm (5) Range of slurry pH values, 5.3–5.7 (2)		49 ppm (1)	no pH
Plants (ppm)								
Spruce	2.5 ppm (2)		—		0.6 ppm (2)		—	
Grass	7 ppm (1)		30 ppm (3)		1.5 ppm (3)		—	
Clover	—		—		0.6 ppm (3)		—	
Willow	16 ppm (3)		37 ppm (6)		0.5 ppm (2)		—	

Material	Upper Blue River 1971 Mo	pH	Upper Blue River 1972 Mo	pH	Lower Blue River 1971 Mo	pH	Lower Blue River 1972 Mo	pH
Water (ppb)	6 ppb (3)	8.0 (2)	3 ppb (1)	8.5 (1)	190 ppb (4)	8.2 (4)	260 ppb (4)	8.3 (2)
Sediment (ppm)	4 ppm (5)		4 ppm (4)		39 ppm (5)		25 ppm (5)	
Soils (ppm)	3.5 ppm (5) Range of slurry pH values, 6.3–7.2 (5)		Upper 2 in., 3.5 ppm (5) Deeper than 2 in., 2.4 ppm (4) Range of slurry pH of surface soils 5.7–7.2 (5) Range of slurry pH of deeper soils 5.4–7.3 (4)		12 ppm (4) Range of slurry pH values, 5.9–7.9 (8)		Upper 2 in., 23 ppm (2) Deeper than 2 in., 19 ppm (2) Range of slurry pH of surface soils 6.8–8.0 (2) Range of slurry pH of deeper soils 7.2–8.0 (2)	
Plants (ppm)								
Spruce	0.5 ppm (3)		0.7 ppm (1)		1.2 ppm (2)		0.6 ppm (1)	
Grass	0.9 ppm (3)		1.2 ppm (2)		1.5 ppm (4)		7.5 ppm (3)	
Clover	—		1.3 ppm (1)		50 ppm (2)		60 ppm (4)	
Willow	1.2 ppm (4)		0.4 ppm (10)		3.3 ppm (3)		1.0 ppm (9)	

[a] Number of separate samples shown in parentheses; several determinations on most samples.

5.0–5.8, the problem would probably have been worse because of greater availability.

A third case of elevated environmental levels of molybdenum associated with uranium processing occurred in Colorado. The uranium mill in question began operation in 1953. In 1965 farmers downslope from the mill began to notice a deterioration in the health of their cattle. The symptoms included diarrhea and weight losses and seemed to be related to the well water which was used for irrigation of pasture. Le Gendre and Katz of our group sampled water, soil, and forage in the area, and, with the cooperation of the company, sampled the tailings ponds. The water in the ponds has concentrations of molybdenum up to 856,000 μg/l. Water behind a retaining dam below the tailings ponds has concentrations of up to 50,000 μg/l. Water from the farmers' wells contained up to 50,000 μg/l, while soils from the farmers' fields contained as much as 50 ppm molybdenum and plants contained as much as 300 ppm molybdenum (on a dry weight basis). It is difficult to extensively sample ground water upslope from the mill; however, the one water sample we have obtained had less than 15 μg/l molybdenum.[18]

Although we have not had any verified reports of livestock problems near copper mines or mills, molybdenum does occur at levels of a few hundredths of a percent in some porphyry copper deposits in the southwestern United States. We have analyzed some wastewaters from beneficiating copper mills and found them to contain as much as 10,000 μg/l molybdenum.

Another industrial source which is not connected with molybdenum production is indicated by a recent report concerning a claypit in Missouri.[19] The clay had been mined for use in the ceramic industry. Surface runoff from the claypiles drained into a stream which, on occasion, floods pastures lying downstream. Livestock (Angus and Charolais) in this area indicated symptoms of molybdenosis. The area was extensively sampled by the USGS. The molybdenum levels reported for the clay were 3–7 ppm for soils, in the vicinity the total molybdenum content was 3–7 ppm, and plants in the area were found to contain as much as 750 ppm molybdenum—dry weight (this value was for sweet clover—in grass and sedges the maximum reported level of molybdenum was 10 ppm). Although the claypiles were clearly a source of elevated molybdenum levels in local environment, the presence of anomalous levels of aluminum, beryllium, selenium, and other elements complicated the diagnosis of the cattle problem. The health of the cattle improved on removal from the affected pastures.

The fact that some coal ashes have elevated levels (10–100 ppm) of molybdenum[20] indicates that fossil fuel plants are a potentially important source of environmental molybdenum. Moreover, our work and others' indicate that the molybdenum in ashes is highly available (the ash is very alkaline) to plants. Indeed, one report indicates that it is as available as sodium molybdate.[21] This work would seem to indicate that large applications of fly ash to soil (the levels used in the report mentioned were on the order of 25 g of ash per kg of soil) is not always an advisable method for disposing of the ash.

The work done by Jorden and Kaakinen[22] of our group involving a mass balance of a coal-fired power plant shows that molybdenum tends to concentrate in the fly ash. This might be expected because the melting point of MoO_3 (the form of molybdenum that is suggested to be the most prevalent in coal) is 1155°C. Temperatures in the furnace often exceed 1400°C. Thus in the hot combustion gases one would expect molybdenum to be primarily in the gaseous phase and then to condense by itself or onto other particles as the combustion gases are cooled on the way to the stack exit which is usually at about 120°C.

Jorden and Kaakinen's work indicates that wet scrubbers are efficient at removal of molybdenum, and it can be inferred that the hot precipitators that some plants are turning to to solve SO_2 problems may increase the output of molybdenum and other volatile metals because of the higher temperatures. The lack of control devices on most oil-fired plants may make these plants an even greater potential source of environmental molybdenum than coal-fired plants. The effectiveness of the scrubber can be seen by noting that the aqueous output from the scrubber had a molybdenum concentration at least 50 times the input water.[22]

In addition to molybdenum mining and milling and the other sources mentioned, there have been several reported cases of livestock problems associated with the processing or use of molybdenum products. These include the area surrounding a molybdenum roasting plant in Pennsylvania where a 1970 report[23] cited the release of 180–240 lb of molybdenum into the atmosphere per day within a quadrant of a circle to the east of the stack (radius 1 mile) being covered with (at the time of the report) 1 lb of molybdenum per acre per month; and the area adjacent to a molybdenum oxide plant in the Netherlands[24] where forage levels of molybdenum were reported to be as high as 80 ppm.

A well-documented case involved the release of a molybdenum-containing catalyst from an oil refinery in England in 1960.[25,26] This catalyst was lost at rates ranging from 6 to 50 tons per month during 1960 and 1961. The catalyst contained about 8.2% molybdenum trioxide (MoO_3). The primary loss of the catalyst, 40 tons in a 2 day period, was the result of equipment malfunction. Animals downwind were affected approximately 5–11 days later. The molybdenum levels in the forage ranged from 10 to 85 ppm with a mean of 28 ppm shortly after the incident. Six years later (the unit was shut down in 1961) the levels in the forage in the area ranged from 0.3 to 2.7 with a mean of 1.2 ppm.

Another industrial case of molybdenosis in grazing cattle was reported by Buxton and Allcroft.[27] This case, also in England, involved pastures downwind from a steel plant which used molybdenum compounds. The molybdenum levels in the forage ranged from 14 to 126 ppm.

The transport away from the source can be either by atmospheric or aqueous pathways. In the case of roasting plants, fossil fuel plants, oil refineries, and steel mills, this transport may be primarily atmospheric (although the aqueous effluent may in some cases be significant). The blowing of tailings dust may be another significant mechanism in some areas. The information on atmospheric transport is rather limited, and our work has been primarily concerned with aqueous transport.

AQUEOUS TRANSPORT OF MOLYBDENUM

Molybdenum is relatively mobile compared to other heavy metals, and this characteristic is used in geochemical prospecting to locate deposits of molybdenum or minerals, such as copper, which are often associated with molybdenum. The concentrations of molybdenum found in water draining deposits varies somewhat. Mallory[28] found the level in a stream below a deposit in Colorado to be about 3 μg/l as opposed to <0.5 μg/l in another nearby stream, and 1.3 μg/l below the point of confluence of these streams. Runnells[11] found a level of 0.4–1.1 μg/l molybdenum in water draining the Mt. Aetna deposit. Brundin and Nairis[29] did an extensive study of over 1000 water samples from northern Sweden and reported that 28 μg/l represents the 95th percentile of molybdenum concentrations in waters from areas with scattered molybdenum mineralization. Sugawara et al.[30] sampled 10 rivers and 19 fresh water lakes in Japan and found molybdenum levels to vary from 0.2 to 0.6 μg/l in rivers and 0.05 to 1.2 μg/l in the lakes. Voegeli and King[31] conducted an extensive survey in surface waters in Colorado and concluded that "surface water containing greater than 5 μg/l molybdenum in the mountainous areas of Colorado may be considered anomalous and may warrant extensive sampling and study to ascertain the source of molybdenum."

The most extensive survey involving aqueous molybdenum is due to Kopp and Kroner.[32] This survey measured several trace elements in water collected from key points on rivers in all sixteen major drainage basins in the United States. The sampling points were, in general, located in the vicinity of (a) people and/or animal concentrations; (b) industrial activity, agricultural activity, heat emission sources; (c) recreational use areas; (d) state and national boundaries; and (e) other areas which for a variety of reasons might be sources of heavy metals.

In the 5 year period 1962–7, covered in the report, molybdenum was detected in 33% of more than 1500 samples from 130 sampling points with a mean level of 68 μg/l. This figure is significantly higher than the median of 0.35 μg/l reported for North American streams, and 1.4 μg/l reported for public water supplies.[15] Table 2 shows the frequency of detection with the mean, maximum, and minimum levels detected for the 16 major drainage basins.

If we recall the proposed irrigation standards of 5 μg/l for continuous use on all soils and 50 μg/l for short term use on fine textured soils,[6] we see that there are many areas where these amounts are exceeded. Of the 130 individual sampling stations, 38 recorded a maximum greater than 100 μg/l. The

TABLE 2
Results of surface water analysis for molybdenum by basin (as reported by Kopp and Kroner[32])

Basin	Frequency of detection (%)	Observed values (μg/l)		
		Minimum	Maximum	Mean
1. Northeast	13.2	4	61	25
2. North Atlantic	32.7	4	168	33
3. Southeast	18.7	2	53	15
4. Tennessee River	38.2	3	67	25
5. Ohio River	28.1	6	473	70
6. Lake Erie	27.7	21	108	68
7. Upper Mississippi	68.8	4	360	88
8. Western Great Lakes	51.5	4	129	28
9. Missouri River	32.0	8	354	83
10. Southwest Lower Mississippi	20.0	11	1100	95
11. Colorado River	37.0	10	444	130
12. Western Gulf	10.6	4	59	24
13. Pacific Northwest	38.9	2	128	30
14. California	37.9	14	124	45
15. Great Basin	57.9	3	338	145
16. Alaska	27.8	2	34	17

TABLE 3
Results of surface water analysis for molybdenum by station (as reported by Kopp and Kroner[32])

Station	Frequency of detection (%)	Observed values (μg/l)		
		Minimum	Maximum	Mean
1. Ohio River below Addison, Ohio (basin 5)	28.1	14	473	88
2. Red River (North) at Grand Forks, North Dakota (basin 7)	90	63	360	210
3. South Platte River at Julesburg, Colorado (basin 9)	30	15	354	181
4. Arkansas River at Coolidge, Kansas (basin 10)	40	200	1100	460
5. Colorado River at Loma, Colorado (basin 11)	81.8	20	444	194
6. Bear River above Preston, Idaho (basin 15)	80	38	338	196

six stations recording the highest maxima are listed in Table 3 which shows the frequency of detection and maximum, minimum and average levels detected.

If we assume that the molybdenum levels reported by authors such as Hem[15] indicate a "background" level ("background" in the sense, perhaps, of Voegeli and King[31] in that neither anomalously high natural occurrences nor industrial activity involving molybdenum is present) of less than 5 μg/l, then it would appear that there are anomalously high values occurring in many rivers which are either due to human activity or to areas of naturally high molybdenum levels. In view of the lack of mineralized areas and the acidic soils in the Midwestern and Eastern United States, it seems safe to assume that elevated levels in these areas indicate industrial sources.

It is of interest that of the six stations having the highest maximum recorded molybdenum values in water, two are in Colorado and one is just across the border from Colorado on the downstream side. More recent measurements of molybdenum levels in the Arkansas River show that the present levels are not as high (<40 μg/l) as those measured by Kopp and Kroner; however, the values for the Colorado and South Platte rivers remain about the same as found by Kopp and Kroner. The levels in the South Platte and Colorado rivers can be tied to either the occurrence of molybdenum deposits or to the mining and milling of molybdenum in the drainage area of these streams. There is some reason to believe that the levels in the Arkansas have decreased, at least in part, because of a change in mining practice at Climax which is also in the Arkansas drainage.

The data by Voegeli and King are particularly interesting because of the extent of the study in Colorado. They analyzed 299 samples from 197 stations. They found 89% of the samples to have 1 μg/l or more of molybdenum, with 87% of the stations having less than 10 μg/l. Of the 39 samples which exceeded 10 μg/l molybdenum, 33 were from streams which drain areas of known molybdenum deposits or areas where molybdenum is being processed. Our own data are consistent with these results.

Roger Jorden of our group has been continuously monitoring the level of molybdenum in Ten-mile Creek which drains the area containing the molybdenum deposit, mine and mill at Climax.[13] The measurements are taken just before the creek enters Dillon Reservoir (Fig. 1). Figure 2 shows that the concentration of molybdenum in the creek is less than 0.5 mg/l for most of the year, but in late spring and early summer of 1972 rose as high as 9 mg/l. It is even more interesting to combine the flow data and concentration data for the two outlets to the reservoir (the other inlet streams, the Blue River and Snake River, have very low molybdenum levels). The results are shown in Fig. 3. We see that the vast bulk of molybdenum was delivered to the reservoir during May and June. This massive slug of molybdenum resulted from a release of tailings water from the Climax tailings ponds. This release was necessitated by a heavy spring runoff which the Climax waste control system, which normally recycles the tailings water, was incapable of handling.

Dillon Reservoir (Fig. 4) then acts as a temporary holding tank for the slug of molybdenum, effectively distributing the output over a longer period of time than the input. The cycling of molybdenum within the reservoir is being studied quite closely by Jorden

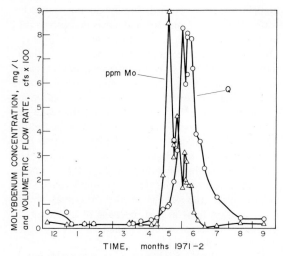

FIG. 2. Molybdenum concentration and volumetric flow rate versus time on Ten-mile Creek (at USGS gaging station near Frisco, Colorado), 1971–2.

and his students. One of the most interesting results of our work on Dillon was the stratification that was noticed by Dr. Winston of our group.[34] He showed that during the summer months molybdenum levels were much higher near the bottom of the reservoir than at the surface. After the fall overturn the distribution becomes more uniform. This is illustrated in Fig. 5, which compares the molybdenum and temperature profiles. A similar result has been reported by Sugawara et al.[30]

For sampling and water management purposes this result is rather significant. The outlets to the Blue River consist of a surface outlet (morning glory) at 230 ft and the Roberts Tunnel subsurface outlet 188 ft below the morning glory, taking water at the depth of about 125 ft. About 50% of the water is discharged through the subsurface outlets. Thus a sampling of the water from these outlets would yield results, in the summer, that were inconsistent with surface measurements.

Briese and Jorden[35] have performed a detailed mass balance calculation on the molybdenum in Dillon Reservoir as a part of the investigation of the cycling of the molybdenum in the reservoir. Table 4 shows their results for the period November 1971 to November 1972. Most of the 252,700 lb of molybdenum that entered the reservoir in May–June of 1972 had left within a 6 month period. Thus, the holding period for the bulk of molybdenum is relatively short.

The cycling in the lake is probably both physical and biological. The summer stratification seems to occur because the cold molybdenum-bearing waters of Ten-mile settle to the bottom of Dillon very rapidly. We are measuring molybdenum levels in the sediment and in detritus to see what roles these have in the cycling of molybdenum. Since molybdenum appears to be essential for nitrogen fixation, there is the possibility of the stimulation of algal growth by elevated levels of molybdenum in the presence of what might ordinarily be limiting levels of nitrogen. We are presently investigating this possibility.

In the drainage system described above, the molybdenum is chiefly carried as a dissolved species (as has been demonstrated by filtration experiments). This species is believed to be the molybdate anion, MoO_4^{--}.[7] This property of molybdenum offers a way to distinguish between point sources and nonpoint sources. Boberg and Runnells[36] suggested

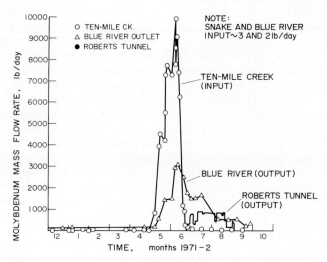

FIG. 3. Molybdenum mass flow rates into and out of Dillon Reservoir, 1971–2.

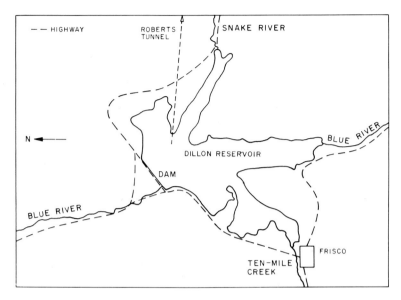

FIG. 4. Dillon Reservoir and tributaries.

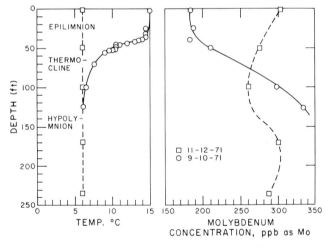

FIG. 5. Temperature and molybdenum concentrations for Dillon Reservoir, 1972.

that it is useful to divide mass flow rate by the drainage area tributary to the particular stream location. For a substance which is conservative, released at a rate proportional to the concentration in solid material in contact with water, and once in solution stays in solution, this provides a tool to investigate point source emission, since the quantity involved should be independent of the area if the release is from a nonpoint source.[37] Figure 6 shows the result of using 138 data points for tributaries not adjacent to molybdenum mining and milling and 40 data points that are adjacent to such activity. A time average was performed to smooth seasonal fluctuations. The slope of the no tributary molybdenum milling curve is not significantly different from zero while that of the tributary molybdenum milling curve is significantly different from zero with a probability $p < 0.001$. This figure also indicates that 10^5 mi^2 is required to "dilute" out the effects of the point sources.

It is expected that in very acid waters containing large amounts of ferric ions a significant amount of molybdenum will be transported as adsorbed ions on colloidal particles of iron hydroxides.[7] This was confirmed by Runnells and his students in laboratory tests, the results of which are shown in Fig. 7.[8]

TABLE 4

Molybdenum mass balance, Dillon Reservoir. November 12, 1971 to November 9, 1972 (units = pounds of molybdenum)

Inputs		Outputs	
Ten-mile Creek	250,400	Blue River	194,000
Snake River	1,100	Roberts Tunnel	45,600
Blue River	1,200		239,000
		Mass storage change	13,100
	252,700		252,700

Standing mass in Dillon Reservoir	
November 9, 1972	224,400
November 12, 1971	181,200
Change	43,200

Percentage error (assuming standing mass correct)

$$\%\text{Error} = \frac{43,200 - 13,100}{224,400} \times 100 = 13\%$$

These results show that as much as 98% of the molybdenum was removed from solution in an industrial wastewater containing molybdenum at concentrations of 1–11 mg/l at a pH of 3.1 by adding $FeCl_3$ which lowers the pH and provides the necessary iron. A similar, but less dramatic result, is obtained by using aluminum instead of iron.

This, then, is a possible removal process and it is interesting that Le Gendre and Runnells have shown that this process occurs accidentally in another stream in Colorado, namely Clear Creek, which drains the area of the Urad mine. The molybdenum concentrations in Clear Creek, range from 200 to 1500 μg/l with pHs of 6.8–7.8. Filtration tests show that essentially all the molybdenum is in solution, except near the outfall from the Argo Tunnel, which drains acid mine water from a mining area and empties into Clear Creek near Idaho Springs. This water is high in iron (2.5 mg/l) and has a pH of about 2.5. The pH of the water in Clear Creek decreases from 7.8 to 3.4 within the first few feet below the Argo Tunnel inlet and rises back to about 6.5 after approximately 300 ft. Le Gendre[38] found a removal of about 40% of the molybdenum from the dissolved fraction in this area of low pH. Sediment below the Argo Tunnel inlet is stained brightly orange for about 150 yd downstream. Within 20 ft of the outlet the sediment contains as much as 700 ppm molybdenum (dry weight) and 2% iron.

We thus see that, under some conditions, molybdenum in water is not primarily in dissolved form. Dr. Roger Jorden of our group and a graduate student are looking at this mechanism to see if it can be used for wastewater treatment. At present, standard water and sewage treatment facilities remove only about 10–15% of the molybdenum put into the system.[39] The results of Berrow and Weber[40] showing a range of 2–30 ppm molybdenum (dry weight) in sewage sludge indicates that it is concentrated in the sludge.

We have also done some work in New Mexico[18] where the Molybdenum Corporation of America operates a molybdenum mine and mill. This operation is located at Questa, New Mexico, and the area

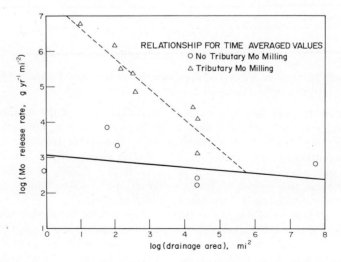

FIG. 6. Relationships of mass release to drainage area for time average values. (Reproduced by courtesy of the Curators of the University of Missouri, Columbia, Mo.)

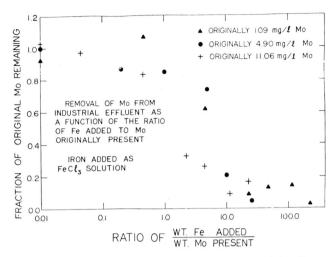

FIG. 7. Removal of molybdenum as a function of added Fe^{+3}.

is drained by the Red River which then joins the Rio Grande. The average molybdenum concentration in a few samples of water from the tailings ponds (as measured by us) is 1520 µg/l. The level of molybdenum in the Red River below the decant pond was measured at 560 µg/l compared to 40 µg/l just above the pond and 6 µg/l a few miles upstream from the operation. The concentration is steadily lowered by dilution so that the Rio Grande below its confluence with the Red River contains 40 µg/l as opposed to 19 µg/l above the confluence.

With some exceptions (such as the small area along Clear Creek mentioned earlier), molybdenum in waters in the Southwest is probably carried almost entirely in dissolved form. This may not be true in the Midwest and East in the vicinity of large industrial areas where local conditions, due to effluent release, could give rise to molybdenum being taken out of solution.

Earlier we noted that there were several rivers in the East and Midwest that were reported by Kopp and Kroner to have molybdenum concentrations above 100 µg/l. More recent data indicates that this situation continues and the EPA STORET data for 1970 and 1972 reveals that several rivers sampled in areas of high industrial activity contained more than 100 µg/l molybdenum. Among these were the Cuyahoga, Maumee, Little Miami, Ohio, Missouri, and Potomac rivers. A sampling of this data is given in Table 5.

In view of the work by Voegeli and King and others,[15,31] the values in the above table should probably be considered anomalously high and would indicate the possibility of various industrial sources in these areas.

TABLE 5
Molybdenum data for selected stations from 1970 to 1972 (compiled by STORET)

	Observed values (µg/l)		
Station	Minimum	Maximum	Mean
1. Mississippi River at Dubuque	28	160	109
2. Ohio River above Greenup Dam, Kentucky	<40	145	102
3. Potomac River at Williamsport	16	115	66
4. Missouri River at St. Louis	40	220	136
5. Missouri River at Omaha	<40	240	149
6. Cuyahoga River at Cleveland	<40	180	108
7. Maumee River at Toledo	<40	320	180
8. Great Miami River at Woodsdale, Ohio	48	420	214

THE TRANSPORT FROM IRRIGATION WATER TO PLANTS

The existence of the suggested irrigation water standard of 5 µg/l for continuous use and 50 µg/l for intermittant use indicates a concern about the transport of molybdenum from irrigation water into plants. This suggested standard is apparently based largely on a study[41] of the relationship of molybdenum levels in soil solution to molybdenum levels in plants. This work indicated that soil solutions containing as little as 10 µg/l were associated with plants containing 10 ppm molybdenum. Since the soil solution is the component from which nutrients are directly removed by plants, the intensity factor or concentration of an element in the soil solution is clearly an important consideration.

There are, however, other factors which need to be considered. The soil solution does not necessarily contain the same level of molybdenum as the irrigation water. Ions can be adsorbed onto various solid constituents of the soil. Moreover, as plants remove a nutrient from the soil solution the replacement of that nutrient depends on the capacity factor of the soil, which is the ability to replenish depleted elements in the solution.[42]

Davies[43] and others[44,45] have characterized the soil components involved as (a) the soil solution, (b) adsorbed nutrients involving exchange sites (sometimes referred to as conditionally available), (c) the organic matter and microorganisms which might be involved in interactions controlling release and uptake of the nutrient, and (d) crystalline minerals and precipitates (relatively unavailable). The interactions between these compartments are highly complex. However, great strides[44] have been made in recent years in the understanding of the basic soil chemistry involved including the effect of the solubility of the solid phases on the ability of the soil to replenish ions in the soil solution upon uptake by the plant from the solution. This understanding has played an important role in the development and improvement of soil testing.

The traditional aims of a soil test are:[46] (a) to group soils into classes for the purpose of recommending management practices, (b) to predict the probability of getting a profitable response to the application of fertilizer nutrients, (c) to evaluate soil productivity, and (d) to determine specific soil conditions that may be improved by a change of management practices. It is immediately apparent that these goals have much in common with the development of irrigation water standards. In fact, in view of the possible impact of fall-out from air pollution, sewage sludge, and fly ash disposal, and reforesting of strip-mined areas and tailings piles, it is important at this point to broaden our view to include these and other alternate delivery paths of metals to soils as well as the transport by irrigation water.

The importance of the soil test is made even more apparent when we note that it is not the total amount of a given substance which is applied to the soil which is important from the point of view of plant and animal health, but the amount which is available for uptake by the plants (although in the case of livestock, soil ingestion can be a significant means of intake—here a different kind of biological "availability" would be important). For example, the application of 70 g Mo/ha to a deficient pasture was adequate for 10 years although the average increase in total soil molybdenum content was, at most, 0.1 ppm.[45] The application of 1 kg/ha on some soils can lead to increases in plant molybdenum on the order of 10 ppm, although the change in total content of the soil is probably, at most, 1.5 ppm. Thus, relatively insignificant increases, compared to "background," in total soil molybdenum can have fairly profound effects if the molybdenum is available. On the other hand, very large molybdenum concentrations in soils can have insignificant effects on the plants if the molybdenum is unavailable. This effect can be noted in Table 1.

I think this point is important to make because of a tendency to, on the one hand, make light of the possible adverse effects if the change in total soil content is negligible, and, on the other hand, to make much of large changes in total soil content. It would seem wiser to investigate the change in the available amount of a given trace metal (or other compound) rather than the change in total amount.

One way to investigate such changes is to analyze plants grown on the soil. This is, of course, an important empirical test. But, one must expect great variability to occur in the results because of the large number of parameters, such as climate, moisture, and root depth, which can affect plant growth. A well-established soil test might help eliminate some of this variability. The use of soil tests may also be helpful in separating natural from disturbed levels of trace elements in soils. This would seem to be particularly feasible in cases where naturally occurring forms are much less available than those due to human activity. The maximal use of this approach would seem to require an understanding of the mineral forms of trace elements which occur in soils.

The development and use of soil tests for molybdenum and other trace metals have been motivated primarily by the existence of areas where there were deficiencies in the available amounts of these metals.[47] Such deficient soils have been found in many parts of the world and the application of as little as $\frac{1}{16}$ oz of molybdenum per acre has increased production by as much as 50–100%.[3] The role of the soil test has been as a diagnostic tool used to locate deficient areas and to prescribe proper treatment.

Several soil tests for available molybdenum have been suggested.[45,46] Among these are water soluble molybdenum, extraction by acid ammonium oxalates, hot water soxhlet extraction, and extraction using an anion exchange resin. The tests are empirically validated by comparing the uptake of molybdenum by plants to the amount of "available" molybdenum measured by a given method. To date there seems to be no method which is reasonably accurate for all soils. However, some of these work reasonably well for given classes of soils. Williams and Thornton[48] have found that ammonium acetate works well with Scottish soils which are high in molybdenum and that EDTA also is a reliable extractant for poorly drained, acidic soils. Lindsay et al.[49] of our group have used anion exchange extraction and have found good correlations with plant uptake for alkaline soils, but poor correlations for acidic soils. Thus, a parameterization of soils by pH, organic matter, and drainage may be needed in conjunction with a soil testing program.

The movement of molybdenum in soils is determined by several influences including leaching, adsorption reactions, and evaporation. The removal from solution can be by:[45] (a) precipitation in mineral forms, (b) binding in organic complexes, and (c) adsorption by anion exchange material (clay minerals, hydrated oxides, etc.). Several soil parameters can influence the availability of molybdenum and among these are the concentration of sulfate, pH, and soil moisture.[41,43,50–53] Molybdenum is generally less available at low pH, high sulfate, and low soil moisture levels. Phosphate often enhances the uptake of molybdenum by plants.[43,52,53]

Most of the work concerning the application of molybdenum to soils has been concerned with the treatment of deficient soils. Molybdenum deficiencies are generally associated with soils which have a high iron content, acidic soils having a high anion exchange capacity, or soils having a very low total content of molybdenum.[3] Molybdenum deficiencies have been corrected by applying levels between $\frac{1}{16}$ oz to 4 lb/acre.[3] The more common levels used are apparently in the range of 2–8 oz/acre.

Some of the work on molybdenum deficient soils, however, does indicate that for some deficient soils an overly generous application of molybdenum can lead to concentrations of molybdenum in plants which are above 5–10 ppm and are therefore potentially toxic to ruminants. Since plants growing on soils where there is insufficient available molybdenum for optimal growth often have very low concentrations of molybdenum (e.g. 0.5 ppm or less in alfalfa),[3,54] the increased plant levels associated with increased productivity by application of moderate amounts of molybdenum are of no concern. However, further application of molybdenum in an available form beyond the amount needed for optimal growth can give rise to increases in plant molybdenum to levels toxic to livestock (sometimes accompanied by no further increase in productivity).[54] Reisenauer[54] found that while the application of 1.6 lb/acre of molybdenum gave an increase in plant (alfalfa) molybdenum from 0.27 ppm to 2.0 ppm with an increase in yield of about 30%, the application of 6.4 lb/acre gave rise to no further increase in yield and increased molybdenum in the plants to 12 ppm. Jones and Ruckman[52] found that the application of 282 g Mo/ha as molybdic acid gave rise to an increase in plant molybdenum from 0.6 ppm to a level which over the course of 6 years remained between 6 and 9 ppm in clover grown on an initially molybdenum-deficient soil. A subsequent reapplication of the same amount of molybdenum at the end of the 6 year period gave rise to 47 ppm of molybdenum in the clover.

Although considerably less work has been done on soils where no molybdenum deficiency exists, the level of application to such soils which gives rise to toxic amounts of molybdenum in plants is reasonably consistent among these reports. Gammon et al.[55] looked at grasses and clover growing on peat and found that the application of 0.2 lb/acre of molybdenum as sodium molybdate increased the white clover molybdenum from 0.5–11 ppm. Jensen and Lesperance[56] reported that the application of 1 kg/ha molybdenum as ammonium molybdate to neutral sandy loam soil gave rise to an increase of 5.2 ppm molybdenum in alfalfa grown on that soil. They did not see any reduction in growth until the alfalfa molybdenum level reached 1500 ppm (after application of 282 kg/ha molybdenum). Barshad[57] found an increase of plant molybdenum (in *Lotus corniculatus*) from 2 ppm to 150 ppm after an application of 10 ppm molybdenum (the equivalent of 20 lb/acre or 20 kg/ha) to a clay loam soil with an initial molybdenum content of 0.3 ppm.

In view of this work, it appears likely that the application of molybdenum via irrigation water can, if applied in sufficient amounts, give rise to significant increases in plant molybdenum. Our concern is with determining what levels of molybdenum in irrigation water various soils can tolerate before reaching the point where plants grown on these soils are potentially dangerous to livestock, wildlife, and humans.

Lindsay et al. of our group[49] are approaching this problem by laboratory, greenhouse, and field experiments. They have taken samples of various types of soils found in the area to use in calibrating the soil test (anion resin extractable). They have monitored the levels of molybdenum in irrigation water applied to field sites and sampled the soils and plants at these field sites over the growing season. They are performing laboratory experiments to determine kinetic constants for exchange between various soil compartments.

The greenhouse work indicates results which are consistent with the results mentioned previously. The application of 0, 2, 4, and 6 ppm molybdenum as sodium molybdate (which is equivalent to 0, 4, 8, and 12 lb/acre when mixed to a depth of 6 in) gave rise to very significant increases in plant molybdenum for the alkaline soils (all calcareous), but to a much smaller increase for the acidic soils as seen in Fig. 8.

The field sampling in the South Platte River Valley north of Denver has yielded molybdenum levels in alfalfa ranging from 2 to 11 ppm with a mean of 6 ppm. The levels in alfalfa in the Lower Blue River Valley ranged between 2 and 81 ppm. Figure 9 shows

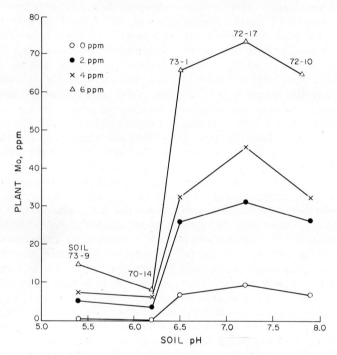

FIG. 8. Results of 0, 2, 4, and 6 ppm molybdenum applied to five Colorado soils.

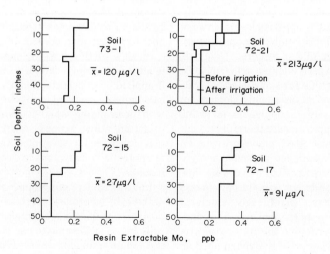

FIG. 9. Mean resin extractable molybdenum in four calcareous Colorado soils irrigated by water containing mean levels of 27, 91, 120, and 213 µg/l. ($\bar{x}$ = mean molybdenum concentration in water.)

the mean levels for resin-extractable soil molybdenum on four sites receiving different levels of molybdenum in irrigation water. One site, that receiving an average of 213 μg/l molybdenum in the water, showed a significant increase ($p < 0.05$) in resin-extractable molybdenum after irrigation. Those receiving lower levels of molybdenum showed no significant change. The alfalfa molybdenum levels on this site changed from 4.3 ppm to 7.2 ppm and this change was also significant ($p < 0.05$). Since the depth of water applied to this site was approximately one-half meter, the amount of molybdenum applied per hectare was approximately 1 kg.

Jackson has found that the molybdenum in the South Platte River Valley soils is highly labile (these are all calcareous soils with pH > 6.5). As much as 65% of the total molybdenum in these soils is solubilized by using a soil–water ratio of 1:50. Thus, unless there is poor drainage, one would expect the molybdenum delivered by irrigation water to these soils to be quite mobile and easily leached from the soils. The main concern would be the immediate response to the molybdenum application.

For other soils which are acidic or high in organic content or have a large anion exchange capacity, the immediate response might be less important than the possible long term accumulation. Jones and Ruckman's work[52] indicates a capacity for a long term effect of a single application of molybdenum to a Pinole loam (pH 5.6–6.1). Carrigan[58] found that 2 lb/acre molybdenum on Leon fine sand resulted in values exceeding 10 ppm molybdenum in the carpet grass 5 years later, while Cunningham and Hogan[58] found that $\frac{3}{4}$ lb/acre applied to an acid, peaty soil increased the plant levels from 2 to 100 ppm the first year with a subsequent decline to 11 ppm over the following 2-year period.

Perhaps the most difficult complication with this type of work is relating greenhouse experiments to what happens in the field. The variability of parameters under field conditions often masks an effect seen very clearly in a controlled greenhouse experiment. It is for this reason that Lindsay et al. have done both greenhouse and field studies.

On the other hand, because we know, within a small range, what forage levels affect livestock, and because we can study the transfer from irrigation water to forage in a fairly detailed way, the establishment of rational tolerance limits for heavy metals in irrigation waters seems a much easier job than the establishment of criteria concerning the effect on humans.

BIOLOGICAL ROLE AND EFFECTS

Molybdenum has been demonstrated to be present in most plants and animals. Molybdenum deficiencies in plants have been demonstrated in several parts of the world,[3] and it is now thought by some to be essential to all plants except green algae. The role of molybdenum in the plants is apparently to stimulate nitrogen fixation and nitrate reduction. At present seven enzymes—nitrogenase, nitrate reductase, xanthine oxidase, aldehyde oxidase, NADH-dehydrogenase, xanthine dehydrogenase, and sulfite oxidase are known to contain molybdenum as a metallic cofactor.[59] Although the evidence is somewhat indirect because of the difficulty of inducing a deficiency in mammals, many authorities now consider molybdenum to be an essential micronutrient.[1]

There is a considerable body of information concerning the biological role of molybdenum in plants. Although the amounts of molybdenum required for optimal plant growth are quite small (ranging from 10^{-5} to 0.1 ppm in plant content),[60] it seems to be definitely required for nitrogen metabolism. The biological role of molybdenum in animal health is considerably less well understood. It is suggested that low levels of molybdenum may lead to chronic copper poisoning since these two elements seem to be biological antagonists.[1,61] There is also some evidence that molybdenum insufficiency might lead to the occurrence of renal xanthine calculi in sheep.[1] Some recent work by Berg[62] indicates an increased occurrence of esophageal cancer in geographical areas containing molybdenum-deficient soils. There have been reports that molybdenum has a beneficial effect on dental caries,[63,64] but this effect is not widely accepted.[65]

The primary toxic effect of molybdenum involves an interference with copper metabolism, inducing a copper deficiency.[1,2] The most sensitive animals are apparently the ruminants, and, indeed, the effect was first noticed in cattle and sheep. There are, in fact, several clinical disorders which are associated with copper deficiency. These include anemia, bone abnormalities, loss of pigmentation, reproductive failure, and cardiac lesions.[1] Not all of these disorders have been associated with molybdenum. This may be due to insufficient evidence or to the particular action by which molybdenum makes copper unavailable. This mechanism is very poorly understood at present.

The most common clinical disorder in livestock is known as molybdenosis or in some areas as "teart" or "peat scours." This disorder is commonly characterized by diarrhea (scouring), discoloration of hair, loss of appetite, joint abnormalities, osteoporosis, reproductive difficulties, lack of sexual interest, testicular degeneration, and occasionally death in cattle and sheep which graze on forage with high levels of molybdenum.[1,66] The disease is usually treated by copper supplementation. The lowest concentrations of molybdenum in forage at which symptoms have been evident is 5 ppm.[66] However, the critical level for onset of clinical symptoms depends on the copper status of the animals. Several workers have tried to characterize the ratio of copper to molybdenum levels in forage at which clinical symptoms occur. Miltimore and Mason[67] have suggested that a copper/molybdenum ratio of 2 represents a critical level and that molybdenosis can be anticipated when the ratio falls below 2. There is not universal agreement on this figure. Indeed, one complication is that the level of sulfate in the diet can have a profound effect on the interference of molybdenum and copper.[1]

At lower levels of molybdenum, subclinical effects have been demonstrated by Thornton et al.[68] They have located herds in areas having mean forage molybdenum levels of 5 ppm and mean forage copper levels of 10 ppm where cattle suffered from decreased weight gains, decreased fertility, and delayed maturity without any clinical indications of molybdenosis. They were able to demonstrate 10–70% weight gains over a 6 month grazing season with respect to controls by giving copper supplementation to these cattle.

Another clinical disorder associated with copper deficiency is "swayback," which is a neonatal enzootic ataxia encountered in lambs.[1,69] This nervous disorder is characterized in congenital cases by partial or complete paralysis, while postnatally acquired cases show a retarded growth rate and posterior muscular incoordination, followed by progressive paralysis. The pathological changes involve demyelinization of the cerebrum in congenital cases and in the spinal cord for postnatally acquired cases.[1,69] The mechanism involved in this pathology appears to be an insufficient production of cytochrome oxidase.[70,71] A similar myelin anomaly has been reported in copper-deficient newborn rats by Di Paolo and Newberne.[72]

Although the disease has been produced in laboratory animals by high molybdenum and low copper diets,[73,74] many workers did not consider that those farms where swayback was found had high enough molybdenum levels in the forage to implicate it as the cause. There is now evidence from England,[75] however, that molybdenum may be a contributing factor in swayback incidences found in the areas of relatively high molybdenum levels in England which have been extensively mapped by Webb et al.[12] These findings by Alloway[75] indicate that mean levels of molybdenum of 3.5 ppm and mean levels of copper of 16 ppm are associated with farms having a history of swayback. The mean copper/molybdenum ratio for swayback farms is 4.7 compared to 8.8 for the controls.

Considerable work has been done on the effects of molybdenum on guinea pigs, rats, mice, and rabbits. In all these species there is a noticeable antagonism between molybdenum and copper. However, the level of toxicity varies from species to species, with nonruminants being much less sensitive than ruminants. As with ruminants, the level at which molybdenum becomes toxic to nonruminants depends on their copper status. At normal copper levels rats may tolerate 100 ppm molybdenum as molybdate without gross effects.[1,76] The form in which the molybdenum is given also affects the toxicity. For example, the highly insoluble MoS_2 is much less toxic than molybdenum given as the highly soluble molybdate.[77]

The symptoms of molybdenum toxicity also vary from species to species.[1] In young rabbits the toxic syndrome is characterized by loss of appetite, loss of weight, baldness, and dermatosis.[1] In rats and guinea pigs, loss of appetite and retardation of growth or loss of body weight are indicators of molybdenum toxicity.[78] Depigmentation has also been noticed in guinea pigs.[78]

Various mechanisms have been proposed to explain why molybdenum displaces copper. One of these is that a biologically inactive molybdenum–copper compound is formed, preventing the utilization of copper for ceruloplasmin synthesis.[79,80]

There have been no reported cases of molybdenum toxicity in humans.[81] This is not surprising since the levels reaching humans in water and food are not likely to be in the range where acute toxicity would occur. The question remains, however, whether long-term exposure to low, but elevated levels (compared to background) might induce chronic disorders.

The study of "recondite" toxicity as it is referred to by Schroeder[82] is one that requires carefully controlled conditions, subtle and accurate measurements, and, as Schroeder points out, "unfailing patience." There are several approaches to this problem. One is to look over many generations for an effect on reproduction. Another is to have a large number of animals, thereby improving statistics;

another is to use very sensitive techniques for measuring small metabolic changes. Still another approach is to subject the animal to a stress on the assumption that the element in question may affect the animal's ability to respond to stress but not its ability (at least for short periods) to function under normal conditions.

Schroeder[82,83] has studied reproductive effects in mice receiving 10 ppm molybdenum as sodium molybdate in their water and 0.25 ppm molybdenum in their food. He found significantly increased rates of young deaths and dead litters in this group after three generations compared to the controls which were receiving the same food and only 1 ppm molybdenum in the water. His findings suggested the following order of toxicity with respect to reproduction: mercury > cadmium > lead > selenium > molybdenum > titanium > nickel > arsenic.

Suttle[84] took another approach to the problem. He fed animals a copper-deficient diet, thereby depleting their copper stores. He then supplemented the diet with copper and various levels of molybdenum and measured the rate of copper repletion. In sheep he found a significantly slower response to the copper supplementation (as measured by plasma copper) in the group receiving 2 ppm molybdenum in their diet as opposed to the controls receiving 0.5 ppm molybdenum. In guinea pigs he found a significant reduction in response in the group fed 4.5 ppm in their diet as opposed to the controls which were fed 0.5 ppm in their diet. Moreover, he found that the influence of molybdenum on copper utilization was not linear. In fact, the first increment of 4 ppm gave as great a reduction in copper utilization as the change from 25 to 100 ppm. Thus, small increments seem to have disproportionately large effects. He suggested that the copper–molybdenum interaction may have a greater significance for nonruminants, including man, than has previously been realized.

Winston[85] of our group has looked at the effects of molybdate in water on rats which were cold stressed. No physiological differences were found between controls and animals drinking 10 ppm molybdate in their water when there was no stress. However, after the rats were stressed by being kept at 2–3°C for 4 days, there was a significant difference (with $p < 0.01$) in weight loss by males drinking molybdated water compared to the controls on deionized water. Both groups received the same diet which consisted of Purina "Lab Chow" containing 1–3 ppm of molybdenum.

Moffit and Murphy[86] have carried out some very interesting work involving the effect of copper deficiency and excess on the metabolism of foreign chemicals such as carbon tetrachloride and parathion to see if the toxicity of these chemicals is related to copper status. Their low copper diet contained 1–2 mgCu/kg diet. Whether one can relate these results to molybdenum is an unanswered question.

One of the most common effects of molybdenum toxicity among different species is an alteration of the bone metabolism. In addition to osteoporosis and joint abnormalities which are seen in cattle and sheep,[1,66] foreleg deformities are noticed in rabbits (which resemble those seen in dogs), and deformities in long bones of rats have been produced by very high levels of molybdenum in the diet.[87] In an experiment by Johnson et al.,[88] rats were found to suffer connective tissue disturbances which were revealed by reduced breaking strengths of femurs and rupture strength of skin samples. The action seemed to parallel that of lathyrism (loss of tensile strength of tissues).

Since some disturbance in bone metabolism seemed common among species, it seemed to us to be a productive area for investigation. Dr. Solomons in our group has developed a very sensitive technique[89,90] for investigating the transport of ions in bones and other tissues. This transport is based upon a model which describes the system in terms of rate constants K_1 and K_2 for movement of the ion out of and into the bone from the surrounding solution, respectively. In this case Dr. Solomons studied calcium transport by using Ca^{45} injected into the rat 24–48 h prior to sacrifice and excision of the ulna which was bathed in Ringers solution. The Ringers solution was sampled and K_1 and K_2 determined by fitting a curve to the data. He found[90] that when there was a constant flow of the bathing solution past the bone K_2, the rate at which calcium is taken into the bone, was significantly decreased ($p < 0.005$) for rats with 10 ppm molybdenum in their drinking water. The rate constant K_1, governing the release of calcium from the bone, was not significantly different from that of controls.[90] This result is consistent with the clinical symptoms of osteoporosis seen with acute dosages of molybdenum.

Dr. Solomons[90] also saw an increased level of molybdenum in the bones of the rats receiving molybdenum in their drinking water (from 0.71 ppm in the ash in controls to 2.5 ppm in the ash of the rats receiving 10 ppm molybdenum in their drinking water). The result is consistent with the work by Davis[91] who saw a similar increase in the molybdenum content in bones when dietary molybdenum was increased from 1 to 30 ppm.

On the other hand, when the experiment was

conducted with no flow of solution past the bone, there was a slight, but not statistically significant, tendency to increase calcium uptake by the bone.[90] This increase would be consistent with the reports[63,64] that molybdenum helps prevent dental caries because the no-flow case simulates teeth, where flow is minimal, while the flow case simulates bone.

When dealing with low level exposures which cause small effects, the isolation of clinical manifestations of these effects, if they exist, is a very difficult job. The effect may be to modify the homeostatic mechanisms involved with other stresses. The effect may only be seen in certain portions of the population which have inadequate copper homeostatic responses because of general health, exposure to other materials, or genetic predisposition. Thus, an epidemiological examination of an exposed population may indicate nothing unless this population is parameterized correctly.

Industrial data are often offered as strong evidence concerning the toxicity of a chemical. Such data are valuable, of course, because they involve information on human exposure which can be used to supplement laboratory experiments. However, the lack of evidence for human toxicity in industrial exposure to an element does not constitute a proof that the general population can be exposed to the same or lower levels with no ill effects. In the first place, the industrial exposure may involve a different form of the element, or a different pathway. In the second place, the population exposed is not representative of the general population, in that the very young, very poor, undernourished, chronically or acutely ill, and very old are underrepresented and this part of the population may be more sensitive than the workers who are being studied.

Hatch[92] recently pointed out that we are becoming increasingly sophisticated in our ability to measure small biochemical, physiological, and psychological disturbances due to various substances. There is often a tendency to say that any such change is dangerous and unallowable. There are, of course, homeostatic mechanisms in living organisms for countering the effects of fluctuations in the surrounding or internal environment. In many cases these mechanisms are capable of responding to an environmental change so that the organism suffers no harm. The problem is compounded, however, by the synergistic reactions which can exist between substances. Thus, by itself, a substance may be at an unharmful level, but if the health status of the organism has been changed by another substance, the first substance might cause a significant effect. In the case of molybdenum we know that the copper status of the individual is important. But, in addition, there are other elements which are antagonistic to copper. Among these are cadmium, zinc, and lead. These elements not only complicate the isolation of effects of molybdenum on human health, but also complicate the setting of tolerance levels.

CONCLUSIONS

I have attempted in this paper to review the important sources and modes of transport of environmental molybdenum and to summarize the present knowledge concerning its biological effects. Because of the vast array of information, some subjects were not treated as fully as they perhaps deserved. For example, the role of molybdenum in biochemistry was not treated in great detail.

I have noted that there are several possible sources of environmental molybdenum. Among these are molybdenum mining and processing,[23,24,33] the use of molybdenum products in alloying and refining,[25-27] copper and uranium mining and milling,[16-18] fossil fuel burning,[20,22] and disposal of fossil fuel ash[21] and sewage sludge.[33] Some of these sources have already been connected to livestock problems or high forage levels; others need to be evaluated in order to more fully appreciate their potential effect. It was noted that one of the most difficult problems is the separation of "natural" sources from "disturbed" sources.[11,14,18]

Although atmospheric transport can be an important means of delivery, only aqueous transport was discussed in detail. It was pointed out that extensive sampling by many workers[15,30,31] would indicate that a "background" level (in the sense that natural molybdenum anomalies and industrial activity are absent) of dissolved molybdenum in water is probably less than 10-20 μg/l. The existence of many measurements[32] of molybdenum at levels above 100 μg/l in several large rivers in the United States indicates the probable presence of industrial sources. The work done in Colorado indicates that molybdenum is transported almost entirely as a dissolved species,[18] that in large lakes it is stratified during the summer months with the greatest concentration at the bottom,[33] and that there is removal from the dissolved component under acidic conditions if there is sufficient iron present.[18]

The biological role of molybdenum has been a subject of great interest to agricultural scientists since the time when it was noted that (a) there are pastures where optimal growth is not achieved because the plants obtain insufficient molybdenum, and (b) forage can concentrate molybdenum to levels (>5–20 ppm, depending on copper and sulfate status) that are toxic to cattle and sheep. Considerable work has been devoted to developing tools to diagnose areas of molybdenum deficiency. One of the most important of such tools is the soil test which attempts to measure the molybdenum in soil which is available to plants.[45,46] Such tests offer much potential for work on trace element contamination because they measure the relevant aspect of the contamination, as opposed to analyses of total soil content. Such tests also can be helpful in developing soil management techniques for treating soils containing excess available molybdenum and avoiding practices (such as liming) which would change molybdenum from unavailable to available forms.

The establishment of tolerance limits for irrigation waters is a complicated task because of the effect of soil parameters, climatic conditions, and farm management practices. Alkaline, organic, or young soils are generally more prone to produce high molybdenum content in plants, whereas acidic soils and soils of high anion-exchange capacity are often found to be associated with deficiencies.[43] High moisture content is also known to be associated with toxic molybdenum levels.[51] Thus different soils will most certainly tolerate different levels, on both a short and a long term basis, of molybdenum in irrigation water. A better understanding of the response of different soils under different conditions to the application of molybdenum will lead to the ability to develop irrigation water criteria which make sense in the context of the types of soils, the climate, and the farm management conditions which prevail in the region of concern.

Molybdenum plays an interesting biological role as a metallic cofactor in seven enzymes.[59] At high levels it apparently interferes with copper metabolism.[1] It does not seem to be as acutely toxic as elements such as cadmium and mercury. The principal environmental concern would seem to be whether there are chronic effects due to long term exposure to elevated environmental levels. Such effects are extremely difficult to detect. The use by various workers of very sensitive tests,[90] response to stress,[85] copper repletion,[84] and long term reproductive studies[83] indicate that rats and mice receiving 10 ppm molybdenum in drinking water or 4.5 ppm in food do have adverse reactions as compared to controls in terms of bone kinetics, reproduction, stress response, and copper repletion. These levels are still somewhat above those actually encountered and further work is required to evaluate the longterm effects of the changes seen and to see if these changes are associated with lower levels of intake.

The role of molybdenum in nature is not unlike that of selenium. At low levels both elements are essential to plant and/or animal life, at high levels both are toxic to animals and/or plants. This dual role distinguishes these elements and others such as chromium and nickel from elements such as cadmium and mercury (although the latter may yet prove to be essential at very low levels). The existence of homeostatic mechanisms imply that there is a range of intake levels over which animals can respond safely to elevated (or depressed) levels. Since molybdenum and selenium are ubiquitous, and since, for some areas, additional amounts of these elements might be desirable, criteria which require no detectable amounts or which apply a single number to all areas are probably unrealistic and undesirable. Although such limits might be established as a tentative measure, it seems desirable to adopt a long term goal of setting standards which take into account the nutritional status, soil and climatic conditions, and farm and wildlife management of the regions where elevated molybdenum levels are of concern.

REFERENCES

1. UNDERWOOD, E. J., *Trace Elements in Human and Animal Nutrition*, 3rd edn., Academic Press, New York, 1971.
2. DE RENZO, E. C., Molybdenum, in *Mineral Metabolism* (Comar, C. L. and Bronner P., eds.), chap. 25, Academic Press, New York, 1962.
3. SAUCHELLI, V., *Trace Elements in Agriculture*, pp. 133–149, Van Nostrand Reinhold Co., New York, 1969.
4. KUKLIS, A., Molybdenum, in *Minerals Yearbook*, p. 753, Bureau of Mines, US Dept. of Interior, 1971.
5. KING, R. U., SHAWE, D. R., and MACKEVETT, E. M., Jr., Molybdenum, in *United States Mineral Resources*, US Geol. Survey Prof. Paper 820, 1973.
6. *Report of the Committee on Water Quality Criteria*, p. 152, Federal Water Pollution Control Administration, US Dept. of Interior, 1968.
7. FLEISCHER, M., An overview of distribution patterns of trace elements in rocks, in *Geochemical Environment in Relation to Health and Disease* (Hopps, H. C. and Cannon, H. L., eds.), New York Acad. of Sci., New York, 1972. KRAUSKOPF, K. B., Geochemistry of micronutrients, in *Micronutrients in Agriculture* (Mortvedt, J. J., Giordano, P. M., and Lindsay, W. L., eds.), Soil Sci. Soc. of America, Inc., Madison, Wisconsin, 1972.
8. ROBINSON, W. O. and ALEXANDER, L., *Soil Sci.* 72, 267–274 (1951).

9. SHACKLETTE, H. T., HAMILTON, J. C., BOERNGEN, J. C., and BOWLES, J. M., *Elemental Composition of Surficial Materials in the Conterminous United States*, US Geol. Survey Prof. Paper 57-D, 1971.
10. WELLS, N., *NZ J. Sci. Tech.* **337**, 482–502 (1956). VINOGRADOV, A. P., *The Geochemistry of Rare and Dispersed Chemical Elements in Soils*, 2nd edn., pp. 187–204, Consultants Bureau, New York, 1959.
11. RUNNELLS, D. D., Detection of molybdenum enrichment in the environment through comparative study of stream drainages, central Colorado, presented at 7th Ann. Conf. on Trace Substances in Env. Health, Columbia, Missouri, June 12–14, 1973. To be published in *Proceedings of Conference* (Hemphill, D. D., ed.).
12. WEBB, J. S., THORNTON, I., and FLETCHER, K., *Nature, Lond.* **217**, 1010 (1968).
13. THOMSON, I., THORNTON, I., and WEBB, J. S., *J. Sci. Fd. Agric.* **23**, 879 (1972).
14. RUNNELLS, D. D., BROWN, D., and LINDBERG, R., Investigation of enrichment of molybdenum in the environment through comparative study of stream drainages, central Colorado, presented at 1st Ann. NSF Trace Contaminants Conf., Oak Ridge, Tennessee, August 8–10, 1973. To be published in *Proceedings of Conference* (Fulkerson, W., ed.).
15. HEM, J. D., *Study and Interpretation of the Chemical Characteristics of Natural Water*, 2nd edn., US Geol. Survey Water-Supply Paper 1473, 1970.
16. CHRISTIANSON, G. A., *Report on Molybdenosis in Farm Animals and Its Relationship to a Uraniferous Lignite Ashing Plant*, North Dakota State Dept. of Health, 1971.
17. DOLLAHITE, J. W., ROWE, L. D., COOK, L. M., HIGHTOWER, D., MAILY, E. M., and KYZAR, J. R., *The Southwestern Vet.* **26**, 47 (1972).
18. RUNNELLS, D. D., LEGENDRE, G., LINDBERG, R., BROWN, D., SMITH, E., HARTHILL, M., and KATZ, B., Geochemistry of molybdenum, in *Transport and the Biological Effects of Molybdenum in the Environment, Progress Report to NSF*, pp. 33–63 (Chappell, W. R., ed.), the Molybdenum Project, University of Colorado, Boulder, Colorado, January 1, 1973.
19. EBENS, R. J., ERDMAN, J. A., FEDER, G. L., CASE, A. A., and SELBY, L. A., *Geochemical Anomalies of a Claypit Area, Callaway County, Mo., and Related Metabolic Imbalance in Beef Cattle*, US Geol. Survey Prof. Paper 807, 1973.
20. ODE, W. H., Coal analysis and mineral matter, in *Chemistry of Coal Utilization*, suppl. vol., p. 226 (Rowry H. H., ed.), Wiley, New York, 1963.
21. DORAN, J. W. and MARTENS, D. C., *J. Environ. Quality* **1**, 186 (1972).
22. KAAKINEN, J. W. and JORDEN, R. M., Determination of a trace element mass balance for a coal-fired power plant, presented at 1st Ann. NSF Trace Contaminants Conf., Oak Ridge, Tenn., August 8–10, 1973. To be published in *Proceedings of Conference* (Fulkerson, W., ed.).
23. *Molybdenum Toxicity in Livestock*, Vet. Service Extension Paper 0–23, State of Pennsylvania Veterinary Service Extension, 1970.
24. Industrial molybdenosis in grazing cattle, in *Air Pollution, Proc. of 1st European Congress on the Influence of Air Pollution on Plants and Animals, 1968*, Center for Agricultural Publishing and Documentation, Wageningen, 1969.
25. GARDNER, A. W. and HALL-PATCH, P. K., *Vet. Rec.* **74**, 113 (1962).
26. GARDNER, A. W. and HALL-PATCH, P. K., *Vet. Rec.* **81**, 86 (1968).
27. BUXTON, J. C. and ALLCROFT, R., *Vet. Rec.* **67**, 273 (1955).
28. MALLORY, E. C., JR., US Geol. Survey Prof. Paper 600B, B-115–B116, 1968.
29. BRUNDIN, N. H. and NAIRIS, B., *J. Geochem. Explor.* **1** (1), 7 (1972).
30. SUGAWARA, K., OKABE, S., and TANAKA, M., *J. Earth Sci. Nagoya*, **9**, 114 (1961).
31. VOEGELI, P. T. and KING, R. U., *Occurrence and Distribution of Molybdenum in the Surface Water of Colorado*, US Geol. Survey Water-Supply Paper 1535-N, 1969.
32. KOPP, J. F. and KRONER, R. C., *Trace Metals in Waters of the United States*, FWPCA, US Dept. of Interior, 1967.
33. JORDEN, R. M., Molybdenum transport in Dillon Reservoir, in *Transport and the Biological Effects of Molybdenum in the Environment, Progress Report to NSF*, pp. 64–93 (Chappell, W. R., ed.), the Molybdenum Project, University of Colorado, Boulder, Colorado, January 1, 1973.
34. WINSTON, P., Movement of molybdenum in Dillon Reservoir and Standley Lake and the effects of excess molybdenum on adult rats, in *Transport and the Biological Effects of Molybdenum in the Environment, Progress Report to NSF*, pp. 42–58 (Chappell, W. R., ed.), the Molybdenum Project, University of Colorado, Boulder, Colorado, January 1, 1972.
35. BRIESE, F. W. and JORDEN, R. M., Analysis of trace metal mass balance for aqueous systems, presented at 1st Ann. NSF Trace Contaminants Conf., Oak Ridge, Tenn., August 8–10, 1973. To be published in *Proceedings of Conference* (Fulkerson, W., ed.).
36. BOBERG, W. W. and RUNNELLS, D. D., *Econ. Geol.* **66**, 435 (1971).
37. JORDEN, R. M. and MEGLEN, R. R., Aqueous release of molybdenum from non-point and point sources, presented at 7th Ann. Conf. on Trace Substances in Env. Health, Columbia, Missouri, June 12–14, 1973. To be published in *Proceedings of Conference* (Hemphill, D. D., ed.).
38. LE GENDRE, G. R., Removal of molybdenum by ferric oxyhydroxide in Clear Creek and Ten-mile Creek, Colorado, Master of Arts thesis, Department of Geological Sciences, University of Colorado, 1973.
39. ZEMANSKY, G. and JORDEN, R. M., Molybdenum removal in conventional water and wastewater treatment plants, in *Transport and the Biological Effects*

of Molybdenum in the Environment, Progress Report to NSF, pp. 128–132 (Chappell, W. R., ed.), the Molybdenum Project, University of Colorado, Boulder, Colorado, January 1, 1973.
40. BERROW, M. L. and WEBBER, J., *J. Sci. Fd. Agric.* **23**, 93 (1972).
41. KUBOTA, J., LEMON, E. R., and ALLAWAY, W. H., *Soil Sci. Soc. Am. Proc.* **27**, 679 (1963).
42. LINDSAY, W. L., Influence of the soil matrix, in *Geochemical Environment in Relation to Health and Disease*, vol. 199, pp. 37–45 (Hopps, H. C. and Cannon, H. L., eds.), New York Acad. of Sci., New York, 1972.
43. DAVIES, E. B., *Soil Sci.* **81**, 209 (1956).
44. LINDSAY, W. L., Inorganic phase equilibria of micronutrients in soils, in *Micronutrients in Agriculture*, pp. 41–58 (Mortvedt, J. J., Giordano, P. M., and Lindsay, W. L., eds.), Soil Sci. Soc. of America Inc., Madison, Wisconsin, 1972.
45. REISENAUER, H. M., WALSH, L. M., and HOEFT, R. G., Testing soils for sulphur, boron, molybdenum, and chlorine, in *Soil Testing and Plant Analysis*, pp. 187–193, Soil Sci. Soc. of America, Inc., Madison, Wisconsin, 1973.
46. COX, F. R. and KAMPRATH, E. J., Micronutrient soil tests, in *Micronutrients in Agriculture*, pp. 13–22 (Mortvedt, J. J., Giordano, P. M., and Lindsay, W. L., eds.), Soil Sci. Soc. of America, Inc., Madison, Wisconsin, 1972.
47. MELSTED, S. W. and PECK, T. R., The principles of soil testing, in *Soil Testing and Plant Analysis*, pp. 13–22, Soil Sci. Soc. of America Inc., Madison, Wisconsin, 1973.
48. WILLIAMS, C. and THORNTON, I., *Plant Soil* **39**, 149 (1973).
49. JACKSON, D., LINDSAY, W. L., and HEIL, R. D., Accumulation of available molybdenum in agricultural soils irrigated with high-molybdenum waters, in *Transport and Biological Effects of Molybdenum in the Environment, Progress Report to NSF*, pp. 133–175 (Chappell, W. R., ed.), the Molybdenum Project, University of Colorado, Boulder, Colorado, January 1, 1973. JACKSON, R., LINDSAY, W. L., and HEIL, R. D., The accumulation of available molybdenum in some agricultural soils irrigated with high molybdenum waters, presented at 1st Ann. NSF Trace Contaminants Conf., Oak Ridge, Tenn., August 8–10, 1973. To be published in *Proceedings of Conference* (Fulkerson, W., ed.). JACKSON, D. R., Effects of high-Mo waters on available Mo in soils, unpublished PhD Thesis, Colorado State University, 1973.
50. BARSHAD, I., *Soil Sci.* **81**, 387 (1956).
51. KUBOTA, J. and ALLAWAY, W. H., Geographic distribution of trace element problems, in *Micronutrients in Agriculture*, pp. 525–554 (Mortvedt, J. J., Giordano, P. M., and Lindsay, W. L., eds.), Soil Sci. Soc. of America Inc., Madison, Wisconsin, 1972.
52. JONES, M. B. and RUCKMAN, J. E., *Soil Sci.* **115**, 343 (1973).

53. WILLIAMS, C. and THORNTON, I., *Plant Soil* **36**, 395 (1972).
54. REISENAUER, H. M., *Soil Sci.* **81**, 237 (1956).
55. GAMMON, N., JR., *Soil Sci. Proc. Am.* **19**, 488 (1955).
56. JENSEN, E. H. and LESPERANCE, A. L., *Agron. J.* **63**, 201 (1971).
57. BARSHAD, I., *Soil Sci.* **66**, 187 (1948).
58. CARRIGAN, R. A., *Proc. Ass. So. Agr. Workers* **47**, 57 (1950). CUNNINGHAM, I. J. and HOGAN, K. G., *NZ J. Sci. Tech.* **31A**, 39 (1949).
59. WILLIAMS, R. J. P., Molybdenum in enzymes, presented at the International Conference on the Chemistry and Uses of Molybdenum, Univ. of Reading, England, September 17–21, 1973. To be published in *Proceedings of Conference*.
60. IVANOVA, N. N., *Agrochimica* **17**, 96 (1973).
61. TODD, J. R., *J. Agri. Sci.* **79**, 191 (1972).
62. BERG, J. W., A second association between low molybdenum and esophageal cancer, unpublished.
63. LUDWIG, T. G., HEALY, W. B., and LOSEE, F. L., *Nature, Lond.* **186**, 695 (1960).
64. ANDERSON, R. J., *Br. Dent. J.* **120**, 271 (1966).
65. HADJIMARKOS, D. M., Trace elements and dental health, presented at 7th Ann. Conf. on Trace Substances in Env. Health, Columbia, Missouri, June 12–14, 1973. To be published in *Proceedings of Conference* (Hemphill, D. D., ed.).
66. DYE, W. B. and O'HARA, J. L., *Nev. Agri. Exp. Sta. Bull.* 208 (1959).
67. MILTIMORE, J. E. and MASON, J. L., *Can. J. Anim. Sci.* **51**, 193 (1971).
68. THORNTON, I., KERSHAW, G. F., and DAVIES, M. K., *J. Agri. Sci. Camb.* **78**, 165 (1972).
69. JENSEN, R., MAAG, D. D., and FLINT, J. C., *J. Am. Vet. Med. Ass.* **133**, 336 (1958).
70. GALLAGHER, C. H., JUDAH, J. D., and REES, K. R., *Proc. R. Soc. Lond.* B, **145**, 134, 195 (1956).
71. FELL, B. F., et al., *Res. Vet. Sci.* **6**, 10 (1965).
72. DI PAOLO, R. V. and NEWBERNE, P. M., Copper deficiency and myelination in the central nervous system of the newborn rat: histological and biochemical studies, presented at 7th Ann. Conf. on Trace Substances in Env. Health, Columbia, Missouri, June 12–14, 1973. To be published in *Proceedings of Conference*.
73. MILLS, C. G. and FELL, B. F., *Nature, Lond.* **185**, 20 (1960).
74. SUTTLE, N. F. and FIELD, A. C., Production of swayback by experimental copper deficiency, in *Trace Element Metabolism in Animals* (Mills, C. F., ed.), E. & S. Livingstone, Edinburgh, 1970.
75. ALLOWAY, B. J., *J. Agri. Sci. Camb.* **80**, 521 (1973).
76. WINSTON, P. W., Effects of dietary molybdenum on the physiology of the white rat, in *Transport and the Biological Effects of Molybdenum in the Environment, Progress Report to NSF*, pp. 217–240 (Chappell, W. R., ed.), the Molybdenum Project, University of Colorado, Boulder, Colorado, January 1, 1973.

77. FAIRHALL, L. T., DUNN, R. C., SHARPLESS, N. E., and PRITCHARD, E. A., *The Toxicity of Molybdenum*, US Public Health Bull. 293.
78. ARTHUR, D., *J. Nutr.* **87**, 69 (1965).
79. DOWDY, R. P. and MATRONE, G., *J. Nutr.* **95**, 197 (1968).
80. MARCILESE, N. A., AMMERMAN, C. B., VALSECCHI, R. M., DUNAVANT, B. G., and DAVIS, G. K., *J. Nutr.* **99**, 177 (1969).
81. LOURIA, D. B., JOSELOW, M. M., and BROWDER, A., *Ann. Intern. Med.* **76**, 307 (1972).
82. SCHROEDER, H. A., Recondite toxicity of trace elements, in *Essays in Toxicology*, pp. 108–199 (Hayes, W. J., Jr., ed.), Academic Press, London, 1973.
83. SCHROEDER, H. A. and MITCHENER, M., *Arch. Environ. Hlth.* **23**, 102 (1971).
84. SUTTLE, N. F., The nutritional significance of the Cu:Mo interrelationship to ruminants and non-ruminants, presented at 7th Ann. Conf. on Trace Substances in Env. Health, Columbia, Missouri, June 12–14, 1973. To be published in *Proceedings of Conference*.
85. WINSTON, P. W., HOFFMAN, L., and SMITH, W., Increased weight loss in molybdenum-treated rats in the cold, presented at 7th Ann. Conf. on Trace Substances in Env. Health, Columbia, Missouri, June 12–14, 1973. To be published in *Proceedings of Conference* (Hemphill, D. D., ed.).
86. MOFFIT, A. E., JR. and MURPHY, S. D., Effect of excess and deficient copper intake on hepatic microsomal metabolism and toxicity of foreign chemicals, presented at 7th Ann. Conf. on Trace Substances in Env. Health, Columbia, Missouri, June 12–14, 1973. To be published in *Proceedings of Conference*.
87. LALICH, J. J., GROUPNER, K., and JOLIN, J., *Lab. Invest.* **14**, 1482 (1965).
88. JOHNSON, R. H., LITTLE, J. S., and BICKLEY, H. C., *J. Dent. Res.* **48**, 1290 (1969).
89. SOLOMONS, C. and HANDRICH, E. M., Skeletal biology of molybdenum, in *Transport and the Biological Effects of Molybdenum in the Environment, Progress Report to NSF*, pp. 241–256 (Chappell, W. R., ed.), the Molybdenum Project, University of Colorado, Boulder, Colorado, January 1, 1973.
90. SOLOMONS, C. C., ERNISSE, D. J., and HANDRICH, E. M., The skeletal biology of molybdenum, presented at 7th Ann. Conf. on Trace Substances in Env. Health, Columbia, Missouri, June 12–14, 1973. To be published in *Proceedings of Conference* (Hemphill, D. D., ed.).
91. DAVIS, G. K., in *Symposium on Copper Metabolism*, p. 216 (McElroy, W. D. and Glass, B., eds.), Johns Hopkins Press, Baltimore, 1950.
92. HATCH, T. F., *Arch. Env. Hlth.* **27**, 231 (1973).

Transport and Biological Effects of Molybdenum in the Environment
(W. R. Chappell)

DISCUSSION by CORALE L. BRIERLEY
Chemical Microbiologist, New Mexico Bureau of Mines and Mineral Resources, New Mexico Institute of Mining and Technology, Socorro, N.M. 87801

INTRODUCTION

W. R. Chappell's paper summarizes much of the research that has been done on the release and transport of molybdenum in the environment and the effects of molybenum on biological entities.

In this presentation I will (1) propose questions which may aid in further studies of molybdenum and related elements, (2) make observations on data presented by the Molybdenum Project Group from the University of Colorado, and (3) interject related data from our laboratory.

MOLYBDENUM IN THE LITHOSPHERE

The major concentration of molybdenum in the earth's crust is found in the form of molybdenite (MoS_2) and wulfenite ($PbMoO_4$).[1] The natural release of molybdenum into the environment is generally attributed to weathering, but studies defining the exact nature of weathering of these minerals are limited.

Although molybdenite, the principal mineral recovered, is quite stable, the velocity of oxidation is sufficient to cause noticeable molybdenum enrichment in natural waters.[2] Often observed near molybdenite deposits is ferrimolybdite, $Fe_2(MoO_4)_3 \cdot nH_2O$, an oxidation product of molybdenum and iron.[3]

The biogenic solubilization of molybdenite has been studied,[4] primarily in an effort to find catalytic leaching methods for recovery of molybdenum from molybdenite ore deposits. Microorganisms are generally found to be ineffectual because of their sensitivity to low concentrations of molybdenum;[5] however, in our laboratory we have isolated[6] and characterized[7] a thermophilic microbe, or symbiotic group of microbes, which oxidize(s) both the molybdenum and sulfide moieties in molybdenite and can grow in concentrations up to 750 ppm of soluble molybdenum. This unusual organism, which appears to lack a cell wall structure, oxidizes reduced, inorganic compounds at temperatures between 45–70°C. It possesses a high degree of thermal stability and functions in an acid medium with an optimum at pH 2.[8] The organism seems capable of growth in both aerobic and anaerobic environments. Although our work is aimed at devising biological techniques to enhance molybdenum recovery from ores and waste material, this entity or a similar biological agent could play a major role in contamination of natural waters by oxidation of molybdenite in mineralized regions.

An earlier study[9] attempted to clarify the role of biogenic factors in migration of molybdenum in the underground waters of rare metal sulfide deposits. The results of this investigation showed that underground waters are populated with bacteria of the thiobacilli type. These workers showed that increased activity of the thiobacilli accompanies an increased molybdenum content in these waters.

As Dr. Chappell's paper indicates, solubilization of molybdenum is enhanced by mining and milling due to the increased surface area exposed for oxidation. An interesting study would be an observation made of various aspects of mining operations separated into areas of (1) mine run-off, (2) run-off from waste piles, and (3) run-off from tailings.

An additional area of environmental concern is the concentration of molybdenum which contaminates fossil fuels. The release of molybdenum in burning fossil fuels has been studied by the group in Colorado as well as others, but of additional interest would be a study of the fate of molybdenum in the Lurgi coal gasification process and a study of methods for recovery of molybdenum and other heavy metals which are accumulated in oil shales.[10]

AQUEOUS TRANSPORT OF MOLYBDENUM

The form of molybdenum in aqueous transport has not been confirmed. The chemistry of molybdenum, complicated by the existence of six valence states and three coordination numbers, has most certainly contributed to the lack of definite information on the subject of aqueous transport.

The major form of molybdenum in transport is thought to be molybdate;[10] however, a complicated series of oxyanions and polymolybdates can be formed in solution. Organo-metallic complexes, as well as molybdenum compounds coordinated with inorganic ligands, are also possible in aqueous systems,[11] and heavy metals are present as colloids in some waters.[12] Dr. Chappell pointed out that mass balance data collected by the Colorado group, indicate that molybdenum is dynamic, e.g. molybdenum is cycled both physically and biologically, but generally remains in solution, particularly in the southwestern region of the United States. Several methods, mentioned in the previous paper, for sedimentation of molybdenum from aqueous systems, are coprecipitation with hydrous metal oxides and precipitation with industrial effluents. The main factors concerning the behavior of molybdenum during sedimentation are (1) the type of medium (pH and Eh), (2) molybdenum concentration in the basin, (3) composition and concentration of other soluble material, and (4) the composition of precipitating material. Considering these factors, molybdenum will behave in one of three ways: (1) Part of the dissolved molybdenum will be precipitated and entrapped by settling material; if the medium is weakly acid and reducing, molybdenum will decrease in the water and increase in bottom sediments. Bituminous coals are formed in this way. (2) No redistribution of molybdenum between solid and liquid phases will occur during sedimentation if the medium is neutral and there is free access of oxygen. (3) The precipitated solid will become impoverished in molybdenum because the molybdenum is leached into solution. This can occur in an alkaline medium of a well-aerated basin. Some limestones, dolomites, and gypsum are formed under these conditions.[13] I would like to speculate upon the removal process described by Dr. Chappell whereby molybdenum is coprecipitated with $FeCl_3$. It is suggested that tailings run-off, high in soluble molybdenum, could be treated in this manner. The tailings water is at pH 7 or greater due to the lime used to depress pyrite in molybdenite flotation. In addition to adding $FeCl_3$, acid may be required to lower the pH of vast amounts of tailings water for *maximum* precipitation of molybdenum. If this water is recycled for use in flotation, the pH would again have to be raised, increasing lime consumption. Therefore, the process would have to be justified on the basis of additional molybdenum recovery. The high lime addition coupled with a recycling program of tailings-pond water would also increase the amount of carbonate and bicarbonate deposits in water lines.

A more probable means of molybdenum recovery from tailings water may be an ion exchange step in which molybdenum from tailings is concentrated. The principles of iron(III) coprecipitation could then possibly be applied in a relatively small treatment area (personal communication, G. Chaun Cadwell, Fluor Utah, Inc., San Mateo, CA).

A natural process which may play an important role in removal of molybdenum from ground water and ocean depths is precipitation of molybdenum with biogenically formed sulfide. Some microorganisms, which oxidize organic compounds and reduce sulfate in the absence of oxygen, produce H_2S as an end product. These sulfate-reducing microbes have been found in underground waters[9] and may be responsible for the formation of metallic sulfide deposits[14] such as molybdenite.

In our laboratory further studies[15] with the unidentified thermophile have shown that in the absence of oxygen this microbe reduces molybdenum(VI) to molybdenum(V), or "moly blue," if provided with a suitable energy source. This reaction has been noted with other organisms,[16] but the environmental significance, if any, of this biological reaction has not been speculated upon. The "moly blue" phenomenon is not well defined, but some evidence supports its being a colloidal material.[11] If this claim is true, it would suggest that biologically mediated formation of molybdenum(V) may play a role in removal of molybdenum from aqueous environments. Further study in transformations of heavy metals by microorganisms may show that molybdenum and mercury are not the only metals altered by biological action, and that microbes may have substantial influence on the availability, as well as toxicity, of metallics to plants and animals.

The control of molybdenum transport from "disturbed" areas may well be classified with other problems of mine drainages.[17] The ultimate, but often impractical solution, is preventing formation of the pollutant. Excluding oxygen would, of course, prevent formation of the oxidized forms of molybdenum found in aqueous environments. Exclusion of air can be accomplished by flooding abandoned mines with water and placing a cover material over exposed minerals to act as an oxygen barrier. The cover material should be vegetated to aid in stabilization and act as an oxygen absorber. The US Bureau of Mines has an extensive program for studying reclamation and stabilization of mine wastes (Dr. J. B. Rosenbaum, Research Director, USBM, Salt Lake City, Utah). Waste dumps could be constructed to prevent air movement within them by placing layers of compacted clay between layers of waste material.[17]

BIOLOGICAL EFFECTS OF MOLYBDENUM

The transport of molybdenum from water to plants is a complex study because of the numerous factors involved in soil chemistry. The pH and Eh, the chelation and ionic exchange properties, and the presence of microbes, all effect plant uptake of molybdenum. Therefore, any investigation of uptake of molybdenum by plants and the interaction of molybdenum with soils is a significant contribution to the overall study.

One consideration which should be made when studying the reaction of molybdenum with microorganisms, plants, and animals is the interchangeability of tungsten with molybdenum. There appears to be an important relationship between tungsten and molybdenum in bacterial nitrogen fixation; tungsten competitively inhibits the normal function of molybdenum in *Azotobacter*.[18] The tungsten inhibition of nitrogen fixation is greatly influenced by pH. There also appears to be antagonism of molybdenum uptake by tungsten in the soil fungus, *Aspergillus niger*.[19] In animals, as in microorganisms, tungsten appears to act as a molybdenum antagonist. Like molybdenum, tungsten is also incorporated into growing bone.[10] When studying the effects of molybdenum on biological systems, one should consider this antagonistic role of tungsten.

There is a great deal of conflicting evidence regarding the toxicity of molybdenum in animals. The copper–molybdenum–sulfate interaction appears to play a major role in both molybdenum toxicity and deficiency; however, the actual interaction and mechanism of these ions have not been defined.

An area of study which has been somewhat neglected in the plant, animal, and microbial kingdoms is the effect of molybdenum at the cellular level. The mechanisms of cellular binding, membrane transport, and intracellular interactions have not been thoroughly studied. Our studies, in which a thermophilic microbe is grown on molybdenite and suspended in a medium containing solubilized molybdenum, indicate that molybdenum did not accumulate within the organism.[20] However, suspension of the organism in concentrations of soluble hexavalent molybdenum exceeding 750 ppm did inhibit growth.[8] Studies with other cell systems may help to clarify the mechanisms of toxicity of molybdenum at the cellular level.

CONCLUDING REMARKS

Because of the complexity of molybdenum chemistry, it is difficult to follow the release and transport of molybdenum in the environment and to study its interaction with the biological world.

The mechanisms for weathering and release of molybdenum into aqueous systems are not well understood. The role of microorganisms in solubilization of molybdenum-containing minerals may be underestimated since current studies in our laboratory indicate that molybdenum can be biogenically leached from molybdenite.

Molybdenum in aqueous systems is transported in the form of molybdate. Removal of this molybdenum is dependent upon a variety of chemical and physical factors with the result being that much of the molybdenum is not removed but transported to the oceans. Methods are being considered for removal of molybdenum in waste waters, but often the proposed methods are economically impractical.

Molybdenum toxicity is difficult to study due to the number of factors which influence uptake and absorption of molybdenum. The copper–molybdenum–sulfate interaction and the interchangeability of tungsten with molybdenum are two factors which are currently recognized.

Only continued studies such as the previous paper described will answer existing questions concerning the hazards of molybdenum to biological entities and confirm or contradict the need for tighter controls on distribution of molybdenum and other heavy metals in the environment.

Acknowledgement—The thermophilic, biological leaching studies described in this paper were supported by US Bureau of Mines Grant No. G0122123.

REFERENCES

1. TOENSING, C. H., Molybdenum metal powder, in *The Metal Molybdenum*, pp. 31–50 (Harwood, J. J., ed.), American Society for Metals, Cleveland, Ohio, 1956.
2. VINOGRADOV, V. I., Some problems in molybdenum geochemistry, in *Materialy Geol. Rudnykh Mestorozhdenii*, pp. 191–204 (Petrog., Mineral. i Geikhim., Akad. Nauk SSSR, 1959).
3. KERR, P. F. et al., *Am. Miner.* **48**, 14–32 (1963).
4. BRYNER, L. C. and ANDERSON, R., *Ind. Eng. Chem.* **49**, 1721–1724 (1957).
5. TUOVINEN, O. H. et al., *Antonie van Leeuwenhoek J. Serol.* **37**, 489–496 (1971).
6. BRIERLEY, J. A., Contributions of autotrophic bacteria to the acid thermal water of the Geyser Springs Group in Yellowstone National Park, Mont. State Univ., PhD thesis, 1966.

7. BRIERLEY, C. L. and BRIERLEY, J. A., *Can. J. Microbiol.* **19**, 183–188 (1973).
8. BRIERLEY, C. L. and MURR, L. E., *Science* **179**, 488–490 (1973).
9. KRAMARENKO, L. E., *Mikrobiologiya* **31**, 694–701 (1962).
10. BOWEN, H. J. M., *Trace Elements in Biochemistry*, Academic Press, London and New York, 1966.
11. KILLEFFER, D. H. and LINZ, A., *Molybdenum Compounds, their Chemistry and Technology*, Interscience Publishers, New York and London, 1952.
12. EREMENKO, V. Y., *Gidrokhim. Materialy* **36**, 125–133 (1964).
13. MIKHAILOV, A. S., *Geokhimiya* **11**, 1171–1181 (1964).
14. ZOBELL, C. E., *Marine Microbiology*, Chronica Botanica Co., Waltham, Mass., 1946.
15. BRIERLEY, C. L., Molybdenite-leaching: use of a high-temperature microbe, presented at the International Conference on the Chemistry and Uses of Molybdenum, Univ. of Reading, England, September 17–21, 1973. To be published in *Proceedings of Conference*.
16. WOOLFOLK, C. A. and WHITLEY, H. R., *J. Bacteriol.* **84**, 647–658 (1962).
17. HILL, R. D., Control and prevention of mine drainage, in *Cycling and Control of Metals*, pp. 91–94, Proceedings of an Environmental Resources Conference (eds. Curry, M. G. and Gigliotti, G. M.), National Environmental Research Center, Cincinnati, Ohio, 1973.
18. TAKAHASHI, H. and NASON, A., *Biochim. biophys. Acta* **23**, 433–435 (1957).
19. HIGGINS, E. S., *et al.*, *Proc. Soc. Exp. Biol. Med.* **92**, 509–511 (1956).
20. BRIERLEY, C. L., *et al.*, *Res./Dev.* **24**, 24–28 (1973).

BIOLOGICAL AND NONBIOLOGICAL TRANSFORMATIONS OF MERCURY IN AQUATIC SYSTEMS

W. P. IVERSON and C. HUEY†

Corrosion Section, National Bureau of Standards, Washington DC 20234

and

F. E. BRINCKMAN, K. L. JEWETT, and W. BLAIR

Inorganic Chemistry Section, National Bureau of Standards, Washington DC 20234

Biological and nonbiological transformations of heavy metals, with present emphasis on mercury, have been and are under investigation in collaboration with Dr. Rita Colwell's group at the University of Maryland.

In our studies of biological transformations, a number of mercury-tolerant bacterial isolates from Chesapeake Bay waters and sediments have been examined for their ability to volatilize mercury from trace concentrations of inorganic or organomercury compounds (e.g. $HgCl_2$ or phenylmercuric acetate). A survey for such aerobic microorganisms has been conducted;[1] these include *Pseudomonas* spp. which constitute about 90% of the phenylmercuric acetate-tolerant bacterial cultures obtained.

Inoculated agar plates containing phenylmercuric acetate (0.3 ppm) were placed in a bioreactor and the atmosphere surrounding the plates analyzed for total and metallic mercury using an atomic absorption (AA) spectrophotometer on-line with a bioreactor[1] (Fig. 1). The results indicated that all of the nine isolates so examined primarily produce metallic mercury. The typical production of $Hg°$ from one isolate (No. 72) is shown in Fig. 2.

Currently we are in the process of examining a number of $HgCl_2$-tolerant anaerobic and facultative bacteria for their ability to produce metallic mercury and other volatile mercury compounds from $HgCl_2$ using a flameless AA spectrophotometer coupled with a vapor phase chromatograph. This approach has been previously described in another connection,[2] but our special application affords a very significant advance in direct separation and characterization of mercury-containing bacterial metabolites at environmental concentrations under con-

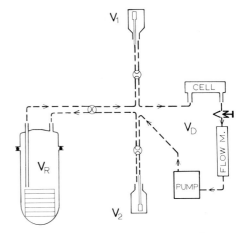

FIG. 1. An on-line atomic absorption spectrophotometer is incorporated into a closed loop fitted with a valving arrangement permitting either flushing or selective switching into a modular bioreactor (V_R, volume 4 liters) containing bacterial isolates on metal-doped synthetic growth media, or into calibration BOD bottle V_1 or V_2 (reprinted from ref. 1).

TABLE 1
Tolerance of Pseudomonas *sp. to metal ions*

Metal	Tolerance (ppm)
Al(III)	1000
As(V)	10–50
Co(II)	50
Cr(III)	10
Hg(II)	50
Mg(II)	<10⁴
Pb(II)	100
Sn(II)	10
Sn(IV)	1000
Te(VI)	10

†NBS guest worker, 1973–74.

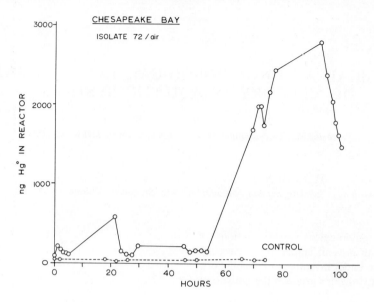

FIG. 2. Gaseous metallic mercury production from a Chesapeake Bay aerobic culture as a function of time, determined by the bioreactor-atomic absorption spectrophotometer apparatus shown in Fig. 1. Diameter of open circles represents ± standard deviation in mercury calibration curves used.

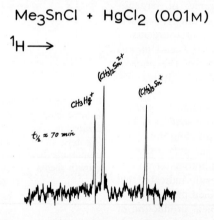

FIG. 3. Proton nuclear magnetic resonance spectrum of intermediate mixture achieved by reacting aqueous solutions containing equivalent amounts of trimethylin chloride and mercuric chloride. Half-life for the complete reaction was about 70 min (ref. 4).

trolled atmospheres. These results are published elsewhere.[3]

One of the isolates, a strain of *Pseudomonas*, demonstrates a tolerance (Table 1) to a variety of other metal cations in agar media. The possibility of volatilization of these metals by this organism is currently under investigation since other transport pathways similar to mercury transformations may be involved.

Related nonbiological pathways for methylation of mercury have been demonstrated with a number of aquated methylated metal cations.[4] These processes are identified and their rates determined by nuclear magnetic resonance and laser Raman spectroscopy (Figs. 3, 4). We view such alternate transmethylation processes for mercury as important to understanding the overall rate of several metals interacting in aquatic media. We are directing

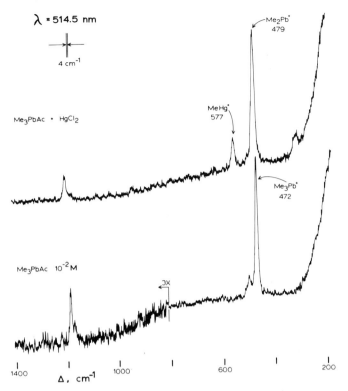

FIG. 4. Comparison of aqueous laser-Raman spectra of reactant trimethyllead acetate solution to that of its reaction mixture with mercuric chloride after several minutes. Product methylmercury ion is readily discernable along with shifted peak for co-product dimethyllead cation (ref. 4).

efforts to establish whether both such biological and non-biological transport paths have kinetic importance among those metals cited in Table 1.

Acknowledgements—The work was supported in part by the Environmental Protection Agency Grants R801002 and R800647 and by the National Bureau of Standards Measures for Air Quality Program. The authors are indebted to the National Science Foundation and the Chesapeake Bay Institute for extended use of the R/V Ridgely Warfield.

REFERENCES

1. NELSON, J. D., BLAIR, W., BRINCKMAN, F. E., COLWELL, R. R., and IVERSON, W. P., Biodegradation of phenylmercuric acetate by mercury-resistant bacteria, *Appl. Microbiol.* **26**, 321 (1973).
2. LONGBOTTOM, J. E., Inexpensive mercury-specific gas chromatographic detector, *Analyt. Chem.* **44**, 1111 (1972).
3. BLAIR, W., IVERSON, W. P., and BRINCKMAN, F. E., Application of a gas chromatograph–atomic absorption detection system to a survey of mercury transformations by Chesapeake Bay microorganisms, *Chemosphere* **4**, 167 (1974).
4. BRINCKMAN, F. E. and JEWETT, K. L., Transmethylation of heavy metal ions in water, Regional Meeting of the Am. Chem. Soc., Washington DC, January 1973, abstr. papers, 52.

rivers that are engaged in the production of ethylmercury for disinfectant purposes and medical use, but no such factories exist along the Shokotsu and Yubetsu rivers in Hokkaido. Thus, it is not beyond possibility that ethylmercury is synthesized naturally within the rivers themselves. This poses a problem for future investigation. The mercury content in the marine fish resources taken from Japanese coastal waters is fortunately not so high, except for some species. For the most part, small marine fish such as sardines, horse mackerel, mackerel, pike, and sillago contain total mercury concentrations of less than 0.1 ppm on an average with maximum values being approximately 0.2 ppm as indicated in Tables 5 and 6. There are, however, some species that show relatively high concentrations of mercury such as *Sebastes baramenuke* and *Beryx splendens* containing more than 0.4 ppm of total mercury and more than 0.3 ppm of methylmercury which are respectively the tentative legal limits of concentration allowed by the Ministry of Health and Welfare as of June of 1973 (Table 6).[10] Skipjack tuna and yellowtail contain approximately 0.2 ppm of total mercury. The distribution of mercury by tissue and organ is indicated in Table 5, and the liver and spleen contain equivalent or higher concentrations of mercury than the muscle tissue. The ratio of methylmercury to total mercury in these fish is 66–100%, while lower values have been reported by other investigators, varying with the species and the water system from which the fish was taken. These values are similar to those derived from fish examined by Ui.[9]

Extensive investigations into mercury accumulation in these types of small to medium size fish was begun in earnest only a few years ago in Japan, and there is not yet any reliable information on background mercury concentration values for these fish to be used as a basis for determining the influence of artificially introduced environmental mercury contamination. Moreover, investigations carried out in Japan are lacking from an ecological point of view, and thus we have little information at our disposal as to the mechanisms of mercury accumulation in the ecosystem, the food chain, and in fish. The following discussion is limited practically to mercury content in the edible portions of fish.

This most troublesome problem is the mercury

TABLE 6
Mercury concentrations in marine fish[10] *(ppm as Hg on wet basis)*

Species	No.	Range (ppm)	Total Hg[a] (ppm)	Methyl Hg (ppm)	Methyl Hg/total (%)
Horse mackerel	10	0.01–0.16	0.047*	—	
Yellowtail	8	0.05–0.13	0.088*	—	
Common mackerel	10	0.03–0.16	0.084*	—	
Pike eel	1	0.32	0.32	—	
Red sea-bream snapper	2	0.27, 0.30	0.285*	—	
Sardine	6	0.01–0.02	0.013*	—	
Black rockfish	20	0.03–0.23	0.127*	—	
Common sea-bass	4	0.04–0.26	0.016*	—	
Sebastes baramenuke	4	a	0.14	—	
		b	0.19	—	
		c	0.56	0.54	96.4
		d	1.10	0.97	88.1
Beryx splendens	11	a	0.22	—	
		b	0.53	0.35	66.0
		c	0.53	0.40	75.4
		d	0.53	0.40	75.4
		e	0.60	0.49	81.6
		f	0.62	0.41	66.1
		g	0.65	0.62	95.3
		h	0.78	0.70	89.7
		i	0.79	0.64	81.0
		j	0.80	0.80	100.0
		k	0.83	0.59	71.0

[a]The numbers with (*) represent the mean level of the species.

accumulation in tuna, marlin, and swordfish. Table 7 shows the values for total mercury in 60 individual specimens of tuna, marlin, and swordfish.[11] The maximum value of 1.34 ppm was detected in a swordfish, and the average concentrations for these fish, except two bluefin caught in Japanese waters, exceeds the level of 0.5 ppm total mercury. Bluefin and bigeye tuna have average values of nearly 1.0 ppm.

The results of an investigation by Yamanaka et al.[12] showed swordfish and shark with high mercury concentrations as indicated in Table 8, while other

TABLE 7
Mercury concentration in tuna, marlin, and swordfish[11] (ppm as Hg on wet basis)

Species	No.	Total Hg (ppm)			Mean ±SD[a]
		$x \geq 1.0$	$1.0 > x \geq 0.5$	$0.5 > x$	
Yellowfin	18	3	7	8	0.53 ± 0.40
Southern bluefin	8	1	5	2	0.71 ± 0.31
Bluefin	12	4	8	0	0.94 ± 0.21
Bluefin[b]	2	0	0	2	0.24 ± 0.03
Bigeye	9	2	5	2	0.83 ± 0.31
Marlin	9	0	8	1	0.61 ± 0.16
Swordfish	2	1	1	0	0.95 ± 0.54
Total	60 (100%)	11 (18.3%)	34 (56.7%)	15 (25.0%)	0.70 ± 0.34

[a] Standard deviation.
[b] These bluefin tuna were caught at Japanese waters.

TABLE 8
Mercury levels in tuna, marlin, swordfish, and shark flesh[12] (ppm as Hg on wet basis)

Subject	Area	Total Hg	Methyl Hg	Methyl Hg/ total Hg (%)
Yellowfin tuna	Indian Ocean	0.25	0.143	57
Yellowfin tuna	Indian Ocean	0.22	0.117	53
Yellowfin tuna	Indian Ocean	0.16	0.093	58
Yellowfin tuna	Indian Ocean	0.21	0.108	51
Albacore tuna	Atlantic Ocean	0.24	0.11	46
Albacore tuna	Indian Ocean	0.34	0.198	58
Albacore tuna	Pacific Ocean	0.32	0.244	76
Bigeye tuna	Pacific Ocean	0.62	0.44	71
Bigeye tuna	Indian Ocean	0.35	0.190	54
Bigeye tuna	Indian Ocean	0.46	0.215	47
Bigeye tuna	Indian Ocean	0.47	0.243	52
Bigeye tuna	Indian Ocean	0.48	0.324	68
Black marlin	Indian Ocean	0.33	0.126	38
White marlin	Indian Ocean	0.46	0.282	62
Swordfish	Pacific Ocean	0.86	0.607	71
Swordfish	Indian Ocean	0.96	0.62	65
Swordfish	Indian Ocean	0.90	0.55	61
Swordfish	Indian Ocean	0.88	0.648	74
Swordfish	Indian Ocean	0.84	0.612	73
Shark	Indian Ocean	1.42	0.828	58
Shark	Indian Ocean	1.86	1.335	72

species showed mercury levels of less than 0.5 ppm. The ratio of methylmercury to total mercury was between 38 and 76%.

As shown in Table 9, bluefin tuna taken from Japanese waters showed high mercury concentration, confirming that it is probably best to consider tuna, marlin, and swordfish as having high mercury content regardless of the water basin from which they are taken.[13]

The ratio of methylmercury to total mercury in tuna is usually high. This ratio was between 58 and 100% according to a report released by Nishigaki et al.[13] as shown in Table 9.

With regard to the source of mercury in tuna, it has been generally accepted that tuna absorb mercury existing naturally in the oceans in the form of dissolved mercury compounds or in the form of protein compounds in plankton and various other sea foods, and that this level of mercury in tuna has no correlation with artificial sources of mercury pollution introduced into the marine environment.

This theory is based on the following facts: (1) The total amount of mercury discharged into the oceans through man's activities is 10^5–10^6 tons, while the oceans contain approximately 10^8 tons of mercury in 1.3×10^{18} tons of sea water.[14] Then the total amount

TABLE 9
Mercury levels of muscle tissue in tuna, marlin, and swordfish[13] (ppm as Hg on wet basis)

Species	Area of capture	Region	Total Hg (ppm)	Methyl Hg (ppm as Hg)	Methyl Hg/ total Hg (%)
Indian Ocean tuna	Indian Ocean	Tail	0.63	0.55	87
	Indian Ocean	Tail	0.70	0.57	85
	Indian Ocean	Tail	0.65	0.58	89
	Indian Ocean	Tail	1.01	0.78	77
	Indian Ocean	Tail	0.59	0.39	66
Big-eye tuna	Pacific Ocean	Tail	0.47	0.47	100
	Pacific Ocean	Tail	0.84	0.84	100
	Pacific Ocean	Tail	0.86	0.50	58
	Kishu Offing	Tail	1.10	1.02	93
Yellowfin tuna	Pacific Ocean	Tail	0.96	0.97	101
	Pacific Ocean	Lean	0.80	0.68	85
	Pacific Ocean	Tail	0.26	0.19	73
	Pacific Ocean	Tail	0.42	0.43	102
	Pacific Ocean	Tail	0.49	0.45	92
Striped marlin	Pacific Ocean				
	(Australia)	Tail	0.58	0.57	98
	(Australia)	Tail	0.59	0.59	100
	(Coral Sea)	Tail	1.08	1.02	94
	(Madagascar)	Tail	0.53	0.46	87
	Kesen-numa	Lean	0.73	0.66	90
	Choshi Offing	Lean	1.13	1.06	94
	Katsuura Offing	Tail	0.88	0.83	94
Swordfish	Pacific Ocean				
	(Peru)	Trunk	1.63	1.46	90
	(Peru)	Lean	1.84	1.39	76
	(Peru)	Tail	1.16	1.08	93
	(Peru)	Tail	1.44	1.39	97
	Pacific Ocean	Lean	1.66	1.61	97
	Pacific Ocean	Lean	0.89	0.78	88
	Pacific Ocean	Tail	0.49	0.50	102
	Katsuura Offing	Tail	1.16	1.15	99
	Katsuura Offing	Tail	0.90	0.82	91
	Katsuura Offing	Tail	1.21	1.22	101
Black marlin	Middway	Head	4.3	0.70	16
	Pacific Ocean	Tail	0.59	0.14	24
	Pacific Ocean	Lean	0.85	0.46	54

of discharged mercury corresponds to only a few percent of naturally existing mercury in the oceans. (2) As a result of mercury analysis of museum specimens of tuna by Miller et al.,[5] it was revealed that these museum specimens caught from 27 to 95 years ago had concentrations of mercury equivalent to more recent specimens.

These arguments, however, are contradictory to the actual state of mercury pollution of the marine environment. First of all the discharged mercury that accumulates in the environment as the result of human activity is not distributed evenly throughout the ocean environment but exists locally in concentrated form. This mercury concentration in sea water and in the ocean bottom sediment along the coast line increases continuously over time with the concentrations in the bottom sediment dissolving into the surrounding water. A considerable amount of this mercury will adhere to organic particles and plankton later being absorbed by small fish, shrimp, and shellfish. Since the coastal waters are the most productive waters in terms of life generation and biological activity, increased concentrations of mercury in these waters enlarge the possibilities for negative influence on the pelagic ecosystem. As an example of the mechanisms involved, small fish generated in coastal waters polluted with mercury will accumulate this mercury in their biomass. Then, as these small fish, at a certain stage in their growth, move into the open oceans, they become prey for the larger fish. Thus the mercury discharged in relation to man's activity becomes a possible basis for increased mercury concentrations in tuna found in the pelagic oceans.

The use of an analysis of the mercury concentration in museum specimens by such as Miller et al.[5] cannot be made the basis for meaningful ecological analysis because the sample size analyzed was so small that ecological comparisons of mercury concentrations in differing tuna populations separated by time span differences such as those represented by the last and present century are not possible.

According to more recent studies done in Japan in relation to mercury concentrations in fish, the general trend is to find higher concentrations of mercury in predatory fish and in larger size fish.

Figure 4 indicates a degree of correlation between body weight and mercury concentration in *Anoplopoma fimbria*. This figure, as well as Fig. 3, gives clear indication that mercury concentration increases with the growth in size and weight. Thus it is reasonable to assume the mechanisms of mercury accumulation as being mainly through the peroral route in the form of prey rather than through direct absorption from sea water. It is important to determine whether this particular mechanism applies to all species of fish or not.

THE PRESENT STATE OF MERCURY ACCUMULATION IN THE JAPANESE POPULATION

The uses of hair, blood, and urine for the determination of indices of mercury accumulation in human beings are all appropriate, but since past analyses used chiefly hair, the following discussion will be advanced in terms of mercury concentration indices related to hair.

It is well known that, generally, the level of mercury in hair of the Japanese people is higher than those levels found in the populations of other countries. Ukita[15] reported that the mean level obtained from the analysis of 74 Japanese hair samples was 6.02 ppm, and, further, that four Japanese students returning from extended stays abroad recorded increases in mercury concentration during the 18 month period after their return to twice

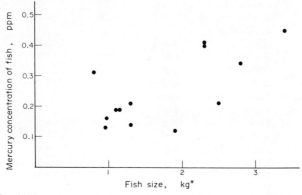

FIG. 4. The relationship between mercury concentration and fish size of *Anoplopoma fimbria* from the Gulf of Alaska. *Body weight of fish from which the head and viscera were removed.

the level recorded during their period abroad.

Aoki[16] analyzed the hair samples of 17 suburb dwellers in Tokyo and derived a mean level of 1.89 ± 1.24 ppm (standard deviation) of methylmercury. The distribution of the levels of mercury found in the hair of 101 citizens of Tokyo of both sexes showed a log normal distribution as indicated in Fig. 5, and the median was 3.85 ppm.[17]

Figure 6 indicates the periodic fluctuation of mercury levels in the hair of seven men and nine women living in Tokyo.[18] These levels are generally higher in men than in women, and the periodic deviation is higher for the hair having a higher mean level.

In an analysis carried out by Nishima et al.[19] the mean value of 104 men (6.9 ppm) was higher than that for 87 women (3.8 ppm). Kozuka[20] reported a mean level of 5.42 ppm for men and 4.62 ppm for women as indicated in Fig. 7. It is characteristic of the Japanese population, it seems, for the men to have higher concentrations of mercury in their hair than women.

On the other hand, Aoki[16] reported a mean value of 1.89 ppm total mercury ranging from 0.1 to 4.7 ppm in the hair of 30 non Japanese. According to an investigation by Perkons and Jervis,[21] the mean level found in 776 persons, mostly Canadians, was approximately 1.8 ppm ranging from 0 to 19 ppm. Coleman et al.[22] analyzed the levels of mercury in the hair of 870 Englishmen of both sexes. The mean values derived by Coleman et al. are similar to values derived from studies of the Japanese, but in the investigation by Coleman et al. the value for men was lower than the value for women, in contrast to the Japanese findings. These sex-related differences in the levels of mercury found from country to country may correlate with differences in dietary life styles.

The most important sources of mercury accumulating in the human population in Japan are rice and fish. Phenylmercury pesticides are still to be detected in unpolished rice at a mean level of 0.134

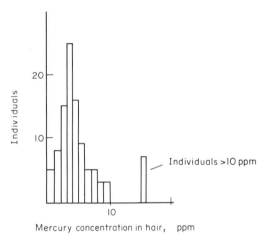

FIG. 5. Mercury levels in the hair of 101 Tokyo citizens.[17]

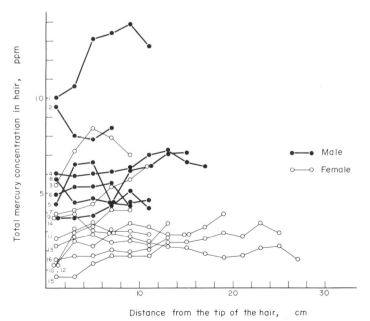

FIG. 6. The mercury levels in the hair of 16 Tokyo citizens.[18]

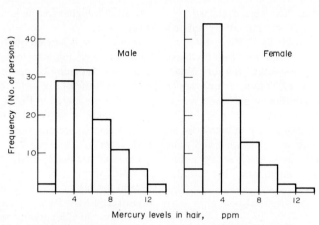

FIG. 7. The mercury levels in the hair of Japanese people.[20]

ppm, ranging from 0.01 to 0.376 ppm[23] in spite of the fact that the use of phenylmercury compounds for agricultural purposes has been prohibited since 1968.

The daily consumption of rice in Japan per person is approximately 200–300 g for men and 200 g or less for women. For the most part the mercury found in rice bran was pesticide-related phenylmercury residue with more than 50% of the mercury pesticide residue found in unpolished rice being removed in the polishing process,[24] intestinal absorption rate for phenylmercury compounds seems to be low, with experimental rats absorbing approximately 40%.[25] Therefore, the actual intake of mercury from rice is considered far less than 0.8–45 μg of mercury per day per person. However, as shown in Table 10, derived from an investigation done by Nishima et al.,[26] the mean level of mercury detected in hair from a group of men who eat boiled rice three times a day was 6.99 ppm, while the mean value was 5.63 ppm in a group of men who make it a habit to eat bread rather than boiled rice. These differences indicate rather clearly that rice is still an important source of mercury accumulation in the Japanese population.

According to Nose and Suzuki,[27] phenylmercury pesticides are rapidly degraded in a rice paddy filled with water and then fixed in the soil in the form of mercury sulfide. Yushima et al.[28] showed that mercury residue accumulations in rice fields may not

TABLE 10
Mercury concentrations in hair and blood of Tokyo citizens according to the consumption of rice and fish[26 from 51]

		Total Hg in hair (ppm)[a]			Total Hg in blood (μg per 100 g)[a]		
			General public			General public	
		Fish consumer	Male	Female	Fish consumer	Male	Female
Daily consumption of boiled rice	Three times	20.75 (34)	6.99 (62)	3.94 (32)	8.45 (30)	2.90 (59)	2.37 (28)
	One–two times	18.62 (45)	6.87 (39)	3.74 (51)	7.92 (42)	2.70 (29)	2.64 (48)
	Mainly bread	34.40 (1)	5.63 (3)	2.90 (4)	—	2.77 (3)	1.70 (3)
Consumption of fish	Prefer	20.52 (70)	7.54 (65)	4.21 (42)	8.41 (65)	3.17 (57)	2.91 (40)
	Not prefer	14.12 (10)	5.79 (38)	3.37 (45)	6.21 (8)	2.26 (33)	2.09 (39)

[a]Mean concentration. Numbers in parentheses represent number of samples.

be absorbed to a great extent by the rice plant unless there was an application of pesticides during the harvest year. However, in this regard, it must be remembered that mercury residue was detected in rice by Wakatsuki,[23] and Kanazawa,[29] making mercury in rice an important probable source of future continued accumulation of mercury in the Japanese population.

As shown in Table 10, the mean level of mercury accumulation in the hair of men who prefer to eat fish is 7.54 ppm, which is higher than the 5.79 ppm value for those whose consumption of fish is on a lower level. These facts seem to suggest that residents of Tokyo generally ingest as much mercury from fish as from rice. Since the ratio of methylmercury to total mercury is larger in fish than in rice, it can be said that with respect to the methylmercury intake of the Japanese population, fish consumption poses a greater threat than the consumption of rice.

In today's Japan, tuna fishermen, sushi shop workers, and fish dealers and retailers have the highest levels of mercury in their hair. These people deal everyday with large amounts of various kinds of fish, especially sliced raw tuna. The high levels of mercury in the hair of the people in these professions became generally known after the United States Government acted to prohibit the export and import of canned tuna with mercury concentrations higher than 0.5 ppm. The largest exporter of canned tuna to the United States is Japan, and the Japanese Government was called upon to prove the safety of the tuna. Analysis of the mercury content found in the hair of 121 tuna fishermen was carried out by the Ministry of Health and Welfare, and unexpectedly high concentrations were found. After this discovery the Ministry of Health and Welfare sent a communication to the fishermen indicating that they should not consume too much tuna but, of course, the fishermen were not about to stop eating their fish. This is due to the very severe working conditions under which tuna fishermen must operate, and long periods at sea lasting anywhere from a few months to more than 2 years preclude the possibility of any other fresh foods being available. Further, it is a great joy for them to eat their fish with their rice wine after daily hard work. Kondo and Takehiro[31] analyzed hair samples taken from 37 fishermen and obtained results that are much the same as those obtained by Yamanaka et al.,[12] which were shown in Table 11. Figure 8 is a compilation of the distribution of mercury levels in the hair of tuna fishermen investigated by Kitamura[31] and Kondo and Takehiro.[30]

In response to these discoveries, researchers of the Tokyo Metropolitan Bureau of Public Health completed several studies of the mercury content in the hair of sushi workers and fish retailers who, in relation to their occupation, consume large amounts

TABLE 11
Mercury levels in hair of crew of tuna fishing boats[12](a)

						Mercury levels in hair		
No.	Age	Career (years)	Working area	Period on tuna meal (days)	Intake of tuna (g/day)	Total Hg (ppm)	Methyl Hg (ppm)	Methyl Hg/ total Hg (%)
1	36	19	Indian Ocean	150	300	45.7	30.7	67
2	36	18	Indian Ocean	150	300	45.3	27.0	60
3	35	9	Indian Ocean	150	300	43.7	33.7	77
4	27	10	Indian Ocean	150	300	43.6	27.6	63
5	32	15	Indian Ocean	150	300	38.3	22.5	59
6	20	3	Indian Ocean	150	150	36.4	24.6	68
7	36	10	Off Africa	180	400	34.8	26.4	76
8	46	29	Indian Ocean	150	150	31.3	21.3	68
9	28	12	Indian Ocean	150	300	31.2	22.2	71
10	29	7	Pacific Ocean	130	50	31.0	17.4	56
11	31	14	Indian Ocean	150	300	27.0	18.5	69
12	39	23	Indian Ocean	150	300	26.1	16.8	64
13	31	14	Pacific Ocean	130	50	25.0	20.0	80
Mean ± standard deviation (n = 58)						19.9 ± 9.9	12.7 ± 7.1	63 ± 11

(a) The upper thirteen values of mercury in hair of 13 individuals out of 58 fishermen were listed in the table in order of total mercury level.

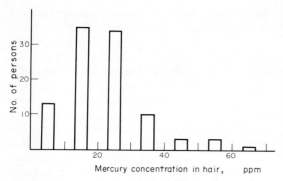

FIG. 8. Mercury levels in the hair of 99 tuna fishermen.[30]

of raw tuna. Doi[11] analyzed 22 hair samples taken from workers in the Central Fish Wholesale Market in Tokyo and obtained a mean level of 10.7 ppm ranging from 2.58 to 25.6 ppm (Table 12). In an analysis of 92 sushi workers, the mean level was 14.8 ppm with the maximum registering 52 ppm.[17] The mean value for 84 fish retailers was 19.3 ppm, and the maximum value was 64.7 ppm, as shown in Table 12.[26]

These people consume, besides tuna, many other kinds of fish, but the high mercury level in their hair is thought to be caused mainly through the ingestion of tuna fish. Many of these people eat as much as from 500 to 1000 g/day. As indicated in Table 10, higher mean levels were observed in men who eat large amounts of boiled rice as compared with men who eat smaller quantities, but the difference in the mean level between the former group and the latter group is only 2 ppm, while the mean level is 6 ppm higher for men who also eat large amounts of fish. In this same group the mercury blood levels are also high as indicated in Table 13, with the mean value reaching 7.9 μg per 100 g with 17 out of the 84 blood samples exceeding the maximum allowable concentration of 10 μg per 100 g of blood, a lower limit value set at the International Committee for MAC held in Stockholm in November of 1968.[32]

In this next section, consideration of the present state of mercury accumulation in the general public and in those who consume large amounts of tuna will be made in comparison to the extent of mercury accumulation in the population living along the Shiranui Sea in Kumamoto Prefecture and along the Agano River in Niigata Prefecture at the height of the Minamata disease outbreak in these areas.

Figure 9 indicates the results of an investigation into mercury levels in the hair of people living along the Shiranui Sea.[33] The hair samples were collected during the period between November of 1960 and March of 1961 just after the worst years for the Minamata disease outbreak. The greatest number of samples were collected in Goshonoura village in the Amakusa district, and the levels show log normal distribution with the levels in men being higher than those found in women. In other towns and villages the distribution of mercury accumulation values is very irregular, and definite correlations are hard to derive from the results perhaps because the hair samples were collected only from members of the fishermen's unions. In any case, it

TABLE 12
The mercury levels in the hair of general population and fish-consumers in Japan[30]

				Total Hg concentration (ppm)		
		Sex	No.	Mean ± SD[a]	Range	Reference
General Japanese	DM	M F	73	6.0 ± 2.9	0.98–23	Hoshino et al.[52]
General Japanese	NAA	M	95	5.42	–47	Kozuka et al.[20]
		F	97	4.62		
Tokyo citizen	AA	M	52	6.35 ± 4.04		Nishima et al.[17]
		F	49	3.9 ± 1.04		
Tokyo citizen	AA	M	104	6.9 ± 2.8	2.6–17.7	Nishima et al.[19]
		F	87	3.8 ± 1.5	1.0–7.8	
Workers in fish market	AA	M F	22	10.7 ± 5.5	2.6–25.6	Doi[11]
Sushi makers	AA	M	92	14.8 ± 6.12	–52	Nishima et al.[17]
Fish dealer	AA	M	84	19.3 ± 10.4	4.7–64.7	Nishima et al.[26]
Tuna fishermen	AA	M	58	19.9 ± 9.9	7.0–45.7	Yamanaka et al.[12]
Tuna fishermen	AA	M	63	24.4 ± 13.2	5.2–69.0	Kitamura[31]
Tuna fishermen	AA	M	37	18.9 ± 9.0	4.8–39.7	Kondo and Takehiro[30]

[a]Abbreviations: SD = standard deviation; DM = dithizone method; AA = cold vapor atomic absorption; NAA = neutron activation analysis.

TABLE 13
The mercury levels in the hair and blood of fish retailers[51]

No.	Age	Sex	Total Hg content Hair (ppm)	Total Hg content Blood (μg per 100 g)	Fish consumption Tuna g/time	Fish consumption Tuna Times/week	Fish consumption Others g/time	Fish consumption Others Times/week
1	49	M	64.7 11.0[a]	9.8 9.0[a]	200	7	1000	7
2	35	M	44.4 6.4[a]	4.9 5.3[a]	100	2–3	80	7
3	59	M	41.2	11.2	100	7	100	7
4	43	M	38.8	19.4	200	7	150	2–3
5	47	M	38.6	11.0	150	2–3	200	2–3
6	42	M	36.5	12.0	150	7	300	7
7	46	M	34.4	10.0	100	7	200	1
8	50	M	33.0	15.4	200	7	100	7
9	61	M	32.5	10.5	100	2–3	100	2–3
10	28	M	31.3	6.9	200	2–3	100	2–3
11	32	M	31.0	10.1	200	7	100	7
12	59	M	29.2	9.8	60	7	70	7
13	57	M	29.1	9.9	100	7	200	7
14	65	M	28.5	14.3	200	1	200	7
15	39	M	28.3	8.5	—	7	—	7
16	70	M	26.7	12.8	100	7	—	—
17	52	M	26.5	6.9	100	2–3	100	2–3
18	61	M	26.1	17.6	300	7	0	0
19	31	M	25.4	11.0	50	7	200	7
20	43	M	25.3	8.1	50	7	50	1
21	35	M	23.6	5.7	100	7	100	7
22	53	M	23.2	8.3	200	7	200	7
23	52	M	22.8	7.9	150	2–3	200	2–3
24	34	M	22.7	4.5	100	2–3	300	2–3
25	45	M	22.3	7.3	200	7	500	7
26	44	M	21.7	8.0	100	1	100	1
27	41	M	21.5	7.5	100	2–3	100	7
28	33	M	21.2	8.1	200	2–3	—	—
29	30	M	20.7	8.6	100	7	100–200	2–3
30	33	M	17.5	15.1	100	7	200	7
31	33	M	—	12.1	300	2–3	150	2–3
32	31	M	13.6	12.6	200	7	70	2–3
33	37	M	—	11.0	100	7	100–150	7
34	46	M	9.9	10.4	100	7	100	2–3

[a] Hair and blood materials were collected first early in March and next in May 1973.

is clear that all of the people in the towns and villages were exposed to severe environmental mercury pollution.

We can understand just how terrible the extent of environmental pollution was in those days when we consider the fact that the dithizone method for mercury analysis was used with its lower sensitivity than the more accurate atomic absorption method and the neutron activation method.

In Niigata, a close correlation between fish consumption, mercury concentration in the hair, and the outbreak of Minamata disease, was established (Fig. 10). According to investigations by Tsubaki,[34] almost all persons with hair levels of mercury of more than 200 ppm developed the symptoms of Minamata disease and were recognized officially as disease victims. The minimum level of mercury in the hair at which the development of Minamata disease symptoms has been considered possible is approximately 50 ppm, but Tsubaki emphasizes that a person with a

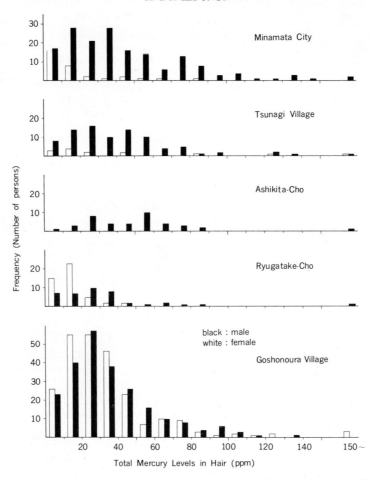

FIG. 9. Mercury levels in the hair of people living along the Shiranui Sea, 1960–1.[33]

level of more than 20 ppm should be required to avail himself of a thorough medical examination. Swedish investigators seem to agree with this conclusion.[35]

HEALTH HAZARDS IN RELATION TO MERCURY ACCUMULATION IN THE JAPANESE POPULATION

What sort of health hazards should we forecast within the context of the pervasive mercury pollution that has been shown to exist in Japan? Shiraki[4] has made a most serious prediction. He suggests that there will be a gradual lowering of mental and neurological faculties in future generations as a result of the pervasive mercury pollution of the human environment in Japan. These problems will be made more manifest in relation to the future of the Japanese people as investigations throughout the country indicate, from generation to generation, the real extent of the problem.

The dreadful story of Minamata disease has impressed on the Japanese people and the people of the world deep fear of methylmercury intoxication, but it is becoming increasingly evident through recent investigations that our knowledge in relation to Minamata disease is, indeed, limited. In 1971 the system for the official recognition of Minamata disease patients was changed in order that much more valuable clinical and epidemiological information could be added to the medical concept of Minamata disease. Many latent patients are now being recognized officially as Minamata disease victims in towns and villages along the Shiranui Sea and along the Agano River. There are now more than 800 officially recognized patients included from both regions, while fewer than 200 persons had been officially recognized as Minamata disease patients before May of 1971. It is estimated that the total number of patients will eventually exceed several thousand.

With regard to the medical aspects of the disease,

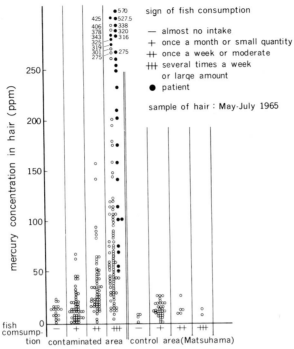

FIG. 10. Correlation between fish consumption and mercury content in human hair.[9 from 52]

various combinations of neuropsychiatric symptoms, such as a type of polyneuritis, mental deterioration, a type of symptom combined with apoplexy and a type combined with hypertension, are now included in the enlarged concept of Minamata disease as well as the classic Hunter–Russell syndrome characterized by such symptoms as visual field constriction, sensory disturbances, ataxia, and dysarthria.[36, 37] It is also considered an important aspect of the problem that there are many patients with disturbed liver function. Thus Minamata disease is not only considered to be a disorder of the central nervous system but rather an intoxication of the entire body.

Almost all of the newly recognized patients contracted the disease during its height, but there are some patients who had been taken ill several years after the most acute period. Harada in 1972 examined the surviving patients who had been examined for mercury levels 10 years ago and was able to find correlations between present symptoms and the levels of mercury in the hair samples taken 10 years ago, as indicated in Table 14. Harada concluded by indicating that there are several patients that are considered to have been taken ill in recent years, and suggests further that there are chronic cases of intoxication with patients maintaining their disease through the ingestion of fish having relatively low levels of mercury concentration over periods of 10 years or longer. Harada also indicated that there were many children with apparent mental disorders in the areas where Minamata disease was most widespread.[38] This fact is considered to have an important bearing on the warning presented by Shiraki[4] above in relation to an expanded medical concept of the disease.

The fact of the third occurrence of Minamata disease was revealed on the May 22, 1973, by the Kumamoto University Minamata Disease Research Group. On this particular occasion the disease was found among fishermen living at Ariake-cho in the Amakusa District on the Ariake Sea. Approximately 10 persons were diagnosed as showing the familiar signs and symptoms of Minamata disease. A chemical plant operated by the Nihon Gosei-Kagaku Co. is considered to be the source of this particular instance of environmental pollution. This factory had been involved in the synthesis of acetaldehyde from acetylene for a 19 year period until 1965, and in the process had discharged a considerable amount of mercury compounds into the Ariake Sea causing mercury pollution of the aquatic biomass and, in turn, the third outbreak of Minamata disease. However the mercury concentrations of fish caught in the Ariake Sea are not as high as those concentrations 10 years ago, and,

TABLE 14
The relationship between hair mercury levels before 10 years and the clinical symptoms at present observed among the people living along the Shiranui Sea[36]

Total Hg level in the hair (ppm):	–10	10–50	50–100	100–
No. of persons examined:	27	79	31	10
Sensory disturbance of peripheral nerve origin	11 (41)[a]	40 (51)	26 (84)	6 (60)
Sensory disturbance of central nerve origin	2 (7)	7 (9)	1 (33)	1 (10)
Ataxic gait	6 (22)	28 (35)	16 (52)	5 (50)
Adiadochokinesis	8 (30)	36 (46)	20 (65)	5 (50)
Dysmetria	8 (30)	23 (29)	15 (48)	4 (40)
Concentric constriction of visual field	9 (33)	23 (29)	13 (42)	4 (40)
Impairment of hearing	10 (37)	37 (47)	24 (77)	6 (60)
Dysarthria	10 (37)	31 (39)	18 (58)	4 (40)
Tremor	5 (19)	17 (22)	10 (32)	1 (10)
Contructure of extremities	5 (19)	12 (15)	5 (16)	2 (20)
Arthralgia, neuralgia	8 (30)	24 (30)	14 (45)	3 (30)

[a]Numbers in parentheses represent percent of the number of persons examined.

further, the present levels of mercury concentrations found in hair samples do not constitute evidence of mercury-related environmental pollution because so many years have passed since the offending factory was shut down. Therefore investigators must find evidence for Minamata disease through clinicoepidemiological examinations which are now under way. Another possible source of mercury contaminations in this area is a chlor-alkali factory operated by the Mitsui-Toatsu Co. in Omuta City. Although inorganic mercury is being discharged, there are several favorable conditions present in which the methylation of mercury could occur, such as larger tide differences and clean oligotrophic water in the Ariake Sea.[39] The discovery of Minamata disease on this third occasion has deeply impressed the Japanese people with the seriousness and pervasiveness of mercury-related environmental pollution as it has been revealed in section after section of the country such as in Tokuyama Bay (Yamaguchi Prefecture) and Toyama Bay (Toyama Prefecture) to mention only two places. Investigations into the seriousness of fish resource contamination and its effect on the human population are in progress.

We have little information on the mercury-related occupational health hazards of tuna fishermen, sushi workers, and fish retailers, especially those who have shown high levels of mercury contamination through hair sample analysis. The authorities in the Ministry of Public Health and Welfare have stated that no neurological disorders have been discovered in the few fishermen with hair mercury levels of more than 50 ppm that have been examined to date, but detailed observational study results related to this human universe have never been published. Medical examinations of 19 laborers working in the Central Fish Wholesale Market in Tokyo showed no neurological symptoms except for one man who had a hair sample mercury level of 25.6 ppm, and this person showed a slight retardation in the Achilles' tendon reflex.[11] The Tokyo Metropolitan Bureau of Public Health examined 34 fish dealers in the Central Fish Wholesale Market who had levels exceeding 20 ppm of mercury in their hair and 10 μg per 100 g of concentration in their blood, and of these 13 men received detailed neurological examinations that indicated no apparent symptoms related to organic mercury poisoning.[26]

On the other hand, it is estimated that there are at least a hundred thousand persons of both sexes who consume large amounts of tuna, and out of this number only about a hundred or so have received sufficient neurological examination. It is important that the warnings of Shiraki and Harada do not go unheeded. Consumers of large amounts of tuna should take these warnings to heart and should subject themselves to periodic medical examinations for at least several years in a row.

There is much argument in regard to a possible correlation between the mercury levels in hair and neurological signs and symptoms of organic mercury poisoning. It is difficult to find such a correlation since the mercury levels in hair and in the blood gradually decrease after the consumption of polluted foodstuffs has been halted. In Japan the maximum allowable concentration of mercury has been set at approximately 50 ppm for hair and 20 μg per 100 g in blood, and these figures have been derived from experience in Minamata and Niigata. An International Committee held in Stockholm recommended maximum allowable levels for professional exposure at 10 μg per 100 g total mercury in whole blood, and stated that normal blood levels for 812 people living in 15 countries ranged from 0 to 30.6 μg per 100 ml with 95% of these people having concentrations of less than 3 μg per 100 ml.[32] As a result of experience with and knowledge of Minamata disease, we feel that Japanese levels should be lowered to correspond to those recommended by the Stockholm conference. Table 15 indicates the mercury levels found in various internal organs of cats that had been living in some

TABLE 15
The mercury levels in the organs of cats living along the Shiranui Sea[40]

		Total mercury concentration (ppm)				
		Liver	Kidney	Brain	Fur	Blood
Minamata disease cats	Natural cases	54.5	12.2	8.08	52.0	10.6
		68.0	—	10.4	39.8	15.8
	Experimental cases	145.5	—	18.1	—	—
		53.5	—	8.05	—	—
		62.0	—	18.6	—	—
		47.6	15.6	10.0	70.0	—
		52.5	15.9	9.14	21.5	—
Healthy cats living along the Shiranui Sea	Yatsushiro City	31.8	2.17	2.43	51.0	1.09
		14.5	4.38	0.71	46.6	0.63
	Shiranui-Cho	23.5	1.25	—	9.8	0.95
	Amakusa-Seto	33.3	3.96	2.6	117.0	2.9
		33.4	3.52	1.08	117.5	5.2
	Ushifuka	9.0	1.83	0.83	33.1	1.4
		20.2	0.9	0.13	17.6	0.3
	Tanoura	75.2	3.64	2.9	134.2	2.12
		12.3	2.75	2.14	86.5	2.34
		29.7	3.64	1.68	39.2	1.4
	Yatsushiro-Shioya	5.4	11.7	—	8.86	0.28
Healthy cats in control area	Kumamoto City	1.18	0.82	0.05	2.2	—
	A fishing village in Oita Pref.	1.28	0.09	0.05	0.51	0.13
		1.56	0.28	0.12	3.34	—
		1.64	0.55	0.13	3.45	—
		0.99	0.25	0.09	1.91	—
	Along the Ariake sea	1.25	0.16	0.04	3.05	0.08
	Arao City	0.64	0.28	0.02	0.8	0.06
	Bay of Misumi	6.58	0.05	0.12	29.2	0.68

of the towns and villages along the Shiranui and Ariake seas as investigated by Kitamura[40] during a few years before 1960. Our attention is attracted to the fact that levels found in the brains of healthy cats living along the Shiranui Sea are higher than those of cats living near the Ariake Sea and Oita Prefecture, though it is questionable whether so-called "healthy" cats were, indeed, in good health in the strictest sense of the word. It would seem, then, to be reasonable to estimate that the levels of mercury in the brains of tuna consumers are higher than the level in the general public. Alkylmercury compounds invade the brain tissue and then remain there because they are not easily excreted by the body. This fact was reported on by Takeuchi et al.[41] and Hirota[42] from the results of autopsy analysis of Minamata disease victims and experimental animals intoxicated with methylmercury chloride. As suggested by Löfroth,[43] what we should expect is an accumulative degeneration of central nervous system cells including very young ones, without possibility of regeneration.

With regard to the accumulation and excretion relationships for methylmercury compounds, Aberg et al.[44] obtained figures which indicated a biological half-life of 70 days in the whole body of three healthy men. Kitamura[45] made estimations as to the correlation between mercury accumulation and poisoning, as shown in Fig. 11, using the biological half-life derived by Aberg et al. Birke et al.[46] examined the mercury levels in the hair and blood of 12 persons who were consuming fish contaminated with methylmercury and obtained half-life figures ranging from 65 to 250 days for hair and a half-life of anywhere from 130 to 270 days for erythrocytes and blood plasma. These values are for longer periods than those obtained by Aberg et al., and differ from individual to individual. The authors of this paper consider it very dangerous to think of biological half-life in terms of fixed and static figures, but agree rather with the orientation of Löfroth calling for a more relative understanding.

In the context of one investigation, several young tuna fishermen were found to be suffering from hypertension; the causes were considered to be unbalanced dietary life coupled with low quality drinking water, habitual drinking, and overly hard work. However, the authors of this paper believe that mercury accumulation related to the consumption of tainted tuna could very well be the cause of the above-mentioned hypertension. In animal experiments carried out by Fukuhara et al.,[47] an increase in blood vessel permeability was discovered in brain tissue following intramuscular injection of methylmercury chloride into rats. Further it must be noted here that hypertension was a rather common symptom of Minamata disease victims.[4]

The genetic effects of methylmercury compounds still remain to be elucidated though the Kumamoto University Minamata Disease Research Group could not find the evidence of genetic effects among the patients of Minamata disease.

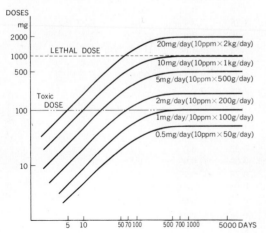

FIG. 11. Estimated accumulation of methylmercury in human body.[45]

ROLE OF WILD AND DOMESTIC ANIMALS AS INDICES OF ILL EFFECT FOLLOWING MERCURY ACCUMULATION IN FISH

It is a well known fact that such animals as cats, dogs, and crows were also victims of the disease as a result of the consumption of polluted fish prior to the outbreak of the disease in the human population in the situation in Minamata City and Niigata. In England and the United States, extensive investigations were carried out in relation to eggshell abnormality, fertility disturbance, and death due to organochloride pesticide residues found in raptorial birds. In Sweden, the death of raptorial birds as a result of the consumption of methylmercury-treated seed grain was also investigated. Careful observation of the abnormalities in wild birds and domestic animals has an important role in the prevention of possible negative effects and disease in the human population.

The authors of this paper are at the present time in the process of investigating a peculiar disease found in the cat population of large cities such as Tokyo. This disease occurs among domestic cats that are more than 3 years old and have been fed a considerable amount of fish. Such cats have been sporadically observed in Tokyo since last year. The characteristic sign of the disease are scales, ataxic gait, spastic convulsions, and paralysis of the extremities. Some of these cats have some form of sensory disturbance while others do not. We analyzed the fur of the cats for mercury level because we thought that the disease was the result of mercury accumulation given the symptoms and history. The results of this analysis of the fur for mercury content, and the correlations between the mercury levels and the ages of the cats are shown in Figs. 12 and 13. It seems that there is no direct relation between mercury in hair and the symptom. After autopsy the mercury levels in various internal organs were analyzed; and the results are given in Table 16. Also, in this case, the levels of PCB and DDT were analyzed as these materials were found in the cerebrum, the fat, and in the liver.

So far we have not been able to reach any definitive conclusions as to whether mercury is the cause or whether it is either PCB or DDT or even the combined action of all of these agents. In any case we consider it very significant and disturbing that there were three kinds of pollutants found in a

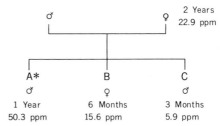

FIG. 13. Total mercury concentrations in the fur of a mother cat and the kittens of different ages. *Cat with CNS signs (August 1973).

domestic cat whose diet was entirely fish provided by its master. Further, we believe that the disease could occur in the human population given the present high level of environmental pollution in Japan. We are continuing our investigations into this problem area.

PCB, DDT, AND OTHER ORGANOCHLORIDE RESIDUES

The coastal areas around Japan are highly polluted not only with mercury, but also with PCB. At

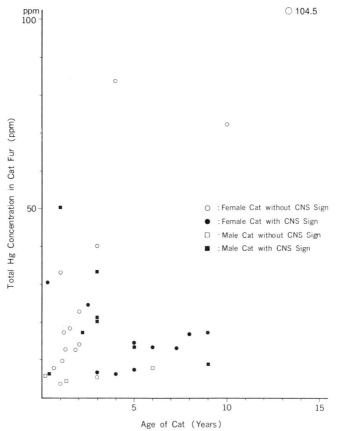

FIG. 12. Relationships between age of cat and total mercury concentration in the fur.

TABLE 16
Mercury, PCB, and DDT concentrations in the organs of a cat
Cat No. 18, ♀, 5 years
Autopsy: June 19, 1973

| Organ | Concentration (ppm) | | | Organ | Conc. (ppm) |
	Total Hg	PCB	DDT		Total Hg
Liver	14.17	1.10	1	Stomach	0.49
Spleen	2.17			Intestine	0.89
Kidney	1.12			Blood (whole)	1.33
Cerebrum	0.49	1.09	1	Muscle	1.35
Cerebellum	0.59			Lung	2.06
Fat	—	33.5	57	Fur (a)	14.7
				(b)	16.9

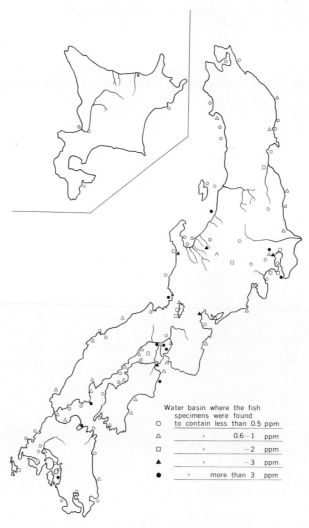

FIG. 14. PCB contamination of marine and freshwater fish in Japan, 1973.[53 from 50]

present the nationwide PCB pollution situation is as indicated in Fig. 14, this information coming out of survey work done by the Fisheries Agency of Japanese Government.

PCB accumulation is found not only in marine fish but also fish-eating birds. Doguchi et al.[48,49] detected a considerable amount of PCB, DDT, BHC, and their metabolites in dead birds found on the shore of Tokyo Bay. In regard to PCB contamination in the Japanese people, fishermen and their families register higher levels of PCB in their adipose tissue than is to be found in people of differing occupations. It is probable that fishermen and their families have higher levels of mercury, PCB, DDT, and related pollutants all in their systems at the same time, but at the time of writing this paper we have discovered no evidence of possible ill effects.

PRESENT ASPECTS OF MERCURY RELATED ENVIRONMENTAL POLLUTION IN JAPAN AND GOVERNMENT COUNTERMEASURES

The Japanese Government disregarded for a long time the occurrence of Minamata disease in both Kumamoto and Niigata prefectures and therefore had no basic policy in relation to mercury pollution of the human environment. It has been only in the last few years that the Government has begun to carry out periodical environmental surveys.

In this regard, the Kumamoto University Minamata Disease Research Group published its findings in relation to the third occurrence of Minamata disease on May 21, 1973, and with it came the possibility that Minamata disease had spread to many other parts of the country. As a result of this pressure, the Government was compelled to begin seriously working out a plan of action in relation to the possibility of pervasive mercury pollution. On June 24 of that year it published guidelines on permissible limits for mercury concentrations in fish along with a tentative schedule of allowable intake of methylmercury tainted fish. Along with this action, on June 25 a decision was made to perform immediate surveys in nine water basins in Japan, i.e. in Minamata Bay, Yatsushiro Sea, Ariake Sea, Tokuyama, Arihama, Mizushima, Himi, Uozu, and the port of Sakata. The results of this survey should be published soon. The Government further decided to carry out extensive surveys for mercury, PCB, cadmium, lead, and BHC by the end of this year at 8000 locations throughout Japan.

In spite of this governmental counteraction there is no reason to expect that the present conditions of environmental pollution in Japanese waters will improve in the near future. Further, there is no reason to believe that the high level of mercury contamination in the Japanese people will be changed in any way simply by the expediency of setting limits on the mercury consumption through guidelines on fish consumption. These guidelines are as follows. (1) Permissible levels for fish are less than 0.4 ppm of total mercury and less than 0.3 ppm for methylmercury. Fish containing mercury above these levels cannot be sold. The exception to this rule is in the sale of tuna, marlin, swordfish, and freshwater fish with little marketability. Recently six other species of fish were added to the exception for the reason that the consumption of these species is quite small though these fishes contain mercury higher than the permissible levels. (2) The allowable level of methylmercury intake for an adult of 50 kg of weight is 170 μg/week.

These limitations and permissible levels have little meaningful significance because most of the fish commonly consumed by the Japanese people are of small to medium size and thus contain mercury levels far lower than indicated in the Government guidelines. Also tuna, marlin, and swordfish are not subject to the application of these governmental standards. The allowable levels of methylmercury intake were determined on the basis of a combination of information derived out of animal experiments with monkeys, allowable intake levels derived from research done by the Kumamoto University Minamata Disease Research Group, and levels of permissible intake published by the World Health Organization. Typical cases of methylmercury poisoning will not occur if all of the people can keep strictly to the allowable intake levels, and through this the higher mercury levels in fishermen, sushi workers, and fish retailers would be lowered. But for the general public these guidelines are meaningless, for they will not have any lowering effect on the mercury contamination level in the general population because the daily intake of fish per man is about 100–150 g of small size fish with relatively low mercury levels, which results in the absorption of about 50–70 μg/week of methylmercury.

Moreover, there have been many instances that make it very difficult for us to believe the statements of the authorities in the central government related to environmental pollution. It is undeniable that the extreme secrecy and endless bureaucracy of the Government have made it difficult to get meaningful information as to basic conditions relevant to environmental pollution in Japan. There were many instances of problems related to restriction on the availability of information in the compilation of this report.

REFERENCES

1. KUMAMOTO UNIVERSITY MINAMATA DISEASE RESEARCH GROUP, *Epidemiological, Clinical and Pathological Studies on Minamata Disease after 10 Years*, Vol. II, 1973.
2. *Annual Report of Consumption of Non-iron Metals*, Ministry of Trade and Industry of Japan, 1972.
3. GRANHALL, *Oikos*, Suppl. **9**, 23 (1972).
4. SHIRAKI, H., Mercury pollution in Japan, *Res. Environ. Disrup.* **2** (3), 1–27 (1973).
5. MILLER, G. E., GRANT, P. M., KISHORE, R., STEINKRUGER, F. J., ROWLAND, F. S., and GUINN, V. P., Mercury concentration in museum specimens of tuna and swordfish, *Science* **175**, 1121–1122 (1972).
6. INTER-AMERICAN TROPICAL TUNA COMMISSION, *Mercury in Tuna, A Review*, I-A TTC Background Paper No. 4, October, 1972.
7. *Annual Report of Fishery*, Ministry of Agriculture and Forestry of Japan, 1972.
8. AOKI, H., Environmental contamination by mercury: II, Inorganic and organic mercury in river bottom mud and river fish, *Jap. J. Hyg.* **24** (5, 6), 546–555 (1970).
9. UI, J., Mercury pollution of sea and fresh water, its accumulation into water biomass, *Rev. Intn. Oceanogr. Med.* **22-23**, 79–128 (1971).
10. *Mercury Levels in Marine Fish*, Tokyo Metropolitan Bureau of Public Health, August 29, 1973.
11. DOI, R., Environmental mercury pollution and its influence in the cities of Japan, *Annual Rep. Tokyo Metropol. Res. Instit. Environ. Protection*, **3**, 257–261 (1973).
12. YAMANAKA, S., SUZUKI, M., KONDO, T., and UEDA, K., Mercury concentrations in tuna and in the hair of tuna-fishermen: II *Jap. J. Hyg.* **27** (1), 117 (1972).
13. NISHIGAKI, S., TAMURA, Y., MAKI, T., YAMADA, H., TOBA, K., SHIMAMURA, Y., and KIMURA, Y., Investigation on mercury levels in tuna, marlin and marine products, *Annual Rep. Tokyo Metropol. Res. Lab. Public Health*, **24**, 239–248 (1973).
14. HAMMOND, A. L., Mercury in the environment: Natural and human factors, *Science* **171**, 788–789 (1971).
15. UKITA, C., Intoxication and public hazard by mercury compounds, *Kagaku* **36** (5), 254–258 (1966).
16. AOKI, H., Environmental contamination by mercury: III, Inorganic and organic mercury in human hair and marine fish, *Jap. J. Hyg.* **24** (5, 6), 556–562 (1970b).
17. NISHIMA, T., IKEDA, S., TADA, U., YAGYU, H., NAGASAKI, M., UDO, R., and KAMIJYO, H., Mercury, lead and cadmium contents in hair of residents in Tokyo metropolitan area, *Annual Rep. Tokyo Metropol. Res. Lab. Public Health* **23**, 277–282 (1971).
18. DOI, R. and SHIMIZU, M., Mercury accumulation in tuna: II, *Kagaku* **43** (7), 436–442 (1973).
19. NISHIMA, T., TADA, U., YAGYU, H., and NAGASAKI, M., Mercury levels in hair and blood, *Jap. J. Hyg.* **28** (1), 58 (1973a).
20. KOZUKA, H., Activation analysis of human head hair: III, Factors having influence on the trace elements contents in hair, *J. Hyg. Chem.* **18** (1), 7–12 (1972).
21. PERKONS, A. K. and JERVIS, R. E., cited from Methylmercury in fish, a toxicologic-epidemiologic evaluation of risks, *Nord. Hyg. Tids.* Suppl. **4**, 222 (1965).
22. COLEMAN, R. F., CRIPPS, F. H., STIMSON, A., SCOTT, H. D., and ALDERMASTON, A. W. R. E., The trace element content of human head hair in England and Wales and the application to forensic science, *Atom Monthly Information Bull. of the United Kingdom Atomic Energy Authority*, **123**, 12–22.
23. WAKATSUKI, T., Pollution by pesticides and insecticides, *Res. Environ. Disrup.* **2** (3), 28–36 (1973).
24. FUJITA, M. and IWASHIMA, K., Residual mercury compounds in rice; *Bull, Inst. Publ. Hlth.* **19** (3), 213–220 (1970).
25. TAGUCHI, Y., Studies on microdetermination of total mercury and the dynamic aspects of methyl mercury compound *in vivo*: Part II, Behavior of low concentrated methyl mercury compound *in vivo, Jap. J. Hyg.* **25** (6), 563–577 (1971).
26. NISHIMA, T., IKEDA, S., YAGYU, H., NAGASAKI, M., UDO, R., KAMIJYO, H., MIYOSHI, C., and NAKAMURA, Y., Studies on the influence of consumption of fish to the total mercury concentration in hair and blood, *Jap. J. Publ. Hlth.* **20** (10), 456 (1973b).
27. NOSE, K. and SUZUKI, T., cited from *Ecosystem and Pesticides* (by Yushima, T., Kiritani, K., and Kanazawa, J.), Iwanami-Shoten Co., 1973, 1967.
28. YUSHIMA, T., KIRITANI, K., and KANAZAWA, J., *Ecosystem and Pesticides*, Iwanami-Shoten Co., 1973.
29. KANAZAWA, J., cited from *Ecosystem and Pesticides*, 1971.
30. KONDO, J. and TAKEHIRO, S., Mercury contamination in tuna, *Jishu-Koza*, No. 24, 1–17 (1973).
31. KITAMURA, M., cited from Kondo and Takehiro.[30]
32. BERLIN, M. H., CLARKSON, T. W., FRIBERG, L. T., GAGE, J. C., GOLDWATER, L. J., JERNELOV, A., KAZANTZIS, G., MAGOS, L., NORDBERG, G. F., RADFORD, E. P., RAMEL, C., SKERFVING, S., SMITH, R. G., SUZUKI, T., SWENSSON, A., TEJNING, S., TRUHAUT, R., and VOSTAL, J., Maximum allowable concentrations of mercury compounds, report of an International Committee, *Arch. Environ. Hlth.* **19**, 891–905 (1969).
33. MATSUSHIMA, G., MIZOGUCHI, S., DOI, S., and CHIYO, S., *An Investigation on the Mercury Levels in Hair with Regard to the Minamata Disease*, Report of the Kumamoto Instit. Health, May 1961.
34. TSUBAKI, T., SHIRAKAWA, K., and HIROTA, K., Clinical and epidemiological investigations of Niigata Minamata Disease, 68th Annual Meeting of the Society of Psychiatry and Neurology, Tokyo, 1971.

35. Methyl mercury in fish: a toxicologic-epidemiologic evaluation of risks, report from an expert group, *Nord. Hyg. Tids.*, Suppl. 4 (1971).
36. HARADA, M., Clinical and epidemiological studies of Minamata disease 16 years after onset, *Adv. Neurol. Sci.* **16** (5), 870–880 (1972).
37. SHIRAKAWA, K., HIROTA, K., KANBAYASHI, K., and TSUBAKI, T., Epidemiological and clinical aspects of Minamata disease in Niigata, *Adv. Neurol. Sci.* **16** (5), 881–891 (1972).
38. HARADA, M., personal communication.
39. JERNELOV, A., The changing chemistry of the oceans, *Proc. 20th Nobel Symposium*, pp. 161–169, 1971.
40. KITAMURA, M., cited from *Minamata Disease*, Minamata report, by Study Group of Minamata Disease, Univ. Kumamoto, 1966.
41. TAKEUCHI, T., ETO, M., KOJIMA, H., OTSUKA, Y., MIYAYAMA, T., SUKO, S., SAKAI, T., SAKURAMA, N., IWAMASA, T., and MATSUMOTO, H., Minamata disease and the pathological changes in Minamata disease victims after the lapse of 10 years, *Nihon Iji-Shinpo*, No. 2402, 22–27 (1970).
42. HIROTA, K., Distribution of methyl mercury in human brain and rat organs and the effect of d-Penicillamine on mercury in rat organs, *Clin. Neurol.* **9** (10), 592–601 (1969).
43. LÖFROTH, G., Methylmercury: a review of health hazards and side effects associated with the emission of mercury compounds into natural systems, *Ecolog. Res. Comm. Bull.*, No. 4, 1–38 (1969).
44. ABERG, B., EKMAN, L., FALK, R., GREITZ, U., PERSSON, G., and SNIHS, J. O., Metabolism of methyl mercury (^{203}Hg) compounds in man, *Arch. Environ. Hlth.* **19**, 478–484 (1969).
45. KITAMURA, M., *Trace Heavy Metals in Foodstuffs: From the Viewpoint of Public Health*, textbook for the 1971 special training course for food hygiene, Ministry of Welfare of Japan, 1971.
46. BIRKE, G., JOHNELS, A. G., PLANTIN, L. O., SJOSTRAND, B., SKERFVING, S., and WESTERMARK, T., Studies on humans exposed to methyl mercury through fish consumption, *Arch. Environ. Hlth.* **25**, 77–91 (1972).
47. FUKUHARA, N., OGUCHI, K., and TSUBAKI, T., On the pathological alteration in the blood-brain barrier in the experimental organic mercury poisoning, *Adv. Neurol. Sci.* **14** (2), 313–320 (1970).
48. DOGUCHI, M., USHIO, F., and NISHIDA, K., Pesticide residues in wild birds in Japan, *Annual Rep. Tokyo Metropol. Res. Lab. Public Health* **23**, 265–270 (1971).
49. DOGUCHI, M., USHIO, F., and ABE, M., Polychlorinated biphenyl in fish and birds captured in Tokyo Bay and its surrounding place, *Annual Rep. Tokyo Metropol. Res. Lab. Public Health* **23**, 271–276 (1971).
50. ISONO, N., Mercury is covering the whole Japan, *Asahi J.* **8**, 24 (1973).
51. Press release on mercury concentration in fish and the health effect of mercury in fish by the Tokyo Dept. of Public Health, June 21, 1973.
52. HOSHINO, O., TANZAWA, K., HASEGAWA, Y., and UKITA, T., Differences in mercury content in the hairs of normal individuals depending on their home environment, *J. Hyg. Chem. Soc. Jap.*, **12**, 90–93 (1966).
53. Press release on PCB concentration in fish, Fishery Agency of Japanese Government, June 4, 1973.

The Distribution of Mercury in Fish and its Form of Occurrence
(Rikuo Doi and Jun Ui)

DISCUSSION by FRANK M. D'ITRI

Associate Professor of Water Chemistry, Institute of Water Research, and Department of Fisheries and Wildlife, Michigan State University, East Lansing, Michigan 48824

and

PATRICIA A. D'ITRI

Associate Professor of American Thought and Language, Michigan State University, East Lansing, Michigan 48824

By offering new information on the subclinical symptoms of mercury poisoning among fishermen, on the continued mercury contamination of rice paddies despite the 1968 ban on mercury-containing fungicides in Japanese agriculture, and on the changing standards for mercury allowed in fish, Drs. Ui and Doi raise questions for other industrialized countries with known mercury problems such as Sweden, Canada, and the United States. After the initial standards were set for the amount of mercury to be permitted, usually based on the mythical average citizen's daily intake of fish, these standards must be critically re-examined in light of the long range subclinical effects of mercury on the Japanese fishermen and fish sellers who eat large quantities of fish as well as individuals with unusual sensitivity to alkylmercurialism. Related problems are how to educate the population to the hazards and overcome the pattern of ignorance, secrecy, and evasion that Dr. Ui has reminded us have frequently been associated with the mercury problem in Japan.[1-3] Dr. Ui has noted the reticence with which people have faced the unpleasant and costly errors of hindsight, and has occasionally offered impassioned pleas for more objective research and evaluation.

Although problems of economic deprivation and bureaucratic evasion such as those that Dr. Ui has courageously pointed out in Japan can, in our opinion, be observed in all industrialized countries, the most significant social issue stems from Dr. Ui's information that latent Minamata disease patients such as those first described by Takeuchi[4] are now being officially recognized. The current official total is more than 800 patients, and the projected number may exceed several thousand, compared with only 200 before May of 1971.

On the basis of Dr. Ui's report, it is appropriate to examine circumstances in the United States and Canada where parallels might be drawn or new questions be raised for examination. Most of the people stricken with Minamata disease were poor fishermen and their families who consumed as many as three fish meals daily, much more than the average American or Canadian. One commercial fisherman on Lake St. Clair announced that he had regularly eaten local fish for many years and, if anything, the mercury had made him more healthy. In fact, however, he probably had a much more varied diet than people in lower economic categories who have access to fish as a diet staple. And among the most economically depressed are numerous Indian tribes that often extract a major portion of their livelihood from the waterways. For instance, the Ojibway Indians who live on the Wabigoon–English River system in western Ontario previously ate substantial amounts of fish. And downstream from the city of Dryden, Ontario, the mercury levels are extraordinarily high. In 1970, burbot tested as high as 24.8 ppm, pike to 27.8 ppm, and walleyes to 19.6 ppm on a wet weight basis.[5] These levels of mercury are comparable with the 20–30 ppm in fish and shellfish taken from Minamata Bay.[6] Yet fish in the English River system were much sought after by both sports and commercial fishermen prior to 1970. Then the Canadian Government set a limit of 0.5 ppm mercury in fish to be released for sale. After a "fish for fun" policy was instituted, some fishing camps on the

Wabigoon–English River system closed, and business decreased dramatically among those that remained open with the encouragement of the Government.

The Ojibway Indians from the local reservation at Grassy Narrows near Kenora, Ontario, had always guided the sport fishermen during the approximately 6 month long summer fishing season. They ate fish at least once a day when they cooked a shore lunch for the fishing parties. Thus, like the Japanese fishermen, their line of work subjected them to an unusually high rate of fish consumption. And at the end of the season, when the Indians returned to the reservation for the winter, the family diet included wild game and meat purchased with unemployment checks. But fish are still the most readily available source of food when the welfare money runs out. The number of game animals and birds to be hunted or trapped have declined substantially in recent years.

Because of their high fish consumption both during and after the fishing season, the blood levels of mercury in the Indians were tested in 1970. But the Ontario provincial government did not notify the Indians of their results. Other blood tests were conducted in the fall of 1972. This time some of the Indians tested were at a hospital in the city of Sioux Lookout, north of the polluted English River chain. The first public confirmation of high mercury levels in a Grassy Narrows Indian came in August 1972, when the Indians finally prevailed upon provincial officials to test the mercury levels in an Indian who had died of a heart attack.[7] The levels were so high that health officials decided the samples must be contaminated. Subsequently, the Indians mounted a campaign to obtain their test results from blood samples taken 3 years earlier. After eating fish caught in local waters containing as much as 50 times the maximum allowable mercury levels—information that had not been readily released either—some of the Indians learned that they had as much as 300 ppb of mercury in their blood.[8] Dr. G. J. Stopps, of the Ontario Ministry of Health, declared that some of them should be dead, based on the fact that some Japanese fishermen who died had blood levels from 144 to 200 ppb of mercury at autopsy. Canadian federal and provincial authorities recognized a 50 ppb safe level of mercury in blood. Not notifying the Indians of their mercury levels for 3 years prompted some unidentified government health officials to charge the Government with a form of "genocide by neglect" and creating "a modern day horror story."[8] To date, only four Cree Indians have been publicly reported to be suffering from mercury poisoning. They lived on a reservation downstream from a pulp mill in Quebec Province.[9]

Like the Indians, guests at the fishing camps are not apt to be fully informed either. Dr. Stopps notified camp owners in 1972 and 1973 that fish from the Wabigoon–English River system definitely should not be eaten. He stated that by eating only one 7 oz fish fillet per day for 3 weeks, the sports fishermen could raise their blood levels of mercury to the range at which people were poisoned at Minamata. And a year after a 2 week fishing trip a guest could still have a 20% increase over normal mercury levels in his blood. Thus, an American fisherman who spent his vacations at a fishing camp could continue to elevate the mercury in his blood from one year to the next even if he had no other exposure in the interim.[10] Despite this, the warning signs that were posted in 1970 have been removed, and camps that remained open under the "fish for fun" policy have received government aid to continue in business. With their increasing debt it is not surprising that camp owners are underplaying the mercury warnings and assuring potential guests that the mercury problem has virtually disappeared. Consequently, once again the shore lunches are being served and eaten by Indians who guide the tourists. And back on the reservation at Grassy Narrows, Ojibway Indian Chief Assin says many of the Indian people will continue to eat the fish because "Some of my people know too much about being hungry and not very much about this mercury;" and the Chief adds: "What else are they going to do?"[8] With millions of dollars jeopardized or lost because of the decline in sports as well as commercial fisheries, it is not surprising that mercury pollution has increasingly tended to be pushed to the back pages of the newspapers.

While it is still possible to say that no Indians have died of mercury poisoning, perhaps because none without a known cause of death have been autopsied, the likelihood of subclinical symptoms cannot be ruled out. While some Indians have experienced changes in personality that led to depression and aggression, these might be attributed to the hopelessness of their situation. And the Indians' greater propensity toward violence and alcoholism cannot be traced to mercurialism either. The rate of death by violence among Indians in the Kenora, Ontario, district has consistently been one of the highest in Canada. In 1971, 13 deaths caused by violence and alcohol were reported at the Grassy Narrows Reservation alone.

As violence and alcoholism may not be related to the elevated mercury levels any more than the milder symptoms of depression and aggression, Kenora also has a higher than average rate of mental disorders among children. Drs. Ui and Doi reported that many children with apparent mental disorders live where Minamata disease was most widespread. Similarly, in Kenora, Ontario, where a new addition to the jail accommodates the increasing numbers of drunken and disorderly Indians from the reservation, schools for the mentally retarded are also being expanded. This city of 11,000 people has facilities for mental illness among persons of all ages since a recent addition for retarded preschool children. And 10 times more infants died of pneumonia and lung disease in Kenora than the provincial average in 1971 while the total number of infant deaths in the Kenora district was double the provincial average—30.5 per thousand compared with an average of 15.8 per thousand.[11] No one knows if the symptoms of mental illness are related to mercurialism, but sufficient evidence exists to warrant autopsies to determine if subclinical symptoms are present. Such pathological studies of brain tissue proved the existence of latent Minamata disease in Japan and could easily be conducted on Ojibway Indians who have died. Moreover, while Takeuchi volunteered to conduct these studies more than a year ago, the local health officials have not made the necessary tissue available.[12]

More serious than identifying this immediate issue is Professor Shiraki's prediction, quoted by Drs. Ui and Doi, that future generations could experience a gradual lowering of mental and neurological faculties as a result of the pervasive mercury pollution in Japan. Shiraki noted that such problems could be made more manifest as investigations throughout the country indicate the real extent of the mercury problem in subsequent generations.[13] This might appear as a grim and far-fetched prediction here, but at this conference Dr. Clarkson stated that the medical implications of the 1972 Iraqi epidemic will not be known for many years. The Iraqis must now live with that reality.

Overt tragedies like Minamata and Iraq cannot be totally suppressed. Nonetheless, secrecy and deception are two recurring themes that are played over and over. And the pattern has many similarities from one country to another. First of all, most of the tragedies have been visited upon the economically deprived—people with the least choice of food sources regardless of whether or not they have been informed of the possible hazards. And the secrecy often begins with the afflicted. Reports from Minamata indicate that the poverty stricken fishermen early kept their illness to themselves in the misguided belief that they suffered from a disease that would bring shame upon the families. And 3 years elapsed before the disease attracted the attention of physicians at the chemical plant in 1956. Similarly, 10 years later the poor farmers of Guatemala were reluctant to admit that poverty had driven them to eat the alkylmercury-treated seed grain. And again in the recent epidemic in Iraq, economics were combined with secrecy and fear.[14,15] Warnings were printed on the bags of treated seed grain in Spanish, not much help to peasants who spoke Arabic and often were illiterate. Moreover, when the extent of the tragedy was discovered, the Government warned that people caught trafficking in treated grain would receive the death penalty. Consequently much of the grain that was not recovered was reported to have been dumped into the River Tigris where it could add to the water contamination. Although the Iraqi Government enlisted the aid of foreign experts to help their victims, they also imposed a news blackout on the press to prevent further publicity. Similarly, some New Mexico farmers continued to sell hogs that had been fed waste alkylmercury-treated grain, even after they were warned of the possible danger, rather than assuming the economic loss if the hogs were confiscated.[16]

Ignorance and poverty combined to cause the tragic mercury poisoning epidemics where treated seed grain was misused as food in Iraq, Pakistan, and Guatemala. The Japanese fishermen and the Alamogordo family were indirectly poisoned by contaminated fish and meat, but secrecy is a common theme in all of these occurrences. Nonetheless, it is too simple to suggest that adequate warnings, or even the assurance of more varied diets, could prevent a recurrence. The victims themselves are only the base of a pyramid of economic and political evasions. As the hog farmers in New Mexico continued to sell their animals even after being warned, sports fishermen do not want to be denied the pleasure of frying their daily catch over the camp fire, and it is only natural that resort owners and commercial fisherman would want to stay in business. When the Tennessee River was closed to fishing, the United Auto Workers Union instituted a lawsuit against the polluting industries on the grounds that the workers had been deprived of recreational opportunities. A similar legal action was begun in Michigan against the governor for denying the fishermen their civil rights to take fish after they had been adequately warned of the danger. Consequently, a compromise was offered. The Michigan Department of Public Health would issue health warnings, and fishing was reinstituted.

Since 1971 the Michigan Department of Public Health has issued a warning that selected species of fish in the waters of the St. Clair river, the Detroit river, Lake St. Clair, and some portions of lakes Huron and Erie contain potentially dangerous levels of mercury and can cause serious illness if eaten regularly over a long period of time. Moreover, children and pregnant women are especially advised to avoid eating any of the listed fish. This health warning was posted at all of the approaches to these polluted waters and was published in the 1972 *Rules for Fishing in Michigan*. This publication contains the fishing laws, and is issued to each fisherman with the purchase of a fishing license.[17] Although this warning from the Michigan Public Health Department continues in effect, many of the signs have been removed from the fishing areas, and the health warning has been deleted from the 1973 and 1974 editions of *Rules for Fishing in Michigan*. While such compromises may be appropriate in some circumstances, they should be made on the basis of a careful evaluation of all of the possible effects including subclinical as well as clinical symptoms of mercury poisoning rather than because of economic and political pressures. The Michigan Department of Public Health is presently conducting experiments to determine the subclinical effects of mercury under the supervision of Dr. Harold Humphrey.

Compromises on a larger scale have been the hallmark of industrial pollution. In Japan Minamata disease was first recognized by a factory health official in 1956, but the information was not made public for several years until a research team from Kumamoto University laboriously developed the answers independently. By then a few isolated cases had developed into an epidemic. And in another political gesture, the workers helped the employers at the Chisso Co. preserve their secret to protect the factory and their jobs. Moreover, once the cause of poisoning was recognized, no precedents were available in environmental law to enable the victims to claim damages. The modest settlements finally obtained were long in coming. The Government has been reluctant to move against powerful industries such as the Chisso Co. that paid half of the local city taxes.

Government inaction and industrial emphasis on profit have begotten similar compromises in the Western Hemisphere. In the United States and Canada the wood pulp industry continued to discharge wastes into the rivers for up to 50 years under the threat that reducing the discharge would be so unprofitable that the plants would have to close and people would be unemployed. Consequently, when the mercury pollution scare was first raised on the Wabigoon–English River systems, the mayor of Dryden, Ontario, downplayed the problem on the grounds that pulp wastes had prevented fishing downstream from Dryden for many years anyway.

And in the United States, where elevated mercury levels were reported in waterways in 1970, both industry and government initially announced a campaign to reduce the mercury discharge to zero. Then the industries quickly decreased their mercury discharges, demonstrating both a current spirit of cooperation and a prior contempt for the environment because the technology was available but had not been utilized. The compromises began in 1970 after the Department of Health, Education, and Welfare requested that the Justice Department take some of the polluting industries to court. Then some industries closed and others announced that eliminating all mercury discharges into waterways could be so expensive that they would have to close and workers would lose their jobs. The litigation was continued while compromises between the companies and the Justice Department permitted a release of 8 oz of mercury into the waterways daily until means of eliminating all pollution could be developed. While the cases were in litigation, another veil of secrecy was promulgated. Congressional subcommittees conducted by Senator Hart and Representative Reuss could not obtain information about mercury polluters and litigation from the Departments of Interior and Justice.[18] Finally, the Justice Department provided the House Committee only with the information that the companies themselves were willing to divulge. At least one company refused to explain its plans for pollution abatement. Therefore, the House Committee presumably was supposed to remain ignorant— hardly an appropriate basis for legislative leadership. Thus, from the ignorant and poverty stricken sufferers to the most important legislative bodies in the United States, a lack of willingness to put the cards on the table concerning mercury pollution has been a distinguishing feature. And it is now appropriate that scientists who have aggressively sought answers to questions about the effect of mercury on the environment and its inhabitants should also critically examine the standards provided to offer a margin of safety. Not only must adverse effects from mercury contamination be prevented now, but aggressive research must continue to find means of restoring and maintaining the environment.

If this discussion seems rather blunt and to the

point, that was intentional. If it stimulates discussion, so much the better. As scientists, educators, members of the industrial community and of regulatory agencies, we are now obliged to work together to find solutions to the numerous problems that relate—not only to mercury pollution—but all forms of pollution. Therefore, all data must be made freely available so that recommendations for action can be based on what is best for all of the citizens all of the time.

REFERENCES

1. UI, J. and KITAMURA, S., Mercury pollution of sea and fresh water—its accumulation into water biomass, *J. Fac. Eng. Univ. Tokyo*, Ser. B, **31**(1) 271 (1971) (in English).
2. UI, J., Minamata disease and water pollution by industrial wastes, *Rev. Intn. Oceanogr. Med. 13/14*, 37 (1969); *Blol. Abstr.* **51**, 14699.
3. UI, J. and KITAMURA, S., Mercury in the Adriatic, *Marine Poll. Bull.* **2**, 56 (1971).
4. TAKEUCHI, T., Approaches to the detection of subclinical mercury intoxications: experience in Minamata, Japan, in *Environmental Mercury Contamination* (Hartung, R. and Dinman, B. D., eds.), Ann Arbor Science Publishers Inc., Ann Arbor, Michigan, 1972.
5. FIMREITE, N. and REYNOLDS, L. M., Mercury contamination of fish in Northwestern Ontario, *J. Wildl. Mgmt.* **37**, 62 (1973).
6. KITAMURA, S., Determination on mercury contents in bodies of inhabitants, cats, fishes and shells in Minamata District and in the mud of Minamata Bay. In *Minamata Disease*, Study Group of Minamata Disease, Kumamoto University, Japan, 1968.
7. BURTON, L., Indians press for inquest, *Kenora Miner and News*, November 23, 1972.
8. LEE, D., Indians threatened, *Winnipeg Free Press*, March 10, 1973.
9. Anon., Quebec mercury poisoning send four Indians to hospital, *Toronto Daily Star*, June 16, 1971.
10. STOPPS, G. J., personal communication to Colin Myles of Doug. Hook's Camps, February 28, 1972.
11. Anon., Kenora birthrate highest but infant deaths soar, *Kenora Miner and News*, September 23, 1971.
12. TAKEUCHI, T., personal communication to Frank M. D'Itri, October 1972.
13. SHIRAKI, H., Mercury pollution in Japan, *Res. Environ. Disrup.* **2**, 1 (1973) (in Japanese).
14. HUGHES, E., Pink death in Iraq, *Readers Digest* **103** (619) 134 (1973).
15. DAMLUJI, S. F. and TIKRITI, S., Mercury poisoning from wheat, *Br. Med. J.* **1**, 804 (1972).
16. ALPERT, D., LIKOSKY, W. H., FOX, M., PIERCE, P. E., and HINMAN, A. R., Organic mercury poisoning—Alamogordo, New Mexico. Viral Diseases Branch, Epidemiology Program Communication, Public Health Service, Report No. HSM–NCDC–Atlanta EPI–70–47–2, March 27, 1970, 10 pp.
17. Anon., *Rules for Fishing in Michigan*, Michigan Department of Natural Resources, Lansing, Michigan, 1972.
18. REUSS, H. S. (Chairman of the Subcommittee) Presiding, Mercury Pollution and Enforcement of the Refuse Act of 1899 (Part 2). Hearings before a subcommittee of the Committee on the Government Operations, House of Representatives, 92nd Congress, First Session, October 21 and November 5, 1971, p. 1132.

DISCUSSION (F. M. and P. A. D'Itri)

COMMENT by Jun Ui

The general tendency of mercury accumulation related to the degree of organic pollution—The discussion in this meeting gave me much information on the role of sediment in the process of accumulation of mercury into the fish. After hearing such valuable discussion, I still feel the comparison in the table below might be somewhat useful to judge the future of mercury and other accumulating pollution. All figures in this table are round numbers for comparison.

The general conclusion from this comparison is that in a cleaner environment with less organic pollution, biological accumulation of mercury pollution can occur much more easily with a small amount of mercury than in a dirty area heavily polluted with organic material. In other words, considering this condition, we must be much more careful when discharging certain pollutants into a clean environment. For instance, Canada and Scandinavian countries have a more delicate environment than Japan and other industrialized European countries. For the United States, the condition may vary from place to place. Of course, this comment does not intend to recommend the promotion of organic pollution in any case.

Place	Hg input (kg/day)	Organic pollution	Hg in fish (ppm)	Disease
Tokyo Bay	200	++	0.1	−
Minamata	20	±	10	+
Stockholm	2	−	1	±
Odda (Norway)	0.5	−	1	?
Ravenna (Italy)	20	++	1–2	−
Indian Ocean tuna	0	−	1	±

ENVIRONMENTAL LEAD DISTRIBUTION IN RELATION TO AUTOMOBILE AND MINE AND SMELTER SOURCES†

GARY L. ROLFE
Assistant Professor of Forest Ecology and Environmental Studies, University of Illinois

and

J. CHARLES JENNETT
Associate Professor of Civil Engineering, University of Missouri–Rolla

INTRODUCTION

Numerous investigators in the past few years have studied the distribution of lead in various environmental components in relation to sources. Typical of these are studies of the distribution of lead in soils and plants in relation to highway sources such as those of Chow,[1] John,[2] Lagerwerff and Specht,[3] Motto et al.,[4] and Page et al.[5] However, few studies have involved the ecosystem concept in which the distribution of lead is studied in an entire system with each component being studied in relation to the remainder of the ecosystem.

The objective of this paper is to discuss two systems studies of environmental contamination by lead: (1) the distribution of lead in an ecosystem with the major source being the combustion of leaded gasoline in automobiles; and (2) the distribution of lead in an ecosystem where the major sources are mines and smelters. The first is an interdisciplinary study involving a team of approximately 30 scientists at the University of Illinois at Champaign-Urbana. The second also involves a large interdisciplinary group of scientists, and is being conducted by a team from the University of Missouri at Rolla. Both projects are supported by the National Science Foundation RANN program.

†Data presented in this paper are preliminary in nature and should be interpreted as such. It is a summary of research completed to date by two interdisciplinary teams of scientists comprised of Drs. L. L. Getz, A. Haney, L. Larimore, H. Leland, P. Price, and G. Rolfe, and Mr. J. McNurney at the University of Illinois, and Drs. J. Charles Jennett, E. Bolter, N. Gale, D. Hemphill, K. Purushothaman, and B. Tranter at the University of Missouri.

AUTOMOBILE SOURCE

Background

The system under study by the Illinois team is an 86 square mile ecosystem based on a watershed located in central Illinois. The watershed lies primarily to the north of Champaign-Urbana and is predominantly rural agricultural in nature. However, the watershed also includes approximately 90% of the cities of Champaign and Urbana, and thus does include a significant urban component. Major crops of the rural area are corn and soybeans. Woodlands, pasture, and wasteland such as railroad right-of-ways and stream margins are also included in the watershed area.

The rural portion of the watershed is drained by the Saline Branch of the Vermillion River and the urban portion by the Boneyard Creek. The streams join on the east edge of Urbana and flow out of the ecosystem to the east.

The objectives of the Illinois study include characterizing the input, accumulation, and output of lead from automobile sources in a typical midwestern ecosystem, understanding the mechanisms controlling fluxes between system components and evaluating potential effects of lead on ecosystem components other than man.

System inputs

Gasoline consumption represents the major source of lead input to the watershed under study. To determine the magnitude and the spatial and temporal variations of gasoline consumption, a traffic monitoring system with road tube traffic counters was established in January 1972. All major roads in the watershed were included in the network.

Combining the accumulated traffic volume information with documented data on automobile gasoline consumption, a reliable estimate of the magnitude as well as the spatial and temporal consumption of gasoline is obtained.

In order to convert the gasoline consumption data to lead emissions in the ecosystem, the following assumptions were made:

(1) 2.5 g of lead per gallon of gasoline.
(2) In urban areas, due to the nature of the traffic flow, 50% of the consumed lead is considered to be emitted from the exhaust system of the automobile.
(3) In rural areas with generally higher vehicle speeds, 80% of the consumed lead is emitted.

Total lead input to the watershed on a yearly basis is shown in Fig. 1. The 12 square mile urban compartment receives approximately 75% of the total lead input, while the 74 square mile rural area receives only 25%.

If the system is divided into zones based on traffic volume, the relative concentration factors for lead inputs to the various zones can be calculated. These are shown in Table 1 with zone I being a 50 m strip of land on both sides of highways with traffic volumes greater than 4000 vehicles per 24 h; zone II being the same width strip along highways with traffic volumes less than 4000 vehicles per 24 h; zone III being rural areas remote from highways; and zone IV being the urban area. The relative concentration factors for Zones I and IV are very high, and indicate areas of concern due to the tremendous lead input per unit area.

TABLE 1
Relative concentrations of lead by zones

	Vehicles/day	% total area	Relative concentration
Zone I	>4000	8.5	70
Zone II	1000–2000	17.0	14
Zone III	Rural, remote	59.5	1
Zone IV	Urban	15.0	140

To substantiate the input data derived from the traffic monitoring program, an atmospheric monitoring network has recently been established to monitor air/lead concentrations and deposition patterns. No data are currently available.

The potential input of lead to the ecosystem in rainwater has also been monitored, and representative data for total lead inputs to urban and rural compartments are shown in Fig. 2. Input from rainfall averages 2% of direct input from automobile emissions.

System outputs

Surface drainage discharge has been monitored continuously at five stream gaging stations located on the Saline Branch and the Boneyard Creek. Total water volume moving past each gage is determined for each 24 h period. This time period corresponds with the water sampling period. Continuous duty, compositing water samplers are in operation at each location. Samples are taken at 15 min intervals and composited for each 24 h period.

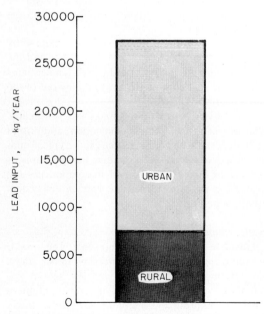

FIG. 1. Total lead input from automobile emissions to the urban and rural compartments.

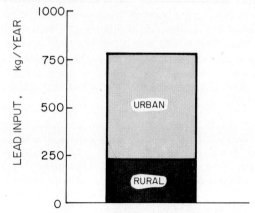

FIG. 2. Lead input to rural and urban compartments of the watershed in rainfall.

The water samples are filtered using a 0.45 μm Millipore filter. The filtrate portion is analyzed by anodic stripping voltammetry and the suspended solids by atomic absorption spectroscopy.

The total output of lead on a daily basis for the urban and rural compartments is shown in Fig. 3. The lead output on a yearly basis for the urban compartment is approximately 600 kg or 75% of the total lead output. Output from the rural compartment totals 210 kg. A comparison between the total output and the total input as shown in Fig. 1 demonstrates that approximately 2% of the total input exits via streamwater. This indicates a considerable accumulation of lead in some component or components of the system.

In consideration of lead output in streamwater, it is important to evaluate the distribution of lead in the filtrate and suspended solids. Figure 4 illustrates the distribution of lead outputs between filtrate and solids for both urban and rural compartments. As would be expected, the majority of lead output is associated with suspended solids in both urban and rural compartments with very little dissolved in the filtrate. The ratio of lead in suspended solids to lead in filtrate varies from 4:1 in the rural compartment to 27:1 in the urban compartment.

Lead in system components

The distribution of lead in soils (0–10 cm layer) in relation to highway traffic volume is shown in Figs. 5 and 6 for low volume and high volume highways, respectively. Values represented in the graphs are averages of 10 transects along the particular highway in question. The distribution profiles are similar

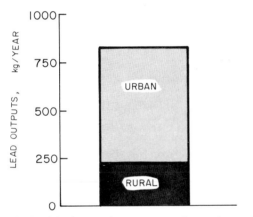

FIG. 3. Total lead output in streamwater from urban and rural compartments.

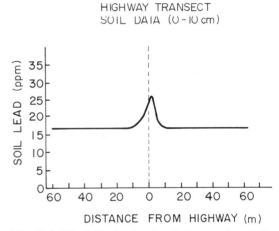

FIG. 5. Soil lead distribution in relation to a low traffic volume highway.

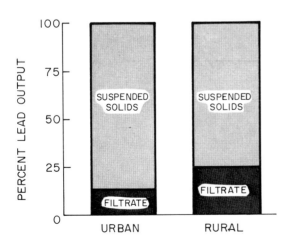

FIG. 4. Lead distribution between filtrate and suspended solids in urban and rural compartments.

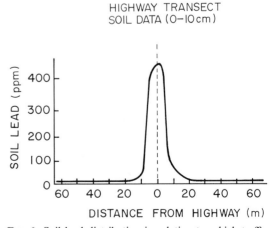

FIG. 6. Soil lead distribution in relation to a high traffic volume highway.

to those reported in the literature with a peak next to the highway, the magnitude of which depends on traffic volume. A sharp decline in concentration within the first 10 m is also evident. Profile distributions of lead in urban (zone IV) and rural soils (zone III) are shown in Fig. 7. Values reported at various depths are averages for 20 profiles in each compartment.

Maximum concentrations are found in the upper 10 cm with a sharp decline in concentration between 10 and 20 cm. Below 20–30 cm the lead concentration is relatively uniform. Considerably higher concentrations are evident in the urban soils than in the rural agricultural area.

The total soil lead pool (urban and rural) is estimated at 800,000–1,000,000 kg or approximately 30 years' input at current automobile emission levels. Approximately 60% of the total soil lead pool occurs in the rural area due to the large total weight of soil as compared to the urban area. The soil system is by far the major reservoir of lead in the ecosystem. Urban (zone IV) and zone I areas contain the largest pools, with zones II and III having soil lead concentrations near background levels.

Vegetation

Plant lead concentrations in relation to highway sources are shown in Figs. 8 and 9 for low volume and high volume highways, respectively. Values

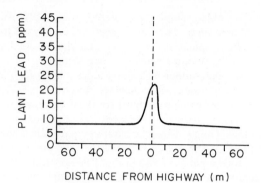

FIG. 8. Plant lead concentrations in relation to a low volume highway.

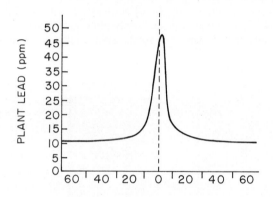

FIG. 9. Plant lead concentrations in relation to a high volume highway.

represented in the graphs are averages of 10 transects along the particular highway in question. The transects are the same as for the soil data reported in Figs. 6 and 7. Data reported are for unwashed plant foliage samples, and thus are a combination of both surface and cellular lead. Again, concentrations peak at the highway and show a sharp decline with increasing distance.

Temporal distribution of lead is shown in Fig. 10 for urban and rural tree ring samples. The urban samples were collected from trees within 10 m of heavily traveled city streets with greater than 10,000 vehicles per 24 h. The rural samples were collected from trees within 10 m of a country road with an average traffic volume of 1800 vehicles per 24 h. Values shown are averages of 20 samples and show a significant increase between 50 year old tree rings and current rings. The present estimate of the total plant lead pool is 3000–4000 kg with approximately 65% in the rural compartment which has low lead concentrations but high plant biomass. The plant

SOIL PROFILE ANALYSIS

	URBAN	RURAL
0–10 cm	390 ppm Pb	16 ppm
10–20 cm	165 ppm	9 ppm
20–30 cm	48 ppm	7 ppm
30–40 cm	40 ppm	6 ppm
40–80 cm	19 ppm	4 ppm

FIG. 7. Profile distribution of lead in urban and rural soils.

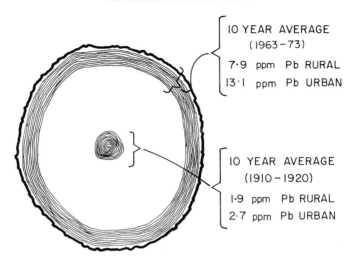

FIG. 10. Tree ring analysis of urban and rural trees.

lead pool accounts for about one-tneth of the average yearly total lead input from automobile emissions.

Small mammals

Samples of small mammals were obtained by snap-trapping from the following situations; within 10 m of a heavily traveled road (15,000 vehicles per 24 h); within 5 m of a medium use road (2500 vehicles per 24 h); and within 5 m of a low use road (200 vehicles per 24 h). Control samples were taken from situations at least 50 m away from any road. Trapping was done seasonally (November, March, June, and August). Animals were washed in deionized water before analysis.

A summary of the results is shown in Table 2 for several species involved in the study. Numbers in parentheses are sample numbers analyzed. Data shown are an average body burden.

All species except *Peromyscus leucopus* had higher lead concentrations in habitats adjacent to the heavy use road. Since the home range of the species averages more than 50 m in diameter, even those individuals caught close to the highway were undoubtedly consuming food from areas remote to the high lead input zone. This likely accounts for the low level of lead in this species.

There was no correlation between lead concentration in the small mammals and the season of the year.

The total pool of lead in the entire small mammal population is estimated to be less than 1 kg. Therefore, there is a very small amount of lead in residency in the total standing biomass at any given time.

Insects. Insect communities were sampled with a portable suction trap on a weekly basis during the 1972 growing season. Sample sites were located in areas known to have either very high lead inputs (zone I) or very low lead inputs (zone III). Insects were categorized according to their type of feeding; sucking plant juices, chewing plant parts, or preying upon other insects.

After collection, composited samples were washed in deionized water on a screen sieve before analysis.

Significant differences in lead content of insects were not observed between sample sites within the same traffic density and vegetation type, or between samples taken at different times of the year. Therefore, data were grouped by insect feeding types and lead input intensity (high versus low).

Lead content in insects was positively correlated with lead emissions ranked in decreasing levels from high emission areas to low emission areas. In all high emission areas there was a trend of increasing lead

TABLE 2
Average lead concentration (ppm) of various small mammals whole body analysis

Species	Heavy traffic	Medium traffic	Low traffic	Controls
Blarina brevicauda	15.2(46)	6.5(71)	3.9(49)	3.6(16)
Cryptotis parva	12.3(13)	5.4(3)	3.2(3)	7.4(7)
Microtus ochrogaster	8.3(40)	4.3(60)	2.6(50)	3.3(13)
Peromyscus leucopus	2.5(16)	2.6(23)	—	2.7(4)
Reithrodontomys megalotis	10.8(32)	3.8(23)	3.1(6)	3.1(20)
Mus musculus	6.9(51)	3.5(150)	4.2(59)	4.6(104)

content from sucking, to chewing, to predatory insects (Fig. 11). This trend was not apparent in low emission areas. Chewing insects likely ingested more lead from surface deposits on leaves than those sucking liquids from internal vascular tissue. Data on predatory insects that feed on lead-containing herbivores suggest that lead is selectively retained in the body, leading to biological concentration of lead in these insect food chains. There was no evidence of biological concentration of lead by herbivorous insects. In the terrestrial portion of the ecosystem, this is the only current example of biological concentration and does give cause for concern over higher trophic levels in the food chain such as insectivorous birds. These are currently under study.

Aquatic sediments. The distribution of lead with depth of sediment is shown in Fig. 12 for the urban compartment and Fig. 13 for the rural. Very high lead concentrations are found in the upper 20 cm of the urban sediment layer. This is likely due to the large quanity of particulates washed from urban impervious surfaces into Boneyard Creek. The high sediment concentrations indicate possible danger to urban aquatic food chains. Less concern is due to rural aquatic chains, but sediment concentrations in the rural area are still exceptionally high in comparison to rural soil lead concentrations.

The total estimated lead pool in stream bottom sediments is 3200 kg with 70% being in urban

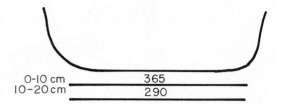

FIG. 12. Stream sediment analysis—urban.

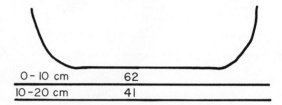

FIG. 13. Stream sediment analysis—rural.

sediments. This is approximately one-tenth of the average yearly total of lead input from automobile emissions.

Aquatic Organisms. Average lead levels (ppm) of components of the aquatic biota are shown in Table 3. The lead accumulation in the biota of the various compartments is irregular with respect to time and location. The differences reflect variations in both the biomass and lead concentrations of the organisms.

TABLE 3
Seasonal average lead levels (ppm) in the aquatic biota of the urban and rural compartments (1972 and 1973)

Compartment	Summer	Fall	Winter
Rural			
Fish	2.4	1.8	NA
Macrophytes	24.1	16.2	—
Benthos	5.4	18.9	14.4
Urban Benthos	139.6	352.3	518.8

The rural compartment has a diverse changeable population. Representatives of 15 orders of invertebrates, 17 species of fish, and a rooted plant have been collected. The concentration of lead in the rural biota does not change significantly with time. Fluctuations in the fish population are slight. The abundance of rooted plants in the summer greatly changes the total lead pool in the rural compartment which closely parallels population growth.

There are only four families of invertebrates present in significant quantities in the urban com-

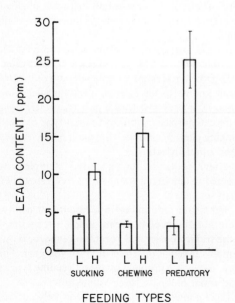

FIG. 11. Lead concentration in grouped insect samples taken from low (L) and high (H) lead emission areas for sucking, chewing, and predatory feeding types of insects.

partment. None of the species are seasonal, and fluctuations in total biomass are small. Concentrations of lead in organisms of this compartment are very high and result in a major lead pool of the aquatic system.

An estimate of the total lead pool in the aquatic system for 1972 is 12 kg with approximately 70% being in the rural compartment due to the greater area and larger biomass. The pool in the urban compartment is restricted to the benthos organisms while a complete trophic structure is present in the rural compartment.

MINE AND SMELTER SOURCE

Background

The second study deals with an evaluation of the sources, transport, and distribution of lead throughout the world's largest lead mining district, the "Viburnum trend" of S.E. Missouri. The "Viburnum trend" or "new lead belt" is almost wholly located within the confines of a national forest in the heavily forested Ozark foothills. The prime lead-emitting operations are mining, milling, ore concentrate haulage by open truck or railroad gondola car, and smelting. The principal sources of lead and other trace metals from these operations are:

(1) *Mine water*—includes high levels of CO_2, carbonates, etc., from the dolomite formation; in addition are minor hydraulic fluid and oil spills, unused blasting materials, and finely crushed rocks.
(2) *Mill waters*—include finely ground dolomite, unrecovered minerals, excess flotation reagents, and dissolved and suspended heavy metals.
(3) *Solids* associated with both the mining and milling operations—known as tails or gangue, these are finely ground (<200 mesh) and are mostly dolomite with varying amounts of heavy metals.
(4) *Concentrate*—approximately 70% lead by weight, this material is the result of the milling process and enters the ecosystem by blowage from storage piles at the mill and smelter or during transport by uncovered trucks and railcars.
(5) *Smelter emissions*—include blowing concentrates, dust from sintering operations, and stack emissions.

Water quality and aquatic studies

Water samples have been taken at 23 sampling sites including control and mine waste-receiving streams on a biweekly basis for the past 3 years to provide the statistical base for establishing changes in water quality with season of the year, mine and mill operating conditions, climatic conditions, etc. Analyses have included lead, zinc, copper, cadmium, and manganese as well as some 20 organic parameters; the analyses were with standard methods. Jennett,[6] has reported that there were few organic differences between control and noncontrol sites. In the case of heavy metals, however, a comparison of filtered versus unfiltered water has shown that large quantities of heavy metals move through the system in association with suspended matter. The filtered samples almost always contain lead in the 4–6 ppb range reported by Bolter et al.[7] to be the background level for the region. Whenever high concentrations of lead or zinc were observed, these were always associated with turbulent stormwater runoff, and the high concentrations were found only in the unfiltered samples. This indicates that one of the major heavy metal transport mechanisms occurs when stream flow and velocity is of a sufficient level to scour the bottom, removing sediments and biological slimes. Unfortunately, this has rarely been observed by regulatory agencies because of the routine filtering of samples before analysis.

The waters from mine and mill waste operations generally do not find their way into the surface stream system until they have received treatment in at least a single tailings pond. Generally, the effluent from these ponds is not chemically different from the receiving streams except for an inordinate amount of suspended solids and high turbidity.[6]

The release of inorganic, and probably organic, nutrients in the mine and mill waters which have not been completely treated by conventional lagoon systems has resulted in seasonal blooms of algae on some of the receiving streams. The dominant forms and extent of growth varies from stream to stream. Gale et al.[8] have reported that these algae trap and hold considerable quantities of heavy metals which have escaped treatment in the tailings ponds. This phenomenon has been observed for at least 3 years with a variety of genera and at all seasons.

The phenomenon of lead uptake by algal forms has been shown repeatedly. Small amounts of *Cladophora* have been found containing 8000 ppm lead, and large amounts have been found containing 5000 ppm. (Zinc levels were approximately 1000 ppm.) *Stigeoclonium*, more tolerant to heavy metals, has been found with associated lead at 74,000 ppm, zinc 2300 ppm, and copper 1200 ppm. Research currently underway has confirmed these figures for a number of algal forms in laboratory studies. It has also been shown that most of this lead is associated with entrapment or fixation on the surface of the algae rather than internal contamination. Lead in fish collected near the lead sources was found to concentrate in the digestive tract and gills on the basis of current limited data. Fish specimens near the tailings dams have shown for higher levels than nonexposed fish, yet very little is known relative to the toxemia of heavy metals in aquatic animals.

Laboratory studies have also shown that carbon is not a limiting nutrient in these waters, but phosphorus is.[6] Manganese and phosphate additions to these waters stimulate algal production. The heavy metals are less toxic in hard waters and in the presence of abundant nutrients.

Air quality

The most obvious source of lead emissions is the region's smelter. Its 200 ft main stack discharges baghouse effluent at an average rate of 28.45 lb/h of total particulate, of which 5.12 lb/h is particulate lead under normal operating conditions.

In order to determine dustfall rates, 20 sampling sites were selected in a 3×3 mile grid surrounding the smelter. Analyses have been made monthly on the material collected.

Deposition rates for lead increased dramatically near the smelter to levels as high as $2 \text{ g/m}^2/\text{month}$, and decreased to background levels of $0.05 \text{ g/m}^2/\text{month}$ within 3 miles. The relation between lead deposition rate and radial stack distance has been found to be

$$y = 8.76 \times 10^6 \times x^{-1.2704},$$

where y is the lead deposition rate (mg/m^2/month) and x the radial distance from the smelter stack in feet. The correlation coefficient for this equation was 0.995.[6]

The distribution of ambient concentrations of particulate lead has been monitored at five sites by making biweekly, 24 hr, high volume sampling runs. In comparing the different stations, particulate lead concentrations are higher closer to the smelter which reflects the effects of trucking, hauling, and windblown dust. By comparing particle size distributions of lead, it is highly probable that elevated ambient lead levels are to a large extent from fugitive sources, since 90% of the stack emissions are less than 12 μm and should travel some distance before settling to the ground or vegetation level. From an analysis of the data, fugitive sources are felt to be the major source of deposited lead and airborne lead levels within the immediate smelter area. Beyond this area, stack emissions are the prime contributor of lead in the air. Further work is being done to precisely measure the individual sources such as roads, tails, etc., and to evaluate the effects of these sources on airborne lead levels.

Geochemical studies

Soil and geochemical studies have been performed in a 300 square mile area around the smelter. Samples were divided into the O_1 fresh or partly decomposed litter, OH-well decomposed litter, 0–2.5 cm, 2.5–7.5 cm, and 7.5–15 cm soil fractions. Analysis was made on the soil fraction passing through an 80 mesh sieve. The background lead range was 17–32 ppm. All anomalous samples are in the vicinity of mines, mills, smelters, or haul roads. The results do not indicate natural surface anomalies. The data have indicated a strong tendency for the leaf litter to retain lead, preventing it from being transported into the soil. Anomalous values for lead, especially in the OH layer, are elongated in a northerly direction (with the prevailing wind) from the smelter (Table 4). The extent of the anomaly is still under study, but OH samples 10 miles north contain 300–450 ppm lead, and 12–14 miles south the same layer is 200–250 ppm.[9]

Profile samples of soil have consistently shown a tendency for the lead to be retained in the top fractions of the 0–2.5 cm samples and nearly always associated with the leaf litter. In a study to determine the absolute content of lead in a unit area, several 0.01 m^2 samples, to a depth of 2.5 cm, were collected, analyzed, and the total lead content determined. The data show several hundred g/m^2 within a quarter mile of the smelter, decreasing to 0.5–2.5 g/m^2 at 5 miles. These data also show the strong ability of the leaf litter to retain the lead and prevent its transport to the soil. Near the smelter, as much as 94% of the lead is retained in the total leaf litter.

Soil samples near trucking roads used for transport of ore concentrates usually show lead concentrations of 100–700 ppm in the 0–7.5 cm soil fraction near the road; anomalous values are generally found up to approximately 300 ft. Similar traverses of the rail transport system were made which also show anomalies to a distance of 300 ft.

Vegetation

Vegetation along the routes over which lead concentrate is hauled from the mills to the smelters has been extensively sampled to determine the extent of contamination. The samples of blueberry,

TABLE 4
Lead content of soil (ppm) within 5 mile radius from smelter[9]

Distance (Miles)	Type	N	NE	E	SE	S	SW	W	NW
0.25	O_1[a]	895	7900		26,000	2760		10,400	8350
	OH	7800	28,000		41,250	15,250		21,500	56,000
	Soil	49	273		46	217		1930	1000
0.5	O_1	605	1900	3880	7750	2760	3220	4550	1220
	OH	1900	12,350	19,750	59,000	16,500	5950	11,750	4400
	Soil	67	83	527	213	53	160	51	49
1.0	O_1	1035	1450	1410		690	1900	590	1095
	OH	5000	6250	4500	2865	1120	6600	2700	4600
	Soil	67	89	31	61	38	72	54	69
2.0	O_1		320	1720		470	198	230	
	OH	2625	1310	2150	1650	2080	490	1400	
	Soil	40	101	53	26	57	24	38	
3.0	O_1		440	890		400	212	156	590
	OH	950	1370	860	600	810	590	260	1980
	Soil	63	33	57	33	31	75	260	28
4.0	O_1			120		164	166	79	287
	OH	810	350	378	270	360	210	360	950
	Soil	59	24	49	34	23	56	29	32
5.0	O_1			99		203	206	86	167
	OH	1050	290	415	500	360	200	220	600
	Soil	29	37	30	18	17	79	16	26

[a] O_1 = slightly decomposed leaf litter. OH = well decomposed leaf litter. Soil = 0–2.5 cm (passing through 80 mesh sieve).

fescue, and purple top were severed 2.5 cm above the soil, visible particulates removed, and the whole plant analyzed unwashed. Table 5 summarizes these data. Soil samples of the 0–15 cm depth, taken at the same time, showed a similar pattern with 800 ppm on the haul roads and 75.6 ppm on the control roads at the immediate shoulder, declining to near background levels at 400 yd from the haul roads.

There are other ways toxic metals can enter the environment from roads, such as the common use of slag on winter roads, in road beds, etc. This material in the blowing concentrate contributes to toxic runoff and dust on plants. Even if this is mere surface contamination, it is part of the food consumed by animals and birds. It can also contribute markedly to human exposure levels of those in close proximity to the highways, and measures should be taken to limit this source of contamination.

A summary of the analysis of all plant and soil samples indicates that vegetation near smelters and, to a lesser extent, near mines and mills, accumulate lead to several times background. The majority of this accumulation is surface contamination. Elevated levels near the New Lead Belt smelter (200 ft stack) carry at least 7 miles, and a study of another smelter, not in the New Lead Belt but having a 612 ft stack, has, in preliminary studies, shown similar elevated levels at least 7 miles out. The level of accumulation is markedly less near mines and mills than near smelters.

Lead mining, milling, and smelting has been carried on in Missouri for over 100 years. In order to evaluate the effects of these operations on growing plants, samples of leaf lettuce, radishes, and greenbeans were taken in the New Lead Belt, the Old Lead Belt (a mining region some 70 miles away and 100 years older), a small town 2 miles from the state's oldest smelter, and a three county control area in north central Missouri which has never shown evidence of lead mineralization. An analysis of the data has shown that the values for the Old Lead Belt region are quite similar to the New Lead Belt except near the old smelter, which has been in operation since 1891 (Table 6). Levels near this old smelter were among the highest recorded during the study. An analysis of these data should stress that these vegetables are not the major part of any human's diet and, therefore, may not contribute a

TABLE 5
Concentrations of lead in vegetation along ore truck routes

Distance from road right-of-way (yd)	All species			
	Micrograms low	Per gram mean	Dry weight high	Number of samples
0	1.0	279.7	4350.0	80
100	1.9	34.2	537.0	58
200	2.4	11.6	35.0	16
300	4.8	8.5	15.0	8
400	4.0	6.5	9.1	6

Concentration of lead in vegetation along control routes

Distance from road right of way (yd)	All species			
	Micrograms low	Per gram mean	Dry weight high	Number of samples
0	2.4	18.1	49.0	27
100	0.15	8.1	20.0	15
200	4.0	7.2	12.0	9
300	2.5	3.9	5.8	5
400	2.3	3.9	5.8	5

TABLE 6
Lead content of vegetable crops (mg/g dry weight)[a]

Crop	Low	High	Mean	No. of samples
Control counties				
Lettuce root	10.0	50.0	20.3	12
Lettuce leaf	6.9	33.9	20.6	12
Radish root	5.0	11.0	7.7	12
Radish top	5.0	32.0	14.0	12
Greenbean root	5.0	20.0	9.1	12
Greenbean pod	<5.0	<5.0	<5.0	12
"New lead belt" counties				
Lettuce root	10.0	66.0	89.7	28
Lettuce leaf	8.9	923.0	114.0	28
Radish root	4.4	107.0	22.3	28
Radish top	13.3	578.0	94.4	28
Greenbean root	5.0	53.0	9.9	28
Greenbean pod	5.0	40.0	8.8	28
"Old lead belt" counties				
Lettuce root	11.7	492.0	68.8	30
Lettuce leaf	10.3	742.0	83.8	30
Radish root	5.0	518.0	33.4	30
Radish top	5.0	117.0	76.7	30
Greenbean root	5.0	67.0	8.6	30
Greenbean pod	5.0	10.0	5.4	30
"Old lead belt" small town with a lead smelter since 1891				
Lettuce root	25.0	3244.0	636.0	8
Lettuce leaf	47.0	1324.0	284.0	9
Radish root	7.4	50.0	22.3	8
Radish top	32.0	385.0	119.5	8
Greenbean root	10.0	704.0	119.3	8
Greenbean pod	5.0	136.0	25.1	8

[a] All samples washed in distilled water immediately after harvest.

great deal to their body burden. However, exceptionally high lead values, particularly in the leafy vegetables, are found where lead is mined, milled, and smelted. Researchers to date have not been able to assess the possible effects of this material.

CONCLUSIONS

At this stage of the studies, the following conclusions seem justified:

(1) Leading mining and milling operations as well as automobile emissions contribute heavy metals to stream sediments and biota. These materials are transported out of the ecosystem to other systems during times of runoff. This is rarely detected by regulatory agencies because of the use of filtered samples for analysis. Preliminary studies do not indicate biological magnification of lead in aquatic food chains.
(2) Air quality studies have shown that the main source of heavy metals in the immediate area of smelting operations is fugitive sources; beyond a few thousand feet the stack emissions are the main contributors of heavy metals. Automobile emissions are the major source in the Illinois study.
(3) The soil anomaly around the smelter extends at least 15 miles north and south. Most of this lead is retained in the leaf litter rather than passed on to the soil. The soil anomaly in the Illinois ecosystem is restricted to a narrow band (±40 m) along roadways.
(4) Concentrate transport contributes to heavy metals exposure near both the roads and railroads used near mines and smelters. Automobile emissions contribute to soil, plant, and animal anomalies along highways.
(5) High levels of lead in or on samples of vegetables, particularly lettuce, have been detected in areas where lead is mined, milled, and smelted. These materials are only a small part of human diets and are not considered too dangerous at this time, though the effects remain to be studied.
(6) Lead levels in the vegetation surrounding the smelter are high for at least 7 miles from the 200 ft stack. Preliminary results of another

study of a Missouri smelter not in the New Lead Belt with a 612 ft stack have also shown elevated lead levels for at least 7 miles.

(7) Major lead pools in the Illinois ecosystem occur in soils, stream bottom sediments, and plants. In terms of magnitude, other lead pools are negligible. However, consideration should be given to these other pools as their importance to the biota may be considerable.

(8) Illinois ecosystem trouble spots include soil-plant zones along major highways and in urban areas and stream sediments in the urban compartment. These areas are receiving emphasis in current research.

REFERENCES

1. CHOW, T. J., Lead accumulation in roadside soil and grass, *Nature* **225**, 295 (1970).
2. JOHN, M. K., Lead contamination of some agricultural soils in western Canada, *Environ. Sci. Technol.* **5**, 1199 (1971).
3. LAGERWERFF, J. V. and SPECHT, A. W., Contamination of roadside soil and vegetation with cadmium, nickle, lead and zinc, *Environ. Sci. Technol.* **4**, 583 (1970).
4. MOTTO, H. L., DAINES, R. H., CHILKO, O. M., and MOTTO, C. K., Lead in soils and plants: its relationship to traffic volume and proximity to highways, *Environ. Sci. Technol.* **4**, 321 (1970).
5. PAGE, A. L., GANJE, T. J., and JOSHI, M. S., Lead quantities in plants, soil and air near some major highways in southern California, *Hilgardia* **41**, 1 (1971).
6. JENNETT, J. C., Environmental transport of lead in relation to mines, mills and smelters, *1st Annual NSF Trace Contaminants Conference, Oak Ridge National Laboratory, 1973*.
7. BOLTER, E., JENNETT, J. C., and WIXSON, B. G., Geochemical impact of lead-mining wastewaters on streams in southeastern Missouri, Presented at the 27th Annual Purdue Industrial Waste Conference, Lafayette, Indiana, and published in the *Proceedings*, 1972.
8. GALE, N., WIXSON, B. G., HARDIE, M. G., and JENNETT, J. C., Aquatic organisms and heavy metals in Missouri's new lead belt, *American Water Resources Association Conference, 1973*.
9. BOLTER, E., BUTHERUS, D. L., and TIBBS, N. H., The impact of lead–zinc mining and smelting on the heavy metal content of soils in the new lead belt of Missouri—a five year case history, *7th Annual Meeting, North Central Section, Geological Society of America*, Literature cited.

Environmental Lead Distribution in Relation to Automobile and Mine and Smelter Sources (G. L. Rolfe and J. C. Jennett)

DISCUSSION by S. R. KOIRTYOHANN
Project Director, Environmental Trace Substances Center, The University of Missouri, Columbia, Missouri

B. G. WIXSON
Professor of Environmental Health, Civil Engineering Department, Environmental Research Center, The University of Missouri–Rolla, Rolla, Missouri

and

H. W. EDWARDS
Associate Professor, Department of Mechanical Engineering, Colorado State University, Fort Collins, Colorado

INTRODUCTION

The paper presented by Rolfe and Jennett consists of separate and distinct reports on environmental lead contamination studies at two different institutions. The first section deals with environmental lead levels in and around Champaign–Urbana, Illinois, where the major lead input is ascribed to the combustion of leaded gasoline. The second section deals with the environmental effects of lead mining, milling, and smelting in the "new lead belt" of Missouri. Environmental lead levels associated with some areas of the Missouri lead processing activities tend to be substantially higher than those values reported for the 86 square mile area of the Illinois study. This presentation of research findings from two different lead sources and types of ecosystem input is an excellent example of outgrowth from cooperative efforts developed between research teams at Colorado State University, the University of Illinois, and the University of Missouri where various areas of environmental lead pollution have been under study for several years.

The objective of the presentation by Rolfe and Jennett were to discuss the studies of environmental contamination by lead in two systems, and unfortunately they were limited by time and length requirements and did not, therefore, provide a section which would integrate the research findings of the two studies. Both authors were also faced with the difficult task of summarizing several years of work by multidisciplinary teams so that many of the details were presented in summary form. However, a more useful comparison of the environmental impacts of the two completely different lead contamination levels would have been helpful.

AUTOMOTIVE SOURCES

The first section concerned with the study area in Illinois speculated, on the basis of extrapolation of limited data, that the accumulation of lead from automotive sources by certain small mammals or aquatic species may eventually result in serious environmental effects. Although the data reported indicated definite elevated lead levels in certain samples of sediments, soils, plants, and animals taken near heavy traffic sources, the authors failed to report specific data in this section on the physiologic effects of lead at these levels or on environmental damage associated with automotive lead. The data reported, however, are generally consistent with the literature: significant lead contamination in environmental samples is normally restricted to a relatively narrow zone near traffic sources. The National Academy of Sciences' report on lead[1] and the Proceedings of the International Symposium, *Environmental Health Aspects of Lead*,[2] for example, are replete with similar data. The authors' conclusion that the lead levels measured in the Champaign–Urbana system may have serious environmental effects is unsupported by the data reported and in disagreement with most of the recent scientific literature concerning the low level effects of lead on biological systems.

The section on system inputs discussed the potential input of lead into the ecosystem via rainwater; however, further explanations would have been helpful in understanding this source. As noted by one of the discussers, the lead in stream flow is very likely only a very small fraction of the total lead output. The major component of the total outflow would more likely be the atmospheric component, e.g., lead-containing particles which survive gravitational settling. The literature indicates that a substantial fraction of the total lead mass in automotive exhaust is associated with particles which, in an aerodynamic sense, possess insignificantly small settling velocities. The mass balance constructed, therefore, is open to question since the atmospheric outflow was disregarded.

The importance of evaluating the distribution of lead in the filtrate and suspended solids was emphasized based on analytical research in the Illinois ecosystem study. The importance of this finding was emphasized again in the Missouri study, pointing out the importance of determining the amount of lead and other trace metals associated with the stream sediments.

Additional information concerning the basis for calculation of the "total soil–lead pool" would be helpful. Furthermore, the calculation may be of questionable value since lead from automotive exhausts is definitely not distributed evenly throughout the system under study. For example, studies near a highway at Colorado State University have indicated that soil lead levels become indistinguishable from background beyond about 100 m downwind from a roadway edge.[3]

The lead concentrations in plants nearby highway sources followed the general findings of prior literature[4] and are limited to narrow zones paralleling the highways. Consideration should also be given to the possibility of elevated lead in dusts from urban atmospheres. Work by Patterson et al.[5] has pointed out contamination problems associated with lead pollution from automobile exhausts and industrial sources in the Los Angeles and San Francisco area which are presently being deposited in Thompson Canyon 250–500 km remote from the congested sources.

Research information on lead concentrations in small mammals was interesting. However, Table 2 would have been more comprehensive if the common name of the mammal had been indicated below the scientific name. For example, *Blarina brevicauda* is a short-tailed shrew, *Cryptotis parva* is a least shrew, *Microtus ochrogaster* is a wood mouse, *Peromyscus leucopus* is a wood mouse, *Reithrodontomys megalotis* is a plains harvest mouse, and *Mus musculus* is a house mouse. This information would serve to assist those without extensive biological training in following the data.

While many animals collected near heavy traffic routes show elevated lead levels, it was interesting to note that the *P. leucopus*, or wood mouse, did not show this effect. Perhaps further studies should consider the possibility of a threshold effect for this species.

The findings on the Illinois area insect study were interesting but raised several questions among the authors of this discussion paper. One of the chief concerns was with the experimental design since insects collected in the trap would continue to accumulate lead from air flowing through the trap and washing with deionized water might remove some of the surface lead accumulated after the insect was trapped. In any case, the amount of lead left on and in the insect might reflect only a part of total lead collected by the following three mechanisms:

(1) Buildup of tissue lead through ingestion of lead-contaminated materials.
(2) Scavenging of lead on exposed insect surfaces before trapping.
(3) Scavenging of lead on exposed insect surfaces during and after trapping.

However, since this difference in lead concentration is based on feeding habits, there may be a divergence of opinion on the insect data since all insects were trapped in the same vacuum system and all would have shown similar lead content if it were due to contamination.

One conclusion reached, that this was the only current example of biological concentration in the terrestrial portion of the ecosystem, may be misunderstood. In other studies a number of reports have been published in the literature which would indicate biological accumulation of automotive lead. The following are examples of such reports: Rabinowitz and Wetherill[6] reported that 50% of the lead found in lead-poisoned horses was from automotive sources; Tornabene and Edwards[7] reported that bacteria can take up substantial quantities of lead from solutions containing lead salts typical of those found in fresh automotive exhaust; Edwards et al.[8] reported that mice collected along Interstate Highway 25 in Colorado have lead body burdens approximately three times higher than those collected from the remote Craig–Hayden area in Colorado; and Gish and Christensen[9] reported that earthworms collected near a roadway accumulated up to 331.4 ppm of lead.

The first section on automotive sources does contain some very interesting environmental lead data, especially for water and sediment samples from urban and rural regions. However, it is important to distinguish between the occurrence of environmental contamination and environmental consequences of such contamination. These and other data leave little doubt that man has substantially contaminated some portions of his environment with lead from automotive exhausts. However, the significance of present levels of environmental contamination by automotive lead is not yet fully understood. Although these and other data point to no drastic environmental effects associated with present levels of lead contamination from automotive sources, the data do indicate a definite need for careful monitoring on a continuing basis. There are substantial economic and petroleum conservation benefits associated with use of lead as a gasoline additive. The decision to omit or to continue to use lead as a gasoline additive is presently in a critical phase and further explanation on the significance of the authors' data to this issue would have been relevant.

MINE AND SMELTER SOURCES

Research findings and supporting data on the mining, milling, and smelting sources were limited by the paper format. The mining and smelter section focused primarily upon the identification of potentially hazardous releases of toxic materials from lead mining, smelting and milling operations in the "new lead belt" or "viburnum trend" of southeast Missouri. Primary sources of lead and other trace substances were identified, and lead concentration data reported for air, water, plant, and soil samples taken as a function of distance from industrial operations. Prior research on the lead industry as a source of trace metals has been reported by Wixson et al.[10] The data demonstrated that significant elevation of lead levels in various environmental media was not highly localized to a region immediately adjacent to the smelting and milling operations but, on the contrary, extended several miles beyond the industrial site.

The sources of trace metal pollution in the aquatic environment by mine water, mill water, tailing materials, blowage of concentrates, and smelter emissions were presented in a straightforward manner but lacked the benefits of additional supporting data on the associated co-products of zinc, cadmium, and copper. However, these parameters have already been noted in prior publications by Jennett and Wixson[4] and Jennett et al.[11]

Environmental lead contamination by application of smelter slag material for snow and ice control has presently been corrected through the mining companies' comprehension of the problem and refusal to give or sell the material to the state or county highway department. The concentrations of lead in vegetation along ore truck routes was found to be considerably more elevated than those reported for research at the University of Illinois due to the difference between the amounts in automobile exhaust and the lead concentrate material blowing off trucks or out of railroad gondolas during transportation from the mine-mills to local lead smelters. The accumulation rates of zinc and copper are additionally important since these materials are usually transported by railroad to other states for further processing. However, the decline of environmental lead distribution with distance from the haulage route is similar to the distribution pattern reported in the Illinois and Colorado State studies for highway traffic except for a difference in the amount of lead detected.

One important feature as noted by both studies, in different ecosystems and at markedly different concentrations, is that a potential contamination problem may occur during periods of storm wash-off whereby trace metals may be transported into area streams. Research evaluations by Bolter et al.[12] indicate that lead is leached from the leaf litter by humic and other acids thereby making it available for run-off rather than percolation into the upper soil layer as reported in other studies. Further research is being continued and additional data should be forthcoming to further confirm these findings.

Stream studies in the "new lead belt" of Missouri have not indicated any significant biological accumulation of lead as noted in herbivorous insects in the Illinois study area. This may be due in part to the types of aquatic life used in the studies. One of the most significant findings reported in both studies was the importance of using unfiltered water samples for the analysis of trace metals. Further work needs to be concentrated on sediment analysis since both the Illinois and Missouri study indicated that the stream, river, or lake sediments constituted a major sink source for lead and other trace metals.

As reported by Purushothaman et al.,[13] the finding is significant that fugitive dust around one lead smelter is the primary source of lead fallout, whereas, beyond this area, stack emissions are the primary source of atmospheric lead. These data also tend to clarify why air and dust lead concentrations near the El Pasco smelter were highly elevated even though a high stack was employed;[14] the importance of good housekeeping for controlling fugitive dust is now clear and should be practised by industries.

In the presentation regarding air quality near one smelter, an empirical formula for the computation of the lead fallout rate was used, given the distance from the stack. The implication may have been given that this information was not direction dependent. Yet there was a very definite direction dependence for the soil–lead data. This apparent confusion may be explained by the type of data reported, since the settleable particulate lead was collected by dustfall buckets and the data reported was based on a 30 day period of collection. This data included the contribution of fugitive dust from the smelter area as well as other materials collected by the dustfall buckets. Limitations of the relationship are still being verified in the collection of data.

New data being evaluated indicates that lead levels in soils around two different lead smelters remain elevated for a distance of 12–15 miles from the smelter source. Further research is being carried out to study this anomaly.

CONCLUSIONS

The paper by Rolfe and Jennett represents a valuable cooperative effort between two universities sharing common research data on the behavior of lead in the environment. This same spirit of cooperation and mutual interest also exists between the University of Illinois, Colorado State University, and the University of Missouri as noted by the comments presented in this discussion. The importance of controlling heavy metal transport in stream sediments has been re-emphasized, and the comparison of contamination from different types of sources has pointed out mutual problem areas.

Further studies are continuing on factors such as determining how the contamination concentrations change with time in order to predict future effects. Additional work is needed to understand the present effects and evidences of biological damage due to environmental lead contamination in different areas. Improved analytical methods developed during these studies should answer questions on the chemistry of lead in sinks in relation to the source type, conditions of mobilization, and what control measures can be employed. The ability to predict the effects of lead and other trace metals will contribute valuable information to agencies and industries alike in protecting the environment through the control of lead and other trace metals.

Acknowledgements—Research reported from Colorado State University, the University of Illinois and the University of Missouri has been supported by the National Science Foundation, RANN (Research Applied to National Needs) Trace Contaminants Program. Grateful appreciation is also extended to the many industries, federal and state agencies for their cooperation and assistance.

REFERENCES

1. NATIONAL ACADEMY OF SCIENCES, *Lead—Airborne Lead in Perspective*, Committee on Biologic Effects of Atmospheric Pollutants, Washington DC, 1972.
2. *Environmental Health Aspects of Lead*, Proceedings of the International Symposium, Amsterdam, October 2–6, 1972.
3. EDWARDS, H. W., *et al.*, *Impact of Man of Environmental Contamination Caused by Lead*, interim report, NSF Grant G1-4, Colorado State University, Fort Collins, Colorado, 1972.
4. JENNETT, J. C. and WIXSON, B. G., Problems in lead mining waste control, *J. Wat. Pollut. Control Fed.* **44**, 2103 (1972).
5. PATTERSON, C. C., *et al.*, *Determination of Biological Discrimination Factors for Metals Within Natural Ecosystems*, Summary Report to NSF, California Institute of Technology, Pasadena, California, 1973.
6. RABINOWITZ, M. B. and WETHERILL, G. W., Identifying sources of lead contamination by stable isotope techniques, *Environ. Sci. Technol.* **6**, 705 (1972).
7. TORNABENE, T. G. and EDWARDS, H. W., Microbial uptake of lead, *Science* **176**, 1334 (1972).
8. EDWARDS, H. W., *et al.*, *Impact of Man of Environmental Contamination Caused by Lead*, interim report, NSF Grant G1-4, Colorado State University, Fort Collins, Colorado, 1972.
9. GISH, C. D. and CHRISTENSEN, R. E., Cadmium, nickel, lead, and zinc in earthworms from roadside soil, *Environ. Sci. Technol.* **7**, 1060 (1973).
10. WIXSON, B. G., *et al.*, The lead industry as a source of trace metals in the environment, in *Proceedings, Cycling and Control of Metals*, EPA, Natl. Environ. Res. Center, Cincinnati, Ohio, **11**, 1973.
11. JENNETT, J. C., *et al.*, Transport mechanisms of lead industry wastes, *Proceedings, 28th Annual Purdue Industrial Waste Conf.*, Lafayette, Indiana, May, 1973.
12. BOLTER, E., *et al.*, Distribution of heavy metals in soils near an active lead smelter, *Proceedings, Environmental Session, University of Minn. Mining Symposium*, Duluth, Minn., January, 1974.
13. PURUSHOTHAMAN, K., *et al.*, Development of an air quality monitoring system to control lead smelter episodes, *Transactions, 23rd Annual Conference on Sanitary Engineering, Univ. of Kansas*, **113**, February 1973.
14. CARNOW, B. W. and ROSENBLUM, B. F., Unsuspected community lead intoxication and emissions from a smelter, preprint number 73-206, APCA 66th Annual Meeting, Chicago, Illinois, June 1973.

INVESTIGATIONS OF HEAVY METALS AND OTHER PERSISTENT CHEMICALS, WESTERNPORT BAY, AUSTRALIA

M. A. SHAPIRO

Professor, Environmental Health Engineering, University of Pittsburgh; Director, Westernport Bay Environmental Study, Melbourne

and

D. W. CONNELL

Marine Studies Coordinator, Westernport Bay Environmental Study, Melbourne

INTRODUCTION

Westernport Bay (Fig. 1) is one of those historical misnomers—it is the eastern of the two major bays in Victoria. It was discovered and named by that intrepid physician, George Bass, who entered the bay in 1798, after sailing south from Botany Bay in an open whaleboat. It was the westernmost bay he found on this voyage of exploration. Just a few miles to the west, over the Mornington Peninsula, he could have found Port Phillip Bay, a larger body of water and upon whose shores the majority of Victoria's population now reside (2.5 million in the Port Phillip Bay catchment and 50,000 on the Westernport Bay catchment).

The Westernport region is noted for its birdlife. In this area (comprising less than 2% of the State) nearly half (220) of the 470 species of birds found in Victoria have been reported. Such a high diversity in a relatively small area is not common in the State. The whole region provides many bird habitats, but the swamps, marshes, mangroves, mudflats, and bay waters are particularly important. French Island alone has been estimated to support more than 1000 pairs of breeding ibis. Also important are ducks, swans, pelicans, short-tailed shearwaters (mutton birds), penguins, and a considerable range of waders and waterfowl.

Phillip Island on the Bass Strait side is one of the most important wildlife areas in Victoria. It contains a large koala population and is the most important single area in the koala management program of the Department of Fisheries and Wildlife. An extensive fairy penguin rookery at Summerland is strictly protected, and it is here at dusk that the famous penguin parade may be seen. The large mutton bird rookeries are also protected. Unrestricted sealing during the nineteenth century decimated the seal population, but, now protected, the most readily accessible and observable seal colony off the Victorian coast is located at Seal Rocks, the southwest corner of the Island.

But it is as a recreation area that Westernport is best known. Although Phillip Island, with its unique fauna, is, perhaps, the only truly international tourist attraction in Victoria, it is the Mornington Peninsula that has become the major recreation outlet for metropolitan Melbourne. Separating Port Phillip and Westernport bays and located immediately to the southeast of the metropolis, the peninsula is an area of great natural attraction containing a variety of recreation resources. Its landscapes form a mosaic of hill and vale, woods, and pasture; its beaches range from the safe, sandy beaches of Port Phillip Bay to the wave-swept surf beaches of the Gunnamatta–Portsea area. A third of the houses on the peninsula are holiday homes, and some 150,000 people use the area as their main source of recreation during the peak summer months.

Both Port Phillip Bay and Westernport are extensively used by amateur fishermen, yachtsmen, swimmers, and power boat enthusiats. Westernport also supports a commercial fishing industry which supplies a large proportion of Melbourne's fish requirements. The bay is fringed with small towns—San Remo, Corinella, Tooradin, Warneet, and Hastings—which have, in the past, been almost totally economically dependent upon the fishing industry.

Westernport Bay has long been recognized as a fine port with deep, sheltered anchorages, but it was adjoining Port Phillip Bay which became Victoria's major commercial and industrial port. Now, with a dramatic increase in deep-draft shipping, the importance of Westernport Bay has increased dramati-

Fig. 1. Westernport Bay and its catchment.

cally. No other harbor on the Victorian coast, and few in Australia, can match its deep, sheltered waters. Almost completely landlocked, its depth at the entrance exceeds 100 ft, dropping to a minimum at low water of 47 ft in the northern arm and 49 ft in the eastern arm.

Other factors add to its value as an industrial port. In the vicinity of Hastings, the deep-water channel is flanked on both sides by large tracts of flat land suitable for heavy industrial development. It is close to the Melbourne–Dandenong–Berwick axis, where the present thrust of Melbourne's growth is greatest. Similarly, it is close to the Melbourne–Gippsland axis with its rich resources. The large labor and consumer market of Melbourne is nearby, and it is also central to the entire market of south-east Australia. The area is also well served by land transport connections.

In the competitive strivings of industry, transportation costs are important. Some industries cannot exist without close access to deep water and the economies of deep-draft bulk cargo shipping. Many of these industries are basic to Australia's economy. Only Westernport on the Victorian coast and a small number of other locations on the entire Australian coast possess the qualities which are essential to these industries. When the factors of markets, labor, deep water, land quality and availability, and raw material sources are analyzed, Westernport is found to be the most suitable site in Australia for the development of an industrial port complex.

Perhaps the most important feature of all of the resources of Westernport is that they do not exist in isolation, but are linked into an ecological system, such that the value of each resource is dependent upon the maintenance of the others. That is why, with multiple pressures for exploitation, it is essential that planning development and conservation go hand in hand.

CURRENT HEAVY METAL INPUTS FROM INDUSTRY

Large and heavy industry located on the catchment is a recent addition to Westernport. In 1963 the BP refinery was constructed. More recently, ESSO–BHP completed a gas fractionation plant, and in September 1973 the John Lysaght cold strip plant, the first phase of an integrated steel mill, went on line. A fertilizer plant was also constructed but is no longer in operation. Original permits to discharge were granted by the Westernport Pollution Committee and administered by the Ports and Harbors Division of the Victorian Department of Public Works. Since the enactment of the Environment Protection Authority Act, the Victorian EPA has the legal responsibility of issuing licenses for discharge of waste to water.

Recent information concerning the composition of discharges from these plants to Westernport Bay indicates that they are highly complex. All contain oil and grease, and the John Lysaght plant, which uses hydrochloric acid to pickle steel and regenerates the spent acid, also contains from 0.1 to 0.2 mg/l chromium and from 1.0 to 2.0 mg/l of zinc.[1]

THE RESEARCH PROGRAM

Materials that are biocumulative and achieve biologically significant concentrations in aquatic areas can be broadly divided into two major groups:

(1) Comparatively stable organic compounds such as organochlorine insecticides, PCBs, and hydrocarbons, etc.
(2) Compounds of heavy metals such as mercury, cadmium, zinc, etc.

These substances can enter the marine environment in a number of ways deriving from natural and manmade sources. On entering an aquatic environment, the substances are usually absorbed by organisms or sediments, or precipitated. Thus, sediments can contain high concentrations of persistent chemicals, e.g. Lake Burley Griffin, Canberra, Australia has sediments containing up to 1500 ppm of zinc.[2]

Since biocumulative substances, such as heavy metals, have been detected in fish and other edible marine organisms in southern Australia, an investigation of selected biocumulants in Westernport Bay was considered necessary. The initial investigation will extend over a period of approximately 12 months and will involve the examination of sediment, zoobenthos, phytobenthos, zooplankton, phytoplankton, fish, birds, mammals and peripheral vegetation. Some of the sampling stations are located so as to indicate the importance of the various input streams, drains, and industrial discharges. The substances to be determined will be a series of heavy metals organochlorine pesticides, PCBs and hydrocarbons. Utilizing these results, further studies will be carried out in particular areas or on organisms which are of critical interest. This will allow the determination of existing concentrations of heavy metals and other persistent substances.

In addition to the field investigation outlined above, a number of laboratory studies are also in progress. One involves an assessment of the rate of uptake of a number of heavy metals (copper, lead, zinc, and cadmium) by phytoplankton and algae (*Ditylum brightwellei* and an *Enteromorpha* species) which occur in Westernport Bay.[3] The organisms are cultured in the laboratory using suitable media and conditions. Similar work has been conducted in a number of laboratories throughout the world which has indicated that some of these organisms are significant accumulators of heavy metals.[4]

Seasonal eximations of the concentrations of heavy metals in bay phyto-plankton also constitute part of this investigation, and some preliminary results are available. Zinc, copper, and lead were found to have concentrations of 80, 35, and 31 ppm (dry weight), respectively, while the concentration of cadmium was below the detection limit.[5] As a comparison, plankton from the Irish sea contained zinc, 282 ppm; copper, 36 ppm; and lead, 900 ppm.[4]

One of the objectives of the study is to determine the effects of possible pollutants on bay organisms and ecology. To develop some understanding of these aspects, a toxicology and physiology research program on fish and invertebrate animals from the bay is underway. The selection of toxic substances for study posed a number of problems. The composition of effluents from the group of industries which may wish to establish on the catchment is extremely diverse. However, in the literature it is reported that certain heavy metals are ubiquitous and can also be biocumulative. As a result, initial experiments have commenced using cadmium and zinc. Lethal concentrations and the effects of sublethal concentrations are being investigated using static tests and flow-through experiments.[6] The test animals have been selected on the basis of availability, suitability for laboratory studies, as well as economic and ecological importance. Species selected up to the present time are juveniles of yellow-eye mullet (*Mugil forsteri*), adults of the marine hardyhead (*Pranesus ogilbyi*), the shrimp (*Macrobrachium intermedius*), and certain crab species. The lethal concentrations of some zinc and cadmium for these animals have been determined. Future work will cover sublethal effects.

CONCLUSIONS

In certain estuaries in Australia, such as the Derwent and Tamar in Tasmania, evidence of abnormally high concentrations of zinc, copper, and cadmium were found in oysters growing in these bodies of water. A recent report by Thrower and Eustace[7] indicates concentrations of zinc as high as 21,000 ppm, cadmium 63 ppm, and copper 450 ppm wet weight. The tragedy in this instance is that the Pacific oyster, *Crassostrea gigas* Thunberg, was introduced into Tasmania from Japan in the late 1940s and proved so successful that commercial leases to grow this oyster were set up in the Tamar river.

No such beds exist in Westernport Bay. However, the bay does contain a remarkable variety of fish and other aquatic organism. The rationale for the studies and investigations briefly described is that obtaining as much basic knowledge about these organisms and their relationships to environmental factors as possible will permit a more rational strategy for planning and development of the catchment, land use, water use, and environmental control.

REFERENCES

1. Private communication, Environment Protection Authority of Victoria, Melbourne, 1973.
2. WEATHERLEY, A. H., Australian National University, private communication.
3. DUCKER, S. and CANTERFORD, S., Westernport Bay Environmental Study, Progress Report 1973, p. 23.
4. RILEY, J. P. and ROTH, I. *J. Mar. Biol. Ass. UK* **51**, 63 (1971).
5. DUCKER, S. and CANTERFORD, S., University of Melbourne, private communication.
6. AHSANULLAH, M. and NEGILSKI, D., Westernport Bay Environmental Study, Progress Report, 1973, p. 18.
7. THROWER, S. J. and EUSTACE, I. J. *Fd Technol. Australia* **25**, 546 (1973).

MERCURY DISTRIBUTION IN THE CHESAPEAKE BAY

F. E. Brinckman, K. L. Jewett, and W. R. Blair
Inorganic Chemistry Section, National Bureau of Standards, Washington DC 20234

and

W. P. Iverson and C. Huey†
Corrosion Section, National Bureau of Standards, Washington DC 20234

As part of the National Bureau of Standards program[1] on heavy metals pollution, the distribution of mercury in the Chesapeake Bay is being surveyed in collaboration with Dr. Rita Colwell's group at the University of Maryland and others.

Water, plankton, and sediment samples were taken at a number of Chesapeake Bay sites which were selected for their accessibility, exposure to technological effluents and availability of biological data. Sediment samples (grab and gravity cores) were frozen to $-80°C$ as soon as they were obtained.

The samples were analyzed for total mercury using flameless atomic absorption techniques (EPA Provisional Method for Mercury in Sediment and Water, January 1972). The values obtained were determined from standard curves constructed by least square fits. Redox potential, pH, and total sulfide ion activity measurements were made at 10 cm depth intervals, on portions of the fresh core samples, alternate to those portions removed and frozen for mercury analysis. Most of the data on the core samples is in the process of evaluation, and the mercury, as well as other heavy metal, concentrations in frozen core samples from the entire Chesapeake Bay still await determination. The results of a preliminary survey of the sediments from the upper portion of the Bay are shown in Table 1.

The concentrations vary from 0.80 ppm, at Colgate Creek near Baltimore, a heavily polluted area, to 0.02 ppm in the middle of the ship channel below Annapolis. Total mercury concentration in plankton (zoo- and phyto-) ranged from 4.9 ± 0.10 ppm to 0.02 ± 0.01 ppm with an average of 0.92 ppm for 15 stations. The total mercury concentration in unfiltered water from 17 stations ranged from 0.00 ppb to 0.49 ppb with an average of 0.24 ppb.

The table illustrates the relationship between the concentrations of total mercury in the sediment, water, and plankton at all stations. As is evident, there is almost a five-thousandfold increase in mercury concentration from the water to the plankton which is in accordance with the findings of others.[2]

No attempts have been made to speciate the mercury in water, plankton, or sediment samples other than the direct field detection of a very small quantities (1–2 ng) of mercury vapor from sediments. In addition to the marked involvement of plankton as a (at least intermediate) "sink" in the transport of mercury through estuarine media, a more recent finding suggests that another "sink" may be significant. Dr. John Walker (Colwell's group) has found mercury to be of significance in petroleum fractions isolated from the sediment in a highly polluted area (Colgate Creek).[3] In association with Dr. Walker, we have found total mercury concentrations approaching 0.1% by weight in some cases which suggests a preferred solvation of certain mercury-containing compounds from the aqueous phase. Preliminary chromatographic analysis indicates the mercury to be eluted along with the C_{16}–C_{18} fragments in the petroleum fractions rather than appearing as free metal or simple methylmercurial species.[4]

To make the situation more complex, we have found elemental sulfur to be present in these same petroleum enriched fractions in concentrations as high as 1% of the sediment (dry weight). A possible mercury trapping mechanism is suggested involving selective partitioning of mercury into the petroleum residues and its subsequent inactivation by sulfur.

---

†NBS Guest Worker, 1973.

TABLE 1
Distribution of total mercury concentration[a] *at selected stations in the Chesapeake Bay*

Station	Position	Sediment[b]	Water[c]	Plankton[d]	Plankton/water
Colgate Cr.	N 39° 15′ 15″	0.90 ± 0.03	0.20 ± 0.01	—	—
	W 76° 36′ 15″	0.67 ± 0.01	—	0.09 ± 0.01	—
		0.65 ± 0.02	0.04 ± 0.02	0.17 ± 0.02	4250
		—	—	0.23 ± 0.00	—
Baltimore	N 39° 11′ 30″	1.12 ± 0.43	0.02 ± 0.01	1.53 ± 0.12	7650
Harbor 7B	W 76° 28′ 54″	0.26 ± 0.01	0.37 ± 0.01	1.65 ± 0.19	4460
		0.25 ± 0.01	0.24 ± 0.01	0.35 ± 0.04	1460
Parson's Is.	N 38° 54′ 05″	0.19 ± 0.00	0.68 ± 0.11	0.39 ± 0.03	570
	W 76° 13′ 35″	0.04 ± 0.01	0.33 ± 0.01	1.20 ± 0.10	3640
		0.09 ± 0.01	0.26 ± 0.01	0.15 ± 0.02	580
		0.04 ± 0.00	0.10 ± 0.02	0.06	600
Sandy Pt. Lt.	N 39° 00′ 00″	0.00 ± 0.00	0.28 ± 0.05	1.07 ± 0.53	3820
No. 900	W 76° 23′ 00″	0.13 ± 0.00	0.08 ± 0.01	0.18	2250
Matapeake	N 38° 57′ 43″	0.20 ± 0.08	0.04 ± 0.01	—	—
No. 858-C	W 76° 22′ 49″	0.20 ± 0.03	—	0.20 ± 0.00	—

[a] All analyses performed in accord with Environmental Protection Provisional Methods (January 1972). Results are cited ± average deviation for three or more duplicate runs.
[b] In ppm or $\mu g/g$ dry weight.
[c] Unfiltered samples; in ppb or $\mu g/l$.
[d] Lyophilized samples; in ppm or $\mu g/g$ dry weight.

The petroleum may, conversely, be an excellent reservoir for mercury contamination since many of the same microorganisms which transform sediment-bound mercury can also degrade the metallized petroleum with possible re-release of mercury. Whether or not the sulfur cycle is ultimately involved in "freezing" mercury to the marine sediment is not yet known, and experiments are underway to clarify this aspect.

Acknowledgments—The work was supported in part by the Environmental Protection Agency Grants R801002 and R800647 and by the National Bureau of Standards Measures for Air Quality Program. The authors are indebted to the National Science Foundation and the Chesapeake Bay Institute for extended use of the R/V Ridgely Warfield.

REFERENCES

1. Cycle of metals in biosphere studied, *NBS Techn. News Bull.* **56**, 282 (1972).
2. DAVIS, J. and FERGUSON, J. F., The cycling of mercury through the environment, *Wat. Res.* **6**, 989 (1972).
3. WALKER, J. D. and COLWELL, R. R., Mercury-resistant bacteria and petroleum degradation, *Appl. Microbiol.* **27**, 285 (1974).
4. BLAIR, W., IVERSON, W. P., and BRINCKMAN, F. E., Application of a gas chromatograph–atomic absorption detection system to a survey of mercury transformations by Chesapeake Bay microorganisms, *Chemosphere*, **3**, 167 (1974).

SESSION VI

SOURCE REDUCTION METHODOLOGY

CANADIAN EXPERIENCE WITH THE REDUCTION OF MERCURY AT CHLOR-ALKALI PLANTS

F. J. FLEWELLING
Canadian Industries Ltd., Montreal

INTRODUCTION

During the $3\frac{1}{2}$ years since early 1970, activities in Canada to reduce environmental pollution by mercury from chlor-alkali plants have been concentrated primarily on the mercury being discharged directly to waterways in effluent streams. Atmospheric mercury emissions have received much less attention, and the matter of the ultimate disposal of mercury-contaminated solids and sludges is still largely unresolved. This paper reviews briefly the steps which have been taken in all three areas, attempts to estimate the extent to which environmental pollution has been reduced, and considers the work still remaining to be done.

The mercury cell chlor-alkali industry in Canada consists for the most part of small plants scattered at wide intervals across the country. Table 1 shows the number of plant locations and the combined daily capacity for the past few years.

TABLE 1
Mercury cell capacity

Year	No. of plant locations	Total chlorine capacity (tons/day)
1969	13	1494
1970	14	1533
1971	14	1681
1972	14	1703
1973	11	1125

The average plant in Canada produces roughly 100 tons/day of chlorine. In this paper mercury losses are normally referred to in terms of lb per 100 tons chlorine production which also represents therefore the daily loss from the average plant.

During 1973, mercury cell operations at three locations were phased out, and as a result the ratio of mercury cell/diaphragm cell capacity in Canada declined from 60/40 to 40/60.

DIRECT DISCHARGE OF MERCURY TO WATERWAYS

The immediate steps which were taken at mercury cell plants during 1970 and 1971 to reduce mercury discharge in effluents involved:

(1) the diversion to temporary ponds of mercury-contaminated solids and sludges which had previously been sewered;
(2) the reduction in process effluent volume through recycling of individual streams to the process where possible, through the elimination of process purge streams, and by reduction in the volume of water used for cleaning purposes;
(3) the replacement of direct contact cooling of mercury liquid and vapors with indirect heat exchangers or recycled water systems.

These initial steps brought about a substantial reduction in the mercury discharge, and levels down to 1 lb/day were achieved at many plants. Unfortunately, actual figures on the losses of mercury to sewer prior to these first abatement measures are very sparse. From process considerations and data on total mercury consumption, an estimate of up to 20 lb/day for an average plant appears reasonable. A reduction then of as much as 95% was achieved at some plants in the initial attack on the problem.

To cope with the mercury remaining in the effluent after the initial measures, facilities were installed at most locations to precipitate mercury in sulfide form. Sulfide precipitation is a highly effective treatment procedure provided that the contaminated effluent, which ranges in volume from 5000 to 25,000 gal/day, is not diluted prior to treatment with the $\frac{1}{4}$ to 2 million gallons of noncontaminated cooling water used in the plants. A lengthy rebuilding of sewer systems to effect this separation was necessary in many cases before sulfide treatment could be adopted.

The experience at many locations where sulfide treatment plants were installed has been that while the level of mercury at the discharge from the treatment plant could be reduced to well below 0.1 lb/day, the mercury content of the plant outfall decreased very slowly and leveled off at a figure well above that from the treatment plant. Table 2 shows the data for a plant in which sulfide treatment was started at the end of 1970.

TABLE 2
Mercury discharge in typical effluent

Year	Mercury discharge (lb per 100 tons chlorine)
1970, 2nd half	0.68
1971, 1st half	0.47
2nd half	0.38
1972, 1st half	0.22
2nd half	0.21
1973, 1st half	0.20

In this particular plant, the effluent from the sulfide treatment unit, about 5000 gal/day in volume, contributes only 0.02 lb per 100 tons to the total figure. The remainder is due in large part to the continued leaching by 500,000 gal/day of water of mercury deposited in previous years in the downstream sewer lines, a process which could continue for decades.

This wide difference in mercury level between the discharge of the treatment plant and the plant effluent at its final discharge point is an important point to consider in setting effluent quality standards. In Canada, federal regulations require that the discharge from a mercury cell plant should not in any one day exceed 0.5 lb for each 100 tons of chlorine produced. While most plants in Canada continue to experience occasional individual days on which the regulation is exceeded, they are able to meet the 0.5 lb per 100 ton figure on a long term average. Data provided by *Environment Canada* show that for a 12 month period from mid 72 to mid 73, nine plants reported a total mercury discharge in effluent of 821 lb which, averaged over 304,000 tons of chlorine production, represented a level of 0.25 lb per 100 tons.

ATMOSPHERIC MERCURY EMISSIONS

Loss of mercury to the atmosphere from mercury cell plants occurs by three routes:

(1) in waste hydrogen which is vented to the atmosphere at most locations;
(2) in the exhaust to atmosphere of air which has been drawn in over the end boxes of the cells to minimize the escape of hydrogen and mercury into the cell room atmosphere;
(3) in the exhaust air from the ventilation of the cell room area.

The extent to which waste hydrogen gas was being cooled before discharge prior to 1970 is not easily established. While most plants were equipped with some form of heat exchanger, the operating effectiveness was often limited by high water temperatures in summer months and by fouling of the waterside surfaces. As an order-of-magnitude estimate, the average discharge of 5 lb per 100 tons of chlorine production is reasonable. In the interval since 1970, improvements in primary cooling at several locations have eliminated excessive losses and brought the discharge down below a 2 lb per 100 ton figure. A few plants have installed secondary chilling to reduce the gas temperature to the 0–5°C range at which a discharge of less than 0.25 lb per 100 should be achievable.

The so-called end-box ventilation exhaust is highly variable from plant to plant. In some plants no such exhaust exists, with a resultant higher level of mercury in the cell room. The levels of mercury in the exhaust are of the order of 1–2 lb per 100 tons. In a few plants, chillers have been installed to reduce the temperature to the 0–5°C range at which the discharge would be comparable to that of the chilled hydrogen stream, i.e. 0.25 lb per 100 tons.

The level of mercury in the cell room atmosphere is governed by the general level of cleanliness of the floors with respect to spilled mercury, by the effectiveness of the end-box ventilation system, and by the types of denuders used. With the relatively small size of plants in Canada, the discharge from the average cell room should be of the order of 1 lb/day if the level in the exhaust air were 100 $\mu g/m^3$. As levels in the cell room are reduced in keeping with the 50 $\mu g/m^3$ threshold limit value, the discharge to atmosphere should also drop accordingly.

For the 12 month period from mid 72 to mid 73, the reported total discharge to atmosphere from nine plants in Canada was 9467 lb of mercury. Averaged over 304,000 tons of chlorine production, the discharge was 3.11 lb per 100 tons, which represents a reduction of 60% in the probable level which prevailed before 1970. With increased application of secondary chilling of waste hydrogen and end-box ventilation, further reduction in this average discharge is achievable. A contribution to ambient air quality of not more than 1 $\mu g/m^3$ (averaged over 30 days) should be possible from Canadian plants.

MERCURY-CONTAMINATED SOLIDS AND SLUDGES

In mercury cell plant operation, spent brine from the cells is resaturated by contacting it with solid salt, and chemicals are added to precipitate calcium, magnesium, and other metallic impurities. Since these steps are carried out in a brine normally containing 2–5 ppm mercury, the insoluble material from the salt feed, principally calcium sulfate, and the precipitated impurities are contaminated with mercury. For most plants, the level of contamination is low, less than 50 ppm; for a few plants levels around 1000 ppm are experienced, apparently as the result of the particular brine purification reagents and system used. Prior to 1970, brine sludges were usually flushed to sewer, representing a mercury discharge of from 0.2 to 4 lb per 100 tons chlorine, depending on the plant.

One of the first steps taken at mercury cell plants in 1970–1 was to divert brine sludges from the sewer system to temporary holding ponds. These materials have now been accumulating for 3 years, and at sites where land is limited they are presenting a space problem. Leaching with sodium hypochlorite solution is an effective means of removing mercury from sludges which are high in mercury, e.g. 1000 ppm, and is being practised successfully at, at least, one location. The level of mercury in the residual solids, 20–30 ppm, is comparable to that in sludges originating at other plants. For these low mercury level materials neither leaching nor high temperature retorting is effective in removing any appreciable fraction of the mercury.

Burial of the brine sludges under controlled conditions is being studied as the only practical means of disposing of the material. A burial pit lined with peat moss has been in use at one location for a year now and samples of subsurface water taken from the immediate surroundings show no evidence of leaching of the mercury from the burial area. Tests without a peat moss lining at a second site indicate that even sandy soil strongly absorbs any mercury moving into it from buried sludge.

Spent filteraid from the filtration of caustic soda, mercuric sulfide precipitates from effluent treatment, and cell room muds have high mercury contents in the range of 2–20%. High temperature retorting can be used to recover the mercury contained in these materials, and leaching with hypochlorite is also effective. At present only a few plants are carrying out mercury recovery operations. The others are accumulating the mercury-rich materials, and consequently their consumption of new mercury does not yet reflect their efforts to reduce environmental pollution.

For the 12 month period from mid 72 to mid 73 the total mercury disposed of in solids from nine plants was 20,900 lb, which, averaged over 304,000 tons production, equals 6.86 lb per 100 tons of chlorine.

MERCURY CONSUMPTION AND MATERIAL BALANCE

Data are available each year from *Statistics Canada* on the amount of mercury used in the production of chlorine and caustic soda. The figures are based on mercury purchases and inventory changes over each year as reported by individual companies and exclude mercury to equip new plants. While not all companies report figures every year, the survey covers 90% of the mercury cell chlorine production. Table 3 shows the mercury consumption data reported for the last 4 years in relation to the chlorine produced by the reporting companies.

TABLE 3
Mercury consumption

Year	Mercury consumption (lb)	Chlorine production (lb)	Unit mercury consumption (lb per 100 tons)
1969	195,239	445,000	43.9
1970	229,715	450,000	51.0
1971	174,190	460,000	37.9
1972	93,860	470,000	20.0

The marked increase in mercury consumption in 1970 over 1969 is surprising since abatement measures at several plants were well under way by the end of 1970. The year 1972 shows a very substantial decrease in consumption. With the planned closing of three mercury cell plants in 1973, it is possible that 1972 purchases were curtailed to some extent in the latter part of that year by depleting the cell room operating stock which is not included in the inventory.

From the data provided by *Environment Canada* for a 12 month period, the plants which closed down in 1973 can be excluded. The mercury purchases by nine plants operating on solid salt feed totalled 65,150 lb or 21.4 lb per 100 tons of chlorine produced. In Table 4 an attempt has been made to reconcile that figure with the total mercury discharged in effluents, gaseous emissions, and in sludges. When allowance is made for the mercury contained in the filtered caustic soda product, only 50% of the mercury purchased is accounted for in the total.

TABLE 4
Mercury material balance

	Nine plants mid 72 to mid 73 (lb per 100 tons)	1969 estimate (lb per 100 tons)
Mercury purchases	21.4	44
Mercury discharged in effluents	0.25	20
Mercury in gaseous emissions	3.11	8
Mercury in solids and sludges	6.86	—
Mercury contained in products	0.27	2
Mercury not accounted for	10.9	14

The problem of arriving at a material balance for the mercury used has been recognized by mercury cell producers for many years. With the higher mercury consumptions in earlier years, the mercury which could not be accounted for was a smaller fraction of the total as shown in the estimates for 1969 in Table 4. Also, with only the more laborious dithizone method available, mercury analyses of waste streams were infrequent, and the significance of the data was questionable. There was always the feeling on the part of many producers that the problem of arriving at a balance was merely one of inadequate sampling and analysis. With the reduction in mercury consumption in recent years, the mercury which cannot be accounted for has become a larger fraction of the total, and industry has become increasingly aware of the gap in the mercury material balance.

There is some evidence that an appreciable portion of the mercury which has not been accounted for elsewhere has found its way through cell room floors to the underlying soil. The cell room structure itself adsorbs some mercury from the cell room atmosphere, and levels up to 1% have been measured in the surface layer of walls and steel columns. However, the total weight of mercury tied up in that manner in one cell room which was checked was less than 500 lb.

The figures in Table 4 show that the amount of mercury discharged to the environment has been substantially reduced in the last 4 years, particularly that part which previously was being discharged directly to water courses. Some further reduction in mercury consumption can be expected with increased recovery of mercury from sludges and further decreases in atmosphere emissions. However, until the route by which 50% of the mercury purchased is being consumed can be identified and remedied, a further major reduction in mercury purchases is unlikely.

Canadian Experience with the Reduction of Mercury at Chlor-Alkali Plants (F. J. Flewelling)

DISCUSSION by FRED T. OLOTKA
Superintendent Environmental Control, Hooker Chemical Corporation, Niagara Falls, New York

INTRODUCTION

First let me state that Mr. Flewelling can quote as an authority on Canada's experiences in mercury reduction. He has represented Canadian industry as a member of our industrial task force on mercury abatement since its formation 3 years ago. The task force was first organized by the Chlorine Institute in 1970, and is comprised of member Canadian and US companies representing mercury cell operators from the chlor-alkali industry. It was the work of this group that engineered the control of mercury loss to our waters within a relatively short period. Some of the original group are continuing these efforts with respect to mercury loss in air and solid waste. A good deal of progress has already been made in this direction.

Unlike our canadian neighbors, the US has only 27% of its chlorine capacity produced from mercury cells. The total output in the US, however, is about seven times as much, or 2.6 million tons of chlorine. It is unlikely, however, that any significant mercury cell expansion will take place in the US, and with an expected increase in diaphragm cell expansion, a drop in the 27% will be evident within the next 2 years.

The "mercury crunch," if I might use the word, has had its impact on US industry. One cell plant was shut down for 2 years due to lack of compliance to regulations, and started up recently under new ownership and commitment to meet all EPA restrictions. A second plant shut down permanently due to adverse public and government pressures, age, and various economic reasons. A third plant is closing its older mercury cells and converting to diaphragm cells. All new cell expansions in the US are diaphragm cell oriented.

The future operation of the existing plants will depend upon their ability to comply with EPA regulations.

MERCURY EMISSION—WATER

In both countries, government pressure was initially applied to mercury loss in waste water, and industry, through its task force efforts, crashed through an abatement program in a remarkably short time to significantly reduce these losses.

The US plants were able to curb their water losses by 95% within one year, and by using similar technology as our Canadian counterparts, mercury loss has now been reduced by 99.9%.

Our experiences in the level-out of mercury are essentially the same as Canada's, with most plants running their abatement systems at 0.02 lb/day mercury loss, and their total plant effluents at 0.15–0.20 lb/day.

You might recall Canadian restrictions are limited to 0.5 lb of mercury per 100 tons of chlorine produced. The US plants ranging from 100 to 700 tons have therefore achieved remarkable abatement, considering their size.

The current EPA effluent limits, as of October 11, 1973, are now aimed at even further reduction.

1. Total dissolved mercury = 0.00014 #/1000 # Cl_2 maximum any one day, = 0.00007 #/1000 # Cl_2 maximum daily average 30 consecutive days.
   These limits are not only far stricter than the Canadian ones, but have not been achieved by any US plant. They are to be met by July 1977.
2. Zero discharge of process waste pollutants.
   This limit is to be met by 1983 for existing plants and as soon as the law becomes effective with respect to newly designed plants. One mercury cell plant estimated its cost to obtain a zero discharge would be equal to $3 per ton of chlorine or 6.0+% of its manufactured cost. This does not include the heavy expenditures yet needed to control air and solid waste effluents.

MERCURY EMISSION—ATMOSPHERE

Generally speaking, air emission limits were imposed earlier on the US plants. To this extent US efforts are more advanced than Canada's. The EPA issued its final emission standards in April 1973. US plants are now given not more than 2 years to meet the necessary requirements, with a total of 2300 g/day of mercury emission as the maximum for any plant, regardless of size. This is broken down as

1300 g for cell room emissions and 1000 g for hydrogen, inlet, and end box. Since cell rooms are extremely difficult to material balance, 18 housekeeping rules have been set up to which each plant must rigorously adhere. A number of US plants have already had EPA conformance inspections. We expect that most of our plants will meet the air standards by the end of 1974.

Unlike water abatement, the technology is more varied for air. Mercury losses depend upon temperature, volumetric flow rates, entrainment, and the degree of saturation in the gas stream.

Abatement designs follow engineering concepts that would:

(1) Cool the gas through indirect-contact heat exchangers at each decomposer as well as additional cooling and refrigeration of gases to 0° and 5°C in the H_2 plant.
(2) Remove entrained mercury using mist eliminators.
(3) Reduce vapor concentration by chemical scrubbing.
(4) Reduce vapor concentration using resin, activated charcoal, and resin-sieve type beds.
(5) Reduce the volumetric flow rate of the gas stream.

Unlike our efforts in water abatement, these concepts are not universally adaptable to all plants. Those located in warmer regions may be limited in the use of water cooling and must back up mercury loss in their hydrogen with scrubbing systems or resin-sieve beds.

Some plants have found that expensive mist eliminator elements sag under the weight of recovered metal and require liquid washes as well as mechanical vibrators. Others have found that particulates must be removed before resin beds and activated carbon towers can be successfully used.

The success in reducing mercury loss from end boxes lies in the ability of a plant to keep several hundred covers tightly on the boxes. If this control is poor, then an expensive abatement installation is needed to curb mercury loss—this could mean the installation of a $70,000 scrubbing unit or a $90,000 resin-sieve bed system.

To be careless with EPA's 18 housekeeping rules might very well involve abatement on the entire cell building. Many of these plants are multistoried and large in size. Their ventilation systems range from 100,000 ft^3/min over 1,000,000 ft^3/min. How do you capture the entire building and pass its ventilation through an abatement system? One plant in Puerto Rico has no walls or roof. Their housekeeping just *has* to be meticulous to stay within the 18 rules.

The impact of the new EPA regulation, which limits atmospheric emissions to 5.1 lb/day, is to reduce mercury loss in the industry by at least 75%.

SOLIDS WASTES

Mr. Flewelling stated that the question of mercury accountability is still unresolved. Our US plants have also found the problem to be as slippery as the metal. Many of us feel that a significant percentage of unaccounted for mercury is tied up in sludge recovered from water abatement operations. Most of the sludge is stored at plant site, some of it is also buried under licensed control. A number of plants have had limited success in recovering mercury using retorts. Some have abandoned the operation due to its low production capacity.

Of interest to the industry is the outcome of a rotary kiln operation sponsored through a Federal grant. The success of this operation will be valuable in proving the contention of some cell operators that the answer to inconsistent mercury accountability is largely due to the idiosyncrasies of sampling and analyzing sludge.

The amount of solids emitted from a cell plant vary widely with the salt source, its purity, and whether graphite or metal anodes are used in electrolysis. These solids consist of graphite, sulfates, calcium, magnesium salts, and other impurities, as well as mercury tied up as a sulfide. Experience in some of the US plants has shown that well over the 6.86 lb reported in the Canadian survey can be tied up in such sludge.

Burial in some states has been permitted under license provided that:

(1) The sludge is kept in covered drums or tied up as concrete.
(2) A weight record be filed with the Government.
(3) Burial must be in a clay compacted area containing a heavy liner.
(4) Burial must be below frost line level.
(5) Drums must be covered with clay immediately upon delivery.
(6) Standpipes must be provided at the site for monitoring purposes.
(7) A record of monitoring must be submitted periodically to the government control agency.

RECLASSIFICATION OF MERCURY FROM HAZARDOUS TO TOXIC

The industry was recently notified that mercury has now been put in a toxic substance category. This now places a concentration as well as a weight limit on allowable mercury discharges. Further, the mercury standard would now be gross rather than net, and relates to discharge into navigable water. A gross limit means that the plant must provide abatement to handle flash floods and rains that might wash mercury-bearing industrial and power plant fallout on its property, excessive mercury in its river water intake, and in the bulk of its operating and raw materials it purchases.

Under review is a 0.2 lb Hg/day maximum limit for each plant regardless of size; and under a "toxic classification" neither economics nor available technology will be permitted as consideration for compliance waiver. The impact of the ruling on the cell industry is significant.

I might point out that one of the cell plants has analyzed 830 river water samples since mid 1971. The mercury content of the river averaged 0.11 ppb and ranged from 0.1 to 0.6 ppb. This plant uses 70,000,000 gal/day of water. In addition, it purchases raw and operating materials to run its operations, all of which by analysis contain levels of mercury ranging from 5 ppb to 2.4 ppm. Most are not related to the cell operation.

Based upon daily composite analysis of its discharges, this plant will not only fail to meet the new criteria, but will have to shut its cell plant down and treat only river water and its raw and operating materials whenever the 0.2 lb/day limit is exceeded. This represents 12% of the year and is equivalent to treating 3 billion, 80 million gal of water, and untold tons of material.

Incidently the plant under discussion was chosen as an exemplary operation by EPA standards.

Its average mercury discharge from abatement process	= 0.015 lb/day
Its average mercury discharge from sewers surrounding the cell plant	= 0.136
Its average mercury discharge relating to non-mercury cell sewers	= 0.046
Its average mercury discharge relating to intake river water	= 0.064
Total	0.261 lb/day average

The toxic classification limit therefore forces the exemplary plant to treat all non-related mercury cell inputs, or face the alternative of operating this plant at "zero discharge."

In addition, depending upon volume and mercury concentration, all joint municipal treatment plants will fall under this restriction. I am not too certain the design of such plants is capable of reducing mercury to within acceptable levels, considering the volumes of flows to be handled. It appears that several EPA agencies may be in conflict with one another at this point, and unfortunately the mercury cell industry will probably be compromised on the ultimate decision.

Of interest to the scientific mind is the fact that nature has liberally endowed the earth with numerous forms of mercury-containing material: ore, rocks, coal, oil, wood, etc. Much of its transport is natural, through volcanic action, vapor pressure, solubility, and burning. Such contributions are natural and we are all certain they do not need man's support to add to their levels. Are we going far enough or are we putting all our abatement eggs in one basket?

By far the greatest pressure of regulatory action in the US has been applied to mercury emanating from mercury cell operations. Comparatively little pressure has been applied in the past 3 years to such mercury users as the electrical and instrument industries, to municipal incinerators, and poorly operated dumpsites handling discarded fluorescent lights, mercury switches, or the vaporization of mercury from millions of buildings painted with mildew-proof paints.

The coal industry has forecast the US coal requirements to jump from 580 million tons this year to 3 billion tons by 1985.

Samples of coal taken by our plant average 0.6 ppm of mercury. After burning, the fly ash which represents 10% of the coal, averages 290 ppb. In western New York the mercury fallout calculates to 28 lb daily from coal burning alone. Oil carries a higher mercury content. Bordering on the Great Lakes and located in a high precipitation area, a good deal of this fallout will find its way into Lake Erie and the Niagara River.

Earlier it was mentioned that one river averaged 0.11 ppb mercury, therefore at a flow rate of 235,000 ft^3/s, it carries 140 lb/day of dissolved mercury.

On a coal plus oil *per capita* basis for the entire Great Lakes basin, fuel consumption probably contributes half of the 51,000 lb of dissolved mercury emptying out into Lake Ontario each year.

And a contribution from 3 billion tons of coal—conservatively approximates a 340,000 lb fallout of mercury per year across the US. I would like to summarize at this point.

(1) Despite the success of Canadian and US water abatement programs, current EPA restrictions for the US are aimed at levels not yet achieved by the trade.
(2) The air program has progressed further in the US than in Canada, and most plants are expected to meet compliance by the end of 1974.
(3) Solid waste handling is still a problem in both countries, needing greater development work, particularly in the recovery of mercury.
(4) Reclassification of mercury from a hazardous to a toxic substance has placed agency regulations in conflict and must be resolved.
(5) An apparent credibility gap is developing in the need to apply added restrictions on the cell industry despite the presence of greater and as yet unregulated sources of mercury emission.

PHYSICAL–CHEMICAL METHODS OF HEAVY METALS REMOVAL

JAMES W. PATTERSON

Associate Professor and Chairman, Department of Environmental Engineering, Illinois Institute of Technology, Chicago 60616

and

ROGER A. MINEAR

Associate Professor of Environmental Engineering, Department of Civil Engineering, University of Tennessee, Knoxville 37916

INTRODUCTION

The toxic nature of many metals, even at the trace level, has been recognized with respect to public health for many years. Lead, cadmium, arsenic, mercury, antimony, and beryllium all fall into this category, and all have been reported as causative agents in accidental industrial deaths.[1] Beyond these unambiguous situations, many metals have been evaluated as toxic to aquatic life above certain "threshold toxicity" levels. Continued release of metal wastes into the environment has been justified on the basis of dilution to undetectable levels or to levels below the "threshold toxicity" value in the receiving water body. That is, the water quality standards could be met upon dilution.

There are current arguments that no such threshold exists.[2] Avoiding involvement in such arguments, one must, nevertheless, concede that metals are not biodegradable in the environment. Nor, unfortunately, are they completely locked into the sediments. This has been dramatically demonstrated in a few instances such as the Minamata Bay incident involving mercury, and the itai-itai byo or ouch-ouch disease resulting from rice fields contaminated by cadmium-containing industrial wastes.[1] There is, in addition, evidence that biochemical transformations similar to those well-documented for mercury, may convert arsenic, selenium, and tellurium to organic compounds much more toxic than the inorganic forms.[3] Cadmium has been shown to be concentrated by shell fish to values reaching 900–1600 times those found in the surrounding water.[1] Other instances of biological magnification of metals similar to the well-documented phenomenon with pesticides appear likely.

The consequence of this is that environmental protection, sought through water quality standards which are in turn based upon instantaneous water concentrations of pollutants, may not be accomplished. The concept of effluent standards is frequently argued for in terms of ease of control and monitoring, and the goal of zero pollutant discharge is gaining wide acceptance. With metals, a conceptual approach (and that which currently prevails in establishing effluent standards) is application of best available technology.

The purpose of this paper is, then, to examine those processes currently available for metals treatment and to identify levels of treatment being accomplished and at what cost. The most widely used approach for metals removal is precipitation, and experience with this process has produced a wide range of treatment efficiencies. We propose to examine a number of reasons for such diversity by considering precipitation theory and translating this into problems encountered in actual practice. Notable among problems associated with precipitation processes are solids separation and ultimate disposal of the solid product (often a high water content, voluminous sludge). The former problem can be related directly to difficulties in achieving high treatment efficiencies based on total effluent concentrations.

SOURCES OF METALS

Metal processing, finishing, and plating industries are obvious sources of metal wastes, and the nature of the waste directly reflects the particular combination of metals and manufacturing processes utilized by a given plant. Generally, the waste problems are related to transfer of dissolved metal from plating baths or metal surface cleaning baths to rinse waters by drag out. Usually, generation of these wastes is continuous. The concentrated plating baths themselves may be periodically discharged on a batch basis due to buildup of contaminant. An extremely wide range of general industries are incorporated in these categories since metal fabrication, cleaning, and plating are frequently integrated into the overall manufacturing process. Similarly, plastics industries may include electroplating of metals onto plastic surfaces.

Other industries may have specific and characteristic metal waste problems, where the particular metal is an integral part of the manufacturing process. Notable examples are the high zinc wastes of viscose rayon manufacturing, groundwood pulp production, and newsprint paper production.[4] Similarly, cadmium and silver may be found in photographic, porcelain and ceramics, and pigment-printing wastes. Table 1 provides some representative values for untreated effluent metals concentrations. A more extensive listing[5] contains a wider range of values, depending upon the exact process and wastewater constituents.

METAL TREATMENT PROCESSES

Table 2 summarizes the general treatment methods employed for the removal of metals from waste discharges. Many of the processes listed have not yet been successfully implemented on a full scale basis for metals treatment. Only precipitation is widely practiced where product recovery from the waste stream is not sought. Consequently, it is the prevalent treatment method in older and smaller operations where waste streams are usually complex and mixed. Even when other processes are used for polishing effluents, the resultant concentrated solutions may be treated by precipitation processes.

For new or more recent installations (either those with proper existing configurations, or those requiring only minor process modification) combined savings in material loss and treatment costs may make use of one of the other techniques economically advantageous. A great deal of current literature is devoted to the use of these processes. When designed into product recovery schemes and completely closed cycle operations, they provide an effective solution to the metal waste discharge problem. When such processes have not been included in the original plant design, the cost of conversion to such systems may be prohibitive, at least on the short term basis. There are many examples of successful conversion but each case must generally be evaluated independently.

Precipitation

Simple equilibrium solubility calculations have few meaningful examples in the real world, particularly in consideration of metal solubility. The simplest case,

$$MA_{x(s)} \rightleftharpoons M^{+x}_{(aq)} + xA^{-}_{(aq)}, \qquad (1)$$

$$K_{so} = [M^{+x}][A^{-}]^x, \qquad (2)$$

TABLE 2
Methods of metallic wastewater treatment

Precipitation	Electrolysis
Hydroxide–oxide	
Carbonate	
Sulfide	
Ion exchange	Activated carbon
Evaporation	Cementation
Freeze purification	Flotation
Reverse osmosis	Electrodialysis

TABLE 1
Representative metal concentrations of untreated industrial wastewaters

	Metal concentration (mg/l)						
Process	Cd	Cr	Cu	Fe	Ni	Zn	Ref
Pickling-dipping	—	—	20–90	$2.0–2.3 \times 10^3$	—	10–35	6
Automatic plating of zinc base die castings	—	5–10	20–30	—	45–55	3–8	6
Automatic plating on plastic base	—	—	10–15	—	30–40	—	6
Automatic band plating and passivation: mixed barrel and rack	7–15	2–25	3–8	—	15–25	5–10	6
Vulcanized fiber	—	—	—	—	—	200–300	7
Cold steel finishing with galvanizing	—	1–6	—	60–150	—	2–88	8

would indicate a concentration of metal in solution wholly dependent upon the concentration of A^-. If A^- is hydroxide ion, then obviously the residual metal concentration is dependent upon the pH of the solution, a continuously decreasing function as pH increases,

$$\log [M^{+x}] = \log \frac{K_{s0}}{K_w^x} - x(\text{pH}) \quad (3)$$

illustrated by line A in Fig. 1. Magnesium hydroxide follows this model. Most metals of industrial concern do not, however, but instead exist in solution as a series of complexes formed with hydroxide and other ions. Each of these complexes is in equilibrium with the solid phase and their sum defines the total metal ion solubility. For the case of hydroxide species only,

$$M_T = M^{+2} + M(OH)^+ + M(OH)_2^0 + M(OH)_3^-$$
$$+ M(OH)_4^{-2} \ldots \quad (4)$$

The theoretical solubility limit then becomes a complex function of the pH, as illustrated by B of Fig. 1. Minimum solubility occurs at some calculated pH, and the total metal solubility increases at higher and lower pH values. Similar considerations for other metals yield characteristic curves as shown in Fig. 2. Obviously, optimum pH values for individual metals removal do not coincide. This provides at least a theoretical basis for optimizing metal removal through segregated treatment.

Figure 3 represents this difference in another way, showing only optimum percent removals as a function of pH for several metal hydroxides. For mixtures of metal ions, a pH which optimizes total metal removal will not provide maximum removal for any given metal in solution.

Theoretical solubility values may not be readily achieved in practice for a number of reasons, a few of which we will examine:

(1) *Wastewater composition* has two effects upon

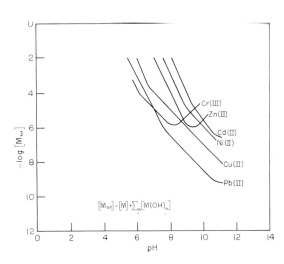

FIG. 2. The solubility of pure metal hydroxides as a function of pH. (Taken from ref. 9.)

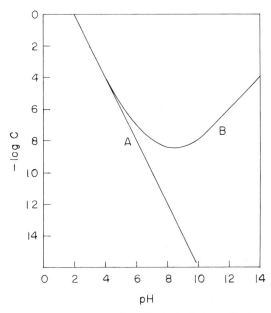

FIG. 1. Theoretical solubility of hypothetical metal hydroxide $M(OH)_2$, with and without complex formation.

A: $\log M_T = \log \left(\frac{K_{s0}}{K_w^x}\right) - x \text{ pH}$, where $K_{s0} = 10^{-10}, x = 2$.

B: $\log M_T = \log [M^{+2}] + \sum_n [M(OH)_n^{+2-n}]$,

where

$K_{s0} = 10^{-10}$, $K_{s1} = 10^{-8}$, $K_{s2} = 10^{-9}$, $K_{s3} = 10^{-4}$, $K_{s4} = 10^{-5}$.

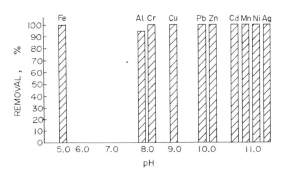

FIG. 3. The pH of optimum percent removal by lime neutralization for selected metals. (Taken from ref. 10.)

the actual solubility. First, the calculations made in generating the relationships in Figs. 1 and 2 are based on stoichiometric or mass concentration. This is not entirely correct, particularly when considering higher charged ion species in the presence of appreciable quantities of other ions in solution. Mass solubilities increase significantly under these conditions since solubility constants are truly constant only for ion activities or thermodynamic concentrations. These values decrease in a nonlinear fashion as ionic strength changes.

$$\alpha_i = \gamma_i [C_i], \tag{5}$$

where,

$$\log \gamma_i = 0.5 Z_i^2 \sqrt{N}, \tag{6}$$

$$N = \text{ionic strength} = 1/2 \sum C_j Z_j^2. \tag{7}$$

Secondly, the presence of organic and inorganic species other than hydroxide, which are capable of forming soluble species with metal ions will further increase the total metal solubility since relationship (2) holds only for the free metal ion while simultaneously the complex ion equilibria contribute to the total metal concentration by

$$M_T = [M^{+x}] + \sum [\text{complex forms}]. \tag{8}$$

Notable examples of inorganic complexing agents which result in very high residual metal concentrations are cyanide and ammonia.

(2) *Temperature variation* can explain deviations between calculated and observed values if actual process temperatures are significantly different from the value at which the theoretical constant has been evaluated. Process temperature variation is to be expected, and accompanying treatment level fluctuations will occur simultaneously.

(3) *Nonequilibrium systems* are likely a major consideration. Thermodynamic calculations express minimal solubility, only if the system has truly reached equilibrium as described by the equations. One consideration, of course, applies to the nature of the solid phase assumed. If solubility products are computed from considerations of crystalline solid phases, the real system may be at metastable solution—solid equilibrium with an amorphous solid form of much higher solubility. Conversion of the solid phase to its most stable form may be very slow and not reliable relative to the treatment system under consideration.

Precipitation kinetics could conceivably be controlling the treatment rate, but particle size rather than completeness of solid phase formation will control the apparent removal effectiveness. The problem then becomes one of solid phase removal by efficient solids removal systems or conversion of small particles to large particles. The data of Table 3 illustrate this problem. Better solids removal improves treatment levels nearly an order of magnitude in each case.

TABLE 3
Comparison of solids removal efficiencies[11]

Residual soluble metal	Precipitation and clarification, plus sand filtration (mg/l)	Clarification followed by paper press filtration (mg/l)
Cd	0.7	0.08
Cu	2.0	0.04
Ni	2.6	0.1

In consequence of these and other considerations, theoretical and laboratory results do not always agree well with results obtained in treatment practice. High temperatures, or dissolved solids made up of noncommon ions, can increase the solubility of the metal. Dissolved solids concentrations at approximately 1000 mg/l may increase the solubility more than 100%, as compared to the solubility in pure water. Dissolved solids concentrations of several thousand mg/l, such as may occur in continuously recycled waters, can have an even more marked effect upon the solubility of the precipitate. Soluble metal concentrations can, however, also be obtained that are even lower than theoretical values, often at pH values that are not optimum. Effects of coprecipitation, and adsorption on flocculating agents added to aid in settling the precipitate, appear to play a significant role in reducing the concentration of the metal ions.

Formation of precipitates other than the hydroxide has potential application for waste treatment. For example, the solubility of cadmium carbonate is approximately a hundredfold less than of the hydroxide.[12] The sulfides of several metals are also much less soluble than their corresponding hydroxides. Sulfide precipitation appears particularly applicable for copper, nickel, and zinc treatment. This technique potentially involves toxicity problems, with excess residual sulfide reagent. However, a system has recently been developed that provides for sulfide precipitation without the toxicity problems.[13] Ferrous sulfide, which has a higher solubility than the sulfides of the metals to be precipitated, is used as the precipitating reagent. The solubility of ferrous sulfide is quite low (10^{-5} mg/l of sulfide ion), so that the toxicity problem is eliminated. Freshly precipitated ferrous sulfide is most reactive and is obtained by adding an excess of a soluble sulfide for precipitating the metals to be removed from the effluent and then adding sufficient soluble ferrous salt to precipitate all excess sulfide ion. The pH is normally adjusted to the range of 7–8, prior to precipitation. Hexavalent chromium which may be present is reduced to chromium(III) by the ferrous iron and immediately precipitated as the hydroxide. Therefore, no extra precipitation steps are necessary to remove the chromium. If the excess ferrous iron in solution is considered undesirable, it may be oxidized to iron(III), which will precipitate as the hydroxide.

Ion exchange

Ion exchange is more properly a concentration than final treatment technique. During treatment, the ion-exchange resin exchanges its ions for those in the wastewater. This process continues until the solution being treated exhausts the resin exchange capacity. When this point is reached, the exhausted resin must be regenerated by another chemical which replaces the ions given up in the ion exchange operation, thus converting the resin back to its original composition, and yielding a concentrated regenerant brine.

The regenerant brine must be treated; often more easily than the original volume of wastewater. In some cases, chemicals can be recovered from the brine. In practice, wastewaters to be treated by ion exchange are generally filtered to remove suspended solids which could mechanically clog the resin bed. Oils, organic wetting agents, and brighteners, which might foul the resins, also must be removed; perhaps by passage of the wastewater through activated carbon filters.

Successful ion exchange treatment has been reported for chromium wastewaters, primarily as the hexavalent chromate and dichromate ions.[14–16] The procedure utilizes an anionic exchange resin, normally regenerated with concentrated sodium hydroxide. The eluate is sodium chromate, from which purified chromic acid may be recovered by removal of sodium in a cationic exchange system. Copper has been successfully treated by cation exchange,[17–19] as have lead,[20] mercury,[21,22] nickel,[23,24] and other metals.

Ion exchange is an attractive method for the removal of small amounts of impurities from dilute wastewaters, or the concentration and recovery of expensive chemicals from segregated concentrated wastewaters. In addition, ion exchange permits the recirculation of a high quality water for re-use, thus saving on water consumption. The limited capacity of ion exchange systems means that relatively large installations are necessary to provide the exchange capability needed between regeneration cycles.

Reverse osmosis

Most of the development work and commercial utilization of the reverse osmosis process, especially for desalination and water treatment and recovery, has occurred during the past 10 years. Reverse osmosis uses a pressure differential across a semipermeable membrane to separate a solution into a concentrated and a more dilute (effluent) solution. It has been utilized in several plating industry installations for chemical recovery, wherein small units of under 300 gal/h capacity have been installed to recover plating bath chemicals and make closed-loop operation of a plating line possible.

Membrane materials for reverse osmosis are fairly limited, and the bulk of the development work has been with specially prepared cellulose acetate membranes, which can operate in a pH range of 4–8 and are therefore restricted to wastewaters that are not strongly acidic or alkaline. Another limitation of currently available membranes is a narrow acceptable temperature range for satisfactory operation. For cellulose acetate-based systems the preferred limits are 65–85°F; higher temperatures will increase the rate of membrane hydrolysis, while lower temperature will result in decreased flux. Certain solutions (strong oxidizing agents, solvents, and other organic compounds) can cause dissolution of the membrane. In addition, fouling of membranes by slightly soluble components in solution is possible, as well as fouling of membranes by wastewaters high in suspended solids. A metal finishing waste treatment plant designed to produce no liquid effluent has been recently installed at the Rock Island (Ill.) arsenal. The system uses chemical precipitation followed by reverse osmosis. Key components in the process are two reverse osmosis units operating in parallel and capable of handling 6800 gal/h of wastewater.

Evaporation

Brief mention should be made of the use of evaporation (or distillation). The primary use of evaporative treatment has been for product recovery. However there has been limited use to treat final concentrated wastewater residues, to dryness. Evaporative treatment may thus be considered, insofar as a treatment process, in terms of end-of-the-line technology, for zero discharge.

There are basically two types of evaporative recovery systems commonly in use: the vacuum evaporator and the atmospheric evaporator. A vacuum evaporator operates at subatmospheric pressures, thus enabling evaporation to take place at temperatures in the range of 130–190°F. An atmospheric evaporator operates at atmospheric pressure and the normal boiling temperature of the solution being processed. These types of evaporators can be utilized in either open or closed loop processing cycles.

Evaporation is a well-established industrial process for recovering plating chemicals and water from plating waste effluents. Commercial units for handling zinc, copper, nickel, chromium, and other metal plating baths have been operating successfully and economically for 10 years or longer. Packaged units for in-plant treatment of plating wastes are available from many manufacturers. Evaporative systems have relatively high capital and operating costs, especially the vacuum units. The economics of distillation, from the standpoint of either investment or operating costs, imposes a constraint on the size range of such systems. Units with a capacity of the order of 300 gal/h are used in practice. They are fairly complex, and require trained personnel to operate and maintain them.

Other technology

There are other technologies with either potential or limited specific applicability in treating metallic wastewaters. Included among these are ion flotation, liquid–liquid extraction, freeze concentration, activated carbon adsorption, electrolysis, and cementation. Only the latter three technologies have been directly utilized in wastewater treatment.

Some success has been reported on chromate removal by activated carbon.[25–27] Most effective treatment results with dilute wastes, containing less than 10 mg/l chromate. To date, the technique has only been applied at the pilot plant level, and applicability for full scale treatment is thus not proven.

An interesting, and apparently effective method of treating copper wastewaters, is cementation onto scrap iron.[28, 29] The principle of the technique is quite simple. It involves the oxidation–reduction reaction of soluble copper(II) with elemental iron as shown below:

$$Cu^{+2} + Fe^0 \longrightarrow Fe^{+2} + Cu^0. \qquad (9)$$

The result is to replace copper with iron in solution; the copper plating onto the solid iron surface. This reaction has been used for many years in the mining industry to recover copper from sulfuric acid leachates. If hexavalent chromium is also present in the wastewater, it may react with either elemental or ferrous iron to yield trivalent chromium:[29]

$$2Cr^{+6} + 3Fe^0 \longrightarrow 2Cr^{+3} + 3Fe^{+2}, \qquad (10)$$

$$Cr^{+6} + 3Fe^{+2} \longrightarrow Cr^{+3} + 3Fe^{+3}. \qquad (11)$$

In the most recent industrial waste application of this technique,[29] scrap iron is suspended in a perforated rotating drum through which the wastewater flows. Copper is cemented onto the iron and scraped off as particulate copper through the abrasive action of the mixed iron scrap as it tumbles within the drum. Chromium(VI) is simultaneously reduced to chromium(III). Effluent soluble chromium, iron, and residual copper are precipitated as the final step in the treatment process. Oil-laden wastewaters cannot be treated by cementation as the oil wets and blinds the iron scrap from reacting.

Electrolysis is primarily a recovery technique, normally yielding an effluent which requires subsequent treatment by precipitation or other means. Direct electrolytic recovery of copper from concentrated waste streams is often both technically and economically feasible. The process may require a preconcentration step, such as ion exchange or evaporation. Recovery of copper or zinc from plating baths normally presents no problem, but the presence of other metal contaminants may interfere by simultaneous or preferential plating at the electrode. For copper cyanide baths, copper recovery by electrolysis can be accomplished in conjunction with cyanide destruction within the cell.

ACTUAL TREATMENT PRACTICES

Pretreatment and precipitation

Precipitation process efficiencies vary for several reasons among which are: nature of the metal(s) to be removed, pretreatment required, volume of waste flow, strength of waste, and cost and availability of treatment chemicals. Lime is the most frequently employed base, primarily due to its lower cost relative to sodium hydroxide or soda ash (sodium carbonate). Limestone slurries and dolomite lime have also been used where substantial process or economic advantages could be demonstrated.

When strong acid solutions are to be treated, such as waste plating or pickling solutions containing sulfuric acid, lime neutralization yields large volumes of calcium sulfate sludge. In some cases this may represent as much as 30% of the original waste liquid volume. Under such conditions, the added cost of sodium bases might be justified provided total dissolved solids is not a criterion of treated effluent acceptability.

Precipitation can be conducted on a batch or continuous basis. The former is usually employed for low flow waste streams or treatment of contaminated pickling, plating, passivating, or other concentrated solutions. Batch systems normally require greater holding volume capacity than continuous systems, but can be operated manually rather than requiring automatic monitoring and control. Concentrated wastes that are discharged intermittently may be assimilated into a continuous, dilute waste stream through sustained bleed-off either directly or from holding tanks, thereby providing a more time-constant flow-contaminant base for system design, but at the expense of rigorous treatment.

Pretreatment prior to precipitation of metal hydroxides is frequently an important part of the overall treatment process. The conversion of soluble chromium(VI) as the chromate or dichromate anion to chromium(III) is mandatory, since only the latter forms the insoluble hydroxide. Equally important is the removal or separation of complexing agents. Cyanide removal is dictated by toxicity considerations. It must also be removed before precipitation of metals due to its strong complexing ability. This complexing action and, especially in the case of nickel and iron, the difficulty with which cyanide decomposes when associated with the metal ion, generally dictates as a first step separate treatment of cyanide wastes. Fluoroborate, fluoride, and pyrophosphate based plating processes have in part solved the cyanide problem. In the case of copper, however, pyrophosphate must be isolated from the cupric copper or reduction of the copper to the cuprous state is necessary prior to base addition. The soluble cupric-pyrophosphate complex is quite strong.

Flow separation offers numerous advantages. Individual waste streams can be pretreated and neutralized to coincide with optimum conditions for a specific metal contaminant. In the case of cadmium and nickel, formation of the carbonate, which possesses better settling characteristics than the hydroxide, would be feasible. Similarly, special methods such as the patented Kastone Process,[30] which seems to offer a distinct advantage for cadmium and zinc cyanide plating wastes, could be applied to these lines only. The major advantage with this process seems to lie in the better solids characteristics and lower solubilities related to formation of the metallic oxides rather than hydroxides. In fact, the oxides are normally the ultimate stable solid form of many metal hydroxides. The transition from more rapidly formed hydroxide to the oxide is generally too slow for meaningful exploitation in treatment. Research into the use of ozone in conjunction with lime[10] may lead to such exploitation. Figure 4 demonstrates the advantage of ozone treatment in terms of shifting the optimum pH for precipitation. These data are from laboratory scale studies only, however.[10]

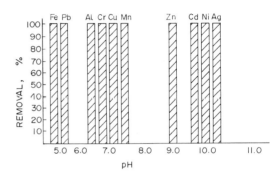

FIG. 4. The pH of optimum percent removal of selected metals treated with ozone after hydroxide precipitation with lime. (Taken from ref. 10.)

Flow separation affords an additional advantage relative to residual solubility. As pointed out by Curry,[31] even for the unlikely case where the pH of minimum solubility for two metals coincided, the final waste concentration is dictated by the solubility relationship, thereby, yielding [M_1] and [M_2] in a flow of Q_T gal/min. If the flows are treated separately, the concentrations will still be [M_1] and [M_2], prior to flow combination. Upon discharge, however, the combined flow metal concentrations will be [M_1]/($Q_1 + Q_2$) and [M_2]/($Q_1 + Q_2$), where Q_1 and Q_2 are the respective separate stream flows and their summation is Q_T.

A form of segregated treatment is the patented integrated treatment processes of Lancy.[32] The process, generalized in Fig. 5, is applied to any metal finishing process where the waste originates through dragout of a process bath solution adhering to the metal part. Rinsing of this part produces a metallic wastewater. The integrated system first conveys the contaminated part to a concentrated treatment chemical bath, immediately after removal from the process bath, and then to the rinse operation. The system can be completely closed cycle, but even if the rise water is wasted, it now contains only

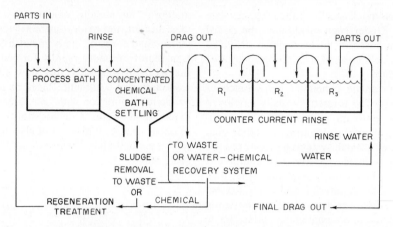

FIG. 5. Generalized schematic of the integrated treatment system.

treatment chemicals, generally not of objectionable nature at the effluent levels. The concentrated treatment bath affords several treatment advantages: (1) pH conditions optimum for a particular metal, (2) where feasible, large excess of treatment chemicals can be used which drives the precipitation reaction to more rapid completion and may enhance the nature of solids formed, and (3) outflow of solids will be only those adhering to metal surface rather than all poorly settling particles which would be carried over the discharge weir in both batch and continuous solids separators.

Actual treatment data for precipitation processes reported in the literature are summarized in Table 4. These values have been selected on the basis of representing treatment that is currently being accomplished in full scale industrial operations. Some of these plants have been in operation for many years and have continued to maintain treatment levels reported in early references.[33,40]

Variations in treatment levels will most certainly reflect variations in process control and reaction kinetics of the nature discussed in the previous sections. As can be seen from the data in Tables 3 and 4, solids separating procedures can be equally important in establishing the apparent treatment efficiency.

Other treatment processes

Among the various processes discussed, a great amount of treatment data is available on precipitation, but much less on other processes. One study, recently commissioned by the US Environmental Protection Agency, surveyed treatment technologies employed by the electroplating industry to treat copper, zinc, chromium, and nickel.[46] The study assessed treatment data for 53 plants, of which 26 were visited. Table 5 summarizes the treatment processes employed for metals (among the four—copper, nickel, chromium, and zinc) associated with each of the 53 plants.

The plants reviewed were selected on the basis of employing "good" treatment technology for at least one of the metals involved. In treating copper, nickel, and chromium, 75–76% of these plants used some form of precipitation. Zinc treatment by precipitation accounted for 64% of the plants with zinc wastes, with evaporation being next most common, at 25%. For the former three metals, treatment other than precipitation was about evenly distributed between ion exchange, evaporation, electrolysis, and reverse osmosis. It must be emphasized that the patterns of treatment described in Table 5 apply only to the selected plants, and do not characterize the treatment technology of the total electroplating industry. Table 6 reports the ranges of effluent and 53 plant median effluent observed in the study.

SLUDGE PREPARATION AND HANDLING

The hydroxide sludges generated by precipitation processes are usually voluminous and difficult to dewater. The removal of these sludges from the liquid phase of the wastewater is normally by gravity settling in a clarifier. Schwoyer and Luttinger[47] have reported that metal hydroxide sludges collected in clarifiers typically contain 0.5–3.5% solids. Freshly precipitated metal hydroxides have the characteristic of a very low specific gravity, which makes effective settling difficult without long clarifier detention times or additional treatment.

Chemical conditioning of the floc may be employed to obtain improved settling characteristics, at increased operating costs. Successful application of

TABLE 4
Selected data representing actual treatment levels accomplished by metal hydroxide precipitation

Metal	Concentration prior to treatment (mg/l)	Effluent concentration after treatment (mg/l)	Comments	Ref.
Cd	—	0.54	1–2 mg/l coagulant added	33
	—	1.0	Integrated treatment	34
Cr	2.9–31.8	0.75		35
	140	1.0		36
	1300	0.06	1–2 mg/l coagulant used in clarifier	33
	—	1.3–4.6	Prior to sand filter	37
	—	0.2–0.5	After sand filter	
	—	0.05	Coprecipitated with $Fe(OH)_3$	38
Cu	—	0.2–2.5	Prior to sand filter	37
	—	0.2–0.8	After sand filter	
	—	21.0	Integrated treatment	34
	33	21.0		39
	204–385	0.5		40
	—	2.2	1–2 mg/l coagulant	33
	14–18	0.05–0.5	Laboratory study	41
	30	0.16–0.3	With sand filtration	41
	—	0.15–9.19		42
	10–20	1–2		43
	—	0.18–0.30	With sand filtration	44
Fe	2–5	0.3		45
	—	0.5		38
	—	0.09–1.90		44
Ni	35	0.4		35
	39	0.1	1–2 mg/l coagulant	33
	—	0.1	Prior to sand filtration	37
	—	15	Final pH 8.1	38
	—	2	Integrated treatment	34
Zn	55–120	1		40
	—	0.5–2.5	Prior to sand filter	37
	—	0.1–0.5	After sand filter	37
	70	3–5		4
	—	0.02–0.23	After sand filtration	44
	—	2	Integrated treatment	34

anionic polyelectrolytes, alum and ferric chloride as sludge conditioners have been reported. An alternate procedure which has proven effective is sludge densification, involving two-stage neutralization, with sludge recycle from the clarifier as the only neutralizing agent in the first stage.[29] The resultant metal hydroxide sludge is reported to have greatly improved settling properties.

Clarification, with or without sludge conditioning, is never completely effective in solid–liquid separation. Some of the smaller floc will escape even the most effectively designed clarifiers. In order to meet increasingly stringent effluent standards, some final polishing step is often necessary. Sand filters are commonly used as polishing devices, with good results. For example, Stone[37] has reported that

TABLE 5
Treatment processes utilized at 53 electroplating facilities[46]

Treatment	Cu	Ni	Cr	Zn
Precipitation				
Batch	4	5	8	4
Continuous	21	22	18	11
Integrated	5	5	12	3
Adsorption	0	0	0	0
Ion exchange	2	3	3	0
Evaporation	2	1	3	7
Electrolysis	2	0	2	2
Reverse osmosis	2	3	2	0
No treatment	2	3	2	1
Plants having specific metal waste	40	42	50	28

filtration reduced a post-clarifier chromium concentration ranging from 1.3–4.6 mg/l down to a range of 0.3–1.3 mg/l. This final residual was primarily in the soluble chromium(VI) form. The same author also reported a reduction of effluent copper concentration of 0.2–2.5 mg/l down to 0.2–0.5 mg/l after sand filtration. Thus sand filtration produced a more consistant quality effluent, although minimum effluent copper values were not reduced. Even filtration will not result in complete removal, particularly when a fraction of the metal remains in soluble form. For example, Nemerow[41] has also reported the results of using sand filtration as a polishing step for treating copper waste, with residual effluent levels of 0.2–0.3 mg/l obtained. This compares well with the results of Stone[37] and others. These results were obtained in gravity sand filters. Somewhat poorer performance has been reported for pressure sand filtration by Chalmers,[11] as shown in Table 3.

Sludge dewatering and disposal represents a major fraction (often 20–40%) of the total cost of metallic wastewater treatment. Dewatering may be accomplished by sand bed drying, lagooning (a thickening process), vacuum filtration, or centrifugation. Ultimate disposal of the dewatered sludge is often to a landfill site. One of the most popular dewatering processes is vacuum filtration, which can achieve final solids concentrations of 100,000–300,000 mg/l. The performance of the filter depends upon the initial sludge solids concentration, plus other conventional vacuum filtration parameters such as use of precoat, and vacuum applied. Effluent from the vacuum filter is normally recirculated back to either the precipitation tanks or the clarifier, depending upon its quality.

ECONOMIC CONSIDERATIONS

To this point we have not considered one of the most important aspects of treating metallic wastewaters—the capital investment and day-to-day operating and maintenance expenses of treatment. Construction costs, as reported in the literature, vary considerably from plant to plant and industry to industry, depending upon numerous factors. These include plant location, degree of waste segregation practiced, water and treatment chemical costs, waste characteristics, type of treatment process employed, and effluent standards to be met. Because of this spectrum of variables encountered, generalizations on capital cost data can be misleading.

Likewise, the operation and maintenance costs associated with treatment reflect the influence of

TABLE 6
Effluent data for 53 plant survey[46]

Treatment process	Effluent concentration (mg/l)				
	Cu	Ni	Cr^{+6}	Cr_T	Zn
Precipitation					
Batch	0.08–1.47	0.25–0.48	0.04–0.16	0.05–0.23	0.03–5.0
Continuous	0.03–3.1	0.06–7.0	0.0–0.6	0.11–10.0	0.08–2.0
Integrated	0.16	0.002–2.0	0.01–0.4	0.03–2.0	0.2–0.45
Ion exchange	1.0–2.0	1.0–5.0	0.01–0.38	0.02–1.0	n.a.
Evaporation	0.08–0.53	1.5	0.05–1.0	0.16–1.5	0.05–18.4
Electrolysis	0.2	n.a.	0.01–0.03	0.32–5.0	2.9
Reverse osmosis	0.29–0.3	0.00–2.5	0.1	0.03	n.a.
Median effluent—53 plants	0.2	0.5	0.06	0.3	0.3

many of the same variables described for capital costs. We can perhaps somewhat narrow this variability, for purposes of demonstration only, by looking at data reported for plants treating similar volumes of equivalent type wastes. Table 7 presents capital investments reported for three plants treating a chromate waste. Two of the plants utilized a process of chromate reduction to chromic ion, followed by pH adjustment and precipitation of chromic hydroxide. The third plant treated chromate waste by ion exchange.

The values reported in Table 7 are typical in several respects. They demonstrate the variability of capital investment experienced, even in constructing the same type of treatment process for wastes of similar character and volume (e.g. precipitation for plants A and B). They do not, in fact, encompass the complete range of costs one finds reported in the literature.[46] Also, typically, ion exchange will represent a greater capital investment than the precipitation plant. This greater cost may be offset in part by product recovery.

Table 8 reports operating costs for the same three plants (A, B, and C) described in the previous table, plus two additional plants utilizing precipitation and one additional plant employing ion exchange. Costs are reported in terms of thousands of gallons of wastewater treated and cost per pound of chromium removed. The latter approach appears to reduce the variability seen when operating costs are based upon wastewater volume, and indicates, for example, treatment costs for reduction plus precipitation of $0.56–0.93 per lb. Variation in operating and maintenance costs appears to be greatly influenced by the degree of instrumentation/automation designed into the plant and the degree of difficulty and associated costs involved in handling and disposing of the sludge.

At present, there appear to be few reliable guidelines available for predicting capital and operating costs incurred in handling metallic wastewaters. Among those generally accepted are that economies of scale on plant size do occur, which may be offset by increased treatment costs if wastes are combined rather than treated on a segregated basis; increased capital investment in instrumentation and plant automation may be largely recovered by reduced chemical and personnel costs; and, finally, product recovery by processes such as ion exchange or evaporative recovery, while not normally economical for the smaller plant or more dilute waste, may offset to a large extent the greater capital and operating costs of these processes when used by larger plants, or for more concentrated wastes.

TABLE 7
Capital investment in chrome-bearing wastewater treatment facility

Plant	Treatment process	Wastewater flow (gal/min)	Chromium(VI) concentration (mg/l)	Capital cost ($)
A	Precipitation	100	50	17,000
B	Precipitation	143	n.a.	44,750
C	Ion exchange	100	50–100	40,000

TABLE 8
Operating costs of chrome-bearing wastewater treatment facilities

Plant	Treatment process	Cr(VI) concentration (mg/l)	Cost ($ per 1000 gal)	Cost ($ per lb Cr removed)
A	Precipitation	50	0.39	0.93
B	Precipitation	n.a.	0.16	n.a.
D	Precipitation	120	1.00	0.62
E	Precipitation	60	1.00	0.56
C	Ion exchange	50–100	0.68	0.40–0.80
F	Ion exchange	n.a.	0.16–0.24	n.a.

REFERENCES

1. *Chem. Engng. News*, p. 29, July 19, 1971.
2. B. P. DINMAN, *Science* **125**, 495 (1972).
3. B. C. MCBRIDE and R. F. WOLF, *Biochem.* **10**, 4312 (1971).
4. D. M. ROCK, Hydroxide precipitation and recovery of certain metallic ions from waste waters, a paper presented at the Annual American Institute for Chemical Engineers Meeting, Chicago, Illinois, December 1971.
5. J. W. PATTERSON and R. A. MINEAR, *Wastewater Treatment Technology*, IIEQ Document No. 71-4, State of Illinois, August 1971.
6. W. LOWE, *Wat. Poll. Control, Lond.* 1970, 270.
7. *Environ. Sci. Technol.* **6**, 880 (1972).
8. E. J. DONOVAN, Jr., *Wat. Wastes Engng.* **7**, F22 (1970).
9. R. NILSSON, *Wat. Res.* **5**, 51 (1971).
10. A. NETZER, A. BOWERS, and J. D. NORMAN, Removal of trace metals from wastewater by lime and ozonation, *Pollution: Engineering and Scientific Solutions* (in press).
11. R. K. CHALMERS, *Wat. Poll. Control, Lond.* 1970, 281.
12. J. W. PATTERSON, unpublished data.
13. J. R. ANDERSON and C. O. WEISS, Method for precipitation of heavy metal sulfides, US Patent 3, 740, 331, June 19, 1973.
14. D. YURONIS, *Plating* **55**, 1071 (1968).
15. S. ROTHSTEIN, *Plating* **45**, 835 (1958).
16. E. W. RICHARDSON, E. D. STOBBE, and S. BERNSTEIN, *Environ. Sci. Techol.* **2**, 1006 (1968).

17. R. Pinner and V. Crowle, *Electroplat. Met. Finish.* **3**, 13 (1971).
18. *Chem. Eng.* **78**, 62 (1971).
19. G. Schore, Electronic equipment and ion exchange for use in activated treatment systems, presented at the 27th Purdue Industrial Waste Conference, W. Lafayette, Indiana, May 2–4, 1972.
20. C. F. Liebig, Jr., A. P. Vanselow, and H. D. Chapman, *Soil Sci.* **55**, 371 (1943).
21. P. N. Cheremisinoff and Y. H. Habib, *Wat. Sew. Wks.* **119**, 46 (1972).
22. W. G. Gardiner and F. Munoz, *Chem. Eng.* **78**, 57 (1971).
23. A. C. Reents and D. M. Stromquist, *Proc. Purdue Indust. Waste Conf.* **7**, 462 (1952).
24. R. F. Heidorn and H. W. Keller, *Proc. Purdue Indust. Waste Conf.* **13**, 418 (1958).
25. Anonymous, *An Investigation of Techniques for Removal of Chromium from Electroplating Wastes*, US EPA Publication 12010 EIE, 1971.
26. D. G. Argo and G. L. Culp, *Wat. Sew. Wks.* **119** (9) 128 (1972).
27. T. Maruyama, S. A. Hannah, and J. M. Cohen, Removal of heavy metals by physical and chemical treatment processes, presented at 45th Ann. Conf., Water Poll. Control Fed., 1972.
28. D. J. Whistance and E. C. Mantle, *Effluent Treatment in the Copper and Copper Alloy Industries*, the British Nonferrous Metals Research Association, No. 33, 1965.
29. T. L. Jester and T. H. Taylor, Industrial waste treatment at Scovill Manufacturing Company, Waterbury, Connecticut, presented at the 28th Purdue Industrial Waste Conference, W. Lafayette, Indiana, May 2, 1973.
30. *Environ. Sci. Technol.* **5**, 496 (1971).
31. N. A. Curry, Philosophy and methodology of metallic waste treatment, a paper presented at the 27th Purdue Industrial Waste Conference, W. Lafayette, Indiana, May 2–4, 1972.
32. L. E. Lancy, *Plating* **54**, 157 (1967).
33. N. H. Hanson and W. Zubban, *Proc. Purdue Ind. Waste Conf.* **14**, 227 (1959).
34. L. E. Lancy, W. Nohse and D. Wystrach, *Plating* **59**, 126 (1972).
35. J. J. Anderson and E. H. Iobst, *J. Wat. Pollut. Control. Fed.* **40**, 1786 (1968).
36. P. I. Aurutski, *Chem. Abstr.* **70**, 206 (1969).
37. E. H. F. Stone, *Proc. Purdue Ind. Waste Conf.* **22**, 848 (1967).
38. C. C. Cupps, *Ind. Wat. Wastes* **6**, 111 (1961).
39. W. A. Parsons and W. Rudolfs, *Sew. Ind. Wastes Eng.* **22**, 313 (1951).
40. O. W. Nyquist and H. R. Carrol, *Sew. Ind. Wastes* **31**, 941 (1959).
41. N. L. Nemerow, *Theories and Practices of Industrial Waste Treatment*, Addison-Wesley, Reading, Mass., 1963.
42. K. S. Watson, *Sew. Ind. Wastes* **26**, 182 (1954).
43. J. A. Tallmadge, Nonferrous metals, in *Chemical Technology*, vol. 2, *Industrial Waste Water Control* (C. Fred Gurnham, ed.) Academic Press, New York, 1965.
44. P. W. Eichenlaub and J. Cox, *Sew. Ind. Wastes* **26**, 1130 (1954).
45. C. S. Cutton, *Can. Munic. Util.* **99**, 19 (1961).
46. Anonymous, *Development Document for Proposed Effluent Limitations Guidelines and New Source Performance Standards—Copper, Nickel, Chromium and Zinc Segment of the Electroplating Point Source Category*, US Environmental Protection Agency Document EPA-440/1-73-003, August 1973.
47. W. L. Schwoyer and L. B. Luttinger, Dewatering of metal hydroxides, a paper presented at the 27th Purdue Industrial Waste Conference, W. Lafayette, Indiana, May 2–4, 1972.

Physical–Chemical Methods of Heavy Metals Removal
(J. W. Patterson and R. A. Minear)

DISCUSSION by JOHN D. WEEKS
US Environmental Protection Agency, National Environmental Research Center, Cincinnati, Ohio 45268

Professors Patterson and Minear are to be complimented on their paper. It is a very concise, yet comprehensive, review of the literature relative to industrial waste treatment techniques. The paper contains a very good review of the theory of hydroxide precipitation and, more importantly, interspersed within the paper are notes of caution which show that the authors are aware of the discrepancy between theory and practice. Furthermore, the authors recognize that successful industrial waste treatment must be related to "good housekeeping" at the point of waste origin. Technique such as counter-current rinsing and air blasting to reduce "drag out" all serve to keep the metal plating solutions in a small volume, high concentration configuration. This permits specialized treatment for specific substances if necessary.

In discussing this paper, it might be well to state some of the problems encountered and to present some preliminary results of a federally funded research project concerned with advanced waste treatment techniques with re-use of the produce water being the objective.

It has become apparent that in certain geographical areas of the United States, notably the southwest, the available water supply is approaching maximum utilization. Various organizations have approached the Bureau of Water Hygiene of the Environmental Protection Agency for guidelines or criteria for producing a potable water from sewage. No such standards exist. The Drinking Water Standards of 1962 presuppose a well-protected and guarded watershed. Such a condition does not prevail in waste treatment plants.

Further, the toxic metals rejection limits specified in the Drinking Water Standards are not pertinent to industrial waste discharges. For these reasons, then, research was undertaken to study the efficiency of various metal removal techniques in advanced waste treatment. To the discusser's knowledge, there is only one plant in the world which is making a direct ingestive re-use of sewage effluents, and this plant is located at Windhoeck, South Africa. In this situation, Stander and van Vurren recognized the burden of industrial metals removal and separated, and treated only domestic sewage. Waste flows from the industrial area of the community are treated separately.

There is a plant producing an excellent quality water from reclaimed sewage at South Lake Tahoe, California. In this case, the product flows into a lake which supports water contact sports and fishing. The lake then drains into an irrigation system. No ingestive re-use was designed into this plant which treats domestic sewage. Thus, there were and are no antecedents of such renovation on a sewage which contains a large industrial component.

The above federally sponsored research project was located on the grounds of the Dallas White Rock Sewage Treatment Plant. Dallas offers an optimum "mix" of domestic and industrial sewage for experimental purposes, and, further, it is located in an area of great concern about water resources.

The advanced waste treatment demonstration plant is an activated sludge treatment plant of a nominal 1 mgd capacity. The pilot plant can be fed from five points in the White Rock plant. These are:

(a) Raw sewage.
(b) Primary sewage.
(c) First stage trickling filter effluent.
(d) Second stage trickling filter effluent.
(e) Final effluent.

For the majority of this research, the demonstration or experimental plant has been fed either first or second stage, or a 50–50 mix of the two. It is felt that reclamation plants will always have some form of sewerage treatment upstream.

Following the activated sludge module is a solids contact or chemical reactor where alum or lime may be added along with various coagulants and polymers to effect metals removal. Following this are multimedia sand filters, carbon contact columns, and a chlorination basin. The final effluent is not used, but rather wasted into the headgate of the White Rock Sewage Treatment Plant.

Each of the above modules are free standing and

treatment configurations can be altered by valving. Therefore this plant offers excellent experimental opportunities at a realistic scale.

The acquisition of metals data started in June 1972. One of the first tasks was to characterize the metals concentration in sewage. Twenty metals are usually studied in a month. Twenty-five metals have been studied in the course of this project. The average concentrations, number of daily composite samples, and the predicted range, mean ±3 standard deviations, are tabulated in Table 1. Many of these metals have been under surveillance for nearly 18 months; some, notably silver, potassium, and magnesium, were dropped. Molybdenum was studied, but the work load forced termination of this element. Noteworthy items in Table 1 are the large number of zero truncated values which occur in statistical analyses and which suggest that the population is being influenced by "slugs" of pollution.

Various short term, high frequency studies have demonstrated such peaks of concentration, for example, cadmium varying from 0.01 to 0.14 ppm within 2 h. Other studies have indicated similar patterns in the concentration time series plots for iron, cadmium, cobalt, and nickel. Lead and manganese have little diurnal variation, while copper is received at the plant in a separate and distinct pattern which indicates a source of copper either removed from the general location of other sources or at a plant which is operating a different work cycle.

To date, three separate treatment configurations have been studied. The first is physical treatment, which is activated sludge, followed by multimedia filtration and two chemical treatments. In the first chemical, or "high lime", procedure, lime is added to the solids contact module to bring the pH to within a range of 11–11.5. Ferric chloride is added as a coagulant, and the effluent is then fed to the sand and carbon columns. In the second chemical treatment process, aluminum sulfate is injected into the chemical reactor along with a small amount of lime to adjust the pH and to insure a good floc formation. The current treatment configuration under study will introduce recarbonation after high lime treatment in order to produce a stabilized water. The results of these studies are depicted in Table 2. From Table 2 it can be seen that the chemicals used in the process can be a source of contamination. The ferric

TABLE 1
Concentrations of metals in Dallas White Rock Sewage

	Mean	Stat. range Max.	Min.	No.
Al	0.56	1.3	0[a]	44
Ag	0.001	0.005	0.000[a]	58
As[b]	12.0	49.0	0[a]	101
B	0.38	0.60	0.16	98
Ba	0.13	0.36	0[a]	118
Ba[b]	0.01	0.05	0[a]	18
Ca	50.0	94.0	7.0	132
Cd	0.014	0.055	0[a]	167
Co	0.03	0.10	0[a]	134
Cr	0.23	0.60	0[a]	167
Cu	0.24	0.97	0[a]	132
Fe	1.13	3.23	0[a]	132
Hg[b]	0.54	212.0	0[a]	68
K	14.0	18.0	10.0	58
Mg	5.2	7.4	3.0	86
Mn	0.07	0.13	0.02	133
Mo	15.0	105.0	0[a]	37
Na	103.0	158.0	48.0	96
Ni	0.10	0.26	0[a]	110
Pb	0.11	0.28	0[a]	110
Se[b]	7.2	31.0	0[a]	164
Si	9.6	17.0	2.2	7
V	5.2	8.2	2.2	13
Zn	0.43	1.87	0[a]	132

[a] Zero truncated if negative.
[b] Concentrations; micrograms per liter, other concentrations in milligrams per liter.

TABLE 2
Train removal (percentage) in physical high lime, and alum treatment trains

Metal	Physical	High lime	Alum
Al	82	—	52
As	22	54	47
B	5	7	6
Ba	64	38	53
Ca	6	+255	15
Cd	56	38	77
Co	25	+25	0
Cr	79	95	93
Cu	65	50	92
Fe	89	94	94
Hg	55	87	45
K	7	7	8
Mg	6	90	6
Mn	43	86	75
No	72	12	85
Na	4	+5	6
Ni	33	75	55
Pb	60	60	64
Se	80	93	90
Zn	77	88	83

chloride in use was found to be contaminated with cobalt, perhaps from storage in unlined stainless steel tanks. Cobalt, once introduced, appears to be resistant to removal. Calcium on the high lime train went to high concentrations as would be expected. Postprecipitation was a never-ending problem in this phase of the study. The increases in sodium are associated with the lime itself.

An interesting point is the degree of removal which can be achieved in a biological–physical process. This process removed more barium than did the alum treatment in which near complete removal was anticipated. This supports Patterson and Minear's caution that theory oftentimes breaks down in processes being carried out near the limit of solubility. To keep my discussion relevant to Professors Patterson and Minear's paper, it might be

TABLE 3
Removal efficiencies in solid contact module

Metal	High lime	Alum
Al	—	+190
As	17	28
B	2	+6
Ba	12	+17
Ca	+295	+28
Cd	23	33
Co	+50	0
Cr	92	75
Cu	44	40
Fe	42	56
Hg	46	+32
K	+7	0
Mg	84	2
Mn	83	20
Mo	+5	21
Na	+11	1
Ni	43	25
Pb	20	17
Se	46	0
Zn	62	+137*28

appropriate to present the removals observed across the solids contact or chemical reactor module. This module may be described as a large cylinder standing upright. Inside the large cylinder is a smaller cylinder suspended vertically which is divided into an upper and lower chamber. The activated sludge effluent and chemicals enter the upper chamber and are mixed. They then flow into the lower chamber for slow speed mixing to promote floc formation. The water then turns and flows into the annular chamber where the upward velocity is quite low. A sludge blanket is maintained so that solid material is removed both by contact with the sludge blanket and subsidence from above. The removals obtained from this unit are demonstrated in Table 3.

The removals shown in Table 3 are from raw data. There has been no rejection of any values since the purpose of this research is more concerned with abnormalities in removal. This permits the inclusion of abnormally high single values which alter the removal figures in some cases. Zinc is such a case. One value in the solids contact effluent averaged out to an unrealistic increase. Arbitrary elimination of this value yields a more reasonable overall 28% removal. The above is true for the alum–mercury value and molybdenum high lime value.

FACTORS WHICH AFFECT REMOVALS

Professors Patterson and Minear's paper mentions the possibility of complexing affecting metals removal. In Dallas we have observed breakthroughs of metals, particularly of copper. This breakthrough has been traced to a shift in the activated sludge module's operation from nitrification to denitrification. The complex ion coefficient for copper ammines is about 10–14. If ammonia is present, it will react with the copper to effectively keep it in solution in spite of the subsequent downstream processes. This same effect pertains to five other metals which were included in this study. The other five metals were present in such a low initial concentration and have complex ion coefficients which are so much higher than that of copper that the effect of complexing is not readily observable.

The possibility of treatment chemicals being a source of contamination was mentioned earlier. The increase in cobalt in the high lime train was traced to the ferric chloride coagulant. It was found that the ferric chloride has been in contact with unlined stainless steel tanks.

Similarly, the increases in sodium were traced to the lime. Therefore, permit this one note of caution. The reagents themselves must be analyzed before attempting to achieve low levels of removal for certain ions.

In general, satisfactory removals can be achieved with satisfactory frequency on all of the metals listed in the Public Health Drinking Water Standards except for lead, cadmium and arsenic. Cadmium, with a mandatory rejection limit of 0.01 mg/l because of its high toxicity, is a problem. Alum treatment effects twice the removal of high lime and would be the method of choice. However, cadmium must be controlled at the source and its discharge to sewers should cease. Here the processes and techniques mentioned in Professors Patterson and Minear's paper will prove valuable.

Lead is a metal which is always present in sewage and of fairly uniform concentration. Like cadmium it is toxic and must be controlled. It occurs more frequently above rejection limits than does cadmium, and removals which would permit sustained ingestion of the effluent cannot be reached. Paradoxically, the best treatment is the biological train, but in this case, 23% of the samples would fail. Theory would indicate a good removal as lead hydroxide in the lime train, but 46% of the samples were above the rejection limit of 0.05 mg/l. Similarly, theory would once again indicate precipitation as lead sulfate in the lime train. However, in actual practice, 34% of the samples were above the rejection limit. It is felt that much of the lead in the sewage is associated with gasoline exhaust emissions. It will be interesting to see if this metal's occurrence will be affected by the "energy crisis" which is developing.

The last metal which requires additional work is arsenic. Alum treatment yields a good removal whereas physical or lime treatment does not. Further, the removal of arsenic compounds in activated carbon was three times better following the lime treatment than the alum treatment. This suggests that the valence and anionic form of the arsenic in the influent may affect its removal in the train. Further, lime treatment may alter the form of organo-arsenical compounds to a form amenable to removal by carbon. This is an area where additional work is most necessary.

Lastly, (Professors Patterson and Minear stress this point also) better removals can only be achieved when better solids separation can be achieved. A high volume throughput, highly retentive filter to replace sand is necessary. Apparently, the more insoluble compounds are escaping the treatment modules as extremely small crystals and then are available for downstream solution. This is the most pressing need in the hardware phase of waste water reclamation.

THE EFFECTS AND REMOVAL OF HEAVY METALS IN BIOLOGICAL TREATMENT

CARL E. ADAMS, Jr., W. WESLEY ECKENFELDER, Jr., and BRIAN L. GOODMAN

Associated Water and Air Resources Engineers Inc.

INTRODUCTION

With the advent of increasingly stringent effluent quality criteria, the reduction or removal of inhibitory or potentially toxic substances has received the primary attention of the regulatory agencies. Heavy metals have received considerable attention in recent years with regard to toxicity to aquatic and human life and, consequently, regulatory authorities are establishing extremely low allowable concentrations of these metals in effluent discharges to receiving waters. Since practically all municipal plants and a great majority of the industrial treatment facilities in the United States employ biological treatment for the reduction of organic materials in the respective wastestreams, it is of significance to realize what effect heavy metals will have upon these biological processes and what removals might be expected in passing through the treatment facilities. By knowing the approximate removals of heavy metals through the biological systems, pretreatment or posttreatment facilities can subsequently be designed to meet the required discharge criteria.

Those heavy metals which have received the most attention with regard to effects on biological processes have been copper, chromium, zinc, and nickel. These metals are associated with the electroplating and other heavy metals industries which tend to discharge into municipal sewerage systems. Also, these metals are found widely in industry and may occur in combination with organic materials so that they will tend to enter a biological treatment process treating only an industrial waste. Consequently, this paper will deal primarily with these heavy metals and will go into greater detail on copper since it seems to have received more attention than other metals. Also, only the activated sludge and anaerobic digestion processes will be discussed since these unit operations appear to be the most common biological treatment processes in existence today. As regulatory standards become more stringent, it is anticipated that the activated sludge process will be the most predominant biological treatment process for organic wastewaters.

COPPER

The heavy metal copper has probably been investigated with regard to effects on biological processes more than any other single metal. This is probably due to the fact that copper is the most common and one of the most toxic of the heavy metals. Copper toxicity has been examined from the standpoint of determining both the toxic threshold concentration and the mechanism of copper toxicity on activated sludge organisms.

Effects of copper on the activated sludge process

Previous investigations have concluded that copper is one of the most toxic heavy metals to biological treatment processes.[1,2] Barth et al.[2] found that copper exerts a much more inhibitory effect on biological processes than chromium, zinc, or nickel, whether the metals were applied as continuous or slug dosages. Continuous studies on activated sludge were conducted with copper,[3] fed as copper sulfate (10, 15, and 25 mg/l as copper), and copper cyanide complex (0.4, 1.3, 2.5, and 10 mg/l as copper). The effects of the copper cyanide complex on effluent chemical oxygen demand (COD) are shown in Fig. 1. After the systems had acclimated to

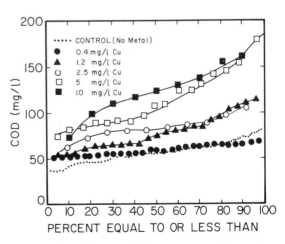

FIG. 1. Effect of copper, fed as copper cyanide complex, on COD effluents.

the effect of cyanide, there did not appear to be any difference in the effects of copper when applied as copper sulfate or copper cyanide. The increase in effluent concentrations shown in Fig. 1 was probably due in part to the increased turbidity, resulting from an increased effluent suspended solids concentration. It was concluded from these studies[3] that the maximum concentration of copper which could be received continuously without producing a detectable effect on the normal waste treatment parameters would be 1 mg/l.

Directo[4] studied the acute effects of copper on activated sludge subjected to 6 h shock dosages of copper at concentrations of 15, 30, and 45 mg/l in the influent sewage. In addition to varying copper concentrations, these studies examined the effect of operational process control parameters, including mixed liquor suspended solids (MLSS) levels of 2000, 3500, and 5000 mg/l, and raw sewage strengths of 200, 300, and 400 mg/l of COD. It was concluded that, at constant copper levels, an increase in MLSS concentration will slightly decrease inhibitory effects of copper. It was confirmed by Preseweckli[5] that tolerance increased with decreasing Me^{++}/MLSS ratio and increasing organic loading. It was also observed that the toxicity effects of copper were more pronounced at higher raw waste concentrations, even though the organic loading was maintained constant.

Additional work on copper toxicity[6] concluded that the deterioration of effluent quality of activated sludge, when subjected to continuous and shock copper dosages, was a result of effluent turbidity due primarily to the presence of individual bacterial cells. It was shown that the effluent COD of the process effluent was directly proportional to the magnitude of turbidity, as measured by optical density.

In studying the effects of copper toxicity at various organic loadings, Salotto et al.[7] concluded that moderate variations of organic loadings of the order of 2–3 : 1 will not markedly effect the toxicity of copper to the activated sludge process. As a result of their investigation, they concluded that the effect of copper, at 1 mg/l concentration, was insignificant with regard to final effluent COD at either high or low loading conditions. They concluded that the effects of higher concentrations of copper are more significant at lower organic loadings than at high loadings.

Finally, the results of previous studies[4,6] were incorporated into an additional investigation[8] in order to determine whether the uptake of copper by sludges was due to a biological or a chemical phenomenon and whether the release of turbidity and COD to the effluent was biological in nature or due primarily to a chemical reaction. Also, it was desired to relate the toxicity effects of copper to the process variables—copper concentration, influent sewage strength, and MLSS concentration. The conclusions from these investigations[1,4,6,8] are summarized as follows:

(1) Apparently copper, which has been introduced into the aeration basin and rapidly mixed, forms complexes with the various available ligands (constituents which will complex metal ions). Consequently, the majority of the copper will be present either as the hydroxide form, depending on the pH of the aeration basin, or as a stable copper–ligand complex. Under a specific set of conditions, a relatively constant amount of the copper reacts to form stable, inactive complexes with active cellular components, e.g., respiratory enzymes. This phenomenon occurs practically instantaneously, and any inhibition or toxicity effects resulting from the formation of these metabolically inactive complexes will be seen immediately. During a constant application of copper, a linear decrease in microbial activity will be observed.

(2) The degree of copper toxicity effect on the process is related to the overall copper equilibrium in the aeration tank. This copper equilibrium is dependent on the three independent variables reflecting either copper or ligand concentrations, i.e., copper concentration, MLSS concentration, and influent sewage strength. The relative concentrations of these variables influences the proportion of active components which are tied up in copper–ligand complexes and, subsequently, affect the extent of copper toxicity.

(3) The metabolically inactive cellular components which have been complexed as a copper–ligand compound may be reinstated to an active state, or the reaction may be reversed, either by dilution or by the addition of uncomplexed ligands. It is highly probable that the respiratory enzymes and other cellular materials may revert to an active metabolic state upon being set free from the copper–ligand complex. Consequently, a decrease in copper dosage, an increase in influent sewage strength, or an increase in MLSS level should decrease the proportion of active cellular components which are tied up as copper–ligand complexes and thereby result in better performance from the activated sludge system.

(4) It appears that the deterioration in effluent quality is immediate and will reach the maximum deterioration in a constant time (10–15 min) regard-

less of the influent concentration of copper. The degree of deterioration, as indicated by effluent COD or turbidity, is a function of influent strength, up to a certain point. Consequently, plots of total effluent COD versus time for different copper concentrations would all peak at approximately the same time; however, the peak rise in COD would be directly proportional to the concentration of copper dosage.

The above studies were all performed with municipal sewage. With an industrial waste, where copper was employed as a catalyst in the process, the effects of copper and cobalt concentrations were examined.[9] A Calgon chelating agent was added in an attempt to determine if complexing the copper would reduce its inhibitory effects on the activated sludge system. The results of copper concentration on COD removal and the relative degradability of the waste as indicated by the COD removal rate coefficient, are shown in Fig. 2. It is evident that the systems receiving Calgon material performed better than those systems receiving only the copper concentration. The correlation of the COD removal rate coefficient, k, also infers that the systems receiving Calgon material performed substantially better than those systems without the complexing agent. The COD removal efficiency dropped from 80% at a copper concentration of 1 mg/l to approximately 55% at a concentration of 2 mg/l. Further reductions in COD removal were not severe. These results inferred that severe effects of copper toxicity resulted at low concentrations, but after a critical threshold concentration of copper (2 mg/l) additional deleterious effects were not evident in proportion to increased copper concentrations. It was also noted that the specific oxygen uptake rate, expressed in grams of oxygen utilized per grams of MLVSS per day, was significantly retarded above copper concentrations of 2 mg/l. Elsewhere, it has been found that a continuous copper dosage as great as 45 mg/l is not adequate to completely terminate biological activity in an activated sludge system,[4] although the effluent COD will deteriorate substantially.

Effects of copper on the anaerobic digestion process

In studies with anaerobic digestion employing copper sulfate,[3] it was found that the digester did not cease to function when fed sludges, both primary and excess secondary, which were generated from raw sewage containing 25 mg/l of copper. Also, slug dosages as high as 410 mg/l to the treatment facility did not result in deleterious performance of the anaerobic digester. The results of the digester operation[3] are summarized in Table 1.

Removal and release of copper through the treatment facilities

In examining a shock dosage of copper at 50 mg/l for 6 h it was observed that only 16.7% of the applied copper had exited the activated sludge system after 50 h.[5] Based on hydraulic considerations, 94.3% should have passed through in 50 h. The majority of the copper in the process effluent was in the soluble or complexed form which passed through an 0.45 mμ millipore fiber. Therefore, a substantial quantity of copper was retained within the activated sludge mass.

Five continuous-flow activated sludge units were subjected to copper concentrations of 1, 10, 15, 30, and 45 mg/l, respectively.[6] The units receiving 1 and 10 mg/l were operated as extended aeration with no sludge wastage. The uptake and release of copper into the effluent is shown in Fig. 3. Effluent COD data indicated an initial upset followed by stabilization near the control (no copper added) values even though no sludge was wasted from the 1 and 10 mg/l

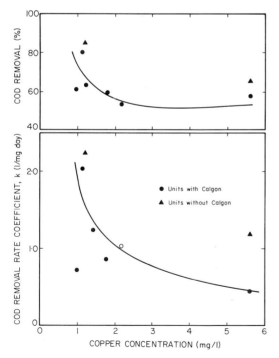

FIG. 2. Effect of copper concentration on activated sludge performance.

TABLE 1
Effects of copper on digestion and accumulation in sludges

		Cu in sludges fed to digesters							
		Primary sludge			Excess activated sludge			Gas production of digesters	
Cu in sewage (mg/l)	Form of Cu fed	Cu		Total suspended matter (mg/l)	Cu		Total suspended matter (mg/l)	Primary sludge	Combined primary and excess activated sludge
		mg/l	% of total suspended matter		mg/l	% of total suspended matter			
5	CuCN Complex	73	0.32	23,000	89	1.8	5000	Normal	Normal
10	CuCN Complex	140	0.76	19,000	—	—	—		
10	CuSO$_4$	280	0.89	32,000	160	6.5	2500	Normal	Subnormal
15	CuSO$_4$	230	0.83	28,000	210	6.2	3400	Subnormal	Subnormal
25	CuSO$_4$	490	2.1	23,000	430	13.1	3300	Subnormal	Subnormal

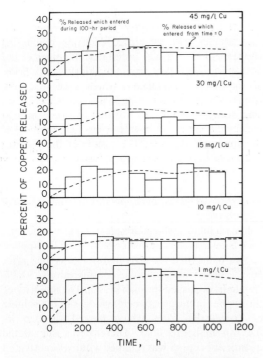

FIG. 3. Percent of copper from activated sludge effluent.

systems and copper accumulated. This inferred that the applied copper can be eventually complexed and accumulated in an inert form which is not subsequently toxic. Each individual bar graph in Fig. 3 represents the percentage of copper released during that 100 h period. The dashed line indicates the percentage of the total copper applied which has been released by that time. The capacity of the units to retain copper, as indicated by the dashed line, was in the range of 80–90% for the 10, 15, 30, and 45 mg/l units. The 1 mg/l unit released a significantly larger proportion of applied copper during the earlier stages, but began to exhibit greater copper retention after about 600 h of operation.

These results indicate that approximately 80–90% of the copper applied to an activated sludge system can be expected to be removed and retained in the sludge. Apparently, the extreme buildup of copper in the systems, as indicated in the 1 and 10 mg/l units, had little effect on the ability of the sludge to take up and retain copper. Thus, it appears that activated sludge can be relied upon as a fairly efficient process to remove and concentrate copper from metal-bearing wastes. It appears that the copper has a tendency to form stable coordination compounds which is the probable cause of its toxic action; however, this capacity of copper for stable formation of complexes may be a defense of the activated sludge against increasing accumulations of copper in the system. It is this characteristic of copper which causes its accumulation and retention in the sludge.[6]

In the earlier mentioned study,[7] where the effects of copper concentrations on high or low organic loadings were examined, copper balances were made at each load condition during the 5 mg/l copper run. The sludges removed from the high organic load unit contained more copper than the sludges removed from the lower organic loaded unit and, consequently, the percent copper removal was higher with the higher organic loading than the lower loading. More copper was lost in the final effluent of the lower organic loaded unit in association with the suspended solids of this effluent which resulted from the endogenous activities of the organisms.

McDermott et al.[3] found that the activated sludge process was approximately 50–79% efficient in removing copper. Approximately 25–75% of the copper in the final effluent was in solution. This is illustrated in Table 2.

The fate of copper in combined primary and excess activated sludges which are fed to anaerobic digesters and its accumulation in the suspended material is shown in Table 1.[3]

CHROMIUM

The heavy metal chromium has been cited as the culprit in many combined municipal-industrial treatment plant upsets. Chromium originates primarily from the electroplating industry and sometimes in large amounts from the leather and tanning industry.

Effects of chromium on the activated sludge process

In a study of five concentrations of hexavalent chromium,[10] ranging from 0.5 to 50 mg/l, it was found that very little deterioration in effluent quality was observed even at the high chromium concentration of 50 mg/l. These results are summarized in Table 3. Even at the highest concentration of 50 mg/l, there was only an increase of 8 mg/l of effluent BOD; however, this increase might be significant in lieu of the present stringent effluent quality criteria.

It had been speculated[10] that the hexavalent chromium would be reduced to the trivalent form which would perhaps serve as a coagulant at the aeration basin pH levels. However, observation of the effluent suspended solids from the control and experimental systems which receive the chromium did not indicate additional sedimentation resulting from a chromium hydroxide precipitate. Consequently, it was hypothesized that the chrome might exist in the form of the hydrous oxide, $Cr_2O_3 \cdot x\,H_2O$.

A significant observation by the authors was the difference between soluble and total effluent chromium for the different influent concentrations. At the 0.5 and 0.2 mg/l dosages, the soluble chromium concentration in the effluent was negligible. At the 5 mg/l dose, the soluble chromium (hexavalent form) was less than 1.5 mg/l. However, at the 50 mg/l dosage, the soluble effluent levels were in the range of 30 mg/l. Since it had been hypothesized that this soluble chrome was in the form of the hydrous oxide, it was felt that the chemically bound oxygen might be utilized as a source of oxygen under anaerobic conditions. Therefore, the primary effluent was mixed with the return sludge from the secondary clarifier in an anaerobic contact basin for a short period of time. It was found that the total effluent chromium from the primary clarifier was reduced from about 30–40 mg/l (in the case of the 50 mg/l influent feed)

TABLE 2
Fate of copper in activated sludge

	Copper cyanide in sewage feed (mg/l)			Copper sulfite in sewage feed (mg/l)			
Type and location of check sample	10.0	15.0	25.0	0.4	1.2	2.5	5.0
Cu fed found in outlet (%)							
Primary sludge	9.0	11.0	12.0	—	12.5	10.7	7.0
Excess activated sludge	55.0	58.0	51.0	—	43.3	25.6	23.0
Final effluent	21.0	21.0	21.0	43.0	25.1	43.3	50.0
Unaccounted for	15.0	10.0	16.0	—	20.0	20.0	20.0
Efficiency of Cu removal (%)	75.0	79.0	79.0	57.0	75.0	57.0	50.0
Soluble $Cu^{(a)}$ in primary effluent (mg/l)							
Total	2.06	1.76	3.10	0.22	0.19	—	—
Reactive	1.12	1.06	1.96	—	—	—	2.65
Soluble Cu in final effluent (mg/l)							
Total	0.53	2.32	1.27	0.12	0.10	—	—
Reactive	0.31	1.12	0.84	—	—	0.67	0.92

[a] Soluble copper is defined as that passing a HA 45 millipore membrane. Total soluble copper is that determined in the filtrate after acid digestion. Reactive soluble copper is that in filtrate which reacts with reagents in absence of prior digestion.

TABLE 3
Effluent quality from activated sludge exposed to hexavalent chromium

Raw waste (mg/l)		Primary effluent (mg/l)		Percent reduction		Final effluent (mg/l)		Plant removal efficiency (%)	
Cr^{+6} fed unit	Control unit	Cr^{+6} fed unit	Control unit	Cr^{+6} fed unit	Control unit	Cr^{+6} fed unit	Control unit	Cr^{+6} fed unit	Control unit
(a) Average BOD									
268	259	180	180	33.5	30.4	14.8	14.7	94.3	94.3
261	288	199	201	22.4	28.3	16.2	18.9	93.2	93.5
311	314	192	173	35.8	44.4	10.9	14.9	96.8	94.9
320	296	193	198	39.8	32.8	15.9	12.7	95.0	95.7
253	263	138	119	45.5	54.8	20.9	13.0	91.7	95.1
(b) Average COD									
452	444	270	266	40.3	40.1	52.0	59.0	88.5	86.7
427	447	297	305	30.4	31.8	70.0	65.0	83.6	85.5
493	496	312	285	36.7	42.5	74.0	75.0	85.0	84.9
458	467	294	277	35.8	40.7	96.0	83.0	79.0	82.2
411	406	234	227	43.1	44.1	67.0	49.0	83.7	87.9
(c) Average suspended solids									
323	303	143	146	52.5	55.1	12.0	13.0	96.3	95.7
242	254	138	130	43.0	48.8	20.0	9.0	91.7	96.5
312	316	157	144	49.7	54.4	12.0	13.0	96.2	95.9
267	267	135	119	49.3	55.4	13.0	9.0	95.1	96.6
277	270	115	114	58.5	57.8	12.0	10.0	95.7	96.3

to an average of 3–5 mg/l. The 3–5 mg/l of chromium was all in the reduced insoluble state.

Slug dosages of 10, 100, and 500 mg/l of hexavalent chromium were investigated by the same authors.[10] The slug concentrations were fed to an activated sludge system, which had no acclimation to chromium, for a period of 4 h. The summary data from these investigations indicate that the 10 mg/l dosage had no deleterious effects on plant performance. However, it is significant to note that the investigators reported a bulky sludge in the final clarifier after 4 days. Other observations indicated that the chromate was more toxic to bacteria than to filamentous organisms. Ingols and Fetner,[11] in studying the effects of the chromic (Cr^{+3}) and the chromate (Cr^{+6}) ions on activated sludge organisms, observed that the chromate ions exert a more toxic effect under both aerobic and anaerobic conditions than the chromic ions. Also of significance was the observation that the chromates were much more toxic to bacteria than to fungi. Therefore, in an activated sludge system, the presence of chromates may tend to stimulate a filamentous, bulky sludge since molds and fungi, such as *Sphaerotilus*, will develop in preference to the bacterial zoogleal masses in the presence of chromate. Consequently, a stimulation of filamentous organisms may have resulted even at the low 10 mg/l concentration which resulted in a temporary bulking of the sludge. The slug dose of 100 mg/l generated an inhibitory effect during the first 24 h of operation after it was added. The effluent BOD doubled and then began to return to normal after about 2–3 days. There was a significant deterioration in effluent quality with the slug dose of 500 mg/l. The BOD and COD of the effluent increased substantially for about 32 h, after which the system began to recover and returned to normal operation in approximately 4 days.

This study[10] also indicated that nitrification was inhibited initially by chromium concentration, even at the lowest level, but the nitrifying organisms apparently acclimated to the chromium and proceeded to nitrify after about 10 days regardless of the chromium concentration fed to the system.

An artificially controlled slug dosage of chromium was added to a full scale combined sewage treatment plant in Bryan, Ohio.[12] A metal plating industry assisted and dumped 50 lb of hexavalent chromium into the sewer. The slug of chromic acid arrived at the treatment plant 2 h and 40 min after being dumped and lasted 1 h in the incoming sewage. During the height of the slug, the sewage had a chromium concentration of 500 mg/l and a pH value of 5.7.

The peak concentration of chromium in the primary clarifier effluent was 65 mg/l with a corresponding pH of 6.8. Eighty-eight percent of the chromium in both the sewage and primary effluent was in solution. Ninety-nine percent of the chromium leaving the plant in the 1-day period following the slug was in a soluble form. Since the return sludge rate (87%) was significantly higher than that of a typical activated sludge plant, a significant quantity of metal was returned back to the aeration basin and was consequently equalized and released in smaller quantities. Three days after the slug entered the plant, the concentration was 10 mg/l in the aeration chamber which was only 3 mg/l less than the peak value several hours after the slug. No significant deterioration in effluent quality was observed. The removal of chromium in a plant with a shorter detention time and a 20% return sludge rate would probably not be as high as in this case, and thus higher concentrations of chromium would be expected in the receiving water.

Effects of chromium on the anaerobic digestion process

Digesters, fed the primary and secondary sludges which had been subjected to a 50 mg/l chromium dosage, did not show any deterioration in performance as indicated by gas production.[10] In order to stress the digester, a slug dose of 300 mg/l of chromium, based on digester contents, was added. This dosage resulted in a temporary cessation of gas production, but after 7 days the digester gradually returned to normal operation. A slug dose of 500 mg/l immobilized the digester from which it did not recover. It was concluded that the digester was resistant to and tolerant of all but the most drastic stresses by chromate. The summary of the aerobic and anaerobic reactions of hexavalent chromium are shown in Table 4.

In the investigation where a slug dose was added to the fullscale operating plant mentioned above,[12] no significant deterioration in gas production was noted in the anaerobic digester.

Removal of chromium through the treatment facilities

The accumulation of chromium in the primary and aeration basin sludges is shown in Table 5,[10] and inferred an increasing concentration and accumulation in the sludges with increasing feed dosage. Table 5 also indicates the relative distribution and recovery of chromium through the primary and aeration portions of the plant. It is significant to note that at the two higher concentrations of chromium, more than 50% of the influent chromium appeared in the final effluent. This quantity included both hexa-

TABLE 4
Reaction of anaerobic digester

Concentration of Cr used	Effect on activated sludge	Short time effects on digester	Sustained damage
50 mg/l (continuous feed to activated sludge plant)	BOD removal efficiency dropped about 3%	—	No damage noted
100 mg/l (slug dose to activated sludge plant)	Plant recovered in about 20 h as measured by BOD removal efficiency	—	No damage noted
500 mg/l (slug dose to activated sludge plant)	Plant recovered within 48 h as measured by BOD removal efficiency	—	No
50 mg/l (fed daily to digester; based on digester contents)	—	Gas production dropped off rapidly. At end of 42 days only 75 mg/l of volatile solids was being produced	Yes
300 mg/l (slug dose to digester)	—	Gas production ceased to 7 days. Digester then gradually recovered	No
500 mg/l (slug dose to digester)	—	—	Yes, digester never recovered

TABLE 5
Distribution and accumulation of chromium in activated sludge

	Cr distribution and recovery							Buildup of Cr in primary and aerator sludges	
Hexavalent Cr (mg/l)	Total Cr fed (g)	Total Cr primary sludge (g)	Total Cr in excess activated sludge (g)	Total Cr plant effluent (g)	Net change of Cr in aerator solids (g)	Cr recovered (g)	Percent accounted for	Average total Cr in primary sludge (mg/g SS)	Average total Cr in aeration tank contents (mg/g SS)
0.5	1.202	0.089	0.743	0.258	+0.064	1.103	92	0.36	4.0
2.0	6.23	0.454	0.862	2.78	−0.05	4.73	76	1.3	8.0
5.0	16.73	0.705	5.32	6.44	+3.17	13.1	78	1.5	26.0
15.0	56.1	1.45	15.4	31.5	+13.9	49.7	89	2.5	36.0
50.0	183.0	2.44	16.8	162.0	−6.9	178.0	97	5.9	66.0

valent and insoluble trivalent chromium. In the case of the 50 mg/l feed, more than 90% was in the hexavalent form. This did not include the reduction of chromium using the anaerobic contact basin for the primary effluent and the return sludge from the secondary clarifier.[10] An observation of the anaerobic digester indicated that practically all of the hexavalent chromium which entered the digester was reduced and accumulated in the sludge in the insoluble form. Therefore, very little soluble hexavalent chromium could be expected in the supernatant from the digester.

ZINC

The heavy metal zinc occurs in many types of industrial wastes, particularly in metal plating and the manufacture of organic constituents, such as acrylic fiber, rayon, cellophane, and special synthetic rubbers. In general, the metal-plating industry tends to discharge into municipal sewers, thereby generating potential problems with subsequent combined treatment. In the production of organic materials, such as viscose rayon, the activated sludge process is the most feasible method of treatment for these wastes if the concentrations of zinc are not deleterious.

Effects of zinc on the activated sludge process

A comprehensive study of zinc and its effects on the activated sludge and anaerobic digestion processes was conducted by McDermott et al.[13] Three concentrations of influent zinc, as zinc sulfate, were examined, 2.5, 10, and 20 mg/l, and in a parallel study zinc cyanide at a 10 mg/l zinc level was used. The average results of these data are shown in Table 6. The data on this table infer that the 2.5 mg/l level of zinc causes no deleterious effects whatsoever in the process effluent while the 10 mg/l unit tended to show slight increases in all effluent parameters with both the complexed and sulfate forms of zinc. These data were taken after the unit had acclimated to the cyanide content of the waste. It was concluded from these data that the maximum level of zinc which will not produce deleterious effects on treatment efficiency is greater than 2.5 mg/l and less than 10 mg/l. However, even at 20 mg/l, the maximum BOD removal efficiency was reduced only about 2%. A study on two viscose rayon wastewaters showed no

TABLE 6
Quality of effluent from control and zinc fed units (final effluents)

Zn in sewage (mg/l)	Form of Zn added	BOD (mg/l, average)	COD (mg/l, average)	Suspended matter (mg/l, average)	Turbidity (standard turbidity units average)
0	Control	13	39	7	18
2.5	$ZnSO_4$	15	40	8	22
0	Control	13	44	10	16
10	$ZnSO_4$	18	49	17	17
10	Complexed Zn	22	57	16	27
0	Control	11	58	7	18
20	$ZnSO_4$	15	68	16	46

effect of zinc on soluble effluent BOD levels up to 125 mg/l fed continuously.[14]

Effects of zinc on the anaerobic digestion process

The effects of zinc sulfate on gas production in the anaerobic process were examined.[13] The zinc was fed in the form of zinc sulfate since it was known that cyanide would exhibit extreme toxicity to the anaerobic process. It was concluded that, for normal digestion of primary or combined sludges, the maximum level of zinc fed in the raw waste sewage is between 10 and 20 mg/l.

Removal of zinc through the treatment facility

The disposition and removal of zinc in the activated sludge process are most effectively illustrated in Table 7. From this it is seen that primary treatment is not too efficient in removing zinc; however, the microbial floc of secondary treatment is extremely efficient. The overall process is apparently 74–95% efficient in removing zinc at feed levels of 20 and 2.5 mg/l, respectively. Data on the concentration of zinc in the primary sludges, excess activated sludges, and digested sludges for three levels of zinc in the investigation[13] are also given in Table 7. Zinc concentrations in the sludges are obviously proportional to the concentration of suspended materials since the zinc is complexed as a part of the suspended material.

NICKEL

Effects of nickel on the activated sludge process

Continuous studies were made with four concentrations of nickel fed as nickel sulfate in the range of 1, 2.5, 5, and 10 mg/l.[15] The average characteristics of the final effluent are presented in tabular form in Table 8 and indicate that the 1 mg/l dosage had no significant effect on the efficiency of the activated sludge process. However, there were significant increases in effluent BOD at the 2.5, 5, and 10 mg/l dosages. These results infer that the effects of nickel on the activated sludge system are not linear with concentration but display decreasing response to increasing concentration. The increased BOD and COD of the final effluents of nickel-fed units in comparison to the control units were about the same for dosages of 5 and 10 mg/l. These same results were also noted with chromium, copper, and zinc.

Slug doses of nickel were also examined at concentrations of 25, 50, and 200 mg/l for 4 h.[15] In each case, the activated sludge was acclimated to a continuous 2.5 mg/l nickel dosage before the slug dose was applied. The slug doses of 25 and 50 mg/l did not seriously upset the system. However, the results for the 200 mg/l slug indicated a significant deterioration in effluent quality but a recovery after approximately 40 h.

It was concluded that nickel ranging from 2.5 to 10 mg/l can be continuously fed to an activated sludge plant with reduced BOD removal efficiencies on the order of 5%. Increased turbidity in the final effluent is the most objectionable characteristic. The maximum level of nickel which will not produce a detectable effect on treatment parameters seemed to be greater than 1 mg/l and less than 2.5 mg/l.

Effects on anaerobic digestion

Combined primary and excess activated sludge from a pilot plant receiving 10 mg/l of nickel digested satisfactorily.[15] In the same study, primary sludge from sewage which had contained 40 mg/l of nickel digested satisfactorily.

Removal of nickel in the treatment processes

The efficiency of primary and activated sludge treatment for removing nickel is shown in Table 8.[15]

TABLE 7
Zinc removal and disposition in activated sludge and anaerobic digestion

Concentration and form of Zn in sewage feed	2.5 ZnSO$_4$	10 ZnSO$_4$	10 complexed Zn	16 complexed Zn	20 ZnSO$_4$
Efficiency of process in Zn removal (%)					
Primary only	13	14	—	8	—
Total activated sludge (including primary)	95	89	96	—	74
Total Zn in sludges (mg/l)					
Primary sludge	64	375	—	548	—
Excess activated sludge	119	328	—	—	—
Digested primary sludge	—	—	—	545	—
Digested combined sludge	—	341	—	—	—

TABLE 8
Effects of nickel on treatment systems

	Concentration of Ni in sewage feed, NiSO$_4$ (mg/l)							
	0	1	0	2.5	0	5	0	10
Average effluent characteristics								
BOD (mg/l)	21	23	13	26	9	13	9	14
COD (mg/l)	48	51	59	63	40	51	40	54
Suspended solids (mg/l)	8	11	5	9	8	16	8	17
Turbidity (stand. units)	25	34	10	29	4	15	4	28
Efficiency of Ni removal								
Primary treatment (%)		5		—		—		3
Complete activated sludge (including primary)		28		42		—		28

	Concentration of Ni in sewage feed, NiSO$_4$ (mg/l)					
	10		20		40	
	Total	Soluble	Total	Soluble	Total	Soluble
Disposition of Ni in sludges						
Primary	62	9.8	—	12.8	308	13.2
Excess activated	89	8.9	—	—	—	—
Digested primary	44	1.6	—	1.9	—	1.5
Digested combined primary plus excess activated	70	1.6	—	—	—	—

and indicates that approximately 30–35% of the influent nickel will be removed in the primary and activated sludge systems. Thus, it appears that nickel does not absorb as well as other metals and will pass through the system in large quantities. In the anaerobic digester it was noted that soluble nickel, introduced with the feed sludges, was converted to an insoluble form during digestion. This conversion was believed due to the long detention times, high alkalinity, sulfide content, and hydroxyl ion concentration which offer conditions favorable to formation of insoluble nickel compounds. Table 8 also shows the amount of nickel which will concentrate in the sludges and in solution at various influent raw waste concentrations of nickel.[15]

COMBINED HEAVY METALS

In the majority of cases, particularly combined municipal treatment facilities, the concern of heavy metals is not related to one or two metals but to a mixture or combination of heavy metals originating from one or several industrial sources. In this regard, it is essential to know if the heavy metals might react synergistically to produce effects which are more deleterious than the individual metals themselves. On the other hand, the sum of the individual metal concentrations may only be similar to an identical concentration of a single metal.

Effect of combined heavy metals on the activated sludge process

Two major studies have been conducted in the United States concerning the effects of mixtures of heavy metals on the activated sludge and anaerobic digestion processes.[16,17] The first study[16] was conducted on a pilot scale with the heavy metal concentrations shown in Table 9. The average characteristics of the final effluents from these systems are shown in Table 10. These data infer that MC 1 and MC 2 had a significant effect on the performance of the system while the effects of MC 3 were very marginal.

The data shown in Fig. 4 illustrate chronological effects of heavy metal concentrations on nitrification. This study concluded that the conversion of nitrite to nitrate was completely inhibited while the oxidation of ammonia to nitrite was somewhat erratic.

The conclusions of this study were essentially that combined heavy metals do not exert synergistic

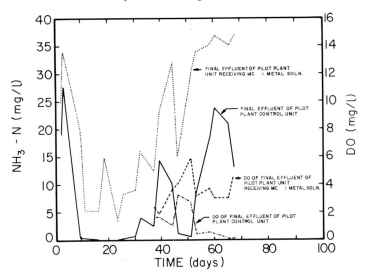

FIG. 4. Effluent ammonia from system receiving MC 1 combination.

TABLE 9
Metal combinations fed to activated sludge process

Metal combination	Metal (mg/l influent sewage)				Total heavy metals (mg/l)	Total CN
	Cu	Cr	Ni	Zn		
MC 1	0.4	4.0	2.0	2.5	8.9	4.3
MC 2	0.4	—	2.0	2.5	4.9	4.3
MC 3	0.3	—	0.5	1.2	2.0	2.0

TABLE 10
Effluent quality from control and metal-fed units

Metal combination fed	BOD (mg/l)	COD (mg/l)	Suspended solids (mg/l)	Standard turbidity units
MC 1	27	66	15	39
Control	18	45	10	26
MC 2	21	63	16	74
Control	21	48	13	32
MC 3	16	57	9	22
Control	21	52	12	16

effects on the activated sludge process and that the combined effect was no more severe than the effect that one metal alone would have at the same concentration. MC 2, with a total metal concentration of 4.9 mg/l, exerted a similar effect as MC 1 with a total metal content of 8.9 mg/l. This same effect had been shown earlier in the nickel investigation[15] at nickel concentrations of 5 and 10 mg/l. Even with a well-operated conventional activated sludge plant, it appears that a considerable amount of metal passes from the plant into the final effluent.

Effects of combined heavy metals on the anaerobic digestion process

The sludges produced by the control unit and the unit being fed the MC 1 sludge were digested anaerobically. The results of feeding the anaerobic digesters are shown in Fig. 5. With primary sludge, it is evident that the accumulation of metals in the digester was low and there was no interference with gas production indicating proper operation of the digester. The data in Fig. 5 reveal an increasing quantity of metal accumulating in the digested sludge which was a result of the higher concentrations of metals entering in the excess activated sludge mixed with primary sludge. A significant observation was that the soluble metal content in the digester was never greater than 1 mg/l for any of the four metals. This confirms earlier investigations which show that anaerobic digestion is efficient in converting introduced soluble metal into an insolu-

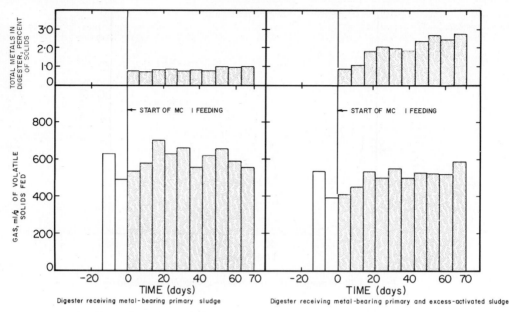

Fig. 5. Effects of combined heavy metals on anaerobic digestion.

ble form. This study[16] concluded that anaerobic digestion of the metal-bearing sludges produced by the pilot-scale unit was satisfactory. Anaerobic digestion of the sludges was not hindered with sludges subjected to the metals chromium, nickel, and zinc, present continuously in the influent sewage at concentrations of 10 mg/l. Copper, continuously present in the influent sewage at a concentration of 10 mg/l, allows normal digestion of primary sludge, and possible difficulty with mixed digestion.

Removal of combined heavy metals through the treatment facilities

A complete material balance for the four metals for the MC 1 run is shown in Table 11. If the influent metal concentration is taken into account, the mixed liquor has a capacity for the heavy metals in the following order in the system studied: zinc, copper, chromium, nickel. In summary, approximately 90% of the zinc, 54% of the copper, 37% of the chromium, and 31% of the nickel were removed from the

TABLE 11
Metal concentrations in process outlets (MC 1)

	Primary sludge	Excess activated sludge	Final effluent	Imbalance (%)	Percent overall removal	Range of observation (%)
Cu						
mg/l	18.2	22.0	0.22		54	32–89
Percent of metal fed	19.0	39.5	45.8	+4		
Co						
mg/l	44.0	97.0	2.6		37	18–58
Percent of metal fed	5.4	20.6	63.0	−11		
Ni						
mg/l	9.8	33.0	1.2		31	12–76
Percent of metal fed	2.3	13.3	69.4	−15		
Zn						
mg/l	59.0	141.0	0.26		90	74–97
Percent of metal fed	11.0	67.5	9.7	−12		

influent sewage. The metals, in combination, behaved independently in their distribution throughout the process.

A survey of four municipal wastewater treatment plants concerning the receipt of heavy metals, distribution of the metals in the various process outlets and effects of the metals on the treatment efficiency[17] has shown satisfactory correlation with pilot plant investigations.[16] The results indicated that the combined heavy metals caused no serious reduction in efficiency of aerobic or anaerobic treatment facilities at concentrations in the range of 1–9 mg/l.

SUMMARY

From reviewing the data of the various studies with individual and combined heavy metals, it appears that essentially all of the heavy metals exhibited the same relative effects on the activated sludge process. In the low concentration ranges, i.e., less than 10 mg/l, there is apparently an increasing deterioration of effluent quality with increasing heavy metal concentration. The efficiency of the activated sludge process seems to be reduced approximately 3–10% at these heavy metals concentrations. Increasing the concentration of heavy metals above these lower levels does not appear to result in any significant additional deterioration of effluent quality, Obviously, there may be some high level concentration of heavy metal at which the activated sludge system would fail completely. However, it appears that this concentration is many order of magnitudes larger than the low levels which cause the initial deterioration in effluent quality, and it also seems that there is not an increasing deterioration of effluent quality with increasing metal concentration up to the threshold failure concentration. A summary of the heavy metals concentrations which resulted in significant effects on the effluent quality from the activated sludge process is given in Table 12.

A summary of the distribution of heavy metals through the activated sludge process is shown in Table 13.[2] Stones[18–21] investigated the fate of copper, chromium, nickel, and zinc through municipal plants and generated results similar to those shown in Table 13. The metals showing the greatest removal through the activated sludge process were copper and zinc. The removal of chromium can vary depending on certain conditions present in and around the activated sludge process. Chromium, introduced into the activated sludge process as hexavalent chromate, may react with reducing compounds to cause precipitation of the trivalent chromium in the primary clarifier. In addition, anaerobic conditions at the front end of the acti-

TABLE 12
Critical metal concentrations

	Concentrations of heavy metals which allowed satisfactory performance of anaerobic digestion		
	Dosage fed continuously in influent sewage		Dosage slugged to influent with no effect on aerobic digester
Metal	Primary sludge digestion	Combined sludge digestion	
Cr(VI)	>50	>50	500
Cu	10	5	410
Ni	>40	>10	200
Zn	10	10	—

	Concentrations of heavy metals which caused significant effects on activated sludge process	
Metal	Continuous dosage (mg/l)	Slug dosage (mg/l)
Cr(VI)	10	>500
Cu	1	75
Ni	1–2.5	>50 <200
Zn	5–10	160

TABLE 13
Distribution of metals through the activated sludge process (continuous dosage) (percent of metal fed)

	Cr(VI) (15 mg/l)	Cu (10 mg/l)	Ni (10 mg/l)	Zn (10 mg/l)
Primary sludge	2.4	9	2.5	14
Excess activated sludge	27.0	55	15.0	63
Final effluent	56.0	25	72.0	11
Metal unaccounted for	15.0	15	11.0	12
Average efficiency of process in removing metal	44.0	75	28.0	89
Range of observations	18–58	50–80	12–76	74–97

vated sludge plant or anaerobic conditions created intentionally may reduce the chromate compounds and absorb the trivalent chromium onto the biological floc. It has been shown[10] that chromium removal can reach 90% under these conditions. In general, the majority of the metal is removed in the secondary sludge by either precipitation or sorption onto the biological flocs. Very little nickel is removed through the aerobic process. With the exception of zinc, copper, chromium, and nickel were present in the final effluent from the secondary clarifier in the soluble form.

The effects of heavy metals concentrations on anaerobic digestion were not similar to those of the aerobic activated sludge process. Apparently the reaction in the anaerobic digester is an "all or nothing" reaction and the digester either performs satisfactorily or ceases to function entirely. The concentrations of heavy metals which allow satisfactory performance of the anaerobic digester are shown in Table 12.[2] The digesters, which received combined primary and excess activated sludges, were subjected to a higher content of metals on a percent basis than those digesters which received only primary sludges. It appears that the effects on digester performance were more pronounced with the combined sludges than the primary sludges at the same heavy metals concentrations.[2]

The anaerobic, reducing conditions in the digester are suitable for converting soluble metals, present in the influent feed to the digester, into insoluble forms which are precipitated out with the sludge. Table 14 illustrates the comparison of the total metal content of the digested sludge in relation to the soluble metal content of the sludges.

Cheng et al.[22] conducted a comprehensive study on the uptake of copper, nickel, cadmium, and lead. They concluded that the formation of metal–organic complexes causes the metal uptake by the biological organisms at low heavy metal concentrations. At higher metal concentrations, precipitation of metal ions will also occur in addition to the biological uptake. It was also concluded that a very rapid uptake of metal occurs within 3–10 min after introduction to the sludge, and that this rapid initial phase of uptake is followed by a longterm, slow uptake by

TABLE 14
Soluble metal content of sludges, compared with total metal content of digested sludge

Metal	Concentration in influent sewage (mg/l)	Soluble metal			Total metal
		Feed sludges		Digested combined (mg/l)	Digested combined (mg/l)
		Primary (mg/l)	Excess activated (mg/l)		
Cr(VI)	50	38.0	32.0	3.0	420
Cu	10	2.0	0.5	0.7	196
Ni	10	10.0	9.0	1.6	70
Zn	10	0.3	0.1	0.1	341

the activated sludge organisms. These investigators concluded that high molecular weight, extracellular materials furnish substantial functional groupings which serve as complexing sites for the heavy metals. These materials would include polysaccharide, protein, ribonucleic acid (RNA), and deoxyribonucleic acid (DNA).[21] It is interesting to note that these conclusions are similar to the mechanisms cited for copper removal by Ayers et al.[8] and mentioned earlier in the section on copper.

These same investigators[22] concluded that the preferred order of uptake of heavy metals by activated sludge was lead > copper > cadmium > nickel. The rate and degree of metal uptake was found to be dependent on several factors, including pH, concentration of organic matter, and concentration of heavy metal. It appears that increasing concentrations of metal ion or mixed liquor suspended solids increases the overall uptake of the metal.[23]

Studies have been conducted on other heavy metals, such as mercury[24] and lead,[25] and their effects on the activated sludge process, although the occurrence of these metals at high enough concentrations to cause harm is relatively infrequent in biological treatment systems.

In sum, it appears that there will be a deterioration of effluent quality at certain low heavy metal concentrations above which no increasing deterioration will be observed. The metals tend to form organic–metallic complexes at low concentrations with certain essential extracellular constituents, thereby inhibiting normal biological activities. These complexes absorb onto the biological flocs and are partially removed from solution along with precipitated metallic ions. A significant portion of the metals can be expected to be removed through the treatment process depending on the heavy metal and the conditions of operation of the treatment facility, such as waste strength, pH, and concentration of microorganisms.

REFERENCES

1. DIRECTO, L. S., Some effects of the copper ion on the activated sludge stabilization process, PhD dissertation, the Ohio State University, 1961.
2. BARTH, E. F., M. B. ETTINGER, B. V. SALOTTO, and G. N. MCDERMOTT, Summary report on the effects of heavy metals on the biological treatment processes, J. Wat. Pollut. Control Fed. 37, 86 (1965).
3. MCDERMOTT, G. N., W. A. MOORE, M. A. POST, and M. B. ETTINGER, Effects of copper on aerobic and biological sewage treatment, J. Wat. Pollut. Control. Fed. 35, 227 (1963).
4. DIRECTO, L. S. and E. Q. MOULTON, Some effects of copper on the activated sludge process, Proc. 17th Industrial Waste Conf. Purdue University, No. 112, p. 95, 1962.
5. PRESEWECKLI, J., PhD thesis, Wroclaw Univ., Poland.
6. MOULTON, E. Q. and K. S. SHUMATE, The physical and biological effects of copper on aerobic biological waste treatment processes, Proc. 18th Industrial Waste Conf. Purdue University, No. 115, p. 602, 1963.
7. SALOTTO, B. V., E. F. BARTH, W. E. TOLLIVER, and M. B. ETTINGER, Organic load and the toxicity of copper to the activated sludge process, Proc. 19th Industrial Waste Conf. Purdue University, No. 117, p. 1025, 1964.
8. AYERS, K. C., K. S. SHUMATE, and G. P. HANNA, Toxicity of copper to activated sludge, Proc. 20th Industrial Waste Conf. Purdue University, No. 118, p. 615, 1965.
9. ADAMS, C. E. and R. M. STEIN, Toxicity of copper and cobalt on the activated sludge process, unpublished industrial report, Associated Water and Air Resources Engineers, Inc., Nashville, Tenn., U.S.A, 1971.
10. MOORE, W. A., G. N. MCDERMOTT, M. A. POST, J. W. MANDIA, and M. B. ETTINGER, Effects of chromium on the activated sludge process, J. Wat. Poll. Control. Fed., 33, 54 (1961).
11. INGOLS, R. S. and R. H. FETNER, Toxicity of chromium compounds under aerobic condition, J. Wat. Pollut. Control Fed. 33, 366 (1961).
12. ENGLISH, J. N., E. F. BARTH, B. V. SALOTTO, and M. B. ETTINGER, Slug of chromic acid passes through a municipal treatment plant, Proc. 19th Industrial Waste Conf. Purdue University, No. 117, p. 493, 1964.
13. MCDERMOTT, G. N., E. F. BARTH, B. V. SALOTTO, and M. B. ETTINGER, Zinc in relation to activated sludge and anaerobic digestion processes, Proc. 17th Industrial Waste Conf. Purdue University, No. 112, p. 461, 1962.
14. Unpublished industrial reports, Associated water and Air Resources Engineers, Inc., Nashville, Tenn., USA, 1972, 1973.
15. MCDERMOTT, G. N., M. A. POST, B. N. JACKSON, and M. B. ETTINGER, Nickel in relation to activated sludge and anaerobic digestion processes, J. Wat. Pollut. Control Fed., 37, 163 (1965).
16. BARTH, E. F., B. V. SALOTTO, G. N. MCDERMOTT, J. N. ENGLISH, and M. B. ETTINGER, Effects of a mixture of heavy metals on sewage treatment processes, Proc. 18th Industrial Waste Conf. Purdue University, No. 115, p. 616 (1963).
17. BARTH, E. F., J. N. ENGLISH, B. V. SALOTTO, B. N. JACKSON, and M. B. ETTINGER, Field survey of four municipal wastewater treatment plants receiving metallic wastes, J. Wat. Pollut. Control Fed., 37, 1101 (1965).
18. STONES, T., The fate of chromium during the treatment of sewage, J. Proc. Inst. Sew. Purific. 4, 345 (1965).

19. STONES, T., The fate of copper during the treatment of sewage, *J. Proc. Inst. Sew. Purif.* **1**, 82 (1958).
20. STONES, T., The fate of nickel during the treatment of sewage, *J. Proc. Inst. Sew. Purif.* **2**, 252 (1959).
21. STONES, T., The fate of zinc during the treatment of sewage, *J. Proc. Inst. Sew. Purif.* **2**, 254 (1959).
22. CHENG, M. H., J. W. PATTERSON, and R. A. MINEAR, Heavy metal uptake by activated sludge, paper presented at 46th Annual Conference Water Poll. Control Fed., Cleveland, Ohio, October 1973.
23. PAVONI, J. L. Fractional composition of microbially extracellular polymers and their relationship to biological flocculation, PhD thesis, University of Notre Dame, 1970.
24. GHOSH, M. M. and P. D. ZUGGER, Toxic effects of mercury on the activated sludge process, *J. Wat. Pollut. Control Fed.* **45**, 424 (1973).
25. STEIN, R. M., C. E. ADAMS, and W. W. ECKENFELDER, Treatability and toxicity of a tetraethyl lead wastewater, *Proceedings of the Second Annual Environmental Engineering and Science Conference, University of Louisville, Kentucky, 1972.*

The Effects and Removal of Heavy Metals in Biological Treatment (C. Adams et al.)

DISCUSSION by E. F. BARTH

Advanced Waste Treatment Research Laboratory, National Environmental Research Center, Cincinnati, Ohio 45268

The paper prepared by Drs. Adams, Eckenfelder, and Goodman was a review, abstraction, and summation of current literature relating to effects of metals on activated sludge processes. Rather than present a review of a review paper, this discussion will indicate some of the *chemical, biological,* and *engineering* parameters that must be considered in performing the type studies the authors have summarized.

Biological factors are important to toxicity studies. Table 1 shows some of the extreme environments that organisms can cope with. It is evident that organisms are tenacious and adaptable, and by usual measures of environmental factors they are tough and resistant. It would be unwise to generalize on toxic environments from a single observation. An environment unsuitable or toxic to one type organism may be quite suitable to another.

Table 2 lists some criteria for determining a toxic concentration of a substance. There are others that can be selected, but the list does point out the variety of parameters that can be chosen. The choice of the parameter can influence the decision of what concentration is considered toxic. Also noted is the fact that purely physical properties of materials can exert toxicity, as is the case of ferric hydroxide floc on organisms in acid mine drainage areas, or the prevention of efficient mass transfer reactions when metallic soaps coat trickling filter rock or excessive oil coats activated sludge particles.

Toxicity studies in biological waste treatment processes are not as straightforward as similar studies with people or fish. In these latter cases, the studies deal with only one genus and limited species, and the study duration is usually only a short portion of their life span. These types of organisms are individually identifiable, and a series of interrelated biological reactions must occur for the individual's survival. Only a single link, such as an enzyme system, need be blocked and the organism falters. The effect of the toxicant is readily observable as life versus death. In contrast to this, Table 3 indicates that in biological waste treatment, toxicity studies must deal with diverse groups of life forms. There is no individual identification, and toxicity criteria must encompass group reactions. Characteristically these microorganisms have short generation times and the phenomenon of acclimation to a toxicant

TABLE 1
Environmental extremes[a]

Factor	Lower limit	Upper limit
Temperature	−18°C (Fungi and bacteria)	104°C (sulfate reducing bacteria in hot springs)
pH	0 (*Thiobacillus*)	13 (*Plectonema nostocorum*)
Hydrostatic pressure	0 atm (spores)	1400 atm (deep sea bacteria)
Salinity	Distilled water (*Pseudomonas*)	Saturated brine (halophytic bacteria)
Oxygen	0 saturation (anaerobes or facultative bacteria)	Supersaturated (algae)
eH potential	−500 (*Clostridium*)	+500 (*Nitrosomonas*)
Total environment span	Ocean bottom (11 km below sea level)	Top of Mt. Everest (8 km above sea level)

[a] From Young, R. S., *Extraterrestrial Biology*, Holt, Rinehart & Winston, New York, 1965.

TABLE 2
Parameters of toxicity

Kills organism
Interferes with growth
Interferes with respiration
Physically interferes with organism
  (a) $Fe(OH)_3$
  (b) Metallic soaps
  (c) Oil

TABLE 3
Active life forms in treatment processes

Process	Phylum	Genera	Species
Aerobic	Several	Hundreds	Thousands
Anaerobic	One	One to ten	One to ten

(Classification: Phylum, Genera, Species)

must be accounted for. Particularly in aerobic systems, due to the diversity of life forms, several genera or species might be adversely affected without any significant effect on the total capability of the system. Anaerobic systems are more susceptible to toxicants because of the narrow spectrum of life forms present. This is consistent with observations at treatment facilities. Anaerobic digesters experience more frequent upsets than aerobic activated sludge or trickling filter units.

Table 4 lists the water solubility of several forms of copper compounds. On the basis of chemical form and solubility, different toxic responses by biological systems might be expected. However, investigation has shown that on the basis of equivalent copper content each compound has the same threshold level of toxicity to aerobic systems as does copper in copper sulfate. Insoluble compounds, such as copper hydroxide in modest amounts, are solubilized by ammonia nitrogen in wastewater by the formation of copper ammine complex. Once a system acclimates to the cyanide portion of copper cyanide complex, the toxic properties are related to the metal component of the complex. Insoluble copper sulfide that enters the oxidizing environment of an aeration tank is converted to copper sulfate,

TABLE 4
Water solubility of copper salts

Metal form	Solubility product
$CuSO_4$	Very soluble
$Cu(OH)_2$	1×10^{-20}
$Cu(CN)_x^-$	Very soluble
CuS	1×10^{-40}

and soluble copper can be detected in the effluent. There are industrial applications of the aerobic oxidation of copper sulfide to soluble copper sulfate. Due to the complex chemical and biological interactions that occur in waste treatment systems, it is difficult to predict the toxicity of a compound based solely on its chemical composition.

The usual concentration range of interest in toxicity studies of metals is 1–10 mg/l. Table 5 shows

TABLE 5
Residual concentration of silver

$$[Ag] \times [Cl] = 1 \times 10^{-10}$$
$$[Ag] = 1 \times 10^{-5}$$
$$\frac{107}{10^{-5}} = 0.00107 \text{ g/l}^{-1}$$
$$= 1.07 \text{ mg/l}^{-1}$$

that a compound such as silver chloride, which is usually considered highly insoluble, has an equilibrium silver concentration of about 1 mg/l. In classical wet chemical analysis, this residual is disregarded as insignificant. However, for metal toxicity studies, the analytical methods have to be selected to cover a broad range of concentration values. For instance, copper introduced to a biological system at 1 mg/l can disproportionate through the system. The effluent concentration can be as low as 0.1 mg/l and the aerator concentration can reach levels of 50–100 mg/l. Choice of the proper analytical method to encompass these variations is of prime importance for interpretation of metal toxicity studies.

Most of the metals of interest in toxicity studies occur in the Periodic Table in the series called the transition heavy metals as shown in Fig. 1. Of the 100 or so elements listed, 75 are classified as metals. At least 10 of these metals are necessary for life processes to function. These are divided into two classes, termed bulk and trace metals. The bulk elements necessary for life are calcium, sodium, potassium, and magnesium. They occur in living tissues in the amount of several grams per kilogram of tissue. The trace elements necessary for life processes are iron, manganese, copper, zinc, cobalt, and molybdenum. They occur in the amount of 1–2 mg/kg of tissue.

The fundamental difference between these two classes is that the bulk metals are nontransitional elements and their ions assume the noble gas configuration, whereas the trace metals are transitional elements with unfilled inner electron shells which favor stable complex formation. The metals

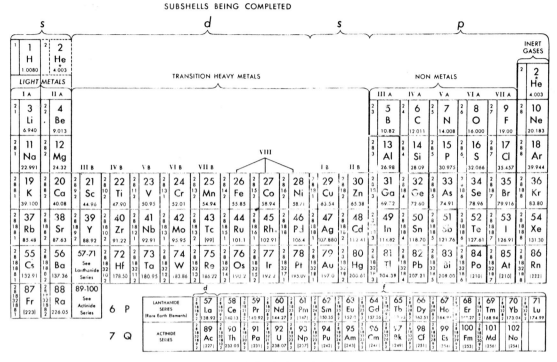

FIG. 1. Periodic Table of the elements.

discussed in the paper by Dr. Adams *et al.*, are copper, chromium, nickel, and zinc, which are members of the transition series. These metals have been shown to form stable complexes with the type of functional groups shown in Table 6. These functional groups are present as polysaccharides, lipids, and nucleic acids in biological protoplasm. Since life processes depend upon reversible reactions, it is not surprising that these metals can exert toxic effects by forming stable complexes which block active sites.

The authors have stressed the fact that the results of blocking these active sites are not quantitatively expressed by the parameter selected to judge toxicity. Figure 2 shows the response of an activated sludge system to various concentrations of copper

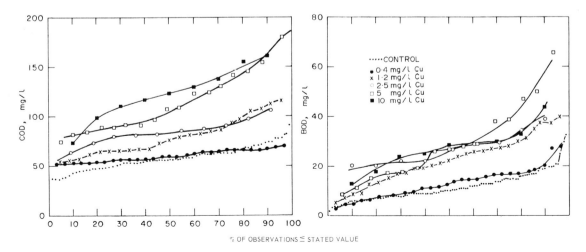

FIG. 2. Effect of copper fed as copper cyanide complex continuously on BOD and COD of final effluents.

TABLE 6
Functional groupings important in metal interactions

—COOH	Carboxyl
—OH	Alcoholic and phenolic hydroxyl
C=O	Carbonyl
—NH$_2$	Primary amino
R | —NH	Secondary amino
| —N—	Tertiary amino
N	Cyclic tertiary amino
—SH	Sulfhydryl
—S—	Thioether
—SO$_3$H	Sulfonate

cyanide complex as judged by the chemical oxygen demand and biological oxygen demand of the final effluent. Although the metal toxicant was increased twenty-fivefold, there was only a twofold decrease in effluent quality in this acclimated system.

From studies of several metals which exhibit this same behavior, a generalized dose-response curve has been prepared as shown in Fig. 3. In general, a dose of 1–2 mg/l in an aerobic process will show a threshold effect, but concentrations 10–20 times this dose do not seriously affect the process. The process does not fail completely until several orders of magnitude beyond the threshold dose are reached. Anaerobic processes are more sensitive and do not tolerate such a wide range of concentration. This is due to the narrower spectrum of life forms present as discussed previously.

Due to the fact that toxicity studies in biological waste treatment processes deal with low concentrations of metals in a heterogenous, randomly oscillating biological system, very long and controlled observations in a stabilized configuration are necessary. Figure 4 illustrates the importance of this fact. A combination of metals totaling 2 mg/l was introduced into an acclimated, activated sludge process and the effluent quality compared to a control unit receiving no metals. If the mean value of the control unit's chemical oxygen demand is extrapolated to the values obtained for the unit receiving the metal combination, the data show that 20% of the time this unit had an effluent quality as good or better than the control unit in terms of chemical oxygen demand.

Figure 5 emphasizes the need to consider acclimation in biological waste treatment toxicity studies. Zinc cyanide complex introduced into an activated sludge process showed an immediate toxic reaction as indicated by turbidity of the final effluent. After 7 days of continuous dosage, the organisms acclimated to the toxic cyanide radical and turbidity of the effluent stabilized. Ionic zinc introduced to another unit had no such immediate effect, and only a slight increase in effluent was noted. Acclimation is

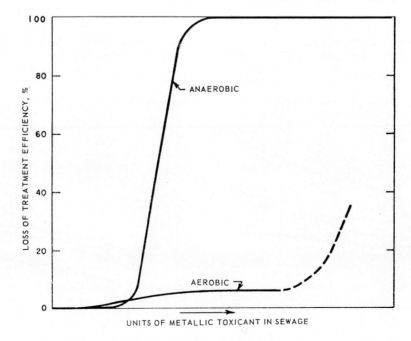

FIG. 3. Patterns of performance depreciation in sewage treatment processes.

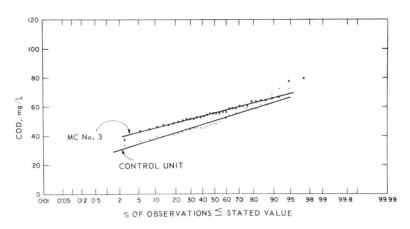

FIG. 4. COD of final effluents.

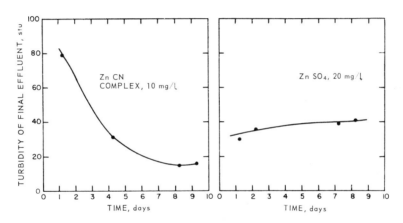

FIG. 5. Comparison of acclimation to complexed zinc and zinc sulfate.

a very important practical consideration, but many elaborate and erudite studies of metal toxicity have neglected this consideration.

This discussion emphasized the points outlined in the paper by Drs. Adams, Eckenfelder, and Goodman, and illustrated the technical considerations necessary for controlling and understanding toxicity studies in biological waste treatment systems.

SESSION VII
CORRECTIVE MEASURES FOR EXISTING PROBLEMS

THE FEASIBILITY OF RESTORING MERCURY-CONTAMINATED WATERS

ARNE JERNELÖV and BO ÅSÉLL

Swedish Water and Air Pollution Research Laboratory, Stockholm, Sweden

INTRODUCTION

High levels of mercury in natural bodies of water are mainly results of man's use of different mercury compounds. Mercury is a very rare element comprising less than 0.3×10^{-10} of the earth's crust, and natural background concentrations in sediments are usually low, 0.01–0.15 ppm.[1] However, in bodies of waters receiving municipal and/or industrial wastewater, concentrations are found in sediments ranging from 0.5 to more than 600 ppm.[1]

Discharge of mercury into water has mainly involved such compounds as: divalent inorganic mercury (Hg^{2+}), elementary mercury (Hg^0), and phenylmercury as acetate ($C_6H_5Hg^+$). Due to the physical and chemical properties of these compounds, most of the mercury discharged to bodies of water is likely to end up in the sediment.[2] From there it can be removed by suspension of the material to which it is attached by forming soluble complex ions such as HgS_2^{2-} or $HgCl_4^{2-}$, or by biological methylation (CH_3Hg^+ and/or $(CH_3)_2Hg$).[3-5] Methylmercury usually represents only a small fraction ($\leq 1\%$) of the total mercury present in an aquatic system. However, more than 90% of the mercury in biota is in the form of methylmercury.[6] This is mainly a result of the high affinity of methylmercury to organic material in general and the efficiency with which it is retained by organisms. Adding to this the fact that, in relation to higher forms of life, methylmercury is one of the most toxic forms of mercury, it is not hard to understand why methylmercury concentrations in fish are of prime concern to man.[7]

In natural aquatic ecosystems the proportion of inorganic mercury annually converted to methylmercury has been estimated to be ~0.1% in some areas in Sweden.[8] Thus, mercury deposits in bottom sediments may continue to release methylated mercury for a very long time unless they are removed or inactivated. This leads to the conclusion that after contamination of a water system, mercury levels in fish may remain high for a considerable length of time, therefore making it desirable to find methods of restoration. Since the primary cause of mercury contamination in fish is the process of methylation, it would seem relevant to concentrate on those methods that best encourage a reduced methylation rate.[9,10]

CONVERSION OF MERCURY TO METHYLMERCURY

The possibilities of finding successful methods of reducing or preventing the conversion of mercury to methylmercury are related to our basic knowledge about the processes and factors involved with methylation. To date, two different biochemical pathways for mercury methylation have been suggested.[11,12] It has also been pointed out that a chemical alkylation becomes possible if alkylating compounds, such as alkyl-lead, are present.[13] It seems, however, that the biological methylation process is, in general, the most important one.

Experimental studies and field observations have shown the biological methylation rate to be influenced by the following factors:

(1) general microbial activity (and factors affecting it, e.g., temperature);
(2) concentration of inorganic mercury;
(3) biochemical availability of inorganic mercury;
(4) redox potential;
(5) pH.

Bisogni and Lawrence[9] have suggested a kinetic model describing the net methylation rate in terms of some of the above mentioned factors.

$$\text{NSMR} = \gamma(\beta \cdot Hg_{total})^n,$$

where NSMR is the net specific methylation rate per amount active sediment,

γ a coefficient related to general microbial activity, and

β a coefficient related to biochemical availability of inorganic mercury;

Hg_{total} the concentration of inorganic mercury, and

n is the order of reaction which is determined by the type of methylation process involved and differs at least between aerobic and anaerobic systems.

It should be pointed out that the model is a simplification in the sense that it describes the *net* methylation rate as a function of some other variables, although this rate might in fact be the result of two competing processes; methylation and *de*methylation. This is implicit throughout this paper whenever methylation rate is mentioned.

A number of restoration methods which might affect the methylation rate in different ways have been suggested, e.g.:

(1) removing of mercury deposits by dredging;
(2) conversion of mercury to mercuric sulfide;
(3) binding of mercury to inorganic material, e.g. silicia minerals;
(4) covering of mercury deposits with mercury-binding or inert material;
(5) raising the pH of a system so that the biological methylation process will give volatile dimethylmercury rather than monomethylmercury.

In order to evaluate the feasibility of the above suggested restoration methods in natural aquatic systems, a number of laboratory and field experiments were performed. Further, a very simplified compartmental model was constructed and analyzed to indicate the relative effects of different restoration methods.

The compartmental model

This model is a description of methylmercury formation and transport in a very simplified aquatic system consisting of (1) three trophic levels of biota, top carnivores (fish, e.g., pikes), carnivores (fish, e.g., roaches) and herbivores (e.g., bottom fauna); (2) two bodies of sediment; and (3) a body of water. The basic structure and the routes of flow of methylmercury are shown in Fig. 1A and the equations for the model are presented in Fig. 1B.

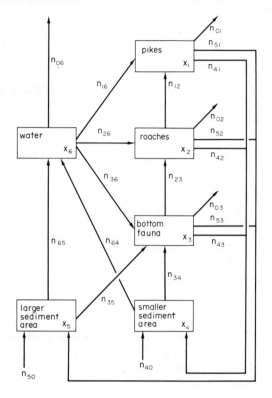

FIG. 1A. Route of methylmercury flow in an aquatic system.

Methylmercury enters the system via methylation of mercury in the sediment (Fig. 1A: n_{40} and n_{50}), the rate being related to inorganic mercury concentration according to the model of Bisogni and Lawrence. The concentrations of inorganic mercury in sediment are considered constant, not being affected by the methylation process. This is valid since the parameters used have been chosen to correspond to an annual conversion rate of 0.1% at a concentration of inorganic mercury of 5 ppm. Methylmercury can either leave the system through metabolic destruction in the biota proportional to the amount present in each trophic level (Fig. 1A: n_{01}, n_{02}, n_{03}) or by outgoing water in proportion to waterflow and methylmercury concentration in water (Fig. 1A: n_{06}).

The rate of exchange of methylmercury from sediment to water (Fig. 1A: n_{64}, n_{65}) is considered proportional to the amount of methylmercury present in sediment. The rate determining constant is chosen in such a way that 99.99% of the amount present at a given time will be lost in one year.

Figures on the standing crop and energy turnover of biota for the three trophic levels in question are taken from H. T. Odum's investigations in Silver Springs, Florida.[14]

$n_{01} = \lambda_{01} \cdot X_1$

$n_{02} = \lambda_{02} \cdot X_2$

$n_{03} = \lambda_{03} \cdot X_3$

$n_{06} = (W/T) \cdot C_6$

$n_{12} = q_{12} \cdot C_2 \cdot f_{12}$

$n_{16} = q_{10} \cdot k_{16} \cdot (1/C_{ox}) \cdot (1/f_{ox,1}) \cdot C_6 \cdot f_{16}$

$n_{23} = q_{23} \cdot C_3 \cdot f_{23}$

$n_{26} = q_{20} \cdot k_{26} \cdot (1/C_{ox}) \cdot (1/f_{ox,2}) \cdot C_6 \cdot f_{26}$

$n_{34} = q_{3,IN} \cdot C_4 \cdot f_{3,IN} \cdot f_{A,4}$

$n_{35} = q_{3,IN} \cdot C_4 \cdot f_{3,IN} \cdot f_{A,5}$

$n_{36} = q_{30} \cdot k_{36} \cdot (1/C_{ox}) \cdot (1/f_{ox,3}) \cdot C_6 \cdot f_{36}$

$n_{40} = \gamma_{40} \cdot (\beta_{40} \cdot C_{Hg,4})^{k_{40}} \cdot A_4 \cdot D_4 \cdot \rho_4 \cdot f_{40}$

$n_{41} = q_{D,1} \cdot C_1 \cdot f_{A,4}$

$n_{42} = (q_{D,2} + q_{12} \cdot (1 - f_{12})) \cdot C_2 \cdot f_{A,4}$

$n_{43} = (q_{D,3} + q_{23} \cdot (1 - f_{23})) \cdot C_3 \cdot f_{A,4}$

$n_{50} = \gamma_{50} \cdot (\beta_{50} \cdot C_{Hg,5})^{k_{50}} \cdot A_5 \cdot D_5 \cdot \rho_5 \cdot f_{50}$

$n_{51} = q_{D,1} \cdot C_1 \cdot f_{A,5}$

$n_{52} = (q_{D,2} + q_{12} \cdot (1 - f_{12})) \cdot C_2 \cdot f_{A,5}$

$n_{53} = (q_{D,3} + q_{23} \cdot (1 - f_{23})) \cdot C_3 \cdot f_{A,5}$

$n_{64} = \lambda_{64} \cdot X_4$

$n_{65} = \lambda_{65} \cdot X_5$

$\dot{X}_1 = n_{12} + n_{16} - n_{01} - n_{41} - n_{51}$

$\dot{X}_2 = n_{23} + n_{26} - n_{02} - n_{12} - n_{42} - n_{52}$

$\dot{X}_3 = n_{34} + n_{35} + n_{36} - n_{03} - n_{23} - n_{43} - n_{53}$

$\dot{X}_4 = n_{40} + n_{41} + n_{42} + n_{43} - n_{34} - n_{64}$

$\dot{X}_5 = n_{50} + n_{51} + n_{52} + n_{53} - n_{35} - n_{65}$

$\dot{X}_6 = n_{64} + n_{65} - n_{06} - n_{16} - n_{26} - n_{36}$

$C_1 = X_1/Q_1$

$C_2 = X_2/Q_2$

$C_3 = X_3/Q_3$

$C_4 = X_4/(A_4 \cdot D_4 \cdot \rho_4 \cdot k_{q,4})$

$C_5 = X_5/(A_5 \cdot D_5 \cdot \rho_5 \cdot k_{q,5})$

$C_6 = X_6/W$

$C_{B,1} = k_{q,1} \cdot C_1$

$C_{B,2} = k_{q,2} \cdot C_2$

$C_{B,3} = k_{q,3} \cdot C_3$

$C_{S,4} = k_{q,4} \cdot C_4$

$C_{S,5} = k_{q,5} \cdot C_5$

FIG. 1B. Equations.

The rate of transport of methylmercury from sediment to herbivores is proportional to the energy intake of the herbivores and to the methylmercury concentration in the sediment (Fig. 1A: n_{34}, n_{35}). In the same manner, the rate of transport of methylmercury between trophic levels is proportional to the energy intake of the predators and to the methylmercury concentration in the prey. The amount of methylmercury which is not assimilated returns to the sediment as does the methylmercury in organisms dying from causes other than predation (Fig. 1: n_{41}, n_{42}, n_{43}, n_{51}, n_{52}, n_{53}).

Finally, the rate of methylmercury flow from water to biota is proportional to the rate of respiration in the trophic levels and to the oxygen and methylmercury concentrations in water (Fig. 1: n_{16}, n_{26}, n_{36}).

The parameters of the model are chosen to correspond to a lake of 6 km² with a mean depth of 5 m. The sediment of the lake is treated as two different compartments, one heavily contaminated with 50 ppm mercury (wet weight) and having an area of 1 km² and the other less contaminated with 1 ppm mercury (wet weight) and having an area of 5 km². The upper 5 cm were chosen to represent that part of the sediment where methylation takes place, i.e., the "active sediment." The mean residence time of water in the lake is set to 1 year, which corresponds to a waterflow of about 1 m³/s.

The rest of the parameter values were collected from various sources (Table 1). It should be pointed out that the parameters were assigned values independent of the model, and, accordingly, no parameter value was estimated by means of fitting this model to real-world data.

Table 2, col. (a), shows the steady state concentrations of methylmercury as calculated from the model with the numerical values used in Table 1. As can be seen, the absolute and relative concentrations are in the order of those empirically found in natural systems. The difference in concentration between pikes and roaches, however, may seem somewhat low. This is an effect of the energy turnover values used (Silver Springs), which gives an extremely high respiration rate for roaches and thus a larger intake of methylmercury from water to this trophic level.

Our model, constructed and based on parameter values from several independent empirical investigations and guesses, has produced methylmercury concentrations which are in accordance with concentrations found in natural aquatic systems. This, we feel, is evidence enough to justify a further analysis of the model in order to qualitatively discuss the results of different changes in the system.

As stated before, the sediment was treated as two compartments with different concentrations of inorganic mercury and different areas. It is interesting to study the contribution of methylmercury from these two areas of sediment since they both have all their parameter values in common except those for area and inorganic mercury concentration. The larger

TABLE 1

Symbol	Physical description	Unit	Numerical value
A_4	Area of the "smaller sediment area"	km²	1
A_5	Area of the "larger sediment area"	km²	5
C_{ox}	Concentration of oxygen in water	g g^{-1}	8×10^{-6}
$C_{Hg,4}$	Concentration of inorganic mercury (smaller area)	g g^{-1}	50×10^{-6}
$C_{Hg,5}$	Concentration of inorganic mercury (larger area)	g g^{-1}	1×10^{-6}
D_4	Depth of "active sediment" (smaller area)	m	5×10^{-2}
D_5	Depth of "active sediment" (smaller area)	m	5×10^{-2}
Q_1	Standing crop of top carnivores (pikes)	kcal	936×10^4
Q_2	Standing crop of carnivores (roaches)	kcal	624×10^5
Q_3	Standing crop of herbivores (bottom fauna)	kcal	156×10^6
T	Mean residence time of water	year	1
W	Amount of water	g	3×10^{13}
f_{12}	Ratio between assimilation efficiency for methylmercury (MeHg) and energy (pikes)	—	15×10^{-2}
f_{16}	Assimilation efficiency of MeHg from water (pikes)	—	75×10^{-2}
f_{23}	Ratio between assimilation efficiency for MeHg and energy (roaches)	—	15×10^{-2}
f_{26}	Assimilation efficiency of MeHg from water (roaches)	—	75×10^{-2}
$f_{3,IN}$	Ratio between assimilation efficiency for MeHg and energy (bottom fauna)	—	6×10^{-1}
f_{36}	Assimilation efficiency of MeHg from water (bottom fauna)	—	50×10^{-2}
f_{40}	Fraction of MeHg produced as mono MeHg (smaller area)	—	1
f_{50}	Fraction of MeHg produced as mono MeHg (larger area)	—	1
$f_{ox,1}$	Assimilation efficiency of oxygen from water (pikes)	—	75×10^{-2}
$f_{ox,2}$	Assimilation efficiency of oxygen from water (roaches)	—	75×10^{-2}
$f_{ox,3}$	Assimilation efficiency of oxygen from water (bottom fauna)	—	50×10^{-2}
$f_{A,4}$	Fraction of sediment treated as smaller area	—	17×10^{-2}
$f_{A,5}$	Fraction of sediment treated as larger area	—	63×10^{-2}
k_{16}	Specific oxygen consumption (pikes)	g kcal^{-1}	2×10^{-1}
k_{26}	Specific oxygen consumption (roaches)	g kcal^{-1}	2×10^{-1}
k_{36}	Specific oxygen consumption (bottom fauna)	g kcal^{-1}	2×10^{-1}
k_{40}	Order of methylation reaction in smaller area	—	3×10^{-1}
k_{50}	Order of methylation reaction in larger area	—	3×10^{-1}
$k_{q,1}$	Specific energy content (pikes)	kcal g^{-1}	1
$k_{q,2}$	Specific energy content (roaches)	kcal g^{-1}	1
$k_{q,3}$	Specific energy content (bottom fauna)	kcal g^{-1}	7×10^{-1}
$k_{q,4}$	Specific energy content sediment in smaller area	kcal g^{-1}	1×10^{-1}
$k_{q,5}$	Specific energy content sediment in larger area	kcal g^{-1}	1×10^{-1}
β_{40}	Biochemical availability of inorganic mercury (smaller area)	—	1
β_{50}	Biochemical availability of inorganic mercury (larger area)	—	1
γ_{40}	Constant relating methylation rate to microbial activity (smaller area)	g$^{1-k_{40}}$ g$^{k_{40}-1}$ year^{-1}	63×10^{-9}
γ_{50}	Constant relating methylation rate to microbial activity (larger area)	g$^{1-k_{50}}$ g$^{k_{50}-1}$ year^{-1}	63×10^{-9}
λ_{01}	Rate constant for metabolic breakdown of MeHg (pikes)	year^{-1}	346×10^3
λ_{02}	Rate constant for metabolic breakdown of MeHg (roaches)	year^{-1}	346×10^{-3}
λ_{03}	Rate constant for metabolic breakdown of MeHg (bottom fauna)	year^{-1}	115×10^{-2}

TABLE 1 (continued)

Symbol	Physical description	Unit	Numerical value
λ_{64}	Rate constant for release of MeHg from sediment (smaller area)	year^{-1}	115×10^{-1}
λ_{65}	Rate constant for release of MeHg from sediment (larger area)	year^{-1}	115×10^{-1}
ρ_4	Density of sediment (smaller area)	g m^{-3}	11×10^5
ρ_5	Density of sediment (larger area)	g m^{-3}	11×10^5
q_{10}	Energy lost in respiration (pikes)	kcal year^{-1}	936×10^5
q_{12}	Assimilated energy (pikes)	kcal year^{-1}	125×10^6
q_{20}	Energy lost in respiration (roaches)	kcal year^{-1}	197×10^7
q_{23}	Assimilated energy (roaches)	kcal year^{-1}	237×10^7
q_{30}	Energy lost in respiration (bottom fauna)	kcal year^{-1}	113×10^8
$q_{3,1}$	Assimilated energy (bottom fauna)	kcal year^{-1}	203×10^8
$q_{D,1}$	Energy lost by natural death (pikes)	kcal year^{-1}	312×10^5
$q_{D,2}$	Energy lost by natural death (roaches)	kcal year^{-1}	281×10^6
$q_{D,3}$	Energy lost by natural death (bottom fauna)	kcal year^{-1}	657×10^7

TABLE 2

Part of model	(a) No treatment	(b) Removal of all Hg from small area		(c) 95% decrease or inactivation of Hg in the small area		(d) 95% decrease of mono MeHg formation in the small area	
	Steady state concentration	Steady state concentration	Decrease (%)	Steady state concentration	Decrease (%)	Steady state concentration	Decrease (%)
Pikes (ppm)	1.67	1.01	40	1.28	23	1.05	37
Roaches	1.63	0.99	39	1.25	23	1.02	37
Bottom fauna (ppm)	0.27	0.16	41	0.20	26	0.17	37
Small sediment area (ppb)	1.26	0.60	52	0.87	31	0.63	50
Large sediment area (ppb)	1.07	0.69	36	0.84	21	0.70	35
Water (ppb)	0.011	0.007	39	0.009	23	0.007	37

area of 5 km^2 with an inorganic mercury concentration of 1 ppm (wet weight) produces ~ 275 g of methylmercury per year, while the smaller area of 1 km^2 with an inorganic mercury concentration of 50 ppm (wet weight) produces ~ 180 g/year. Thus, although the amount of inorganic mercury in the smaller area is 10 times greater than that in the larger area, it contributes less than 40% of the total methylmercury produced. This demonstrates a consequence of using the methylation model proposed by Bisogni and Lawrence in which the biochemically available concentration of inorganic mercury and the coefficient related to the biochemical availability of mercury are raised to a power less than unity. Bisogni and Lawrence estimated this power to be ~ 0.3 and ~ 0.15 for aerobic and anaerobic systems, respectively. Thus, even if all inorganic mercury in the smaller area, which constitutes $\sim 90\%$ of all mercury in the system could be removed from the system the methylation rate would still not be reduced more than 40%. Figure 2, line a, shows the concentration of methylmercury in pikes as a function of time after an instantaneous removal of all inorganic mercury from the smaller area, while Table 2, col. (b) gives the final steady state concentrations and their relative changes for all parts of the model.

The main feature of Fig. 2, line a, is that the new steady state methylmercury concentration in pikes is attained after ~ 7 years and that this new steady state value lies just above 1 ppm. Although it is not realistic, an instantaneous removal of all inorganic mercury from the sediment would be the most efficient restoration procedure. Thus, for restoration procedures performed in the smaller area, the results of this simulation can be considered as a limiting case. The simulations of the effects of the different restoration methods that follow have been made under the assumption that the restoration measures are confined to the smaller, highly contaminated area and should therefore be compared to this limiting case.

One should also consider what the effects on methylmercury concentrations in biota might be if the amount of mercury present in the sediment were reduced.

One aspect of this that we have studied in laboratory experiments involves the effect of stirring on the biological methylation rate and suspending of sediments. In order to measure the biological methylation rate of mercury, added as $HgCl_2$ at a concentration of 10 ppm (wet weight), two types of lake sediment were incubated in 250 ml Erlenmeyer flasks for 20 days. One group of flasks was placed on a shaking table and another group was kept stable. The results are given as mean values of three replicates in Table 3. As can be seen in Table 3, the methylation rates were considerably higher in the groups where the mercury-contaminated sediment was kept in suspension. Thus, in cases of dredging, attempts should be made to reduce stirring and turbulence to a minimum.

Studies of mercury transport in a municipal sewage treatment plant[15] showed that compara-

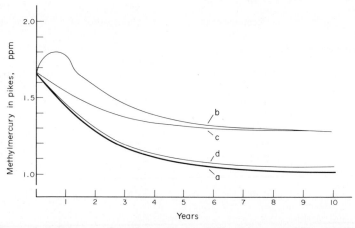

FIG. 2. (a) Removal of all mercury from the small area. (b) Dredging operations with 95% removal of mercury from the small area. (c) 95% decrease or inactivation of mercury in the small area. (d) 95% decrease of monomethylmercury formation in the small area.

FEASIBILITY OF DIFFERENT RESTORATION METHODS

Dredging

As said before, the most obvious method for restoring mercury-contaminated water systems involves the removal of mercury from the system. As mercury is stored mainly in the sediment, dredging may seem to be a useful method. There are, however, many aspects of this method that have to be taken into consideration, e.g., the vertical dispersion of sediment with the subsequent exposure of more mercury in underlying sediment layers and the problems related to the disposal of dredging spoils.

tively high concentrations of mercury exist on small particles that are not removed from the water phase during the settling process but are removed only after precipitation with aluminum sulfate.

In 1970 a dredging operation in Lake Trummen was performed[16] and the dredged sediments were placed in the first of two consecutive sedimentation ponds. Mercury concentrations in "water" were studied in the lake and in the two sedimentation ponds. The mercury concentration in the lake water was within the range 0.05–0.15 ppb both before and during the first dredging operation. In the first

TABLE 3
Effect of shaking on the rate of biological methylation of mercury

Type of sediment	Increase in CH_3Hg amounts during 20 days (ng/g)		Increase of methylation rate in shaking table factor
	Shaking	No shaking	
Eutrophic lake (Lake Långsjön)	555.0	22.2	25
Oligotrophic lake (Lake Djupkärra-Dammsjö)	57.0	9.2	8

sedimentation pond, the mercury concentrations in the water showed a mean of 0.6 ppb, but after settling and before precipitation the average concentration was down to that in the lake water (~ 0.1 ppb). It remained at that level during precipitation and flocculation until the water was back in the lake. Thus, the mercury concentration in the water had already decreased to about background level after the first sedimentation process, and no further decrease in mercury concentration was detected as an effect of precipitation. This was in contrast to the findings of Westermark and Ljunggren,[15] and can be seen as an example of the importance of the type of particle structure of the sediment.

The Lake Trummen project revealed another problem; disposal of the mercury-contaminated sediment from the sedimentation ponds. The mercury concentrations of the drying dredged sediment decreased during the first months on land ($\sim 30\%$). Two processes could be responsible: mercury transport by drainage water or evaporation of volatile mercury compounds. Analyses of the drainage water did not support the former explanation. Laboratory experiments were then performed to study the latter process. These studies revealed a rapid formation of volatile dimethylmercury when wet mercury-contaminated organic sediment was exposed to air ($\sim 10\%$ in 5 days).[8,17]

In order to help find out what effect dredging has upon methylmercury concentrations in biota, a simulation was performed using the previously described model. The conditions for the simulation were as follows: at the beginning of the operation the system was in the steady state of Table 2, col. (a); the dredging operations were limited to the smaller, highly contaminated area and carried out for a 3 month period; during the dredging period the amount of "active" sediment was five times the normal; during the same time the amount of mercury in the smaller, highly contaminated area was linearly decreased to 5% of the previous amount.

Thus, the overall result of the restoration operation was a 95% reduction of the concentration of inorganic mercury in sediment. The resulting new steady state methylmercury concentrations and their relative changes are given in Table 2, col. (c). Figure 2, line *b*, shows methylmercury concentrations in pikes, as a function of time, for the simulated dredging operation.

From Table 2, col. (c), it can be seen that the steady state concentrations in biota decreased $\sim 25\%$ as a result of dredging. These decreases can be compared to those of $\sim 40\%$ in the limiting case (Fig. 2, line *a*), in which *all* inorganic mercury was instantaneously removed from the smaller, highly contaminated model area. Figure 2, line *b*, reveals the interesting facts that methylmercury concentrations in pikes increase shortly after dredging, begin to decrease after ~ 1 year and take approximately 8 years to reach a new steady state. The increase in methylmercury concentrations is a result of the stirring and turbulence of the sediment, simulated by an increased volume of "active" sediment during the dredging period.

CONVERSION OF MERCURY TO MERCURIC SULFIDE

Mercuric sulfide is well known for its extremely low solubility. The theoretical solubility product is $\sim 10^{-53}$, and, even if this figure has little relevance for natural waters with competing organic complexes, competing metal ions, and oxidizing conditions, it has been demonstrated that there is little mercury available for biological methylation when mercury is bound as mercuric sulfide. To obtain the same yield of methylated mercury from mercuric sulfide as from divalent inorganic mercury, the concentrations of mercury in the former form would have to be $\sim 10^3$ times that of the latter.[18]

A large number of aquarium experiments have been performed to test different *in situ* methods of converting divalent inorganic mercury in the sediment to mercuric sulfide. The effectiveness of the methods has been measured by using the relative rate of accumulation of methylmercury in fish. The following methods for the production of mercury sulfide were tried:[10]

(1) Anaerobic conditions have been created by adding glucose—an easily degraded, oxygen-consuming organic substance—to water. Sulfate is then reduced to sulfide.
(2) Sulfide ions were also directly added (20 mg/l) and prior to the introduction of fish the aquaria were aerated till no more free sulfide was detected.
(3) Mercury enriched sediment was covered with FeS (50 g/dm^2).
(4) Mercury enriched sediment was covered with the mineral FeS$_2$ (100 g/dm^2).

The results of the experiments are summarized in Table 4. In all series a reduced accumulation of methylmercury in the fish was obtained. The best results were achieved by adding sulfide. The reduction here was ~98%.

Field studies of Lake Gårlången, prior to and after a winter of anaerobic conditions and free sulfide present in the deeper water, have shown that the biological methylation of mercury in the sediments was reduced by ≥96%.

As mentioned earlier, the conversion of mercury to mercuric sulfide decreases the biochemical availability of mercury. A simulation was performed to demonstrate the effect of 95% instantaneous decrease in the biochemical availability of mercury in the smaller, highly contaminated model area. The initial conditions of the simulation were those steady state conditions given in Table 2, col. (a). The resulting new steady state concentrations and their relative changes are the same as for the simulated dredging operation (Table 2, col. (c)) because, according to the methylation model,[9] a 95% decrease

TABLE 4
Methylmercury accumulation in fish after different treatments of the aquaria with mercury-rich sediments

Substance added to cause the formation of mercuric sulfide	CH$_3$Hg in fish at the end of the experiment (ng/g)	
	Control	Treated
Glucose	510	130
S^{2-}	610	30
FeS	980	160
FeS$_2$	1030	410

In all series the methylmercury concentration in the fish was <20 ng/g at the start of the experiment.

in biochemical availability is equivalent to a 95% decrease in mercury concentration.

Although the resulting steady state methylmercury concentrations obtained by the two methods (dredging and decrease of biochemical availability) are the same, the period of time to attain steady state is different, as illustrated by the methylmercury concentrations for pikes in Fig. 2, line c. A faster decrease is achieved in the latter case since the initially increased methylation rate, the result of stirring and turbulence during dredging, does not have to be taken into account.

COVERING OF MERCURY DEPOSITS

General studies concerning both the chemical and biological reactions in sediment columns and the exchange processes in the sediment–water interface show that microbiological activity is highest in the very upper layer of the sediment and that exchange reactions occur much more quickly between the water and upper layer of sediment than between the water and the deeper layers. Since the formation and release of methylmercury from deposits of sediments containing inorganic mercury clearly follow this process, mercury deposits—if covered—might be expected to release less methylmercury to the water phase.

Naturally, if the material used to cover the mercury-rich sediment also in itself had a mercury binding capacity, the effect would be even larger. With reference to the restoration method previously discussed—binding of mercury in the sulfide form—organic sediments covered with, e.g., sand or clay, have a higher probability of turning anaerobic due to the reduced water and oxygen exchange. Thus, in this case, sulfide ions may be formed and mercury may be bound as mercuric sulfide.

In laboratory experiments fiber sediments contaminated with phenylmercury, as well as other types of organic sediments containing inorganic divalent mercury, were covered by layers of fine sand and freshly ground silica, respectively. A substantial reduction of methylmercury uptake in aquaria fish was thereby obtained (95–70%).

To determine covering efficiency, field studies were carried out (Lake Gårlången and Lake Mellanfryken) where certain mercury-containing sediment areas were covered with mine tailings and sand respectively. It was concluded that a covering efficiency of ~95% could be obtained.

A simulation of a covering operation in the smaller, highly contaminated model area was also performed. Although the covering of sediment in a real-world system might affect many factors, such as amount of "active" sediment, microbial activity, exchange of methylmercury from sediment to water, biochemical availability, and so on, this operation was simulated to represent an instantaneous decrease of 95% in the rate of methylmercury formation. The resulting, new steady state concentrations and their relative changes are given in Table 2, col. (d).

The steady state decreases of methylmercury concentrations in biota are in nearly the same order as those of the limiting case, ~37% compared to ~40%. The timecourse of the change of methylmercury concentrations in pikes, as shown in Fig. 2, line d, is also almost the same as for the limiting case (Fig. 2, line a).

Raising of pH

Several field investigations have indicated that a negative correlation exists between levels of methylmercury in fish and pH.[19-21] Laboratory experiments have indicated that one of the bases of this correlation is the changeover from mono- to dimethylmercury as a net result of the processes of biological methylation of mercury[22] (cf. Fig. 3). More explicitly, this changeover is likely to be a combination of two factors: (1) dimethylmercury is converted to monomethyl mercury in an acid environment, and (2) the composition of microorganisms changes with pH levels. Higher pH favors those microorganisms producing dimethylmercury while lower pH favors those forming monomethylmercury as the net result of the methylation process.[23]

While dimethylmercury might evaporate when formed, monomethylmercury is more likely to accumulate in water-living organisms. Experiments were performed where fish (*Lucioperca lucioperca*) were kept for up to 2 months above a layer of fiber sediment containing phenylmercury added as acetate. In one series of experiments, $CaCO_3$ was added to increase pH from 6.7 to 7.1. At the lower pH, the average concentrations of methylmercury in fish muscle increased from 130 to 390 ng/g. At the higher pH, the increase was from 130 to 180 ng/g.[24]

Further tests with organic sediment contaminated with inorganic mercury have given similar results, showing a negative correlation between pH and the accumulation of methylated mercury in fish, the strongest effect being around pH 7.[8]

The simulation previously shown (Table 2, col. (d) and Fig. 2, line d) exemplifies what happens when 95% of the methylated mercury is in the form of dimethylmercury. However, this study was not concerned with the possible problems dimethylmercury might cause outside the model system.

DISCUSSION AND CONCLUSIONS

The simulations performed, in order to indicate how different restoration methods affect methylmercury concentrations in biota, can be considered a kind of sensitivity analysis of the constructed model. In order to judge the significance of the simulation results, one must accept the hypothesis that there is a definite possibility that the corresponding alterations can be achieved in a natural system. In other words, if we believe that it is possible to, e.g., achieve a 95% decrease in the methylation rate by covering the sediment of a lake, then the corresponding simulation has an indicative value, specifying the best method to be chosen in a practical case.

Unfortunately, many questions still remain unanswered concerning the practical use of different restoration methods. Nevertheless, it seems possible to affect the biochemical availability of mercury to a much larger extent than in the simulation, especially under anaerobic conditions. The effects of sediment covering on methylation rate have also been demonstrated to be high, at least in a limited-time perspective. It seems that the removal of mercury by means of dredging must be quite efficient in order to have a significant effect on the methylmercury concentrations in biota.

However, the simulations showed that restoration operations in general would have little effect if they were limited to only a part of a water system, even if that part contained, as in this case, the main amount (90%) of the mercury in the system. This brings about another aspect of the inhomogeneous distribution of mercury in the simulated system. What would happen if the mercury were to be spread from the smaller, highly contaminated area in such a way that a homogeneous distribution of mercury were

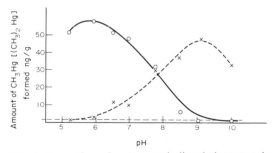

FIG. 3. Formation of mono- and dimethylmercury in organic sediments at different pH during 2 weeks with a total mercury concentration of 100 ppm in the substrate. ○ CH_3Hg^+ found in water and sediment. × Volatile CH_3Hg-compound trapped in Hg^{2+} after aeration. $(CH_3)_2Hg$.

achieved in the sediment of the entire lake? The resulting steady state concentrations of methylmercury in biota after such a simulated homogeneous distribution are shown in Table 5, col. (a). The corresponding values of a homogeneous distribution of mercury in the sediment, after a 50% loss of mercury from the system, are given in Table 5, col. (b).

As can be seen from Table 5, cols. (a) and (b), a sediment transport process resulting in a homogeneous distribution of mercury, covering all the lake sediment, might cause increased methylmercury levels, even if 50% of the total amount of mercury in the "active sediment" were lost during the "transport period." In addition to such sediment transport mechanisms, there are other long term processes that might affect the methylation rate in a natural water system. It is obvious, then, that the effect of restoration operations must not be compared solely to today's mercury levels but also to those levels that would have resulted from such a natural process. In other words, one must also consider the ability of a restoration operation to prevent negative effects of natural processes.

A very important factor in judging the effects of restoration operations is the water exchange in an aquatic system. Water exchange has a great influence both on steady state concentrations of methylmercury and on the time courses of their changes after an operation. For example, Table 6 gives the steady state concentrations for simulated mean residence times of water. These times are 1 month and 5 years, respectively. The approximated time taken to reach a new steady state, after restoration operations of the kind previously simulated, would be 2 years and 25 years, respectively. The River Mörrumsån in southern Sweden, with an average water transport of ~30 m^3/s, is an example of a water system where the methylmercury concentrations in fish have decreased rapidly after the outlets to the river were reduced in 1966.[10]

TABLE 6
Steady state concentrations of methylmercury

	Mean residence time of water	
	1 month	5 years
Pikes (ppm)	0.18	4.3
Roaches (ppm)	0.17	4.2
Bottom fauna (ppm)	0.04	0.7
Small sediment area (ppb)	0.38	2.8
Large sediment area (ppb)	0.19	2.6
Water (ppb)	0.002	0.03

REFERENCES

1. KONRAD, J. G., in *Environmental Mercury Contamination* (ed. R. Hartung and B. Dinman), Ann Arbor Science Publishers, Ann Arbor, Michigan, 1971.
2. JERNELÖV, A., *Vatten 1*, **68**, (1968).
3. JENSEN, S. and JERNELÖV, A., *Biocidinformation* **10**, 4 (1967).
4. JENSEN, S. and JERNELÖV, A., *Biocidinformation* **14**, 3 (1968).
5. JENSEN, S. and JERNELÖV, A., *Nature* **223**, 753–754 (1969).
6. JERNELÖV, A., *Nobel Symposium 20, The Changing*

TABLE 5

Part of model	(a) Homogeneous distribution of Hg in the whole lake sediment Steady state concentration	(b) Homogeneous distribution of Hg in the whole lake sediment, 50% of the amount being lost Steady state concentration
Pikes (ppm)	2.36	1.92
Roaches (ppm)	2.31	1.88
Bottom fauna (ppm)	0.38	0.31
Small sediment area (ppb)	1.57	1.28
Large sediment area (ppb)	1.57	1.28
Water (ppb)	0.016	0.013

Chemistry of the Oceans, pp. 162–169, Almqvist & Wiksell, Stockholm, 1972.
7. WESTÖÖ, G., *Acta chem. scand.* **20**, 2131–2137 (1966).
8. JERNELÖV, A., LANN, H., and LORD, M., *Vatten* **2-71**, 234–239 (1971).
9. BISOGNI, JAMES J., JR., and LAWRENCE, A. WILLIAM, Cornell University, Techn. Rep. No. 63, Ithaca, NY, 1973.
10. JERNELÖV, A. and LANN, H., *Environ. Sci. Technol.* **7**, 712–718 (1973).
11. WOOD, J., KENNEDY, F., and ROSÉN, C-G., *Nature* **220**, 173–174 (1968).
12. LANDNER, L., *Nature* **230**, 452–453 (1971).
13. BEIJER, K., JERNELÖV, A., and RUDLING, L., Report from Swedish Water and Air Pollution Research Laboratory, 1970.
14. ODUM, H. T., *Ecol. Monogr.* **27**, 55–112 (1957).
15. WESTERMARK, T. and LJUNGGREN, K., Report to the Swedish Applied Research Council No. 4952, 1968.
16. BJÖRK, S. VÄRLDEN, Vattnet och vi. Conference, Jönköping, 1970.
17. SKOGLUND, P-O., Report to Swedish Environmental Protection Board No. 300871, 1971.
18. FAGERSTRÖM, T. and JERNELÖV, A., *Wat. Res.* **5-71**, 121 (1971).
19. OLSSON, M., *Nord. Hyg. Tskr.* **50**, 179 (1969).
20. LARSSON, J.-E., Swedish Environmental Protection Board, Stockholm, private communication, 1972.
21. JERNELÖV, A., *Environmental Mercury Contamination*, pp. 174–177, Ann Arbor Science Publishers Inc., Ann Arbor, Michigan, 1972.
22. FAGERSTRÖM, T. and JERNELÖV, A., *Wat. Res.*, **6-10**, 1193–1202 (1972).
23. JERNELÖV, A., *Nord. Hyg. Tskr.* **50**, 174–178 (1969).
24. LANN, H., Report to Swedish Environmental Protection Board No. 060370, 1970.

The Feasibility of Restoring Mercury-contaminated Waters
(A. Jernelöv and B. Åséll)

DISCUSSION by FREDERICK G. ZIEGLER
Director of Air and Water Resources, Associated Water and Air Resources, Inc., Nashville, Tennessee 37204

and

PETER A. KRENKEL
Director of Environmental Planning, Tennessee Valley Authority, 268–401 Building, Chattanooga, Tennessee 37401

INTRODUCTION

Two general observations should be made with respect to restoring mercury-contaminated waters or, to be more specific, in preventing the release of mercury bound to contaminated sediments from entering natural waters. The first observation is that most of the work in evaluating restoration of sediments concentrates on the specific problem of removing or covering highly contaminated sediments localized in the vicinity of industrial discharges. It should be emphasized that this is only one of three obvious forms of sediment disturbance which may create increases of mercury concentrations within natural waters. The two additional situations by which contaminated sediments may be extensively disturbed are both forms of dredging and are listed below:

(1) Dredging for the mining of sand and gravel used within various phases of the construction industry.
(2) Dredging of ship channels to improve and maintain commerce.

Obviously, each of these forms of dredging is crucial to the economy of our nation and is often approved on the basis of an acceptable environmental impact. In many instances, disapproval of these dredging techniques results in a more severe economic impact on a region than the restoration of sediments resulting from contamination by industrial sources.

The second general observation is that the situations under which mercury contamination exists within the environment are as numerous as the number of mercury-contaminated discharges. Discharges through swamps or shallow ponds permit sedimentation of high concentrations of mercury which may or may not be covered by water. This situation is often accompanied by severe erosion during periods of the year which are characterized by rainy seasons. Hydraulic shocks may be more easily regulated if wastewater is discharged through holding ponds. Still other circumstances similar to those described by Jernelöv and Åséll result in direct discharges to large reservoirs or river systems which may, in turn, result in various situations for which remedial action will depend on natural conditions. Each of these situations may call for different treatment techniques for restoration of the sediment and the prevention of mercury release to the natural environment. Thus, some of the techniques described by Jernelöv and Åséll may be more or less acceptable depending on the specific situation, although application of the model may aid in evaluation of the various treatment alternatives.

MODEL DEVELOPMENT

No exception is taken with the model development presented in this paper. However, it should be noted that the rate of methylation described as being approximately 0.1% per year is in all likelihood the result of measurements conducted in the relatively cold climate in Sweden. As has been found by Shin,[1] the methylation rate is quite dependent on temperature. Thus, methylation rates in the southern United States may be significantly higher than those described.

METHODS OF RESTORATION

Numerous techniques have been proposed for the restoration of mercury-contaminated sediment. These techniques may apply to any of the three forms of dredging as well as in-place sediment isolation. It is instructive to note the potential application of the various dredging and covering techniques in preventing severe contamination by sediments.

Two primary forms of dredging are presently employed, which are hydraulic dredging and mechanical dredging. Mechanical dredging normally is accomplished by draglines or clam shells which are barge or land based and are often used to remove relatively small quantities of sediment from around piers or channels. Mechanical dredging disturbs and resuspends sediments at the dredging site to a greater extent than does hydraulic dredging. Additionally, this dredging technique often requires the handling of dredging spoils more than once. For example, dredging of piers often requires that the sediment be deposited in close proximity to the dredging operation to permit drying prior to retransportation to an acceptable disposal site. During drying, washing by rainfall and draining of the water contained in the spoil may result in a discharge of mercury-contaminated sediment back to the environment. Furthermore, as described by Jernelöv and Åséll, the release of dimethylmercury from drying sediments has been shown to be quite significant and may result in contamination of the environment. Mechanical dredging does result in a spoil which contains a significantly lower water content than that resulting from hydraulic dredging. Nevertheless, high resuspension of sediment is characteristic at the site during the dredging operation.

On the other hand, hydraulic dredging normally results in less contamination and resuspension of sediment at the dredging site. This can be suppressed even further with the omission of the cutter head, which is only necessary when large objects such as tree stumps are found in the sediment. The primary objection to hydraulic dredging is the high water content of the spoil which, if disposed of on land, results in a relatively high discharge of water back into the natural environment. It should be noted that the small (6 μm or less) particle fraction contains the majority of the mercury found in the contaminated sediment and may be colloidal in nature.[2] Thus, sedimentation alone may not be adequate to prevent mercury from re-entering the environment. The particle distribution will vary significantly with each situation and may require analyses of the settling characteristics of the particular spoil prior to dredging. This basic testing will reveal the nature of the sediments and dictate the requirement for coagulants and coagulant aids which may be required in addition to plain sedimentation.

Irrespective of the dredging method used, disposal of dredging spoil becomes a critical phase of the operation which must be identified and proven acceptable prior to initiation of the dredging operation. Disposal of dredging spoils may be achieved in a number of ways. Some of these techniques are listed below:

(1) Lining of spoiled disposal sites with either natural or artificial liners. Artificial liners, i.e., polyethylene will only retard the transport of inorganic mercury. Organic mercury, and in particular methylmercury, may still contaminate the groundwater.[3]
(2) Mixing of the spoil with binders which stabilize the mercury within the sediment. Substances such as sulfides and clays (illite) have been reported to inhibit the release of mercury to the environment.[4,5]
(3) Disposal or discharge of the spoils into an environment which will be less conducive to methylation or transport of methylmercury to aquatic life. An example of such a technique might be to permit disposal of contaminated spoils in a relatively shallow ocean environment. Presently, disposal is permitted only at depths of greater than 100 fathoms. It has been shown by Reimers and Krenkel[5] and Shin,[1] that methylation and mercury uptake reduced in a saline environment. Additionally, the high dissolved solids will increase the settling rate of the solids and thus aid in reducing the time of exposure of organisms to the contaminated spoils.

MERCURY MINING OF SPOILS

Numerous methods for extracting mercury from contaminated spoils have been proposed within the last few years. The feasibility of such techniques is dependent upon the location, mercury concentration, and economics of the particular situation. Some of the methods are described here because of the possibility of their use under certain situations.

Sieving[2] has been investigated as a possible means of separating mercury-contaminated sediments from noncontaminated sediments. This method has not proved feasible because mercury is often contained on the smaller particles which are not easily controlled during the sieving process. In addition, the distribution of mercury throughout a wide range of particle sizes lowers the efficiency of this method. Similar comments apply for gravity separation (sedimentation) unless followed by chemical coagulation.

Flotation[2] has been shown to be feasible under

certain conditions. This would be expected inasmuch as flotation techniques are employed in certain mining operations for heavy metals. The relative cost of equipment and hardware to perform flotation mining of contaminated sediment is rather high, however, and may preclude the use of flotation because of the minor benefit of the removal and recovery of this mercury. Nevertheless, if environmental considerations are included in this economic decision, flotation may prove to be a feasible technique.

Pyrometallurgy[2] has also been offered as a possible means of extracting mercury from sediment. This method becomes exceedingly expensive because significant amounts of energy are required to dry the sediment prior to removal of the mercury. These additional expenses will, in most instances, preclude the use of pyrometallurgy for the removal of mercury from contaminated spoils.

Leaching with sulfuric or hydrochloric acid has been investigated.[2] It has been shown that hydrochloric acid but not sulfuric acid is a feasible reagent for the leaching of mercury. Leaching with hydrochloric acid creates an ionic form of mercury, tetrachloromercury II, which will adsorb on carbon or be precipitated with alum. Leaching techniques have been employed for both dredging spoils and undisturbed sediments. Obviously, the implementation of leaching techniques in undisturbed sediments requires numerous safeguards to prevent pH alteration of the natural environment. Leaching techniques may prove to be the least costly of all the alternatives, including covering and overlaying with organic film. Obviously, the numerous requirements for control of the leached material and the use of acid will hinder the use of leaching techniques.

Hypochlorination may also provide an acceptable leaching technique.[2] It should be noted that leaching methods are in general less costly than the other methods described.

In summarizing the various aspects of dredging, it would seem from the work described by Jernelöv and Åséll and others that the removal of contaminated sediments may not significantly change the concentration of the mercury in the biota over a short interval. In this regard, a lower limit of mercury contamination of the sediment which will trigger methylation was found by Shin.[1] Thus, surveying contaminated sediments and delineating the areas in which mercury is at a level high enough to support methylation, if any, may permit suppression of methylation either by removal or covering of a relatively small volume of material.

COVERING METHODS

Another technique for segregating mercury-contaminated sediments is depositing layers of uncontaminated sand or gravel on top of the contaminated sediments, although the technique may be relatively expensive, ranging in price from $3000–$4000 per acre.[6] Transport and distribution accounts for most of the cost, although studies by Langley[7] showed the method to be feasible.

As shown by Jernelöv,[8] sediment transport and destruction of the sediment cover by bottom organisms may reduce the effectiveness of the cover.

Other methods proposed included artificial coverings such as films of nylon polymer, which appear to be effective in suppressing both inorganic and organic mercury transport.[3] High density polyethylene has also proved to be effective with inorganic mercury but less effective with organic mercury. Problems involved in employing these artificial coverings include the difficulty of depositing the films and holding them in place because of the movement of water over the sediment and film and the creation of gases within the anaerobic layer under the film. Leaded weights, sand or gravel deposited on top of the film, or more sophisticated anchors might significantly increase the cost of this method.

Distribution and weighting the films may cost between 1.5 and 3.3 cents/ft^2.[3] The least costly method of distribution involved alcohol-mixed organic films which solidify on contact with water. However, this technique does add organic compounds to the water which will increase the BOD and thus reduce dissolved oxygen concentration. Hot-mixed techniques have also been proposed, although the cost is probably prohibitive. Preformed films have been placed successfully on a small scale but the cost is, again, quite high.

Other coverings which have been investigated and appear to have potential on a small scale include:

(1) Cotton nets covered with sulfide.[4]
(2) Iron and sand mixtures. The cost of this distribution system is approximately the same as that of covering the sediment with sand.[2]
(3) The use of wool.[9]

It may be concluded that, while the investigations described herein are interesting, the cost-to-benefit ratio will probably not be conducive to their use. Furthermore, with few exceptions, application to a field-type system has not been attempted. Prior to the serious consideration of any of these methods of sediment decontamination, the real hazard existing from mercury in sediments, if any, should be delineated.

Finally, it is hoped that this discussion of sediment decontamination investigations in the United States has completed the excellent summary of Swedish experience presented by Jernelöv and Åséll.

REFERENCES

1. SHIN, E. B., Methylmercury uptake by fish and reaction mechanisms of mercury biomethylation under controlled laboratory conditions, PhD dissertation, Vanderbilt University, 1973.
2. FEICK, G., JOHANSON, E. E., and YEAPLE, D. S., *Control of Mercury Contamination in Fresh Water Sediments*, Environmental Protection Technology Series EPA-R2-72-077, October 1972.
3. WIDMAN, M. A. and EPSTEIN, M. M., *Polymer Film Overlay System for Mercury Contaminated Sludge-Phase I*, Water Pollution Control Series 16080 HTZ, May 1972.
4. SUGGS, J. D., PETERSEN, D. H., and MIDDLEBROOK, J. B., JR., *Mercury Pollution Control in Stream and Lake Sediments*, EPA 16080 HTD, March 1972.
5. REIMERS, R. S. and KRENKEL, P. A., *J. Wat. Pollut. Control Fed.* **46** (2) 352 (1974).
6. BONGERS, L. H. and KHATTAK, N. M., *Sand and Gravel Overlay for Control of Mercury in Sediment*, Water Pollution Control Series 16080 HTD, January 1972.
7. LANGLEY, D. G., Mercury methylation in an aquatic environment, *J. Wat. Pollut. Control Fed.* **45** (1973).
8. JERNELOV, A., Release of methylmercury from sediments with layers containing inorganic mercury at different depths, *Limn. Ocean.* **15** (6) (1970).
9. FRIEDMAN, M., HARRISON, C. S., WARD, W. H., and LUNDGREN, H. P., Sorption behavior of mercuric and methylmercuric salts on wool, Division of Water, Air and Waste Chemistry Meeting, ACS, Los Angeles, California, 109, March 29–April 2 1972.

EXPERIENCE WITH HEAVY METALS IN THE TENNESSEE VALLEY AUTHORITY SYSTEM

W. R. NICHOLAS
Chief, Water Quality Branch, TVA

and

BRUCE A. BRYE
Assistant Chief, Water Quality Branch, TVA

INTRODUCTION

During a 5 day period—July 9 through July 13, 1968—more than 500,000 fish were killed in the Watauga arm of Boone Reservoir. The cause of this fish-kill was difficult to track down, and it was not until all the more common causes of fish-kills were eliminated that toxic mercury compounds originating from barrels were pinpointed as the cause. This was the first time in the Tennessee Valley that mercury and compounds of mercury were identified as causing an environmental problem.

In 1970, after the Food and Drug Administration established a guideline for allowable mercury concentration in fish flesh, TVA investigations revealed two locations in the Tennessee Valley where the mercury levels in fish flesh exceeded the guideline of 0.5 $\mu g/g$. Both of these locations were immediately downstream from the discharges of chlor-alkali plants.

The first location was Pickwick Reservoir, an impoundment of the Tennessee River in northwestern Alabama, northeastern Mississippi, and southwestern Tennessee. This reservoir was closed to both commercial and sport fishing in July 1970. In recent months, the restrictions on sport fishing were relaxed, but the states issued warnings to restrict the dietary intake of fish taken from this reservoir. Commercial fishing is still banned.

The second area was the North Fork Holston River in southwestern Virginia and northeastern Tennessee. A 70 mile reach of this river was closed to sport fishing in September 1970. It remains closed.

Investigations of Chickamauga Reservoir in southeastern Tennessee, which also receives the discharge from a chlor-alkali plant, did not show mercury concentrations in fish flesh or water that exceeded the guidelines. Consequently, it was not closed to fishing.

TENNESSEE VALLEY AREA

The drainage basin of the Tennessee River encompasses a landlocked area of nearly 41,000 square miles in the southeastern United States (Fig. 1). Portions of seven states—Virginia, North Carolina, Georgia, Tennessee, Alabama, Mississippi, and Kentucky—lie within its boundaries. It is a watershed of contrasts. Rugged mountains and forests dominate the eastern portion of the Valley; rolling hills, open fields, and woodlands lie to the west. From Mount Mitchell, towering 6600 ft near the eastern boundary of the watershed in North Carolina, the topography of the Valley ranges downward to Paducah, Kentucky, its western extremity, only a little over 300 ft above sea level.

The Tennessee River system has its headwaters in the mountains of western Virginia and North Carolina, eastern Tennessee, and northern Georgia. Two rivers, the Holston and the French Broad, join at Knoxville, Tennessee, to form the Tennessee. From Knoxville the river flows southwesterly through Tennessee, then westerly through northern Alabama, and at the northeast corner of Mississippi turns north, recrosses Tennessee, and continues to Paducah, Kentucky, where it enters the Ohio River. Water from tributaries in the northeast corner of the Valley flows more than 900 winding miles to the mouth of the river at Paducah.

Since 1933 the Tennessee Valley Authority has been developing the water resources of the Tennessee River Valley by the construction of several large multipurpose reservoirs. In the TVA system, reservoirs (as shown in Fig. 1) are operated primarily for purposes of flood control, navigation, and the production of electricity. Two basic types of reservoirs are in use. Large, deep, headwater storage reservoirs are located on the principal tributaries of the Tennessee River. In these pools, water levels usually vary over a wide vertical range

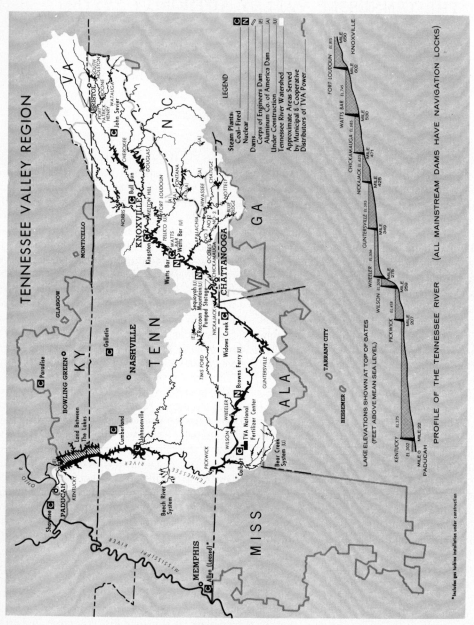

FIG. 1. Tennessee Valley region.

during the course of a year. (A difference of 100 ft or more between high and low levels is not uncommon.) On the main Tennessee River, navigable depths must be maintained at all times throughout the full length of each of the nine contiguous pools. This means drawdown must be limited, and as a result the vertical range of pool-level fluctuations is relatively narrow. The physical characteristics of the TVA reservoirs are considerably different from those of natural lakes. In essence, they blend the characteristics of a free-flowing stream and a natural lake. The physical characteristics of the

TABLE 1
Summary of physical characteristics of selected TVA headwater storage reservoirs

Reservoir	Normal summer pool				Normal[a] reservoir drawdown (ft)	Mean annual discharge[b] 1965 (ft³/s)	Mean detention time (days) for various percents of reservoir volume[b]		
	Length of reservoir (miles)	Depth at dam (ft)	Surface area (acres)	Volume (acre-ft)			100%	80%	60%
Blue Ridge	10.0	150	3320	200,800	101.0	510	199	159	119
Boone	17.3[c] 15.3[d]	125	4520	196,700	55.0	2470	40	32	24
Chatuge	13.0	120	7150	247,800	68.0	440	284	227	170
Cherokee	59.0	163	30,200	1,504,000	93.0	4385	173	138	104
Douglas	43.1	122	30,600	1,452,000	80.0	6395	114	92	69
Fontana	29.0	450	10,670	1,444,300	185.0	4010	182	145	109
Hiwassee	22.0	245	6120	422,000	109.5	1860	114	92	69
Norris	72.0[e] 56.0[f]	200	34,200	2,047,000	90.0	4150	249	199	149
Nottely	20.0	170	4290	184,400	90.0	380	245	196	147
South Holston	24.3	245	7580	638,200	113.0	968	332	266	199

[a] Drawdown from normal summer pool to normal minimum pool.
[b] Based on average year—1965.
[c] On South Fork Holston River.
[d] On Watauga River.
[e] On Clinch River.
[f] On Powell River.

TABLE 2
Summary of physical characteristics of TVA mainstream reservoirs

Reservoir	Dam location (Tennessee river mile)	Normal summer pool				Normal[a] reservoir drawdown (ft)	Mean June–Sept.[b] 1965 discharge (ft³/s)	Mean detention time (days) for various percents of reservoir volume[b]		
		Length of reservoir (miles)	Depth at dam (ft)	Surface area (acres)	Volume (acre-ft)			100%	80%	60%
Kentucky	22.4	184.3	85	158,000	2,713,000	5.0	40,300	34.0	27.2	20.4
Pickwick	206.7	52.7	91	42,700	937,000	6.0	34,600	13.7	10.9	8.2
Wilson	259.4	15.5	101	15,930	642,500	3.0	32,800	9.9	7.9	5.9
Wheeler	274.9	74.1	66	67,100	1,131,000	6.0	33,000	17.3	13.8	10.4
Guntersville	349.0	75.7	76	69,100	987,500	2.0	28,300	17.6	14.1	10.6
Nickajack[c]	424.7	46.3	64	10,900	242,800	2.0	—	—	—	—
Hales Bar	431.1	39.9	94	6420	147,700	2.0	27,200	2.7	2.2	1.6
Chickamuga	471.0	58.9	76	34,300	585,000	7.5	27,300	10.8	8.6	6.5
Watts Bar	529.9	72.4	104	38,600	970,000	6.0	22,900	21.4	17.1	12.8
Fort Loudoun	602.3	55.0	89	14,450	356,000	6.0	12,400	14.5	11.6	8.7

[a] Drawdown from normal summer pool to normal minimum pool.
[b] Based on average year—1965.
[c] Nickajack replaced Hales Bar on December 16, 1967.

selected TVA headwater reservoirs and the nine mainstream reservoirs are summarized in Tables 1 and 2.

BOONE RESERVOIR FISH-KILL

Boone Reservoir is a headwater storage reservoir created by Boone Dam on the South Fork Holston River at river mile 18.6. The confluence of the two main inflows to Boone Reservoir, the Watauga River, and the South Holston River, is about 1.3 miles upstream from Boone Dam. Early on July 9, 1968, the Tennessee Game and Fish Commission informed the TVA Fish and Wildlife Branch that a fish-kill was in progress on the lower end of the Watauga arm of Boone Reservoir. A biologist and a bacteriologist from the Water Quality Branch investigated the situation early on the morning of July 10. They observed dead and dying fish in a reach of the Watauga arm of Boone Reservoir extending from about mile 4 to about mile 6.5. The water in which the fish were dying had a noticeably brown color that resembled that which could have been produced by a high concentration of diatoms. The dead and dying fish were primarily gizzard shad (possibly 99% of the total), carp, catfish, and sunfish. Some of the fish affected were observed to swim spasmodically near the surface, eject air bubbles orally, hemorrhage about the dorsal fin, lose sensitivity, gradually decrease their swimming efforts, lose equilibrium, and, finally, die. Most of the fish simply lost equilibrium and died.

The map of Boone Reservoir (Fig. 2) indicates the site of the kill. The "midreservoir" location of the dead and dying fish strongly indicated a local source of toxic material. However, because of known industrial waste sources of zinc, copper, and ammonia further upstream, water samples were collected and analyzed for these parameters. Determinations were also made of dissolved oxygen concentrations, temperature, and pH. Some of the results of analyses of water samples collected on July 11 from various locations and depths in the Watauga arm of Boone Reservoir are shown in Fig. 3. During the investigation, water samples were also analyzed for cyanide and many metals including mercury, lead, chromium, cadmium, and nickel. Because some species of algae are known to be toxic, algae that could have caused the brown discoloration of the water were also suspected, and the plankton population was investigated.

Fish studies

Dying and dead fish were collected and specimens were sent to the following laboratories for various examinations: (1) the Federal Water Pollution Control Laboratory at Athens, Georgia, (2) the United States Fish and Wildlife Service Parasite Disease Laboratory at Auburn, Alabama, and (3) the United States Fish and Wildlife Service Pesticide Laboratory at Columbia, Missouri. The Auburn Parasite and Disease Laboratory identified *Aeromonas liquefaciens* in bacterial cultures from water and fish. This is a notorious carp-killing bacteria that produces liquifying lesions in the kidneys and muscles but acts slowly and would have to build to concentrations of more than one million cells per milliliter to kill fish. No lesions were observed in any fish, and a 2 h kill of previously unexposed fish was not compatible with a bacterial infection.

All reports from all cooperating laboratories were negative as to detectable causes of the kill.

Plankton studies

During the period of the fish-kill, a "normal" population of phytoplankton existed in Boone Reservoir. The total number of cells per milliliter throughout the upper 20 ft of water progressively increased in a downstream direction from the riverine habitat to the lake habitat. Total cell numbers increased from about 200 cells/ml at Watauga River mile 13.0 to more than 3000 cells/ml at Boone Dam. The diversity of genera decreased toward the dam, as is expected with the progressive loss of heavy, formerly attached diatoms (*Melosira, Gomphonema, Diatoma, Synedra*, etc.) that settle after being washed into the pool. Near the dam few genera but large numbers of diatoms existed, and the diversity of forms was replaced by green algae such as *Ankistradesmus, Scenedesmus, Chlorella, Chlorococcales*, and *Tetrahedron*, all of which are small planktonic forms that have maximum abundance at some intermediate depth in the euphotic zone.

An unusual occurrence of large numbers of a dinoflagellate tentatively classed as a *Glenodinium* sp. was found throughout the reservoir during the period of the fish-kill. Average numbers of these in the epilimnion exceeded 300 cells/ml; the highest concentrations were found about Watauga River mile 6.0. Concentrations were high at all depths. The abundance of dinoflagellate was also indicated by unusually large amounts of chlorophyll c in test results.

Though *Glenodinium* sp. are dinoflagellates like *Gymodinium breve*, which cause "red tide" conditions in the sea when present in concentrations of about 100 cells/ml, *Glenodinium* sp. are not known

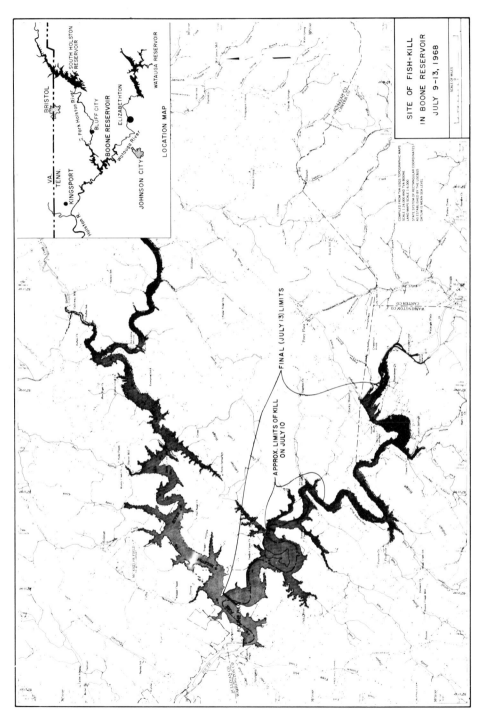

FIG. 2. Site of fish-kill in Boone Reservoir, July 9–13, 1968.

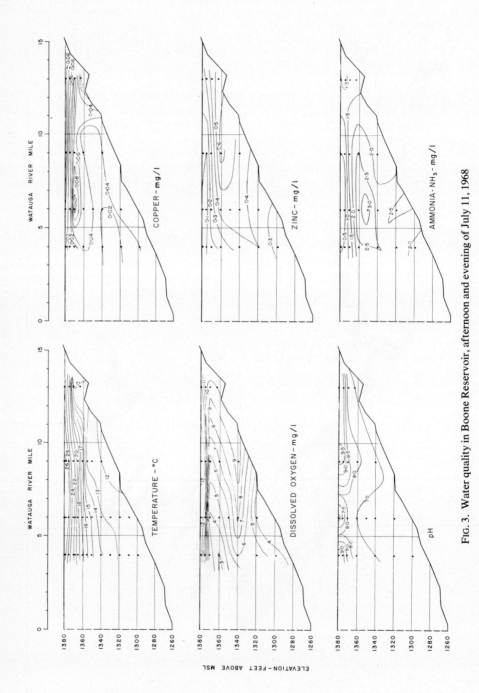

FIG. 3. Water quality in Boone Reservoir, afternoon and evening of July 11, 1968

to release toxic products. A normal "bloom" of dinoflagellates in the ocean produces toxic conditions for other algae, zooplankton, and larger forms by release of soluble carbohydrates, amines, and other compounds. In this reservoir, however, no indications of effects on other algae were observed, and zooplankton were extremely abundant.

It should also be pointed out that toxic algae (blue–greens or dinoflagellates) produce the following symptoms in affected fish: frenzied activity, loss of equilibrium, shallow and rapid breathing, an extended period of slow opercular (gill covering) movement, and then violent bursts of activity and death with the mouth and opercula open. These were not the symptoms of the dying fish observed in Boone Reservoir.

In addition, every toxic algal product has been found to be soluble in water or alcohol. The toxic material sought in this fish-kill was neither—it adsorbed onto membrane filter pads, indicating either a large molecular structure or possibly an electric charge, neither of which is the case for known toxic-algae products.

Initial bioassays

For the first series of field bioassays, redear sunfish and reservoir water that had been filtered through a 1.2 micrometer membrane filter were used to test for soluble toxicants. There were no mortalities in 24 h.

In the next bioassay, several dilutions of unfiltered water from the affected area of the reservoir killed redear sunfish in proportion to the concentration.

Additional bioassays showed the material backwashed from the filters to be toxic. However, water passed through a Foerst plankton centrifuge (17,000 rev/min) was as toxic as unfiltered water. Force-feeding of fish with algal concentrates from the centrifuge produced no toxic symptoms. Plankton collected by net tows were also nontoxic.

Fish used in the bioassays were examined for abnormalities. Gills from both living and dead fish appeared normal. No accumulations of mucus or discolorations on the gills of the test fish were evident.

Water quality observations

Field and laboratory analyses were made on several hundred samples of water collected from the reservoir and the Watauga River during the period of the kill. As shown in Fig. 3, dissolved oxygen concentrations were high (10–13 mg/l) throughout the warm epilimnion and were adequate for fish survival throughout the entire depth of the Watauga arm of the reservoir. The pH ranged from about 7 at various depths below the surface to more than 9 at the surface.

Tests were made for various compounds and metals that could possibly have been present and that are known to be toxic. As shown in Table 3, the concentrations of several toxic metals and ammonia were sufficiently high in some samples from the Watauga arm of Boone Reservoir to kill some fish—if tolerance concentrations reported in the technical literature are used as criteria. However, it is well recognized that the concentration of a potentially toxic material is, alone, not a reliable criterion. Other concurrent physical and chemical conditions such as temperature, pH, and hardness are also very important.

TABLE 3
Toxic compounds and metals

Pollutant (in ionic form for metals)	Range of concentrations found in Watauga River Arm of Boone Reservoir July 11–15 (mg/l)
Ammonia	0.12–3.52
Cyanide	<0.01
Copper	0.01–0.15
Zinc	0.03–0.63
Mercury	<0.01–0.18
Lead	<0.01–0.06
Chromium (total)	0.01
Cadmium	<0.01–0.03
Nickel	<0.05

A bioassay was designed to reconstruct the water chemistry of Boone Reservoir at the time of the fish-kill and thereby to test the toxicity of the metals present. The reconstructed water chemistry is shown in Table 4. Pure oxygen was bubbled through the test containers to produce a range of concentrations of dissolved oxygen from 10.2 to 14.4 mg/l. Test solutions were held within a pH range of 8.8–9.7.

Although this test came close to duplicating concentrations of metals that existed in Boone Reservoir at the time of the fish-kill, there were no mortalities or indications of distress throughout the 48 h test period. However, as learned much later, this test did not even come close to duplicating toxicity conditions related to mercury. Mercuric nitrate was used as the source of mercury for the bioassay. Ionization calculations show that with the

TABLE 4
Bioassay designed to test toxicity of metals present

Vessel	Average pH	Concentration (mg/l)											
		Pb	Hg	Cu	Zn	NH_3-N	NO_3-N	NO_2-N	Average dissolved oxygen	Total alkalinity	Ca	Mg	SO_4
Control	9.3	0.00	0.00	0.00	0.00	0.00	1.5	0.05	12.5	60.0	20.0	6.0	40.0
T_1	9.3	0.00	0.00	0.10	0.20	0.40	1.5	0.05	12.5	60.0	20.0	6.0	40.0
T_2	9.3	0.00	0.00	0.05	0.10	0.40	1.5	0.05	12.5	60.0	20.0	6.0	40.0
T_3	9.3	0.00	0.00	0.02	0.05	0.40	1.5	0.05	12.5	60.0	20.0	6.0	40.0
T_4	9.3	0.00	0.00	0.10	0.00	0.40	1.5	0.05	12.5	60.0	20.0	6.0	40.0
T_5	9.3	0.00	0.00	0.00	0.20	0.40	1.5	0.05	12.5	60.0	20.0	6.0	40.0
T_6	9.3	0.00	0.00	0.00	0.00	0.40	1.5	0.05	12.5	60.0	20.0	6.0	40.0
T_7	9.3	0.00	0.20	0.10	0.20	1.5	1.5	0.05	12.5	60.0	20.0	6.0	40.0
T_8	9.3	0.00	0.20	0.10	0.20	0.40	1.5	0.05	12.5	60.0	20.0	6.0	40.0
T_9	9.3	0.06	0.20	0.10	0.20	0.40	1.5	0.05	12.5	60.0	20.0	6.0	40.0

form of the other chemicals used in the bioassay test solution, less than 1% of the total mercury was present in the form of the very toxic Hg^{++} ion.

By August 10, 1968, the field and laboratory investigations had not revealed the cause of the kill. On August 11 and 12, TVA staff revisited the area to look for possible local sources of toxic material. During the investigation, many 55 gal steel drums were noted floating loose in the reservoir and stranded along the shorelines. Because a toxic material was suspected, the investigators began to examine these drums.

A barrel bearing the name "Tesol" was found to contain 2–4 gal of liquid. Continued investigation revealed a green and white barrel bearing the inscription "Buckman Laboratories, Inc., Industrial Microorganism Control," floating about two-thirds submerged in the water. This barrel contained an estimated 27 gal of liquid. Bioassays conducted at a local boat dock proved the material to be highly toxic—goldfish were killed within 6 min.

More of the reservoir was examined, and many additional barrels were noted, most from chemical and petroleum campanies. Many were riddled with bullet holes while others were rusted through. About 200 loose barrels were found in the water or in the zone of water fluctuation. An additional 229 barrels were counted onshore above the waterline. Seventeen old houseboats and 16 docks with barrel floats in various stages of disrepair were noted. Of the 200 loose barrels, 29 were sampled. Of the 29 sampled, 28 provided one or more gallons of liquid. Of the 29 sampled, nine were found to be toxic, and two of the nine were extremely toxic. One of these drums was that which contained the "industrial microorganism control" material. The toxic components of this material were later determined to be phenylmercuric acetate and 2,4,6-trichlorophenol. (One of the decomposition products of phenylmercuric acetate is diphenylmercury.) The other contained about 5 gal of a highly concentrated syrupy liquid, which turned out to be an industrial detergent.

Laboratory bioassays were performed with standard solutions of phenylmercuric acetate, diphenylmercury, and 2,4,6-trichlorophenol obtained from chemical suppliers. Diphenylmercury was found to be acutely toxic at concentrations between 20 and 50 ppb at pH and temperature conditions similar to those observed during the fish-kill (pH = 7.2–9.2; temperature = 83°F). However, algae and zooplankton were not harmed by concentrations of diphenylmercury as high as 10 mg/l.

A water sample that had been collected on July 14, 1968, from the kill area was then thoroughly analyzed. Diphenylmercury was isolated and identified by concentrating the solids on a glass-fiber pad, extracting with organic solvent, and determining the gas chromatographic behavior. At the same time, 2,4,6-trichlorophenol was also isolated and identified.

The most important sample was collected with a carbon filter fabricated onsite during the field bioassays on July 13, 1968. Approximately 10 l of water was filtered through the cartridge. Diphenylmercury was isolated from the carbon by solvent extraction, purified, and tentatively identified by thin-layer chromatography and gas chromatographic behavior in two different phases, and, finally, quantitated by gas chromatography. Efforts were unsuccessful to determine the extraction coefficient. However, it was determined that at least 225 µg/l of diphenylmercury had been in the water at the time of filtration. At least 14 µg/l of 2,4,6-trichlorophenol were also in the water that was filtered. Based on the evidence of these investigations, it was concluded that the cause of the fish-kill was the release of toxic forms of mercury compounds from metal drums used for flotation purposes.

Drum removal program

In 1967, the TVA had adopted rules and regulations applicable to houseboats and boathouses on the Tennessee River system and issued the following notice:

> Because of danger to small boats and water skiers from loose metal drums floating or submerged in the lake, floating facilities and craft *placed on TVA lakes after January 1, 1968,* may not use empty drums for flotation. Drums filled with plastic foam are acceptable, as are flotation materials such as pontoons and plastic foam logs.
>
> Metal drums used for flotation of facilities and craft *already on TVA lakes before 1968* must be replaced with acceptable flotation materials by January 1, 1972.

Because the 1968 fish-kill on Boone Reservoir was traced to derelict metal drums, in March 1969 TVA urged that (1) all derelict drums be removed from Boone Reservoir, and (2) all drums then in use be considered potentially dangerous and be removed unless it could be *positively established* that their contents are not hazardous. Guidelines for proper handling of the drums were issued, and the TVA arranged to pick up and dispose of drums removed from Boone Reservoir before July 1, 1969.

During the period March to July 1969, about 5000 metal drums (either derelict or identified as previously containing toxic materials) formerly used to float docks and other structures on Boone Reservoir were picked up and disposed of. An estimated 18,000 to 20,000 drums were still in use for flotation on Boone Reservoir.

Prior to 1968, the Watauga River arm of Boone Reservoir had almost annually experienced fish-kills of varying magnitude. During the period May 1961 to July 1968, eight serious and unexplained fish-kills had been reported for this area. Since the drum removal program in the spring of 1969, there has not been a report of a fish-kill on Boone Reservoir.

MERCURY IN THE TVA SYSTEM

The problem

In the fall of 1969, the TVA environmental engineers found trace amounts of mercury and chlorine in the Hiwassee River, a tributary of Chickamauga Reservoir. As a result of these findings, the Tennessee Water Quality Control Board in November ordered a chemical company to control its chlorine and mercury discharges so that effluent concentrations would not exceed specified limits.

In May 1970, after the Food and Drug Administration had established the first specific mercury guidelines ever to be suggested in this country, the TVA began checking mercury levels in fish, water, and sediment samples from its impoundments and streams. By September 1970 the TVA Water Quality Laboratory had performed some 1300 analyses of fish, water, and sediment samples collected from 26 lakes and four river reaches. By April 1971 the number of analyses had increased to 2900. The samples were from 28 reservoirs, 34 river reaches, and 24 farm ponds. In addition, several industrial and municipal waste treatment plant effluents had been analyzed for mercury.

Results of these analyses indicated that the flesh of fish from Pickwick Reservoir and the North Fork Holston River contained mercury in amounts exceeding the Food and Drug Administration guideline of 0.5 ppm. (It was later established that the mercury problem in Pickwick Reservoir extended into Kentucky Reservoir, the next downstream Tennessee River impoundment.) These results were sent to the Environmental Protection Agency and appropriate state agencies. Early in July 1970 these agencies closed Pickwick Reservoir to commercial fishing and recommended that fish caught by sport fishermen not be consumed. Similar action was taken in September 1970 for the North Fork Holston River. Although there was a mercury discharge to the Hiwassee River, the amount of mercury found in flesh of fish taken from the Hiwassee River and Chickamauga Reservoir was found to be less than the Food and Drug Administration guideline.

In almost all cases during this study, water samples were found to contain less than the recommended limit of 5 µg Hg/l for drinking water. However, one sample collected from the North Fork Holston River at mile 70.2 (immediately downstream from a chlor-alkali plant) contained 17.0 µg/l, and another water sample collected from the North Fork Holston River at mile 23.9 contained 8.6 µg/l.

Sediment samples were collected during this study with an Ekman dredge. Mercury was found in all sediment samples collected. A considerable range of mercury concentrations (dry-weight basis) was determined—from less than 0.05 µg/g to 73 µg/g in sediment collected from Pickwick Reservoir at Tennessee River mile 257.8.

The principal known sources of mercury in the Tennessee Valley were a few large chlor-alkali plants. In most instances, their manufacturing processes were such that they could substantially reduce mercury discharges into streams in a short time. State agencies took immediate action to reduce mercury discharges in accordance with specific schedules. The company on the North Fork Holston River ceased operation, and the other two companies reduced discharges in accordance with set standards and schedules. There have been no reports from the respective agencies that would indicate the failure of these two companies to comply fully with the standards and schedules.

Continued surveillance

Since the mercury problem in the Tennessee Valley was identified in early 1970, the problem areas have been kept under continued surveillance. The analytical procedures used by the TVA Water Quality Laboratory for mercury analysis are summarized in Fig. 4. In a March 1971 meeting with representatives of interested Federal and state agencies, the TVA accepted the role of coordinating the mercury surveillance program in the Tennessee Valley. Initially, fish samples were collected monthly from Pickwick and Kentucky Reservoirs, and quarterly from Chickamauga Reservoir and the North Fork Holston River. In July 1972, sampling of Pickwick and Kentucky Reservoirs was reduced to bimonthly. Samples are collected from two stations on Pickwick Reservoir, two stations on Kentucky Reservoir, two stations on Chickamauga Reservoir, and two stations on the North Fork Holston River. For each collection, an attempt is made to obtain at least 10 gamefish (sauger, largemouth bass, smallmouth bass, white bass, yellow bass, spotted bass, crappie, walleye, bluegill, redear sunfish, chain pickerel, and longear sunfish) and 10 other fish (smallmouth buffalo, redhorse, carp, spotted sucker, blue catfish, channel catfish, gizzard shad, hogsucker, mooneye, skipjack, drum, and herring). The samples are divided for analysis by the TVA Water Quality Laboratory at Chattanooga, Tennessee, and the Tennessee Water Quality Control Division Laboratory at Nashville, Tennessee. For quality control, about 10% of the samples from each lot are divided into two portions, one of which is sent for analysis by the Environmental Protection Agency Southeast Water Laboratory at Athens, Georgia. Results of analyses are sent periodically to appropriate Federal and state agencies and to others who desire the information. The results of the fish surveillance program for Pickwick and Kentucky Reservoirs are summarized in Figs. 5 and 6, respectively.

Bottom sediments in Pickwick Reservoir

A task force composed of representatives of the TVA, the Environmental Protection Agency, the Oak Ridge National Laboratory, and Vanderbilt University recommended that more definitive sampling for mercury in the bottom sediments of Pickwick Reservoir should be undertaken to evaluate more thoroughly the problem and the possible need for corrective measures. Following the task force recommendation, the TVA made a rather extensive survey in November 1971 to assess the areal and vertical distribution of mercury in bottom sediments throughout Pickwick Reservoir.

Sampling stations were selected in Pickwick Re-

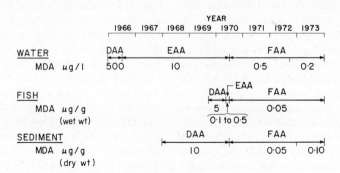

FIG. 4. Analytical procedures used by TVA Water Quality Laboratory for mercury analysis. MDA, minimum detectable amount. DAA, direct atomic absorption. EAA, MIBK extraction, atomic absorption. FAA, flameless atomic absorption.

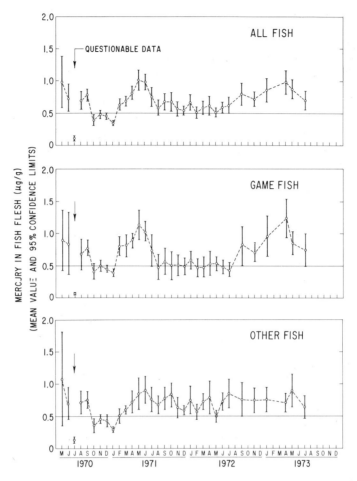

FIG. 5. Summary of mercury concentrations in fish flesh, Pickwick Reservoir, TVA, Water Quality Branch.

servoir at Tennesse River miles 207.2, 211.2, 217.2, 224.5, and 229.8; Yellow Creek embayment mile 2.1; and Bear Creek embayment mile 2.4 as shown in Fig. 7. Station selection was tied in with ranges previously established by the TVA to monitor silt accumulation in the reservoir. In addition to placing a marker buoy accurately at each sampling station, investigators made an up-to-date depth sounding along each range. Distance was measured by metering a wire as it was played out from one end of the range, and alignment was maintained by transit with radioed instructions. Depth soundings along each range were made with a recording sonar unit. Previous siltation data had been obtained in 1938 (the year Pickwick Dam was closed), 1946, 1951, 1956, and 1961.

Core samples were collected manually by hard-hat divers. Three cores (two 3-in-diameter cores and one 5-in-diameter core) were collected at each of the 12 locations in Pickwick Reservoir. An Ekman dredge sample was also collected at Tennessee River mile 204.0 below Pickwick Dam (station K-1). When the core samples were brought to the surface, the sample was dewatered by syphoning off the overlying column of water and then sealed with a plastic cap. After proper identification, the sample was then placed on dry ice. At the end of each sampling day, the samples were frozen in a commercial food freezer establishment.

For analysis, the cores were removed from the sampler while still frozen by running hot water over the sampler. The cores were then cut with a band saw into 1-in-thick slices. These slices were then analyzed for total mercury, solids content, specific gravity, chemical oxygen demand, organic carbon, and organic nitrogen. Results are shown in Table 5.

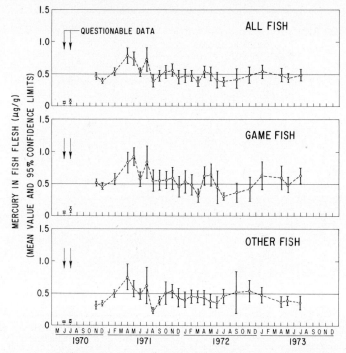

FIG. 6. Summary of mercury concentrations in fish flesh, Kentucky Reservoir, TVA, Water Quality Branch.

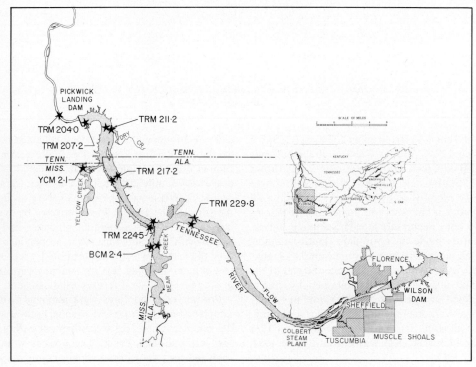

FIG. 7. Sampling locations, bottom sediment core samples, Pickwick Reservoir, TVA, Division of Environmental Planning, Water Quality Branch.

TABLE 5
Compilation of selected data dredge and core samples of bottom sediments in Pickwick Reservoir, TVA Water Quality Branch, November 1971

Core slice (in)	Sampling stations[a]												
	K-1	P-1	P-2	P-3	P-4	P-5	P-6	P-7	P-8	P-9	P-10	P-11	P-12
Total mercury, μg/g, dry-weight basis													
0–1		5.3	6.7	2.9	1.2	9.6	7.6	12.0	9.9	1.8	1.6	12.0	4.3
1–2		3.5	10.0	3.8	1.3	9.4	8.9	14.0	9.1	1.9	1.7	8.3	2.9
2–3		2.2	10.0	3.3	1.2	6.8	7.1	15.0	5.6	2.3	1.1	4.2	1.3
3–4		0.54	8.6	3.3	0.92	3.9	5.0	9.9	4.3	1.4	0.48	3.8	1.3
4–5		0.21	6.8	2.6	0.73	2.7	4.2	5.9	6.2	1.8	0.41		1.2
5–6		0.06		2.4			3.6	11.0	6.1				1.3
6–7		0.07					2.6	7.0					0.59
7–8		0.06						4.8					
8–9								3.0					
Ekman dredge	<0.05	4.0	11.0	4.5	1.2	11.0	9.0	10.0	9.1	1.7	1.6	7.5	4.1
Specific gravity, wet-weight basis													
0–1		1.35	1.10	1.20	1.17	1.16	1.32	1.29	1.17	1.25	1.18	1.44	1.52
1–2		1.46	1.33	1.36	1.25	1.27	1.51	1.39	1.41	1.32	1.30	1.55	1.57
2–3		1.67	1.36	1.33	1.23	1.26	1.41	1.36	1.67	1.34	1.28	1.92	1.70
3–4		2.10	1.26	1.29	1.21	1.21	1.50	1.55	1.36	1.43	1.55	1.89	1.68
4–5		1.75	1.37	1.53	1.11	1.36	1.40	1.50	1.44	1.43	1.74		1.79
5–6		1.83		1.45			1.54	1.58	1.40				1.76
6–7		2.00					1.66	1.57					1.78
7–8		2.02						1.39					
8–9								1.59					
Ekman dredge	1.93	1.64	1.25	1.31	1.44	1.27	1.38	1.29	1.31	1.51	1.31	1.58	1.53
Organic carbon, mg/g, dry-weight basis													
0–1		11.0	18	15	21	19	18	18	16	14	20	15	12
1–2		8.5	17	14	20	17	16	19	14	13	20	11	11
2–3		6.7	17	13	20	15	14	16	14	14	18	6.2	9.8
3–4		3.9	14	13	20	16	14	17	12	19	20	6.8	9.8
4–5		3.0	15	13	21	15	13	12	14	16	20		10
5–6		2.4		14			14	15	13				12
6–7		3.0					13	16					8.7
7–8		2.3						13					
8–9								14					
Ekman dredge	4.0	7.4	16	12	21	13	13	13	12	13	14	13	8.6
Solids content, % (opposite of moisture content)													
0–1		49.5	28.6	33.0	19.7	35.0	31.2	38.3	26.3	27.5	33.2	46.2	55.6
1–2		59.3	39.8	43.6	35.8	42.0	49.4	47.7	44.6	38.6	46.0	57.7	62.2
2–3		69.2	41.0	46.5	36.9	42.9	50.6	48.6	45.6	39.8	48.3	72.2	68.0
3–4		76.4	41.6	48.4	34.3	40.0	52.3	50.3	49.5	40.2	62.0	78.2	69.0
4–5		88.0	41.3	49.5	25.0	42.7	53.2	55.3	46.2	42.1	56.0		69.8
5–6		81.9		49.0			53.9	51.3	38.8				69.9
6–7		77.8					57.3	53.4					72.2
7–8		75.7						53.8					
8–9								57.7					
Ekman dredge	75.9	62.5	30.8	39.8	51.0	36.6	41.2	39.3	39.6	52.5	38.3	58.3	55.9
Chemical oxygen demand, mg/g, dry-weight basis													
0–1		26	45	40	59	45	35	46	44	36	64	27	27
1–2		23	40	36	54	42	42	48	37	36	79	24	28
2–3		12	40	35	53	36	40	42	39	33	55	19	27
3–4		12	39	36	55	40	35	45	33	42	47	27	28
4–5		9.2	39	36	57	38	37	33	37	23	37		28
5–6		6.5		36			36	41	42				26

TABLE 5 (continued)

Core slice (in)	Sampling stations[a]												
	K-1	P-1	P-2	P-3	P-4	P-5	P-6	P-7	P-8	P-9	P-10	P-11	P-12
6–7		7.6					34	36					27
7–8		6.4						36					
8–9								37					
Ekman dredge	8.0	22	42	33	55	38	38	41	34	42	44	27	29
	Organic nitrogen, mg/g, dry-weight basis												
0–1		1.3	2.0	1.5	3.0	2.0	1.9	1.7	0.82	2.6	1.9	1.1	1.0
1–2		0.93	1.8	1.7	2.7	2.0	1.9	2.2	1.9	2.0	2.1	1.0	1.0
2–3		0.57	1.9	1.7	2.7	1.8	1.8	1.6	1.9	2.0	1.9	0.70	1.0
3–4		0.50	1.8	1.6	2.4	2.0	1.7	1.5	1.8	1.8	1.6	1.0	1.4
4–5		0.26	1.8	1.7	2.8	1.9	1.6	1.0	1.7	1.4	1.6		1.0
5–6		0.32		1.6			1.6	1.6	1.8				1.0
6–7		0.40					1.6	1.3					1.1
7–8		0.37						1.2					
8–9								1.2					
Ekman dredge	0.44	0.89	1.9	1.9	2.5	2.6	2.1	1.7	1.6	2.0	1.8	1.7	1.1

[a] K-1, TRM204.0R; P-1, TRM207.2L; P-2, TRM211.2M; P-3, TRM211.2R; P-4, YCM2.1L; P-5, TRM217.2M; P-6, TRM217.2R; P-7, TRM224.5L; P-8, TRM224.5M; P-9, BCM2.4M; P-10, BCM2.4R; P-11, TRM229.8M; P-12, TRM229.8R. L, M, R, refer to left overbank, main channel, and right overbank.

TABLE 6
Analysis of samples collected quarterly, water quality monitoring network

Temperature	Aluminum
Dissolved oxygen	Arsenic
Coliforms (fecal and total)	Barium
pH	Beryllium
Alkalinity	Boron
Hardness	Cadmium
Color (true and apparent)	Chromium
Turbidity	Copper
Conductance	Iron
Solids, suspended and dissolved	Lead
Oxygen demand, biochemical	Lithium
Oxygen demand, chemical	Manganese
Carbon, total organic	Mercury
Nitrogens (organic, ammonia, nitrite plus nitrate)	Nickel
	Selenium
Phosphorus	Silver
Sodium	Titanium
Potassium	Zinc
Calcium	
Magnesium	
Chloride	
Fluoride	
Sulfate	
Silica	

No corrective measures have been undertaken as regards the mercury-laden silt deposits in Pickwick Reservoir. Studies show that silt has accumulated in Pickwick Reservoir since 1938 at an average rate of 0.3 in/year. At this average rate of sedimentation, about 5–7 years would be required to accumulate the $1\frac{1}{2}$–2 in of new mercury-free sediment that research indicates is necessary to seal existing deposits.

EXPERIENCE WITH OTHER HEAVY METALS

Other than the mercury problem, there have been no specific environmental problems identified in Tennessee Valley waters as regards heavy metals. However, many of the fish, water, and sediment samples collected during the study of the mercury problem were analyzed for other heavy metals, including arsenic, beryllium, cadmium, chromium, copper, lead, mercury, nickel, selenium, silver, and zinc.

Early in 1973 the TVA began collecting samples once each quarter from about 50 water quality "trend stations" located throughout the Valley. These samples are analyzed for many water quality parameters including heavy metals (Table 6).

THE USE OF SYNTHETIC SCAVENGERS FOR THE BINDING OF HEAVY METALS

George Feick,† Edward E. Johanson, and Donald S. Yeaple‡

JBF Scientific Corporation, 2 Ray Avenue, Burlington, Massachusetts 01803

INTRODUCTION

Much of the recent work on heavy-metal contamination of water bodies is concerned with the environmental effects of mercury. Because of its extreme toxicity, its widespread use, and its ease of detection in trace quantities, mercury provides an ideal subject for study. One feature of mercury chemistry which is absent with most other heavy metals is its biological methylation into compounds which are more toxic than the mercuric ion itself. The work on which the present paper is based was directed mainly toward the restoration of mercury-contaminated water bodies by the application of chemical binding agents or scavengers, which hold the mercury in insoluble form in the bottom sediments and keep it out of the water column. An additional feature of some of these scavengers is that the insoluble forms of mercury (e.g., mercuric sulfide) may be less subject to biological methylation than the more soluble mercury species. These scavengers may also be useful for binding other heavy metals.

Among the alternative methods for restoring metal-contaminated water bodies is dredging of the contaminated sediments or covering them with a layer of clean fill. Dredging involves problems of cost, disturbance of the environment, dispersal of the contaminants, and disposal of the contaminated spoil. The use of an inert cover involves location of a source of suitable cover material and the costs of transporting and distributing large tonnages of fill. A recent summary of field experience in Sweden with mercury-control methods is given by Jernelöv and Lann.[1] The experience of these authors illustrates some of the difficulties of this kind of experimentation. In one lake, the natural generation of sulfide ions during a winter period of anoxic conditions was much more effective in reducing the mercury uptake of fish than the inert cover which was being tested.

In the present paper we consider first the requirements for practical metal scavenging agents and then discuss the results of an experimental program designed to evaluate some promising materials in the light of these requirements. These laboratory tests show that the long-chain alkyl thiols (R—SH; where R is an alkyl radical containing about 10–14 carbon atoms) are very effective mercury-binding agents, which appear to have some advantages over the simple sulfides.

This work was supported in part by the US Environmental Protection Agency. Further details may be found in Report No. EPA–R2–72–007.[2]

REQUIREMENTS FOR METAL SCAVENGING AGENTS

The following list of requirements is proposed as a basis for evaluating the effectiveness and potential usefulness of metal scavengers:

(1) The equilibrium constant for the formation of the metal complex should be as high as possible.

(2) The resulting metal complex should be extremely insoluble in water.

(3) The complex should be stable toward oxidation, reduction, hydrolysis, biological action and the presence of dissolved salts such as chlorides.

(4) The rate of reaction with mercury at very low concentration should be reasonably rapid.

(5) The Scavenging agent should not adversely affect the quality of the water or the bottom sediment for their intended uses. This includes effects on fish and bottom biota in areas where such considerations are important.

(6) The material should be readily convertible into a dense, granular form that will sink quickly and will not readily be dispersed into the water.

(7) The cost of the material, in place at the bottom of the water and per unit of mercury complexed, should be as low as possible.

---

†*Present address*: Consulting Chemical Engineer, 144 Fair Oaks Park, Needham, Mass. 02192.

‡*Present address*: General Electric Co., River Works Utilities Operation 1100 Western Ave., Lynn, Mass. 01901.

EQUILIBRATION EXPERIMENTS

The first four of the above requirements can be assessed in a general way for the case of mercury by experiments measuring the equilibrium distribution of mercuric ion or methylmercuric ion between water and various sedimentary materials. Sediment samples weighing 100–300 g were equilibrated in closed bottles with about 500 ml of solutions of $HgCl_2$ or CH_3HgCl by continuous agitation for 7 days at 25°C. Preliminary experiments showed that most samples came close to equilibrium within this time.

After equilibration, pH and dissolved oxygen were measured and the solids were filtered off on a 0.45 μm membrane filter. The filtrates and solids were analyzed by the cold vapor atomic absorption procedure of Hatch and Ott.[3] Prior to analysis, the solid samples and all solutions containing CH_3HgCl were digested by refluxing with HNO_3–H_2SO_4 for about 3 h. All samples were acidified and treated with $KMnO_4$ to a permanent color in order to destroy organic matter. In most cases, the mercury analyses balanced within ± 15%. The mercury content of all solid samples is reported on an oven-dry basis.

The results are reported in terms of the partition coefficient K which may be defined as

$$K = \frac{\text{ppm Hg in solution}}{\text{ppm Hg in dry solids}}.$$

The partition coefficient affords a convenient numerical measure of the effectiveness of the solids in removing mercury from the water column.

The results of some partitioning experiments are shown in Table 1. The Acton sand and Acton peat were taken from the bottom of a freshwater reservoir called Nagog pond in Acton, Mass. The sand contained less than 1% and the peat about 44% of organic matter. Both contained traces of sulfides decomposable to H_2S by dilute H_2SO_4. The partition coefficient of the Acton sand increased by about two orders of magnitude when it was aged in an open container for about 5 weeks. The increase is probably due to oxication of sulfides since the aged sediment no longer released H_2S on acidification. This result suggests that mercury-laden sediments may release mercury if they are dredged and exposed to air or oxygen-rich water on a landfill.

As might be expected of an organic material, the Acton peat gave a very low value of the partition coefficient. Despite the high concentration of mercury in the solid, the dissolved mercury was below our detection limit of about 0.02 ppb. This sediment reacted rapidly with dissolved oxygen, probably because of the presence of $Fe(OH)_2$ or other reduced iron species. The strongly reducing conditions appear to be necessary to maintain the mercury in the insoluble sulfide form. When aged in air for 5 weeks, the Acton peat lost some of its reducing ability, and the partition coefficient increased by about two orders of magnitude.

Georgia kaolin was chosen as representative of a common mineral constituent of sediments. The sample used was a commercial material sold for ceramic work. Its partition coefficient and that of the ground silica were relatively high at these high concentrations of mercury. The above results may be compared with those of Smith et al.[4] who report

TABLE 1
Partition coefficients for $HgCl_2$ with various materials at 25°C

Sediment type	Concentration of Hg in sediment (ppm)	Coefficient	pH	Dissolved oxygen (ppm)
Acton sand (fresh)	13.7	3.4×10^{-4}	6.6	4.5
Acton sand (aged)	30.0	0.014	6.4	3.5
Acton peat (fresh)	1430.0	$< 1.4 \times 10^{-8}$	5.2	0.0
Acton peat (aged)	1335.0	2.3×10^{-6}	4.8	0.6
Georgia kaolin	90.0	0.069	5.2	—
Silica (240 mesh)	29.6	1.1	7.4	8.0
Kaolin plus additive:				
3% coarse pyrite	193.0	0.16	5.0	9.0^+
5% pyrite (−325 mesh)	300.0	8.3×10^{-6}	4.5	7.1
5% ZnS (pptd.)	300.0	1.8×10^{-6}	5.1	7.0
0.8% NDM[a]	1000.0	1.5×10^{-7}	4.4	9.0^+
5% $CaCO_3$	314.0	0.037	7.4	7.5
5% $CaCO_3$ plus 0.8% NDM	1000.0	2.0×10^{-8}	6.8	9.0^+

[a] NDM = normal dodecyl mercaptan.

mercury analyses for water and solids in the River Thames near London. Their data lead to partition coefficients in the range of 10^{-7}–10^{-4}, which are comparable to our results for the Acton sediments.

The lower part of Table 1 shows the effect of various chemical additions to Georgia kaolin, which was chosen as a substrate because of its low natural mercury-binding capacity. The inorganic sulfides, such as pyrite (FeS_2) and ZnS, were chosen because mercuric sulfide (solubility product around 10^{-50}) is one of the least soluble compounds of mercury.[5]

The coarse pyrite was a mill byproduct from Climax, Colorado, with 96% by weight falling in the range between 35 mesh (500 μm) and 325 mesh (44 μm), and 4% below 325 mesh. The partition coefficient of 0.16 and the appearance of the pyrite after exposure to the solution indicated that little reaction had taken place and the formation of HgS was probably kinetically limited. The behavior of coarse ferrous sulfide (FeS) was similar with a partition coefficient around 10^{-4}.

The use of finely divided sulfides improved the situation somewhat. Finely ball-milled pyrite (all through 325 mesh) and a sample of precipitated ZnS (laboratory reagent) gave values of 8.3×10^{-6} and 1.8×10^{-6}. Ferrous sulfide precipitated *in situ* (form $CaS + FeSO_4$) gave similar values, all results being obtained in the presence of 4–10 ppm of dissolved oxygen.

It thus appears that highly reducing conditions, such as are provided by the Acton peat, are necessary to develop very low values of the partition coefficient with the inorganic sulfides. Free oxygen probably promotes progressive hydrolysis of HgS by oxidizing the H_2S as it is formed.

An alternative to the inorganic sulfides is found in the long-chain alkyl thiols or mercaptans, RSH, where R denotes an alkyl group containing 12 or more carbon atoms. In contrast to the lower members of the series, the long-chain thiols are highly insoluble in water and are relatively nonvolatile and nontoxic, and some have relatively mild odors. The stability constant for the formation of the Hg^{++} complex is high. (For mercaptoacetic acid, according to Sillen and Martell,[5] it is 10^{44}.) With a molecular weight of about 202, n-dodecyl mercaptan (NDM) should theoretically be capable of binding about half its weight of mercury as a complex of the form $Hg(SR)_2$. Under anaerobic conditions, no biological degradation should take place.

As shown in Table 1, NDM with kaolin gives a partition coefficient of 1.5×10^{-7}. This value is reduced by about an order of magnitude when the mixture is buffered to a pH of 6.8 with $CaCO_3$, which does not produce the effect alone. The buffered value approaches the value obtained for Acton peat. A possible advantage of the long-chain alkyl mercaptans is that the mercury complexes may remain insoluble under alkaline conditions, in contrast to the simple sulfide ion which forms water-soluble anionic complexes under these conditions. It is to be noted that the above values were obtained in the presence of over 9 ppm of oxygen, hence the mercaptides appear less susceptible to oxidation than the sample sulfides.

Neither the sulfides nor the mercaptides give the very low values of mercury concentration to be expected from the stability constants of the respective mercury complexes. The difference may be due to incomplete removal of suspended mercury by the 0.45 μm membrane filter or to side reactions such as oxidation which convert some of the complexed mercury into soluble species. In any event, the mercaptans appear to approach their theoretical potential more closely than do the sulfides.

Preliminary experiments showed that the mercaptides are much more stable in the presence of dissolved chlorides than are the simple sulfides.

TABLE 2
Partition coefficients for CH_3HgCl with various materials at 25°C

Sediment type	Concentration of Hg in sediment (ppm)	Partition coefficient	pH	Dissolved oxygen (ppm)
Acton sand (fresh)	12.0	0.076	7.1	7.6
Acton peat (fresh)	1470.0	6.8×10^{-4}	5.2	0.2
Kaolin plus additives:				
5% pyrite (−325 mesh)	300.0	0.125	4.1	2.8
5% ZnS (pptd.)	300.0	1.5×10^{-3}	5.4	9.0
5% $CaCO_3$ plus 0.8% NDM[a]	300.0	8.0×10^{-4}	7.0	9.1

[a] NDM = normal dodecyl mercaptan.

The partition coefficient of NDM on kaolin buffered with $CaCO_3$ was increased from 2.0×10^{-8} to 2.0×10^{-7} in the presence of 35 g/l of NaCl. In the case of FeS, a similar concentration of salt produced an increase of about six orders of magnitude.

In all the naturally contaminated sediments that we have studied, the methylmercuric ion (CH_3Hg^+) accounts for less than 1% of the total mercury present. It is of interest to compare the binding of CH_3Hg^+ to that of Hg^{++}. The results of some preliminary distribution experiments with CH_3HgCl are given in Table 2. As would be expected for a monovalent ion, the binding of CH_3Hg^+ is much weaker than than of Hg^{++}. Comparison with Table 1 shows that the materials which bind Hg^{++} strongly are those which also bind CH_3Hg^+ most strongly, but the partition coefficients are three to five orders of magnitude greater in the case of CH_2Hg^+.

TOXICITY OF THIOLS

There is only a limited amount of data on the toxicity of thiols to fish. Van Horn et al.[6] estimate that a safe concentration of butyl mercaptan for bluegill sunfish is around 2.5 ppm. As the chain length of the thiol is increased, its toxicity will decrease. Shugaev[7] concludes that the toxicity of the long-chain thiols is more or less equal to that of hydrocarbons of equal chain length. The solubility of n-dodecyl mercaptan in water is estimated at less than 0.013 ppm: hence no toxic effect on fish is to be expected.

This expectation was confirmed by aquarium experiments in which 2 in goldfish were exposed to mercury-contaminated sediments treated with NDM at a rate of about 0.025 lb/ft^2 for a 2 in depth of sand. No toxic effects were observed up to 30 days of exposure.

These aquarium experiments showed that under favorable conditions the mercaptan was effective in reducing both the mercury content of the water and the rate of uptake by fish.

METHODS OF APPLICATION AND COSTS

The thiols are normally oily liquids with a density less than that of water. Some means is therefore required to sink and hold these materials at the bottom of the water. We have found that sand treated with long-chain alkylamines is readily wetted by the mercaptans and that the treated sand is stable for periods of immersion of at least 40 days. A possible method of treating a sandy lake would be to dredge a limited amount of fill from the bottom, treat it aboard a barge with the required chemicals, and return the treated material to the bottom. Alternatively, the required carrier sand could be brought in from an outside source.

The cost of NDM is around 80 cents/lb in drum lots. Each pound is theoretically capable of combining with about $\frac{1}{2}$ lb of mercury. We have obtained good results with about a fourfold excess of mercaptan. The chemical cost for treating a water body should therefore be about $6.40 per lb of mercury to be treated. This cost will probably be almost negligible in comparison to the cost of placing the material.

COMPLEXING OF OTHER HEAVY METALS

Most metals which form insoluble sulfides also form stable mercaptides. This is true of copper and lead and to a lesser extent of cadmium and zinc. The long-chain alkyl thiols may be expected to be useful scavengers for these metals as well as for mercury.

REFERENCES

1. JERNELÖV, A. and LANN, H., Environ. Sci. Technol. 7 (8) 712–8 (1973).
2. FEICK, G., JOHANSON, E. E., and YEAPLE, D. S., Environmental Protection Technology Series Report EPA–R2–72–077, October 1972.
3. HATCH, W. R. and OTT, W. L., Analyt. Chem. 40, 2085–7 (1968).
4. SMITH, J. D., NICHOLSON, R. A., and MOORE, P. J., Nature 232, 393–4 (1971).
5. SILLEN, L. G. and MARTELL, A. E., Stability Constants, Special Publication No. 17, The Chemical Society, Burlington House, London, 1964.
6. VAN HORN, W. M., ANDERSON, J. B., and KATZ, M., Trans. Am. Fish. Soc. 79, 55–63 (1949).
7. SHUGAEV, B. B., Khim. Seraorg. Soedin., Soderzk. Neftyakls Nefteprod. 8, 681–6 (1968). See Chem. Abstr. 71, 99956s (1969).

The Use of Synthetic Scavengers for the Binding of Heavy Metals (G. Feick et al.)

DISCUSSION by D. G. LANGLEY
I.E.C. International Environmental Consultants Ltd., Toronto, Canada

It is fair to title the paper just presented "The use of synthetic scavengers for mercury binding." All data presented deal with mercury. I will, therefore, restrict my remarks to mercury as a heavy metal and its potential environmental hazards. The pragmatic approach which the scientists have taken to the "mercury dilemna" must be commended. I wholly support their approach of practical rehabilitation of a mercury-contaminated ecosystem, but much more applied research is required.

The microbial methylation of mercury in an aquatic environment is the most hazardous pathway through which mercury may contribute to environmental pollution. The experimental support for this statement was furnished by Jensen and Jernelov.[1] Microbial methylation or biotransformation can be simply shown as follows:

$$\text{compounds of Hg (in sediment)} \rightleftharpoons Hg^{++} \xrightarrow{\text{bacteria}} CH_3HG^+ \text{ and } CH_3HgCH_3$$

Methylmercury concentrates progressively in aquatic organisms and eventually in higher trophic levels where it penetrates the blood–brain barrier and may result in severe damage to human nerve cells at very low concentrations. A total body burden of 80 mg of methylmercury corresponding to 0.8 μg/g in whole blood has been found to be fatal.[2] Fish as an important dietary source of protein form the major source through which methylmercury may be transferred to man.

It is noteworthy to stop and reflect briefly on the significance of the research to date. Research studies have rigorously defended the toxicity of alkyl mercurials to man. There are many documented cases of human poisonings attributed directly to alkyl mercurials. These range from the ingestion of treated grain to the discharge of methylmercuric chloride at Minamata, Japan, and subsequent contamination of fish. However, I have been unable to discover any direct evidence of human death directly linked to the biological methylation of inorganic mercury compounds in an aquatic environment. There was a case of poisoning in North America through the consumption of swordfish containing methylmercury eaten by a dieting woman who ate swordfish twice daily for more than a year.[3] If any participants at this conference have any further information on poisoning from ingestion of methylmercury from fish I would be pleased to hear their comments.

In the paper presented, the researchers shook sediment samples with water containing very high concentrations of $HgCl_2$ and CH_3HgCl. The concentrations were in excess of what might normally be found in the aquatic environment, viz. 0.02–1 ppb. To summarize, their studies showed that normal dodecyl mercaptan (NDM) reacted effectively with mercury in solution and behaved as a scavenger to purge mercury from water. It was concluded that such scavenging action might be considered an effective rehabilitation measure for mercury contaminated waters. This research raises several important considerations.

(1) Current data do not demonstrate that mercury contaminated waters require rehabilitation. The concentration of mercury in receiving waters is extremely low and nearly always below the acceptable drinking water standard of 5 ppb. The mercury which was formerly discharged by industry is now tied up in bottom sediments. The methylmercury produced from its biotransformation is estimated from methylation studies to be less than 0.1% mercury/year. It is my belief that the disturbance of bottom mercury-containing sediments by pumping to a barge and then treating with a synthetic scavenger will result in a far greater environmental impact than the present 0.1% which may be methylated annually.

(2) The "scavenging" of mercury from water does not mean the methylation will cease. In fact, the addition of copious quantities of an organic chemical such as NDM will enhance aquatic microbial activity through its natural bacterial decomposition. Research conducted

in our laboratory has shown that the presence of increasing amounts of organic material available for decomposition will enhance methylation potential. This was confirmed in studies by Beak[4] where bleached kraft mill effluent (containing mercaptans) was introduced at 5% by volume to waters overlying sediments containing mercury from a chloralkali plant. In this research the methylation rate increased 100%.

(3) The toxicity of mercury mercaptiles and alkyl mercaptans has not been thoroughly investigated. The introduction of large quantities of NDM (8000 ppm) to an ecosystem might prove far more hazardous than the present low levels of mercury which in themselves are not toxic to aquatic organisms. The toxicity results presented in this paper are acute tests. For example, it is known that mercaptans inhibit cell cleavage. I would, therefore, hope that chronic and sublethal toxicity data be obtained before copious additions of NDM were made to an ecosystem.

(4) And, finally, the uptake and excretion of alkyl mercaptans and mercury mercaptiles by the food chain is not known. The ability of mercury mercaptiles to undergo oxidation and subsequent methylation is not known.

(5) It would have been most useful if some studies had been performed on real-life samples of mercury infested sediments.

SUMMARY

I strongly urge that two fundamental issues be resolved before we start tampering with the natural environment. The issues are:

(1) What is the real relationship between total and methylmercury in fish in North America?
(2) Does the 0.5 ppm acceptance limit for fish consumption have real scientific support based on present toxicological data?

Several restoration methods have been suggested by scientists. These include raising pH to liberate dimethylmercury, dredging, covering with inert materials, conversion of bottoms to anaerobic conditions producing relatively inert HgS, and binding chemically with inorganic or organic agents. Certainly, more research is required before we should accept artificial rehabilitation of mercury-laden bodies of water with NDM. Nature, for the time being, appears to be the most efficient rehabilitator!

Mr. Feick's research does demonstrate that NDM may be a promising effluent treatment chemical for mercury removal from contaminated discharges. It may be possible to use NDM in a controlled treatment process with a sand column to eliminate mercury discharge to receiving bodies of water.

REFERENCES

1. JENSEN, S. and JERNELÖV, A., Biological methylation of mercury in aquatic organisms, *Nature (G.B.)* **223**, 5207 (1969).
2. BERGLAND, F. M., *et al.*, Methylmercury in fish: a toxicological epidemiological evaluation of risks, *Nord. Hyg. Tidskr.*, Supplementum (1971).
3. LAMBOU, V. W., *Problems of Mercury Emissions into the Environment of the United States*, EPA Report, 33 (1972).
4. Unpublished.

SESSION VIII
LEGISLATION, STANDARDS, SURVEILLANCE, AND MONITORING

CURRENT REGULATIONS AND ENFORCEMENT EXPERIENCE BY THE ENVIRONMENTAL PROTECTION AGENCY

CARL J. SCHAFER
Chief

and

DR. MURRAY P. STRIER
Chemical Engineer

Permit Assistance Branch, Permit Assistant and Evaluation Division, Office of Water Enforcement, EPA, Washington DC 20460

ENFORCEMENT BACKGROUND

Prior to the formation of the Environmental Protection Agency on December 2, 1970, federal environmental programs placed only slight emphasis on enforcement, focusing primarily on R&D demonstration projects, field investigations and other studies, approval of state standards, state program grants, technical assistance, and grants for construction of municipal waste treatment facilities. The principal enforcement activity involved public-hearing-type conferences. Rarely was court action brought to compel compliance with pollution control requirements.

EPA policy

The policy of EPA has been to engage, directly and forcefully, in a full range of enforcement actions, the majority of which have dealt with problems of water pollution. The objective of the EPA enforcement policy is to be fair, but firm. A dominant aspect of the enforcement program has been its concentration on individual cases of major environmental abuse, typically involving large national corporations or big cities. For example, one out of three civil injunctions initiated by EPA under the Refuse Act involved plants owned by companies ranked among *Fortune Magazine's* "500 Largest Corporations in the United States." A group of approximately 2700 "major dischargers" has been identified for priority efforts.

The new law

The Federal Water Pollution Control Act Amendments of 1972, enacted on October 18, 1972, have transformed the basis for enforcement activities concerning water pollution. The new law has eliminated the traditional enforcement conference, and has replaced the 180 day notice procedure with an administrative permit procedure and provision for prompt court action against violators.

In enacting the new legislation, Congress stated that it is the national goal that the discharge of pollutants into navigable waters be eliminated by 1985. An interim goal, where attainable, is to achieve by July 1, 1983, water quality which provides for the protection and propagation of fish and shellfish, and provides for recreation in and on the water. In order to achieve such goals, the Act requires establishment of effluent limitations based on technology for the achievement of "best practicable control technology currently available," by July 1, 1977, and "best available treatment economically achievable," by July 1, 1983. These effluent limits apply to point sources discharging to navigable waters.

DEFINITIONS

The term "point source" is defined by the Act as "discernible, confined, conveyances" which encompass any pipe, ditch, channel, tunnel, conduit, discrete fissure, container, rolling stock, vessel or cattle feedlot. The Act defines "navigable waters" as "waters of the United States." This has been interpreted to include all interstate waters, and any interstate non-navigable river, stream, or lake utilized by industries in interstate commerce, or used by interstate travelers for recreational or other purposes. Actually, this latter definition applies to virtually all surface waters of the United States.

DISCHARGE PERMIT SYSTEM

A major facet of this law was the establishment of a new national permit system, called the National Pollutant Discharge Elimination System (NPDES). Under this program, industrial, municipal, and other point source dischargers must obtain permits setting forth specific limitations on the discharge of pollutants into the navigable waters of the United States.

This new permit system is national in scope involving both federal and state participation. The objective is a state administered permit program but the basic requirements such as final implementation dates and effluent limits are applied consistently nationwide. Every discharge qualifying under the point source and navigable water definitions must apply for a permit if not already done so under the 1899 Refuse Act.

Application

If an application form has not already been filed, it should be done immediately. In March, new application forms to be used for the NPDES permit were issued. Forms are specific for agriculture, industry, commercial, and municipal discharges. Both a standard form and a short form are available. The latter is a single registration applying to discharges of 50,000 gal/day or less which do not contain any toxic or hazardous substances. These short forms are now available. The Standard Form C is also available and to be used for all other manufacturing and commercial organizations. Applications previously submitted to the Corps of Engineers under the Refuse Act are considered as applications under NPDES.

For a discharge beginning after July 15, 1973, application must be filed 180 days prior to the start of such a discharge.

Permit content

The permit that will be issued under the NPDES will have a basic format of effluent limits, a compliance schedule, and monitoring requirements. The 1977 and 1983 dates are targets to be viewed as the outside limits for compliance. The Act envisions that in meeting effluent limitations, there will be stages of compliance described as the attainment of levels of substantial improvement even before these legislative dates.

The compliance schedule may contain dates for achieving certain levels of progress such as preparation of engineering reports, final construction plans, beginning and completion of construction, and, finally, operation of facilities. Interim dates and requirements are specified in the permit as a means of monitoring progress and minimizing time slippage. For each interim date, the permittee must submit a written notice of compliance or noncompliance with the interim requirements. Where such interim dates may not be appropriate, reports of progress will be required at least every 9 months.

The permit may therefore have more than one set of effluent limits. These limits are to be attained on dates conforming with the compliance schedule. The effluent limits will be in quantitative terms such as kilograms (and pounds) per day, except for parameters like pH and temperature, where this is not relevant. Limitations may be expressed as concentrations where the rate of pollutant generation is not directly linked to the production effort, such as in mining, or where other requirements, such as removal of toxicity, apply. The effluent limits would be described in terms of average and maximum kilograms (or pounds) per day. They are based on effluent limitations or water quality standards, whichever are more stringent. Generally, the pH limit is in the range of 6–9 unless the water quality of the receiving stream requires some variation.

It is to be noted that the new permit system applies only to those organizations that discharge their wastewaters directly into the waters of the United States. Such wastewaters include not only process wastewaters but sewage, utility wastewater such as boiler blowdown and water treatment waste, and cooling waters. Every discharge pipe must be included in a permit. Manufacturing plants discharging to a municipal or publicly owned treatment works will be required to meet certain pretreatment requirements. In essence, these pretreatment requirements are established for the purpose of assuring that pollutants reaching the receiving waters will not be in excess of that which would be allowed if the discharge were direct.

The self-monitoring requirements contained in the permit are developed on an individual basis with consideration given for the type of treatment, the impact of the proposed treatment facility on the receiving waters, and the parameters to be measured. The purpose of the monitoring program is to establish that the treatment facility is consistently meeting the effluent limitations imposed in the permit. Data must be recorded and retained on file by the permittee for at least 3 years. The frequency of reporting monitoring results will be specified in the permit. A uniform reporting form has been developed and will be provided to the permittee.

Permits are to be issued for fixed terms. The maximum duration of a permit will be 5 years, and it is expected that the majority of permits will be written for that period since it will involve a commitment to a long term abatement program.

State programs

States may apply to EPA for authorization to issue national permits. Such authorization would be granted where the state has the "capability of administering a permit program which will carry out the objectives of the Act." It is a major objective of the Act to encourage states to administer the permit program.

Current enforcement

Except for spills and a few other limited situations such as failure to apply for an NPDES permit, the law forbids initiation of new enforcement cases before December 31, 1974, against any discharger until its application for a permit has been processed.

The enforcement program in water pollution will therefore be substantially curtailed for the immediate future. Virtually the full efforts of the water enforcement staff are channeled into the permit program until sufficient permits have been issued to provide a basis for renewed enforcement activities. Although this change will temporarily suspend the pressure for effective control which springs from the threat of enforcement action, the new permit program will provide a complete and systematic review of existing abatement requirements for all dischargers and will permit a precise definition of their obligations.

STATUS OF PERMIT PROGRAM WITH REFERENCE TO HEAVY METALS

The Agency presently has applications filed by companies of various industries. These applications are processed by our regional personnel in conjunction with the applicable state regulatory agencies. Permits will be issued to dischargers who are located in areas where water quality standards will require more stringent effluent limits than what is envisioned as best practicable control technology or for those industries where effluent guidance for best practicable control technology can be defined sufficiently well to give a high degree of confidence that permits so written will not be materially inconsistent with Effluent Guidelines to be issued in the near future.

TREATMENT CONSIDERATIONS

Space would not allow a thorough discussion of heavy metal waste treatment practices that would afford meeting requirements of the Act. We shall devote our remaining space to the electroplating industry which is a major contributor to heavy metal discharges. It may be said that wastes from electroplating processes contain alkalis, acids, solvents, oils, greases and most significantly, heavy metal salts.

The three most significant routes by which wastes originate in the electroplating industry are:

(1) concentrated plating bath;
(2) spills;
(3) dragout.

Spills and dragout represent the major sources of water. The four most commonly used and best developed treatment processes in use are precipitation-chemical, integrated process, evaporative recovery, and ion exchange.

(1) *Precipitation-chemical treatment process.* Involves oxidation of cyanide and ammonia complexes to CO_2 and N_2 and precipitation of metals at appropriate pH. If chromium is present, it must first be reduced from Cr^{+6} to Cr^{+3} prior to precipitation. From the standpoint of maximum efficiency of precipitation and recovery of metals values, this method is most advantageous when streams are segregated and precipitation is performed at the optimum pH for separate metals.

(2) *Integrated process.* The design concept of the integrated process is that the dragout film and droplets containing plating bath chemicals are first rinsed in a chemical treatment solution to eliminate these compounds before the work pieces are rinsed. The chemical treatment solution is specific for the kind of chemicals which are being treated. There is a separate method for cyanide treatment, for the treatment of chromium chemicals, copper, and other metals. The treatment solution is recirculated between the various treatment wash tank stations and a reservoir tank where the precipitated metal hydroxides are settled. In essence, the integrated process is a closed loop chemical destructive system which offers possibility of water recovery to the extent of 30–90%.

(3) *Evaporative recovery.* When the process solution is operated at room temperature or the quantity of rinse water exceeds evaporation losses, the

reconcentration of the rinse water by evaporating the excess water is possible. Vacuum evaporation reduces the heat input requirements otherwise needed to evaporate the volume of water required for return of rinse water. Usually, the distillate is used in these systems as the rinse water in the process, providing improved purity of rinse water as an added gain. This is a closed loop method which would permit return of the concentrate to the plating bath. Needless to say, this method is most effective when countercurrent water conservation rinse practices are used, thereby decreasing the amount of energy required for evaporation.

(4) *Ion exchange.* The use of ion exchange to remove dissolved solids, including the toxic or harmful plating bath components, returning pure water and the solids removed from the waste stream in high concentration to the plating bath is conducive to a smaller size and therefore less costly waste treatment system. Specific uses of ion exchangers to provide purified process solutions which would otherwise be discharged or to provide continuous or periodic metal recovery to maintain plating bath concentration, could have profound economic value. The joint use of ion exchange in connection with evaporative recovery may also be highly beneficial, eliminating chances for accumulation of impurities. The major advantage of ion exchange is that it is amenable to recycling of plating bath components.

Optimization of waste treatment practices and minimizing costs may involve any one or more of the waste treatment practices described. Once a particular effluent discharge requirement must be met as established by municipal, state or federal standards, one must decide upon the particular treatment technology to be used. Connected with this decision, are such factors as nature, concentration and volume of pollutants, space, and economics. Precipitation techniques are encumbered by water costs. For example, the cost of the treatment system will be proportional to the volume of water used. This volume of water can be reduced significantly by exercising water conservation practices during rinsing of parts upon removal from the plating bath. A useful practice is that of countercurrent rinsing in a multistage operation. Each additional rinse tank through which the water flows in a countercurrent manner allows a theoretical reduction of water use by one-tenth. In a practical way, double countercurrent rinse tanks are used quite extensively.

Another problem with precipitation techniques is that of sludge handling and drying. Finally, the dry or partially dry sludge resulting must be disposed of in an appropriate manner so as to minimize its environmental pollution potential.

The use of ion exchange has its greatest potential in the recovery of water from dilute and countercurrent rinses. The recovery of concentrated heavy metal residues is possible when the streams are segregated. The key to advantageous evaporative recovery lies in use of procedures to concentrate the rinse so that the volume of water that must be evaporated is kept to a minimum.

The choice of the type of waste treatment systems and combinations thereof used in the electroplating industry is obviously a complex process for which there is ultimately only a subjective answer. For a particular set of electroplating conditions, the treatment selected should be that which lends itself to minimal water use and optimal recovery of plating bath chemicals. The net costs advanced in publications on actual plant operations indicate that the amount would not be more than in the 5–10% range of value added by the finishing operation when the daily water use is over 20,000 gallons.

SUMMARY

In summary, NPDES permits are required for point source discharges to the waters of the United States, and will contain effluent limitations, schedules of compliance, and monitoring requirements. Permits may be based on water quality requirements or best practicable control technology currently available, defined on the basis of engineering judgment until formal Effluent Guidelines are issued, but by the Effluent Guidelines for those parameters specified in the guidelines after they are issued. Such guidelines for the various industries are being proposed by the Environmental Protection Agency for public comment.

The waste treatment practices for the Electroplating Industry, a major source of heavy metals discharge, has been discussed in general. Waste treatment practices for this industry such as chemical precipitation, integrated treatment, ion exchange, and evaporative recovery have been indicated as being useful for meeting effluent limitations at least for 1977. Any practice requiring minimal water use and maximum recovery of materials is in best keeping with the true spirit of the law.

RISK–BENEFIT ANALYSIS AND THE ECONOMICS OF HEAVY METALS CONTROL†

GEORGE PROVENZANO

Research Economist, Institute for Environmental Studies, University of Illinois, Urbana, Illinois 61801

In the past few years a good deal of research has been directed toward achieving an understanding of the characterization, movement, and effects of heavy metals in the environment. Research has also been directed toward the equally important objective of explaining the technical and economic reasons that our economic system utilizes large quantities of heavy metals. Information from both types of research is essential for the design and implementation of efficient strategies for reducing the hazards of heavy metals in the environment. Decision makers must consider the tradeoffs between the economic benefits and the environmental risks of utilizing heavy metals. The purpose of this presentation is to define a framework for benefit–risk decision making for the control of heavy metals.

Heavy metals impart economic benefits to virtually every kind of economic activity. The advanced technology of our mass production economy is highly dependent on the physical and chemical properties of these materials. Each year American industry demands large quantities of copper, zinc, lead, chromium, and nickel, and lesser amounts of beryllium, cadmium, and mercury. As Table 1 indicates, industry has steadily increased its utilization of these and several other metals with known toxic properties in the postwar period.

The kinds of economic benefits which stem from the use of heavy metals are as diverse as their physical and chemical properties. For example, arsenic and mercury compounds have been used for their toxic capabilities in pesticides and fungicides; zinc galvanizing and cadmium plating are employed for their materials-saving properties; and lead antiknocks are used for their energy-saving benefits.

The intensive rate of utilization of heavy metals also results directly in the generation of large quantities of waste materials which contain heavy metals and heavy metal compounds. During production activities heavy metal inputs are partly converted into final goods, which may contain heavy metals, and into heavy metal waste materials. Once the services of final goods have been depleted, their material contents also become wastes. Consequently, after the economic benefits of utilizing heavy metals have been realized, their material substance remains and must be either recycled or discharged into the environment.

This conclusion is an obvious reflection of the basic principle of the conservation of matter. Economists have recently integrated the materials balance approach into their analysis of environmental problems.[2] This approach has economic validity as well as scientific validity in understanding the problems of controlling heavy metals.

Because of their toxic properties, the direct discharge of heavy metal wastes into the environment creates a risk of biological damage. The term "risk" refers specifically to the probability, which may be very small, of damage to living organisms.

Heavy metal wastestreams may be treated in order to reduce the risk of damages in the environment. But because it is impossible to ultimately eliminate heavy metals, efforts to reduce the risks of damage in one area of the environment may merely transfer them to another. For example, the removal of heavy metals from industrial wastewater discharges may ultimately create a land disposal problem for solid wastes containing heavy metals.

Because of their persistence in the environment, heavy metals may accumulate in environmental sinks such as the soil or concentrate within the components of the food chain. Again, because of their toxicity, any buildup of heavy metals within any component of the environment may pose a threat to man and to other forms of life. These facts are indicative of the risks of using the natural environment as a place of disposal for wastes containing heavy metals.

---

†This research has been supported by a National Science Foundation Grant, NSF GI31605.

TABLE 1
Estimated US consumption of selected metals, 1950, 1969, 1970, and 1971[a]

Metal	Total estimated consumption[b] (in tons)				Annual rate of increase, 1950–69
	1950	1969	1970	1971	
1. Arsenic (As_2O_3)	32,104	18,170	18,763	16,406	−2.9
2. Barium (barite)	786,131	1,604,742	1,408,626	1,355,397	3.6
3. Beryllium (beryl)	3007	8483	9496	10,373	5.3
4. Cadmium	4773	7531	4532	5416	2.3
5. Chromium	980,369	1,411,000	1,403,000	1,093,000	1.8
6. Copper	1,424,434	2,142,218	2,043,303	2,019,507	2.1
7. Lead	1,237,981	1,389,358	1,360,552	1,431,514	0.6
8. Manganese ore	1,650,429	2,181,333	2,363,937	2,155,454	1.4
9. Mercury	1870	2940	2337	1994	2.3
10. Molybdenum	13,015	25,811	22,669	20,475	3.5
11. Nickel	98,904	141,737	155,719	128,816	1.8
12. Selenium	546	988	755	529	3.0
13. Vanadium	n.a.	6154	5134	4802	—
14. Zinc	967,134	1,814,167	1,571,596	1,650,585	3.2

[a] Source: ref. 1.
[b] Includes stocks released to the open market by the federal government and imports; does not include exports.

What do economists mean by the *benefits* of utilizing heavy metals? A basic assumption of economics states that the benefits of any product are evaluated in terms of the product's ability to satisfy an individual consumer's desires. The consumer expresses his willingness to purchase goods and services in the market place, and he makes those purchases which maximize his level of satisfaction subject to his income and to market prices. In other words the consumer's level of expenditure for each particular product provides a measure of the product's benefits; the benefits of a product are determined at the point of final consumption.

Because consumers generally do not make direct use of heavy metals, the above principle cannot be readily applied. Although the presence of a heavy metal may be essential for the performance of a product, the heavy metal does not provide the motivation for buying the product. Take the storage battery or leaded gasoline, for example. Therefore, an alternative means other than consumer satisfaction must be established for measuring the benefits of heavy metals.

For the most part, heavy metals enter as intermediate inputs into the production of final goods and services that consumers enjoy. The value-in-use of any input, including heavy metals, can be measured in terms of the output it generates during the various stages of production. This is the concept of productivity. In practice it is virtually impossible to isolate the productivity of one single input because several inputs are usually combined during the production process.

In addition to productivity, producers also select a combination of inputs on the basis of cost. Producers will make additional purchases of resource inputs as long as the cost of an additional unit of input is less than the value of the additional services it provides.

If the implementation of an anti-pollution control forces a producer to eliminate or to reduce the use of a particular input, e.g. a heavy metal pesticide, then the producer must substitute an alternative input, such as an organic pesticide, in order to accomplish the same end. Any increase in cost for the substitute input represents a net replacement cost of not being able to use the original input. If no substitute inputs are available, then the market value of any decline in production represents the foregone economic benefits of not being able to use the original input. The concept of net replacement cost or the concept of foregone economic benefit provides a useful means of measuring the economic benefits-in-use of heavy metals. The USEPA has used this measure in at least one instance in measuring the economic impacts of removing lead additives from gasoline.[3]

What do we mean by the *risks* of heavy metal

pollution? A recent colloquium on benefit–risk decision making grouped risks into three categories.[4] These are:

(1) *Risks determined by individual option.* This category included risks that are taken voluntarily; sports, smoking, flying in a private plane. Everyone has a propensity for taking risks in order to achieve benefits that outweigh any perceived hazards.
(2) *Risks determined by individual option but limited by social action.* This category included risk–benefit analyses that are made by individuals subject to limitations imposed by society. An example is the use of drugs in medicine.
(3) *Risks determined by social action decisions.* Certain kinds of hazards are pervasive; they may affect large segments of the population. In addition the individual usually has no immediate control over the causal agent. The hazards are public "bads," and they are subject to government control.

Environmental pollution from heavy metals clearly falls into the third catagory of risk.

We may illustrate the risks and benefits associated with the use of heavy metals by means of the environmental–economic systems diagram in Fig. 1. The following paragraphs deal with lead as an example for illustrating what the diagram means.

Lead is one of the most intensively used industrial metals, and, as a direct consequence of its widespread use, lead and lead compounds are discharged into the environment in extremely large quantities. In 1971 alone, industries in the United States utilized over 1.43 million tons of lead:[1] lead ranked fifth, behind iron, copper, aluminum, and zinc, in terms of total tonnage consumed.

This widespread use can be attributed to the versatile chemical and physical properties of lead and to the favorable economic factors involved with obtaining and processing the metal in the mineral state. Lead is the heaviest and the softest of the common metals; it is abnormally resistant to chemical corrosion, and it alloys very easily with several other metals. Lead also has low melting and high boiling points, and it has very useful energy absorption and transmission qualities. The usefulness of lead in the fabrication or composition of any product is, to a large extent, a function of one or more of these characteristics.

The high frequency of occurrence of economically viable lead ore deposits and the relative geographic convenience of these sources has facili-

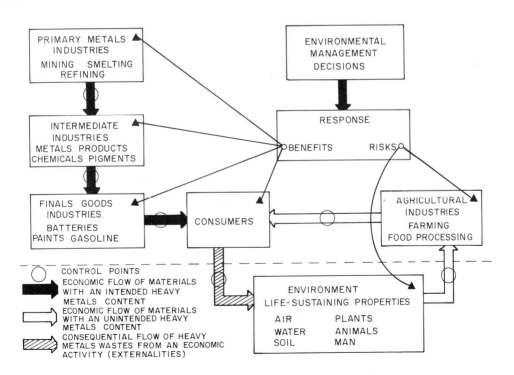

FIG. 1. Economic–environmental system for heavy metals.

tated meeting industrial demands for lead metal at a very low cost. For example, the large galena ore deposits in southeastern Missouri have made that region one of the most productive lead mining regions in the world. In 1971 the Missouri lead belt, which is conveniently located for transporting lead to markets, accounted for 74% of the total domestic lead in ore.[1]

The combination of these two factors—a large number of desirable properties and the relatively low costs of extraction—have encouraged intensive use of the metal. In the absence of any regulations that prohibit the manufacture and distribution of products that contain lead, the relatively low price for this material will continue to be an incentive for its widespread use.

Figure 1 illustrates the interindustry production and utilization of lead. Lead production begins, of course, at the mine. Mined ore is then crushed and dressed in a number of operations to effect mineral separation. The resultant lead concentrate is then smelted and refined into lead metal of varying degrees of purity.

In 1971 domestic mine production of recoverable lead contributed 88% of the total US supply of primary metal. Imported ores, concentrates, the pig-lead accounted for the remainder. Primary lead accounted for about 60% and secondary or recycled lead output represented about 40% of the total US market supply. The total domestic supply of lead, primary, secondary, and imports, amounted to 1.46 million tons.[1]

As Fig. 1 illustrates, primary and secondary lead output is then sold to intermediate industries for fabrication and processing into other commodities. In other words, lead is an intermediate commodity that becomes an ingredient in many of the final goods that eventually reach the consumer.

The major demands for products that contain lead are related to uses in transportation. For example, the storage battery and gasoline additives industries absorbed approximately 45 and 20%, respectively, of industry's total consumption of lead in 1971. The data in Table 2 provide a more detailed view of interindustry utilization of lead in the US during the past decade. In this period the amount of lead utilized in pigments and metals products has declined while the quantities used in ammunition, batteries, and gasoline additives have increased rapidly. The 10 year average annual rate of growth of lead used for gasoline additives is twice as high as the growth rate for all lead consumption.

It is within the interindustry network that the benefits of lead and lead compounds are realized. The patterns of demand and supply reflect industries' selection of materials according to their prices and their properties. Tetraethyl lead (TEL), for example, is used as an inexpensive substitute for higher octane, more expensive blending components in the production of gasoline. The addition of a few cubic centimeters of TEL enables refiners to produce high octane motor fuels with less crude oil and with less severe processing than would be required in the production of unleaded gasolines.

The flow of lead emissions that enter the environment is integrally related to the economies of lead production and utilization in the following manner. Lead enters the ecosystem as a constituent of

TABLE 2
Selected trends in the largest categories of industry's utilization of lead, 1963–72, in percentages of total lead consumed[a]

Manufacturing category	1963	1964	1965	1966	1967	1968	1969	1970	1971	1972[b]	Average annual rate of growth
Storage battery grids, pastes, and oxides	37.8	35.8	36.7	35.7	37.0	38.6	41.9	43.6	47.5	45.5	4.0
Gasoline additives	16.6	17.0	18.1	18.6	19.6	19.7	19.5	20.5	18.4	19.5	4.6
Pigments	8.5	8.6	8.8	9.1	8.2	8.3	7.3	7.2	5.7	6.3	−1.3
Ammunition	4.3	4.7	4.6	5.9	6.3	6.2	5.7	5.4	6.1	5.9	5.3
Other metal products	29.7	30.6	28.4	27.3	25.8	24.1	22.5	20.5	19.5	20.3	−3.4
Miscellaneous	3.1	3.3	3.4	3.4	3.1	3.1	3.1	2.8	2.8	2.3	−0.3
Total lead	100.0	100.0	100.0	100.0	100.0	100.0	100.0	100.0	100.0	100.0	2.3

[a] Source: ref. 5.
[b] Preliminary.

various solid, liquid, and gaseous waste products. These wastes are discharged as a direct consequence of using lead or products that contain lead in certain economic—transportation, production, consumption—activities. The amount of waste that is discharged is a direct function of the level of economic activity. For example, the combustion of leaded gasoline, which is the primary fuel for our land transportation system, has accounted for approximately 96–98% of the airborne emissions of lead.[6]

Any environmental management decision that is directed toward reducing the flow of lead that is currently entering the environment will obviously reverberate back through the lead and lead products industries and possibly through several other supplying industries such as mining equipment, construction, and transportation as well. The extent and intensity of these effects may be quite diverse and will depend on several factors including the size of the reduction, its timing, regional concentration of the affected industry, economic conditions in the industry, and the overall economic importance of the restricted product.

Consider the energy–economic implications of the USEPA's proposals for eliminating the use of lead additives in gasoline. The tetraethyl lead industry absorbs nearly 20% of the total annual industrial consumption of lead in the United States. This is primary lead for the most part and therefore a decline in the use of lead additives will bring immediate losses in income and employment to the primary lead mining and smelting and tetraethyl lead industries. Although the EPA has proposed a phased-reduction plan that will take several years to complete, the lead-mining and smelting industries are highly concentrated in Missouri, Utah, and Idaho, and the economic losses to those areas are likely to be severe. The fact that a large part of the lead industry in Missouri consists of modern, low-cost operations, however, may enable some mining and smelting companies to ride-out cutbacks in domestic consumption by expanding sales in foreign markets.

As indicated above, the addition of tetraethyl lead results in energy-saving benefits in the production of high octane gasolines. High octane gasolines (94–100 RON) are necessary for the efficient operation of engines with compression ratios in the 9:1 to 10:1 range. All things being equal, these engines achieve better fuel economy than engines with lower compression ratios.

The EPA's overall strategy for controlling automobile air pollution will, in all likelihood, require the use of a catalytic emissions control device. Lead emissions from the combustion of leaded gasoline have a deleterious effect on the operation of these devices. Consequently, in 1971 Detroit began producing automobiles with lower compression ratios that run on lower octane unleaded gasoline in preparation for the introduction of the catalytic converters in 1975 and 1976.

The compression ratio drop in the post-1970 automobiles has had an adverse effect on fuel economy. The addition of the catalytic emissions control systems may also produce additional adverse effects.

In a study that was completed for the EPA, Bonner and Moore Associates estimated that the Federal Government's strategy for controlling automobile emissions will result in an increase in the demand for gasoline of 9.2% or over 10 billion gallons per year by 1980.[3]

To conclude the example, tetraethyl lead provides rather significant economic benefits which, in view of the present energy crisis, may be too valuable to forego. At the same time, the risks of allowing increasing amounts of lead to enter the environment warrant some kind of controls on the use of tetraethyl lead. The problem that decision makers must solve concerns where—between the extreme positions of total restriction and no restriction on the use of lead in gasoline—to place the level of control.

As a first step in making this kind of decision, decision makers must have an analytical framework for organizing the components of the problem they want to resolve. In this regard, benefit–risk analysis provides a practical method for comparing the relevant economic benefits with the environmental risks of using heavy metals.

As the name implies, benefit–risk analysis is somewhat akin to benefit–cost analysis; both forms of analysis provide a systematic method for assembling information about the prospective benefits and costs of a government program decision. The name "benefit–risk" analysis may be somewhat of a misnomer because "benefit" refers to the foregone benefits or the costs of not being allowed to use a particular material. "Cost–risk" analysis may be a more appropriate term to use.

As Fig. 2 illustrates, a benefit–cost ratio is the ratio of the discounted stream of benefits to the discounted stream of costs that accrue to the public from a particular government program. A benefit–risk ratio is the ratio of the discounted stream of economic benefits gained (lost) to the discounted stream of the expected values of the risks of

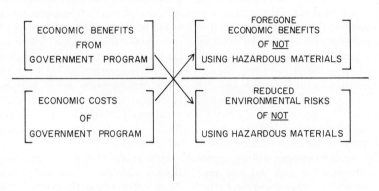

FIG. 2. Benefit–cost ratio versus benefit–risk ratio.

environmental damages which increase (decrease) with the benefits of using (not using) a heavy metal. By discounted stream we mean the present value of prospective benefits and expected values of the risks of environmental damage. The present value of benefits, for example, is expressed as

$$\text{P.V.} = \sum_j \frac{B_j}{(1+r)^j}, \quad (1)$$

where B_j is the dollar value of benefits that accrue in the jth time period and r is the appropriate rate for discounting.

The benefits of using a heavy metal can be estimated directly using price–quantity information and either the net replacement cost approach or the foregone economic benefits approach discussed above. The expected values of the risks of environmental damages from the use of heavy metals, however, can only be estimated indirectly from monitoring information and from data on the health, plant, and animal effects of ambient concentrations of heavy metals in the environment.

Benefit–cost analysis has long been used to evaluate public expenditure programs of various kinds.[7] The most notable examples include river basin developments, urban renewal projects, and public health programs. In addition benefit–cost analysis is a useful tool for evaluating the economic consequences of a regulatory decision.[8] In both of these situations, however, the benefits and costs are estimated with reasonable certainty. If there is considerable uncertainty in quantifying costs and benefits, an *ad hoc* probability factor may be used to give a range of estimates instead of a single estimate.

For benefit–risk analysis, the importance of probability is really the heart of the analysis. The probability that certain events will lead to a specified outcome, i.e. the probability that an ambient environmental concentration of heavy metal will have an effect on an organism, is the basis for assessing risk. The crux of risk–benefit decision making lies in balancing the probabilities of damaging effects to organisms against the reasonably certain economic benefits.

The problem of estimating the improvements in public health that result from a program of disease control is analogous to the problem of estimating the reduction in environmental risks that results from a program of pollution control. In the disease control program the measure of improvement is a function of the reduction in the incidence of a particular disease, and in the pollution control program the measure of improvement is the value of environmental damages that society avoids *vis-a-vis* the implementation of the control program. A comparison of the measurement of improvement in public health to the measurement of improvement in environmental conditions will serve to illustrate the fundamental importance of probability in risk–benefit analysis.

The measure of improvement is a function of the difference between a baseline measure of the incidence of disease or environmental damage and the anticipated results of the control measures. The following equation summarizes the measure of benefits for the ith segment of the population

$$M_i[E(Y_i) - E(\hat{Y}_i)] = B_i, \quad (2)$$

where $E(Y_i)$ is the expected number of illnesses or

deaths due to disease in the absence of a disease control program and $E(\hat{Y}_i)$ is the projected number of illnesses or deaths after the program has been implemented. M_i is an arbitrary average cost factor for a single illness or death. The conceptual framework for estimating the cost factor M_i is well established for disease control programs. These costs include two components: (1) the direct costs of medical care, and (2) the indirect costs of losses in productivity due to death or illness.[9]

The incidence Y_i of disease is measured directly from data collected by the public health authorities. It is important to note that diseases with low incidence rates pose a special disadvantage in terms of lack of data. This is essentially the same "can of worms" that plagues the measurement of health or biological damage due to heavy metals in the environment.

The incidence of damage to the health of the general population due to heavy metals in the environment is not directly observable in the same sense as the occurrence of disease is. Many of the effects of an increase in heavy metal pollution are subclinical and therefore go undetected. Some of these effects may be latent and turn up well after the toxic stress is no longer present. Finally, the effects of a heavy metal are dependent on concentrations reaching the individual, and these, in turn, are dependent on stochastic variables such as the weather conditions and distance from the source of emissions. Because of these stochastic factors the number of occurrences of a particular effect cannot be measured directly. The number of occurrences must be estimated using the following kind of probabilistic damage response function.[5]

Equation (3) defines the expected value of a probabilistic response function for specified ambient levels of pollutant,

$$E(x) = \sum_j \sum_i [x_i f(x_i|c_j)] f_c(c_j). \quad (3)$$

In this equation the expected response $E(x)$ is the percentage of the population that is affected by various concentrations c_j of pollutant.† Expected response is a function of the probability $f_c(c_j)$ that ambient concentration c_j will occur at a receptor point and the relative frequency $f(x_i|c_j)$ that a random event x_i occurs in the sample population. The random event in this instance is the percentage of the sample that exhibits a particular biological effect at concentration c_j. For example, $f(x_i|c_j)$ is the relative frequency that 10% of the sample population will have a blood lead concentration of 20 μg per 100 ml of blood after being exposed to ambient airborne concentrations of 1 μg/m^3 of lead for one month.

The expected number in the ith population group affected is obtained by multiplying the size of the ith group Y_i by the expected response $E(x)$:

$$E(Y_i) = Y_i \left[\sum_j \sum_i [x_i f(x_i|c_j)] f_c(c_j) \right]. \quad (4)$$

The expected change in the incidence of these damages in the ith population group is defined in terms of the difference in the expected number of occurrences of the pollution effect at a baseline or no-control level of pollution and a target level of control. The expression on the left-hand side of the minus sign in eqn. (5) is the expected level of damage at the no-control level of pollution, and the expression on the right is the expected level of damage at the targeted level of heavy metal pollution control. The difference between these two levels is the basis for measuring the benefits of pollution control,

$$B_i = M_i \left\{ Y_i \left[\sum_j \sum_i [x_i f(x_i|c_j)] f_c(c_j) \right] \right.$$

$$\left. - \hat{Y}_i \left[\sum_j \sum_i [\hat{x}_i f(\hat{x}_i|\hat{c}_j)] f_c(\hat{c}_j) \right] \right\}. \quad (5)$$

The circumflexed terms denote values which must be predicted if the benefit–risk analysis is made *ex ante*. Predictive simulation models of environmental processes provide the means for predicting changes in ambient concentrations and changes in biological

---

†A set of joint observations on two variables, concentration c, and effect x, may be organized in the form of a joint frequency distribution $f(x, c)$. For each (i, j) pair of values, $f(x_i, c_j)$ is the relative frequency with which it occurs in the sample set of observations. If in the joint distribution, one of the variables x_i is conditional on the outcome of the other, c_j, then the conditional frequency distribution is written as $f(x|c_j)$, and $f(x_i|c_j)$ is the relative frequency x_i occurs in the subset of variables c_j. Because each conditional frequency is the ratio of a joint frequency to the frequency of the conditional variable, we have

$$f(x|c) = f(x, c)/f_c(c) \quad \text{or}$$
$$f(x|c)f_c(c) = f(x, c).$$

Finally, the expected value $E(x)$ of a sample of observations is the mean of the sample. For a conditional frequency distribution the expected value of one variable for a given value of the second is expressed as

$$E(x) = \sum_i x_i f(x|c_j).$$

responses for specified changes in the level of heavy metal pollution.

Equation (5) suggests that in order to estimate risk, considerable research is needed to define the following parameters:

(1) The probability distributions for ambient concentrations at receptor points.
(2) The conditional probability distributions for damages to all receptor populations.

Monitoring information can be used to construct the probability distributions for ambient concentrations. The conditional response probability distribution is much more difficult to obtain; these distributions must be constructed from expensive and time-consuming epidemiological studies.

Further problems involved in constructing this kind of conditional probability transformation arise because the pollutant may not have the same effect on different target groups within the population; the effects of the pollutant may be dependent on factors other than ambient conditions. In addition, the occurrence of harmful physiological effects at low levels of exposure may be questionable.

In conclusion, risk–benefit analysis may serve as the basis for rational decision making to control heavy metal pollution. A great deal of research will be needed, however, in order to perfect this analytical tool.

REFERENCES

1. US BUREAU OF MINES, *Minerals Yearbook*, US Government Printing Office, Washington DC (various years).
2. KNEESE, A. V., AYRES, R. U., and D'ARGE, R. C., *Economics and the Environment: A Materials Balance Approach*, John Hopkins Press, Baltimore, 1970.
3. BONNER and MOORE ASSOCIATES, INC., *An Economic Analysis of Proposed Schedules for Removal of Lead Additives from Gasoline*, prepared for USEPA, June 25, 1971.
4. NATIONAL ACADEMY OF ENGINEERING, *Perspectives on Benefit–Risk Decision Making*, National Academy of Engineering, Washington DC, 1972.
5. SCHWARTZ, SEYMOUR, Probabilistic models for calculating air pollution damage, *J. Environ. Systems*, June, 1971, pp. 111–132.
6. ENGEL, R. E., HAMMER, D. I., HORTON, R. J. M., LANE, N. M., and PLUMLEE, L. A., *Environmental Lead and Public Health*, US Environmental Protection Agency, Research Trianlge Park, North Carolina, 1971.
7. PREST, A. R. and TURVEY, R., Cost benefit analysis: a survey, *Econ. J.*, December 1965.
8. SCHMID, A. ALLEN, Effective public policy and government budget: a uniform treatment of public expenditures and public rules, in *The Analysis and Evaluation of Public Expenditures, The PPB System*, Joint Economic Committee, Congress of the US, 91st Congress, 1st Session, vol. 1, 1969, pp. 579–591.
9. KLARMAN, HERBERT E., Syphilis control programs, in *Measuring Benefits of Government Investments* (Robert Dorfman, ed.), pp. 367–414. Brookings Institution, Washington DC, 1965.

AUTHOR INDEX

Adams, C. E., Jr. 277
Allersma, E. 85
Åséll, B. 299

Barth, E. F. 293
Bennett, T. B. 63
Bisogni, J. J., Jr. 113
Blair, W. R. 193, 251
Brierley, C. L. 189
Brinckman, F. E. 193, 251
Brye, B. A. 315
Burrows, W. D. 51

Chappell, W. R. 167
Clarkson, T. W. 1
Connell, D. W. 247

de Groot, A. J. 85, 103
D'Itri, F. M. 223
D'Itri, P. A. 223
Doi, R. 197

Eagle, M. 117
Eckenfelder, W. W., Jr. 277
Edwards, H. W. 243

Feick, G. 329
Finger, J. H. 63
Flewelling, F. J. 253

Goldwater, L. J. 13
Goodman, B. L. 277
Goyer, R. A. 163

Hem, J. D. 149
Huey, C. 193, 251
Humphrey, H. E. B. 33

Iverson, W. P. 193, 251

Jenne, E. A. 131
Jennett, J. C. 231

Jernelöv, A. 299
Jewett, K. L. 193, 251
Johanson, E. E. 329
Jonasson, I. R. 97

Katz, M. 25
Koirtyohann, S. R. 243
Krenkel, P. A. x, 117, 311

Laitinen, H. A. 73
Langley, D. G. 333
Lawrence, A. W. 133
Lee, G. F. 137

Marsh, D. O. 1
Miettinen, J. K. 155
Minear, R. A. 261
Mount, D. I. 31

Nicholas, W. R. 315
Nomiyama, K. 15

Olotka, F. T. 257

Patterson, J. W. 261
Provenzano, G. 339

Reimers, R. S. 117
Rolfe, G. L. 231

Schafer, C. J. 335
Shapiro, M. A. 247
Skogerboe, R. K. 81
Smith, J. C. 1
Strier, M. P. 335
Sumino, K. 35

Timperley, M. H. 97
Tragitt, G. 117
Turner, M. D. 1

Ui, J. 197, 229

Weeks, J. D. 273
Westöö, G. 47
Wixson, B. G. 243
Wood, J. M. 105

Yeaple, D. S. 329

Ziegler, F. G. 311

SUBJECT INDEX

Activated sludge process
  effects of metals
    chromium 281
    combined heavy metals 286
    copper 277
    nickel 285
    metal removal 273
Adsorption capacitance tests 117
Aluminium, concentration and removal by sewage treatment 274
Anaerobic digestion
  effects of metals
    chromium 283
    combined heavy metals 287
    nickel 285
    zinc 284
Antimony, in European river sediments 88
Arsenic
  concentration and removal by sewage treatment 27
  content of river sediments 88
  estimated US consumption 340
Atmospheric emissions
  lead 45, 343
  mercury 254, 257

Bacteria
  as methylating agents 106, 114
  as molybdenum leaching agents 189
  mercury tolerant 193
Barium, estimated US consumption 340
Beryllium, estimated US consumption 340
Biological half life
  mercury 159
  methylmercury 41
  lead 164
Biological oxygen demand (BOD) of activated sludge effluents
  effect of metals 282, 285, 295
Biological treatment
  active life forms involved 294
  effects and removal of metals 277
  mercury transport 304
  White Rock Plant 273

Cadmium
  absorption and elimination
    by fish and molluscs 158
    by man 157
  content of river sediments 88
  diagnostic techniques 21
  estimated US consumption 340
  mechanisms of accumulation and excretion 15
Chemical oxygen demand (COD)
  comparison with dry combustion estimates of inorganic carbon 135
  of sediments 119
  of sewage effluents 277, 293
Chromium
  content of river sediments 88
  effects on, and removal by, biological sewage treatment 281
  estimated US consumption 340
  solubility 152
Chromosomal abnormalities 1, 13
Cinnabar 124
Civil engineering projects, influence on behaviour of heavy metals in sediments 91
Clay minerals, role in heavy metal transport 144
Complexation of heavy metals 142
Copper
  accumulation by sulfide precipitation 100
  as catalyst in ferrous sulfate oxidation 141
  content in river sediments 88
  effect on biological sewage treatment 277
  estimated US consumption 340
  factors controlling levels in Lake Monoma 143
  solubility of salts 294
Co-precipitation of heavy metals 137

DDT
  contamination of Japanese animals
    cats 218
    fish 199
Dimethylmercury
  formation 106, 307, 322
  photolysis by UV light 110
Dredging 304, 311, 329

Effluent Guidelines 338
Erosion muds 88, 103

Ferrites 152
Fish
  annual Japanese catches 201
  avoidence threshold to heavy metals 27
  consumption 33, 198, 224
  effects of heavy metals 13

Fish (*cont.*)
  mercury distribution 197, 229
  mercury levels 201
  methylmercury accumulation 302
  mortality in Tennessee Valley 315
  recovery in clean water 28
  toxicity bioassays 26, 321
Flameless atomic adsorption spectrometry 51, 119

Gasoline, as-source of lead 231, 343

Heavy metals
  analytical techniques 73
  binding 100
  co-precipitation with hydrous metal oxides 137
  content of river sediments 88
  control, risk–benefit analysis 339
  correlation with endemic disease in fish 27
  discharges to Westermost Bay 249
  economic–environmental system 341
  effects
    on biological sewage treatment 277
    on fish and aquatic organisms 25
  estimated US consumption 340
  mobilization processes 93
  removal
    by biological sewage treatment 288
    by physical–chemical methods 261
  sorption on suspended matter 93
  sources 261
  transport in sediments 85, 137
  uptake by marine algae 249
Hydrous metal oxides
  role as co-precipitation agents in heavy metal transport 137
  source in natural water systems 139
Hydroxide sludges 268

Inorganic mercury
  effect of abiotic parameters on uptake by natural sediments 124
  in sediments 300
  metabolism and absorption by man 159
  methylation 43
Instability partition coefficients
  effect of abiotic parameters 124
  mercuric chloride
    with long-chain alkyls 120
    with natural sediments 123
Iron, organic-color precipitates 142
Irrigation water 167
Itai-itai disease 15, 261

Kumamoto University Minamata Disease Research Group 197, 213

Lanthanum 87, 90
Lead
  absorption and elimination in man 160
  contents of river sediments 88
  distribution
    in man 164
    in relation to automobile sources 231
    in relation to mine and smelter sources 237
  emissions from gasoline combustion 343
  estimated US consumption 340
  largest categories of US industrial utilization 342
Legislation 335

Manganese
  content in sediments 85
  estimated US ore consumption 340
Mercury
  binding 300, 306, 329
  biological transformation 193
  cellular effects 163
  concentrations in river sediments 88
  conversion to mercuric sulfide 305
  cycle 110
  daily loads 229
  distribution in Chesapeake Bay 251
  estimated US consumption 340
  extraction from contaminated spoils 312
  Japanese consumption and recovery 198
  levels
    in Canadian population 207
    in English population 207
    in Japanese cats 215
    in Japanese fish and shellfish 199, 223, 229
    in Japanese population 206
    in Japanese rice 208, 223
    in Ojibway Indians 224
  mineralization 98
  pollution
    Japan 197
    Tennessee Valley 315
  reduction at chlor-alkali plants 253, 257
  removal from industrial effluents 261
  sorption characteristics 117
  sources
    Japan 197, 207
    USA 261
  vaporization 105
Metabolic interconversions
  arsenic 112
  mercury 111
Methylation 43, 106, 195, 299, 333
Methylmercury
  accumulation in fish 306
  analysis 35
  analytical reproducability 39, 50
  biological half-life in rat organs 41
  comparison with total mercury levels
    freshwater fish 201

Methylmercury (cont.)
   human organs 41
   marine fish 204
   differential uptake and elimination by organs 14
   dose–response relationships in humans
      Iraq population 3
      Japanese population 2
      population with high fish intake 6, 13, 33
   extraction techniques
      fish flesh 36, 48
      mud 39
   levels in Japanese population 41
   metabolism in man 159
   model for formation and transport 300
   sorption 125
   synthesis by methylation
      biological 106, 114, 299, 333
      non-biological 43, 195
Metallothionein 155
Minamata Bay 1, 35, 197, 223
Minamata disease 15, 35, 197, 211, 223
Mineralization enriched by heavy metals 97
Mobilization of heavy metals 90, 103
Molybdenum
   biological role and effects 181, 191
   deficiencies 179
   distribution 167, 189
   domestic use 167
   estimated US consumption 340
   removal
      by addition of $FeCl_3$ 176, 190
      by co-precipitation with biogenically formed sulfide 190
   transport 172, 177, 189
Molybdeniferous rocks 168, 189
Molybdenite, bacterial solubilization 189
Molybdenosis 171, 181

National Pollutant Discharge Elimination System (NPDES) 336
Natural water coloring matter 142
Neutron activation analysis 55, 63, 78, 87
Nickel
   concentration in river sediments 88
   effect on, and removal by, biological sewage treatment 285
   estimated US consumption 340
Normal dodecyl mercaptan (NDM) 331, 333

Organic carbon, values for bed sediments 131
Organic compounds
   adsorption of mercuric chloride 123
   concentration, composition and reactive groups 131

PCB contamination
   Japanese population and animals 217

Westermost Bay 249
Periodic table 295
Phenylmercuric acetate 322
Phenylmercuric pesticides 197, 207
Physical–chemical treatment methods
   evaporation 265, 338
   integrated system 268, 337
   ion exchange 265, 338
   precipitation 262, 337
   reverse osmosis 265

Redox potentials, significance 149
Reproduction in mice, relative toxicity of heavy metals 183
Rice, contamination
   by cadmium 261
   by phenylmercury 197, 207

Samarium 90
Scandium 87, 90
Sediments
   distribution of lead 236
   mercury concentrations 88, 251, 300
   organic content and relationship to sorption of mercury 117
   origin and transport 85
Selenium
   estimated US consumption 340
   influence on absorption of methylmercury 156
Shellfish
   annual Japanese catches 201
   cadmium concentration 261
Stack emissions as source of atmospheric lead 245
Standards
   effluent limits 336
   for mercury
      drinking water 333
      US chlor-alkali plants 257
Sulfide preparation of mercury 253
Synthetic scavengers 329

Tetraethyl lead (TEL) 342
Thiols 332
Titanium, relative toxicity in respect to reproduction in mice 183
Total mercury
   review of analytical methods
      fish tissues 64
      industrial wastes and waters 63
      reproducability 64
      sediments 65
Toxic elements, classification according to toxicity 105
Toxicity bioassays 26, 293, 321
Transport of heavy metals 85, 137

Uranium, transport and content in Mackenzie River 98

Vanadium
  estimated US consumption 340

Whole-body counting technique 157

Zinc
  content in river sediments 88
  effect on, and removal by, biological sewage treatment 284, 296
  estimated US consumption 340
  toxicity to aquatic organisms 142